Foundations of Statistics

Foundations of Statistics

D. G. Rees
Department of Computing and Mathematical Science
Oxford Polytechnic

LONDON NEW YORK
Chapman and Hall

First published in 1987 by
Chapman and Hall Ltd
11 New Fetter Lane, London EC4P 4EE
Published in the USA by
Chapman and Hall
29 West 35th Street, New York NY 10001

Printed in Great Britain by
J. W. Arrowsmith Ltd., Bristol

ISBN 0 412 28560 6

British Library Cataloguing in Publication Data

Rees, D. G.
Foundations of statistics.
1. Mathematical statistics
I. Title
519.5 QA276

ISBN 0-412-28560-6

Library of Congress Cataloging in Publication Data

Rees, D. G.
Foundations of statistics.

Including index.
1. Mathematical statistics. I. Title.
QA276.R39 1987 519.5 87-15770

ISBN 0-412-28560-6 (PBK.)

Contents

Preface

This book is aimed at the following categories of students:

(a) students taking statistics at A-level either as a full subject, or as a half-subject (e.g. pure mathematics with statistics);
(b) students taking A-level papers in mathematics, applied mathematics or further mathematics which contain some compulsory statistics questions or a statistics option;
(c) students in universities, polytechnics and institutes of higher education taking introductory courses in statistics.

In order to gain most benefit from this book, students should have taken or be taking a course in pure mathematics beyond O-level.

My experience over many years, first as a student of statistics, next as an industrial statistician, and now as a teacher of statistics to both non-mathematicians and mathematicians, has led me to think of statistics as a science. While it is true that statistical methods for analysing data often rely on an underlying theory, partly based on pure mathematics, nevertheless the concepts of data analysis, for example, setting up hypotheses and using data to test them, have much in common with the applied sciences of biology, agriculture, medicine and experimental psychology. The reason why practising statisticians need to study certain aspects of mathematics is so that they can correctly select the appropriate statistical methods for analysing data, and at the same time be fully aware of the assumptions and limitations of these methods. Sometimes, new or modified methods need to be developed to deal with new situations not covered by existing methods, and in such cases a strong mathematical background is vital. This book, in describing some of the most important statistical theory and methods in a logical and clear manner, will, I hope, provide a firm foundation for the budding statistician.

In this book I have covered a large and substantial proportion of the latest GCE syllabuses, partly by studying these syllabuses, but mainly by tackling past and specimen examination papers myself. Two points quickly became apparent:

1. Although there is some material common to all syllabuses, there are nevertheless wide variations in syllabus content.
2. Although Appendix 3 of 'Common Cores at Advanced Level' (GCE Examining Boards of England, Wales and Northern Ireland, Autumn 1983) makes a step in the right direction, there is not much common statistical notation. The same applies to statistical tables.

The topics covered, and the statistical notation and tables adopted, are my attempt to present a consistent and coherent coverage of statistics at A-level.

Because of the broad range of topics covered (necessitated by the variations in A-level syllabus content already mentioned) this book should also be suitable for many first-year 'foundation' courses in statistics in universities, polytechnics and institutes of higher education. Indeed certain applied statistics topics have been included which, although they appear in relatively few A-level syllabuses, are often taught to students in higher education.

Chapters 1 to 3 deal with descriptive statistics while Chapters 4 to 8 cover probability and probability distributions. Populations and samples are defined in Chapter 9 which naturally leads to a discussion of point estimation, while interval estimation is covered in Chapter 10. Hypothesis testing is introduced in Chapters 11 to 14, while Chapters 15 and 16 deal with correlation and regression analysis. Elements of experimental design and analysis are covered in Chapter 17, while Chapter 18 is concerned with quality control charts and acceptance sampling.

There are roughly 200 worked examples to illustrate new concepts and methods as they arise. At the end of each chapter there is a set of exercises (a 'Worksheet'), each set being divided into subsets which are referenced by the appropriate section in the chapter. There are more than 500 end-of-chapter exercises, roughly half of which are from recent past or specimen papers of the GCE examining boards. Answers to all exercises are given in Appendix A.

All the data used in the worked examples and in the non A-level end-of-chapter exercises are fictitious, unless otherwise stated.

Acknowledgements

I wish to thank the following for permission to use questions from their A-level examination papers:

The Associated Examining Board (AEB)
University of Cambridge Local Examinations Syndicate (Cambridge)
Joint Matriculation Board (JMB)
University of London School Examinations Board (London)
University of Oxford Delegacy of Local Examinations (OLE)
Oxford & Cambridge Schools Examination Board (O & C), (SMP), (MEI)
Southern Universities' Joint Board (SUJB)
Welsh Joint Education Committee (WJEC)

I also wish to thank the Institute of Statisticians (IOS) for permission to use questions from their examination papers. Acknowledgements for the use of various statistical tables are given in Appendix C.

A number of teachers and lecturers reviewed part or all of the manuscript in draft form. Their helpful comments were all considered and hopefully have resulted in an improved final version. In particular, I would like to thank Mr S. Barry, Mr J. Chapman, Mrs E. Collins, Mr D. Z. Holmes, Mr E. Marsh and Mr M. Stamp. I am also grateful to Mr P. D. Hinchliffe, former Chief Examiner for the statistics paper of the JMB's Pure Mathematics with Statistics A-level, for his many helpful suggestions and for meticulously checking my answers to the end-of-chapter exercises. Naturally any errors which remain are my responsibility.

Most of all, thanks to my wife, Merilyn, for producing the typescript and for her support and encouragement throughout.

1

Diagrams and tables

▼

1.1 INTRODUCTION

In this chapter we will look at simple ways of representing a set of statistical data in diagrams and tables. These will help us to get some idea of the main points of interest of the data.

1.2 DATA AND AN EXAMPLE OF A DATA SET

The *Oxford English Dictionary* definition of the plural noun *data* is 'facts or information especially as a basis for inference'. Consider the data set shown in Table 1.1 which was obtained from 40 undergraduate students who attended an introductory statistics course at Oxford Polytechnic.

Table 1.1 is an example of *raw* data in the sense that the data have not been grouped or summarized. The six variables recorded provide examples of three types of variables as defined below:

1. A *continuous variable* can take any numerical value in a given range.
2. A *discrete variable* can take only certain distinct numerical values in a given range.
3. A *categorical variable* can take 'values' which denote non-numerical categories or classes.

'Height' and 'distance from home' are examples of continuous variables. Continuous variables are usually *measured* in some units.

Table 1.1 Data set for 40 students

Student reference number	*Sex* *M* = *male* *F* = *female*	*Height* (cm)	*Number of siblings*	*Distance from home to Oxford* (km)	*Type of degree* *A* = *BA* *S* = *BSc* *N* = *BA or BSc* (see note (i))	*A-level count* (see note (ii))
1	M	183	1	80	S	3
2	F	163	2	3	A	16
3	F	152	2	90	A	11
4	F	157	3	272	S	6
5	F	157	1	80	S	6
6	F	165	3	8	S	9
7	M	173	1	485	S	7
8	M	180	2	176	S	4
9	F	164	2	10	S	3
10	F	160	3	72	N	9
11	F	166	0	294	S	8
12	F	157	1	22	N	6
13	F	168	0	144	S	6
14	F	167	2	160	S	6
15	F	156	1	50	S	5
16	F	155	1	64	N	6
17	M	178	1	224	S	4
18	F	169	3	480	S	5
19	F	171	5	56	S	3
20	M	175	3	141	S	4
21	M	169	2	259	S	2
22	F	168	4	96	S	3

23	F	165	1	104	S	6
24	F	166	1	90	S	4
25	F	164	3	72	S	11
26	F	163	1	37	S	4
27	F	161	1	208	S	3
28	F	157	2	40	S	9
29	M	181	2	120	S	5
30	F	163	1	400	S	5
31	F	157	2	208	N	5
32	F	169	2	160	S	6
33	F	177	2	410	A	8
34	F	174	1	90	S	5
35	M	183	1	80	S	5
36	M	181	2	278	S	4
37	M	182	1	240	S	3
38	M	171	9	192	S	12
39	M	184	2	35	N	7
40	M	179	1	45	N	5

Notes

(i) Students combining an arts subject with a science subject may opt for either a BA or a BSc during their final year.

(ii) Grades are converted to numbers using A = 5, B = 4, etc. The A-level count is the total over all subjects passed.

'Number of siblings' (number of brothers and sisters) and 'A-level count' are examples of discrete variables. Discrete variables are usually *counted* and have no units.

'Sex' and 'type of degree' are examples of categorical variables. Categorical variables are qualitative (not quantitative).

1.3 TABLES AND DIAGRAMS FOR CONTINUOUS VARIABLES

Consider the continuous variable 'Distance from home' in Table 1.1. The 40 observed values may be represented in a *cross-diagram* as in Fig. 1.1.

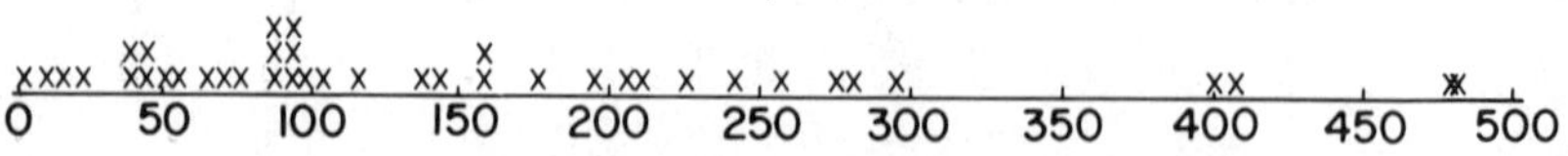

0 50 100 150 200 250 300 350 400 450 500

Fig. 1.1 Cross-diagram for Distance from home (km)

The purpose of this diagram is to show in a simple way how the variable is distributed. Clearly there is an uneven spread of data over the range 0 to 500 km. In fact we can quickly decide that roughly half the 40 students live closer than 100 km from Oxford, while only four (10%) live further than 300 km away.

Another method of handling 20 or more observed values of a continuous variable is to form a *frequency distribution* table and use this table to draw diagrams such as the *histogram* and the *frequency polygon* (see Table 1.2, Figs 1.2 and 1.3).

Table 1.2 Frequency distribution for Distance from home

Distance from home (km)	*Number of students (frequency)*
less than 100	20
100 upto but not including 200	8
200 „ „ „ „ 300	8
300 „ „ „ „ 400	0
400 „ „ „ „ 500	4

When forming a frequency distribution the following guidelines should be followed:

1. Each observed value must go into one and only one group (or class). So groups should not overlap, nor have gaps between them (e.g. 0–90, 100–190, and so on).
2. There should normally be between 5 and 15 groups. Generally the more values we have, the more groups we require.

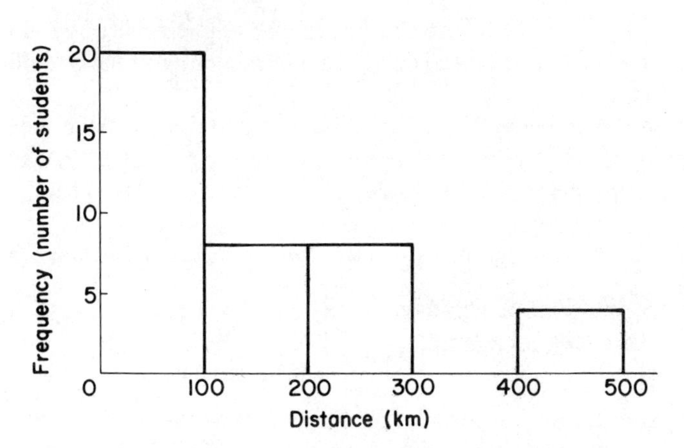

Fig. 1.2 Histogram for Distance from home

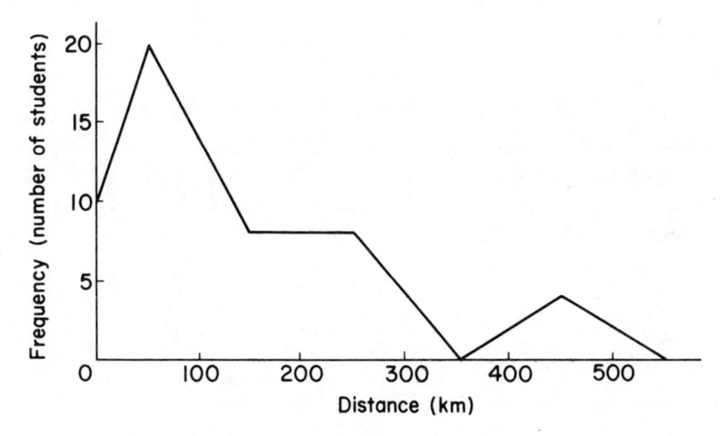

Fig. 1.3 Frequency polygon for Distance from home

3. The group width (100 km in the example above) should normally be the same for all groups, unless there is a good reason for making them different (see Table 1.5 for an example).

The histogram (Fig. 1.2) shows the data from each group of the frequency distribution represented as a rectangle. The area of each rectangle *must* be proportional to the corresponding frequency. If, as in this example, the groups are of equal width, then it follows that the height of each rectangle must also be proportional to the frequency, and so we can label the vertical axis 'frequency'. We cannot do this if the group widths are not all the same (see Table 1.5 and Fig. 1.6 later in this section).

The main purpose of the histogram is to give a quick visual two-dimensional image of the distribution of the observed values (compare the one-dimensional image of Fig. 1.1).

The frequency polygon (Fig. 1.3) can be obtained directly from the histogram by joining up the mid-points of the top sides of the rectangles. We also assume that the groups below the first group and above the last group have zero frequency.

Sometimes the histogram is drawn with *relative frequency* (instead of frequency) as the vertical axis, where

$$\text{relative frequency} = \frac{\text{frequency}}{\text{total frequency}}$$

From Table 1.2 we can form Table 1.3 and draw the histogram of Fig. 1.4.

The shape of Fig. 1.4 is identical to that of Fig. 1.2. The only additional information is that we can obtain percentages by multiplying the relative frequencies by 100. (The idea of relative frequency will be important when we discuss probability in Chapter 4).

Table 1.3 Relative frequency distribution for Distance from home

Distance from home (km)	*Relative frequency*
less than 100	0.50
100 upto but not including 200	0.20
200 ,, ,, ,, ,, 300	0.20
300 ,, ,, ,, ,, 400	0.00
400 ,, ,, ,, ,, 500	0.10
Total	1.00

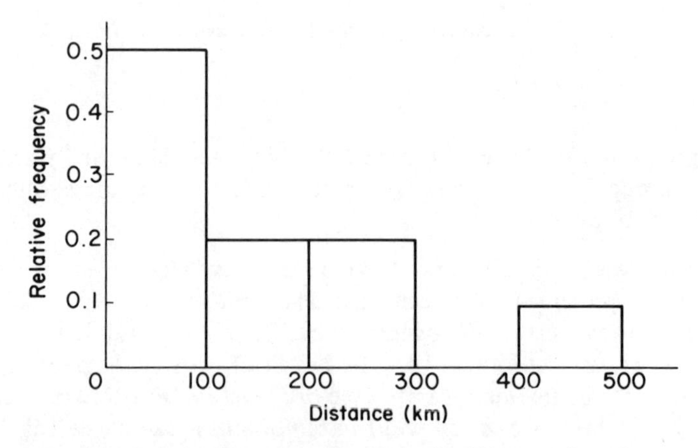

Fig. 1.4 Relative frequency histogram for Distance from home

The *cumulative* frequency distribution and the *cumulative* frequency polygon (sometimes called the ogive) are alternative and often useful ways of displaying data.

From Table 1.3 we can form Table 1.4 and draw Fig. 1.5.

Table 1.4 Cumulative frequency distribution for Distance from home

Distance from home (km)	*Cumulative frequency*
less than 0	0
,, ,, 100	20
,, ,, 200	28
,, ,, 300	36
,, ,, 400	36
,, ,, 500	40

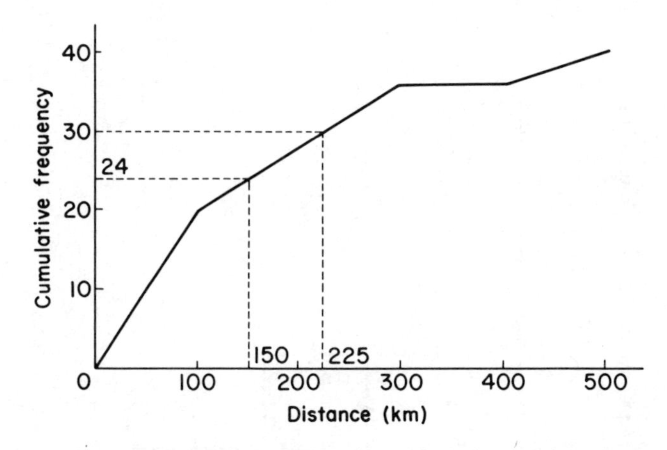

Fig. 1.5 Cumulative frequency polygon for Distance from home

Note that the numbers in the first column of Table 1.4 are the lower end-points of each group, and *not* the group mid-points.

Each row of Table 1.4 contributes one point to Fig. 1.5, and the points are joined by straight lines. We can use Fig. 1.5 in two ways. We can answer questions like 'What percentage of the students live within 150 km of Oxford?' and 'Within what distance from Oxford do 75% (30 out of 40) of students live?'. The answers are 60% (24 out of 40) and 225 km respectively; see Fig. 1.5.

Finally in this section we discuss frequency distributions and histograms with unequal width groups. Using the raw data from Table 1.1, Table 1.5 has been formed with group widths of 50, 50, 100, 100 and 200 respectively.

The right-hand column of Table 1.5 has been calculated using a 100 km group width as a standard (we could have chosen any other reasonable standard, e.g.

Table 1.5 Frequency distribution, unequal width groups

Distance from home (km)	*Frequency*	*Frequency per* 100 *km group width*
less than 50	8	$8 \times \frac{100}{50} = 16$
50 upto but not including 100	12	$12 \times \frac{100}{50} = 24$
100 " " " " 200	8	$8 \times \frac{100}{100} = 8$
200 " " " " 300	8	$8 \times \frac{100}{100} = 8$
300 " " " " 500	4	$4 \times \frac{100}{200} = 2$

50 km). If we draw our histogram with 'frequency per 100 km group width' on the vertical axis, then this diagram will have the property (which all histograms must have) that the area of each rectangle is proportional to the corresponding frequency; see Fig. 1.6.

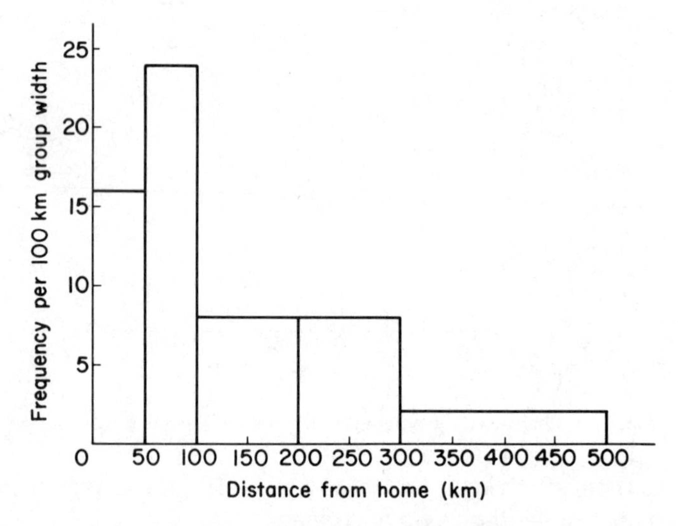

Fig. 1.6 Histogram with unequal width groups

The effect of unequal width groups on the other tables and diagrams, namely Figs 1.3, 1.4, 1.5, and Tables 1.3 and 1.4, is left as an exercise for the reader.

1.4 TABLES AND DIAGRAMS FOR DISCRETE VARIABLES

Consider the discrete variable 'A-level count' in Table 1.1. A frequency distribution table can be drawn up (Table 1.6).

Table 1.6 Frequency distribution for A-level count

A-level count	*Number of students (frequency)*
2	1
3	6
4	6
5	8
6	8
7	2
8	2
9	3
10	0
11	2
12	1
13	0
14	0
15	0
16	1

Notice that the A-level counts have not been grouped since there are only 15 different values to consider. With more values, grouping would have been advisable.

A diagram which can be used to represent discrete data is the *line chart*; see Fig. 1.7.

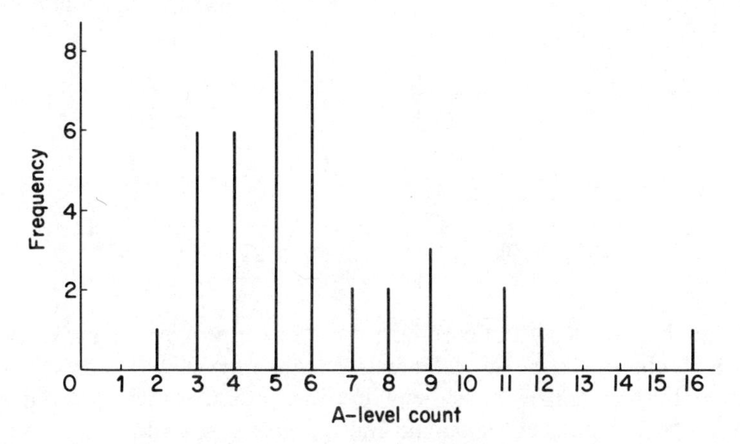

Fig. 1.7 Line chart for A-level count

Note the gaps between the vertical lines, which emphasize that we are dealing with a discrete variable.

We could also draw a relative frequency line chart by dividing each frequency by the total frequency. Its shape would be identical to Fig. 1.7. For discrete data cumulative tables and diagrams are not very useful. Many of the kind of questions of importance can be answered by direct reference to Table 1.6 or Fig. 1.7:

Question What percentage of students have an A-level count of at least 10?

Answer $\frac{0+2+1+0+0+0+1}{40} \times 100 = 10\%$

Question What A-level count is bettered by 50% (20 out of 40) of the students?

Answer 47.5% (19 out of 40) have an A-level count of 6 or more.

1.5 TABLES AND DIAGRAMS FOR CATEGORICAL VARIABLES

Consider the categorical variable 'sex' in Table 1.1. A frequency distribution table can be drawn up (Table 1.7).

Table 1.7 Frequency distribution for Sex

Sex	*Number of students (frequency)*
Female	27
Male	13

A diagram which can be used to represent categorical data is the *bar chart* (Fig. 1.8).

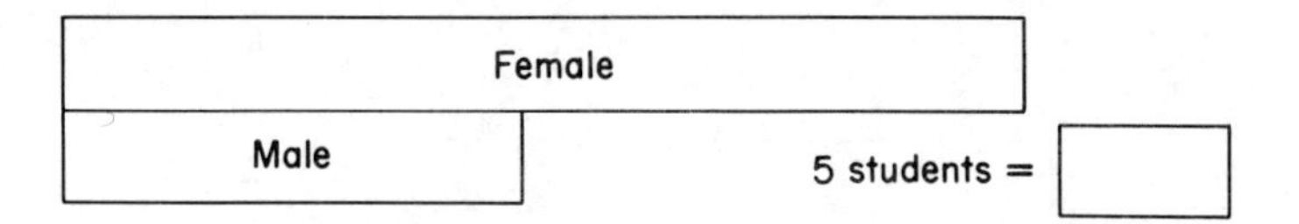

Fig. 1.8 Bar chart for Sex

In a bar chart the length of each horizontal bar is proportional to frequency. The variable changes in the vertical direction to distinguish the bar chart from the histogram. In the example above the main conclusion we can draw is that about two-thirds of the students are female and one-third are male.

1.6 SUMMARY

When we collect statistical data the variables may be continuous, discrete or categorical.

With sufficient data we can draw up frequency distributions for each variable. Examples of diagrams we can draw for continuous variables are the histogram, frequency polygon, relative frequency histogram and cumulative frequency polygon. For discrete variables we can draw line charts, and for categorical variables we can draw bar charts.

WORKSHEET 1

1. Suppose you had to collect data from each of the following:
   (a) 100 cars in a car park
   (b) 100 dogs at a dog show
   (c) 100 'pop' groups
   (d) 100 North Sea oil rigs
   (e) 100 countries
   (f) 100 hotels
   (g) 100 schools

   In each case think of two examples of a continuous variable, two examples of a discrete variable, and two examples of a categorical variable (e.g. for cars, the fuel consumption varies from car to car and so is an example of a variable. But which type?).

2. Using the data from Table 1.1, consider the variable 'Height'. Using groups 149.5 up to but not including 154.5; 154.5 up to but not including 159.5; and so on, draw up a frequency distribution table, a histogram and a cumulative frequency polygon.
   (a) Which group occurs most often?
   (b) By looking only at the histogram, guess the mean height (nearest whole number).
   (c) 50% of the students have a height less than X cm. What is X?
   (d) Y% of the students have a height less than 170 cm. What is Y?
   (e) Z% of the students have a height greater than 165 cm. What is Z?

3. Using the data from Table 1.1, consider the variable 'Number of siblings'. Draw up a frequency distribution table and a line chart.
   (a) Which number of siblings occurs most often?
   (b) By looking only at the line chart, guess the mean number of siblings (not necessarily a whole number).

4. Using the data from Table 1.1, consider the variable 'Type of degree'. Draw up a frequency distribution table and a bar chart.
   (a) What is the most popular type of degree for these students?
   (b) Why can't a mean be calculated for the variable 'Type of degree'?

5. Starting with Table 1.5 draw up a relative frequency distribution table. Hence draw up a relative frequency histogram, and compare its shape with Fig. 1.6.

6. The times in minutes for 50 workers in a factory to complete a standard task were summarized as follows:

Time (minutes)					*Number of workers*
from 8.5	upto	but not	including	9.5	3
9.5	"	"	"	10.5	5
10.5	"	"	"	11.5	13
11.5	"	"	"	12.5	12
12.5	"	"	"	13.5	9
13.5	"	"	"	14.5	6
14.5	"	"	"	15.5	1
15.5	"	"	"	16.5	1

►

Draw a histogram and a frequency polygon in one diagram. Also draw a cumulative frequency polygon.

(a) Approximately, what percentage of workers completed the task in:

(i) less than 10 minutes
(ii) between 10 and 13 minutes
(iii) more than 13 minutes?

(b) Half the workers completed the task in less than X minutes. What is X?

7. The weights of coffee in a random sample of 70 jars marked 200 g were measured in grams and were summarized as follows:

Weight (g)	*Number of jars*
less than 200	0
,, ,, 201	13
,, ,, 202	40
,, ,, 203	58
,, ,, 204	68
,, ,, 205	69
,, ,, 206	70

Form a frequency distribution table and draw a histogram.

8. One hundred seeds were selected from each of 50 consignments of seeds, and kept under carefully standardized conditions favourable to germination. After 14 days the number of seeds which had germinated (in each set of 100) were counted and recorded as follows:

86	95	92	89	92	92	91	88	88	94
92	89	90	86	88	93	92	91	87	91
92	92	91	92	88	94	93	91	92	91
86	93	91	89	90	90	89	84	89	88
90	86	93	88	93	86	94	90	87	93

Draw up a frequency distribution table and a line chart for these data. For what percentage of consignments was there:

(a) at least 90% germination
(b) more than 90% germination?

9. The numbers of goals scored in league matches in a season by the highest scoring players of the 44 teams in Division 1 and 2 of the Football League were as follows:

Division 1	34	29	28	27	27	26	25	25	24	23
	22	21	20	20	20	19	18	18	17	16
	16	14								
Division 2	38	37	37	36	35	34	33	33	31	30
	30	29	29	29	28	24	23	21	18	15
	14	11								

►

(a) Draw up two frequency distribution tables (one for Division 1 and one for Division 2) using groups 10–14, 15–19, and so on.
(b) Explain why a line chart would be inappropriate for these data.
(c) What number of goals is exceeded by half the highest scorers in:

(i) Division 1 (ii) Division 2?

Comment on your answers.

2

Measures of location

2.1 INTRODUCTION

We are all familiar with the term 'average'. This is a statistic which is an attempt to summarize a set of univariate (one variable) data by a single 'middle' value. For example, we might say 'the average age of students in our class is 16 years and 7 months'. Statisticians prefer to use the term 'measure of location' instead of 'average'. There are many different ways of defining such a measure. In this chapter we will consider the following measures of location:

mean
median
mode
geometric mean
weighted mean (including index numbers)

For the mean, median and mode, ungrouped and grouped data will be considered separately.

2.2 MEAN OF UNGROUPED DATA

The *mean*, or to be strictly accurate the arithmetic mean, is simply the total of a set of observations of a variable divided by the number of observations. It follows that, if we use $\bar{x}$ to represent the mean of n observations $x_1, x_2, \ldots, x_n$, then:

$$\bar{x} = \frac{x_1 + x_2 + \ldots + x_n}{n}$$

This formula may be written more economically in Σ (sigma) notation as:

$$\bar{x} = \frac{\sum_{i=1}^{n} x_i}{n}$$

where $\sum_{i=1}^{n} x_i$ means 'sum all terms obtained by first putting $i = 1$ then $i = 2$, and so on up to $i = n$'. (Σ notation is discussed in the Appendix to Chapter 3). If it is clear that there are n observations we write simply:

$$\bar{x} = \frac{\sum x_i}{n} \tag{2.1}$$

If the n observations are a sample from a larger set (population) of observations, $\bar{x}$ is called the 'sample mean'.

Example 2.1

For the first 5 students in Table 1.1, the 'distances from home' are 80, 3, 90, 272, 80 km.

$$\bar{x} = \frac{\sum x_i}{n} = \frac{80 + 3 + 90 + 272 + 80}{5} = 105.0$$

The mean distance from home is 105.0 km. □

2.3 MEAN OF GROUPED DATA

If we have so many values that we decide to form a frequency distribution table (see Section 1.3), we use the following formula for the mean:

$$\bar{x} = \frac{1}{n}\sum f_i x_i, \qquad \text{where} \quad n = \sum f_i \tag{2.2}$$

Here x_i is the mid-point of the ith group, and f_i is the frequency for the ith group. Σ means 'sum over all the groups' (rather than the individual values). Also, since Σf_i is the total frequency, this is the same as n, the total number of observations.

Example 2.2 for continuous data

The weights of coffee in 70 jars are shown in Table 2.1.

$$\bar{x} = \frac{1}{n}\sum f_i x_i = \frac{1}{70} \times 14{,}137.0 = 202.0 \text{ g} \quad (1 \text{ d.p.})$$

Table 2.1 Mean of continuous grouped data

Weight (g)	*Group mid-pt.* x_i	*Frequency* f_i	$f_i x_i$
202 upto but excluding 201	200.5	13	2,606.5
201 " " " 202	201.5	27	5.440.5
202 " " " 203	202.5	18	3,645.0
203 " " " 204	203.5	10	2,035.0
204 " " " 205	204.5	1	204.5
205 " " " 206	205.5	1	205.5
		$n = \sum f_i = 70$	$\sum f_i x_i = 14{,}137.0$

Notes

(i) The column $f_i x_i$ is formed by multiplying corresponding values of f_i and x_i.

(ii) For a particular group, $f_i x_i$ represents the total weight of coffee for that group, assuming that all values occur at the mid-point. So 2,606.5 g is the total weight of coffee for the 13 jars in the first group, and 14,137.0 g is the total weight of all 70 jars. We can see from this example a justification for formula (2.2). □

Some examination boards include coding methods of obtaining the mean in their syllabuses. This makes the arithmetic simpler, but the formula is a little more complicated:

$$\bar{x} = \frac{c}{n}\sum f_i d_i + a, \qquad \text{where} \quad d_i = \frac{x_i - a}{c} \tag{2.3}$$

The proof that formula (2.3) for $\bar{x}$ is equivalent to the formula (2.2) is given in the Appendix to Chapter 3. In (2.3), a is any assumed mean, but it is best to choose one of the group mid-points, for example the mid-point corresponding to the highest frequency. Also in (2.3), c is the group width (sometimes called the class interval), which must be the same for all groups – formula (2.3) does not apply if we have unequal width groups.

Example 2.3

Using the data from Table 2.1, we note that $c = 1$. If we choose $a = 201.5$ we obtain Table 2.2.

$$\begin{aligned}\bar{x} &= \frac{c}{n}\sum f_i d_i + a \\ &= \frac{1}{70} \times 32 + 201.5 \\ &= 202.0 \text{ g} \quad (1 \text{ d.p.})\end{aligned}$$

Table 2.2 Mean of continuous grouped data using coding

Group mid-pt. x_i	$d_i = \dfrac{x_i - 201.5}{1}$	f_i	$f_i d_i$
200.5	−1	13	−13
201.5	0	27	0
202.5	1	18	18
203.5	2	10	20
204.5	3	1	3
205.5	4	1	4
		$n = \sum f_i = 70$	$\sum f_i d_i = 32$

This agrees exactly (as it should) with the answer we obtained without coding.

With the increased use of calculators there is perhaps less justification for this method of coding. □

If we wish to find the mean of *discrete* grouped data, and the groups consist of single values of the variable, as in Table 1.6, then these values are clearly the group mid-points (the groups have zero width), and we can use formula (2.2).

Example 2.4

For the data in Table 1.6, find the mean A-level count.

$$\bar{x} = \frac{1}{n}\sum f_i x_i = \frac{1}{40} \times 239 = 6.0 \quad \text{(1 d.p.)}$$

Table 2.3 Mean of discrete grouped data

A-level count x_i	*Frequency* f_i	$f_i x_i$
2	1	2
3	6	18
4	6	24
5	8	40
6	8	48
7	2	14
8	2	16
9	3	27
10	0	0
11	2	22
12	1	12
13	0	0
14	0	0
15	0	0
16	1	16
	$n = \sum f_i = 40$	$\sum f_i x_i = 239$

The method of coding could be used taking $a = 6$ and $c = 1$, but the arithmetic would not be simpler in this case. □

For categorical data we cannot calculate a mean since the 'values' of the variable are not numerical.

2.4 MEDIAN OF UNGROUPED DATA

The median is defined as the middle value of a set of univariate data once the observations have been ranked (from lowest to highest or vice versa). Suppose we have n observations:

If n is odd, the median is the $\left(\frac{n+1}{2}\right)$th value

If n is even, the median is the mean of the $\left(\frac{n}{2}\right)$th and the $\left(\frac{n+2}{2}\right)$th value.

Example 2.5

Find the median of the observations 80 3 90 272 80.
Ranking, we obtain 3 80 80 90 272.

$$n = 5, \quad \frac{n+1}{2} = 3,$$

and so the median = 80 □

Example 2.6

Find the median of the observations 80 3 90 272.
Ranking, we obtain 3 80 90 272.

$$n = 4, \quad \frac{n}{2} = 2, \quad \frac{n+2}{2} = 3,$$

and so the median $= \dfrac{80+90}{2} = 85$ □

2.5 MEDIAN OF GROUPED DATA

For *continuous* grouped data the median can be found by a graphical method from the cumulative frequency polygon. Since the definition of the median (Section 2.4) implies that half the values are less than the median, it follows that the median is the value corresponding to half the total frequency. The median can be read directly from the cumulative frequency polygon, or calculated more 'precisely' using similar triangles and interpolation.

Example 2.7

The weights of coffee in 70 jars were as shown in Table 2.4.

Table 2.4 Cumulative frequency distribution for the weights of coffee in 70 jars

Weight (g)	*Cumulative frequency* (*number of jars*)
less than 200	0
,, ,, 201	13
,, ,, 202	40
,, ,, 203	58
,, ,, 204	68
,, ,, 205	69
,, ,, 206	70

The cumulative frequency polygon for these data is shown in Fig. 2.1. Half the total frequency equals $\frac{1}{2} \times 70 = 35$. Using the construction shown in Fig. 2.1 we read that the median weight of coffee = 201.8 g (approx.).

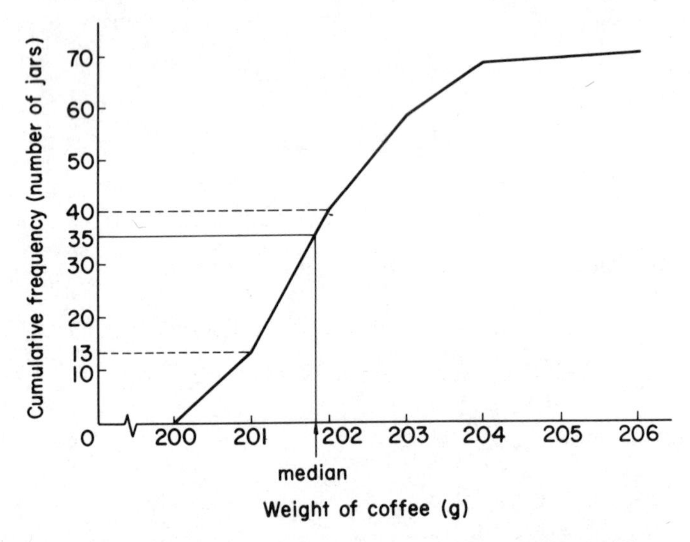

Fig. 2.1 Cumulative frequency polygon for weight of coffee

For a more precise value, we note that the median occurs within the group 201–202, the cumulative frequencies being 13 at 201 and 40 at 202. Consideration of similar triangles (see below) leads to:

$$\text{median} = 201 + \frac{35-13}{40-13} \times (202-201) = 201.8 \text{ g} \quad (1 \text{ d.p.})$$

This follows from consideration of similar triangles:

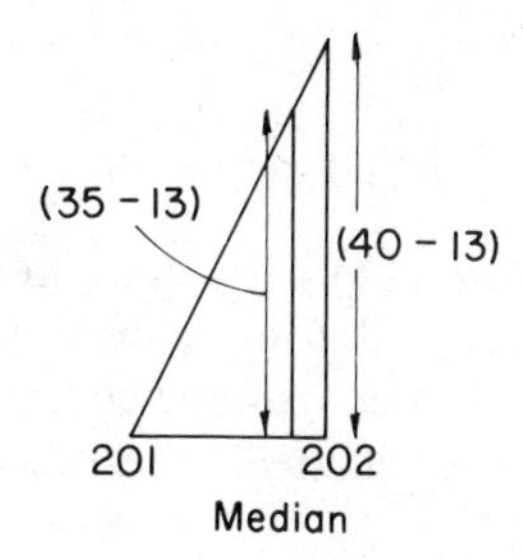

By similar triangles,

$$\frac{\text{median} - 201}{202 - 201} = \frac{35 - 13}{40 - 13}$$

The result above follows.

For most practical purposes the first method of obtaining the median is preferable because it is simpler and precise enough. The precision of the second method is to some extent illusory since the straight lines of the cumulative frequency polygon assume that the weights of coffee are evenly distributed within each group. □

For *discrete* grouped data we can obtain the median by inspecting the frequency distribution table, if each group consists of only a single value of the variable, as in Table 2.3. For these data, half the total frequency is $\frac{1}{2} \times 40 = 20$. The median A-level count is therefore the mean of the 20th and 21st value when these are arranged in rank order. If we write the 40 values of A-level count in rank order we obtain:

```
                                            20th  21st
                                             ↓     ↓
2 3 3 3 3 3 3 4 4 4 4 4 4 5 5 5 5 5 5        5     5    6 and so on
```

The 20th value is 5, so is the 21st, so median A-level count = 5.

For *categorical* data we cannot calculate a median since such data cannot have a ‘middle value’.

2.6 MODE OF UNGROUPED DATA

The *mode* is defined as the value with the highest frequency.

Example 2.8

Find the mode of the observations 80 3 90 272 80.

The mode is 80, since its frequency is 2 while the other three values have a frequency of 1. □

Example 2.9

Find the mode of the observations 80 3 90 272.

All observations occur with a frequency of 1, so we either conclude that each one is the mode, or that there is no mode. This example highlights one of the disadvantages of using the mode as a measure of location – there are others as we shall see later.

2.7 MODE OF GROUPED DATA

For *continuous* grouped data the mode can be found by a graphical method from the histogram. We use the modal group, that is the group with the highest frequency, and the group on either side of it. Using a simple construction the mode can be read directly from the histogram, or calculated more 'precisely' using similar triangles and interpolation.

Example 2.10

The weights of coffee in 70 jars are shown in Table 2.5.

The histogram for these data is shown in Fig. 2.2.

Table 2.5 Frequency distribution for the weights of coffee in 70 jars

Weight (g)	*Frequency (number of jars)*
200 upto but excluding 201	13
201 " " " 202	27
202 " " " 203	18
203 " " " 204	10
204 " " " 205	1
205 " " " 206	1

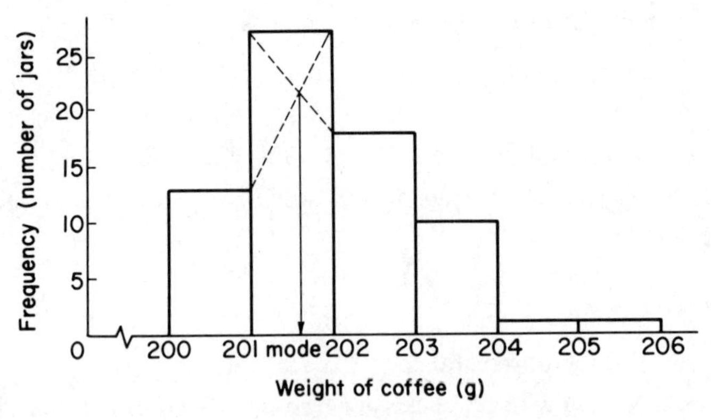

Fig. 2.2 Histogram for weight of coffee

The modal group is 201–202, and where the two dashed lines cross corresponds to the mode. From Fig. 2.2 we read that the mode $= 201.6$ g (approx.).

For a more precise value,

$$\text{mode} = 201 + \frac{(27-13)}{(27-13)+(27-18)} \times (202-201) = 201.6 \text{ g}$$

This follows from consideration of similar triangles:

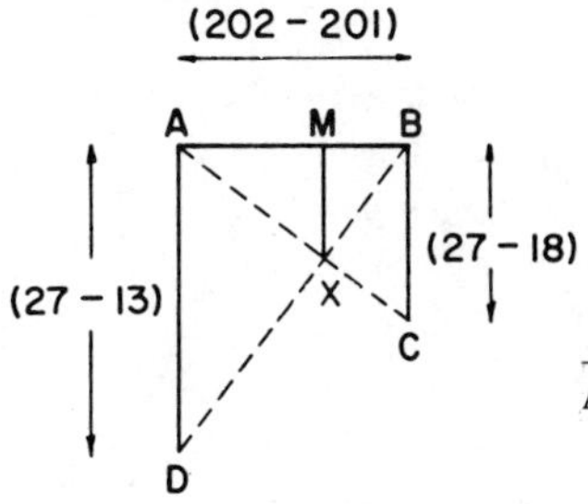

$$\frac{\text{AM}}{\text{AB}} = \frac{\text{MX}}{\text{BC}} \quad \text{by similar triangles AMX, ABC}$$

$$\frac{\text{BM}}{\text{AB}} = \frac{\text{MX}}{\text{AD}} \quad \text{by similar triangles BMX, BAD}$$

$$\therefore \quad \frac{\text{AM}}{\text{BM}} = \frac{\text{AD}}{\text{BC}} \quad \text{dividing equations.}$$

$$\frac{\text{AM}}{\text{AB}-\text{AM}} = \frac{\text{AD}}{\text{BC}}$$

$$\text{AM} \cdot \text{BC} = (\text{AB}-\text{AM})\text{AD}$$

$$\text{AM} = \frac{\text{AD} \cdot \text{AB}}{\text{AD}+\text{BC}}$$

The numerical result above (mode $= 201.6$) follows since mode $= 201 + \text{AM}$.

For most practical purposes the first method of obtaining the mode is preferable because it is simpler and precise enough. □

For *discrete* grouped data we can obtain the mode by inspecting the frequency distribution table, if each group consists of only a single value of the variable, as in Table 2.3. Here there is not a unique mode, since the values 5 and 6 both have the highest frequency of 8.

For *categorical* data, the mode is simply the category with the highest frequency. So for the data in Table 1.7, the modal sex is female.

2.8 WHEN TO USE THE MEAN, MEDIAN AND MODE

In order to decide which of the three measures of location to use in a particular case we need to consider the shape of the distribution as indicated by the histogram (if our variable is continuous) or the line chart (if our variable is discrete). For categorical 'variables', the mode is the only measure which is defined.

If the shape of the histogram or line-chart is roughly symmetrical about a vertical centre line and is unimodal (single peak), then the mean is the preferred measure of location. Such is the case for the 'coffee data' shown in Fig. 2.2. We have already calculated the mean, median and mode for these data (Sections 2.3, 2.5 and 2.7) and found that

$$\text{mean} = 202.0 \text{ g}; \qquad \text{median} = 201.8 \text{ g}; \qquad \text{mode} = 201.6 \text{ g}.$$

These are in close agreement, a difference of 0.4 g between the mean and mode being small in comparison with the 6 g difference between the highest and lowest individual weights ($6 = 206 - 200$). So why should we prefer the mean to the median and mode, when they all give more or less the same value? The answer to this will be fully discussed in Chapter 9. For the moment you are asked to accept that the mean is more precise in the sense that, if lots of samples of 70 jars of coffee were taken, the means of these samples would vary less than the medians or the modes of the samples.

If the shape of the distribution is not symmetrical it is described as *skew*. Figures 2.3(a), (b) and (c) are sketches of unimodal distributions exhibiting symmetry, positive skewness and negative skewness respectively.

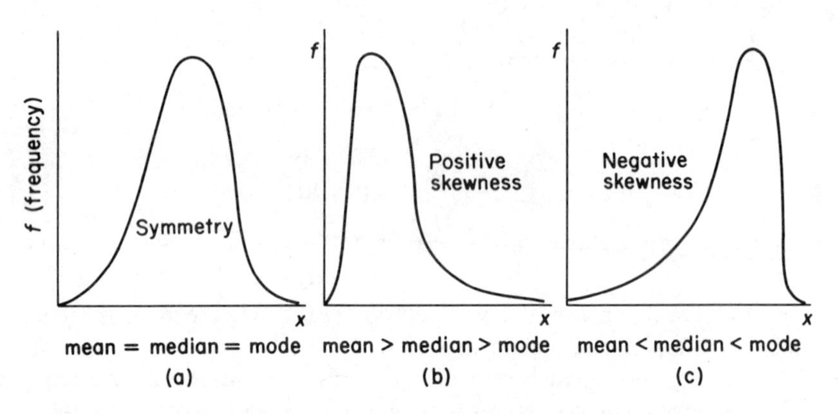

Fig. 2.3 One symmetrical and two skew distributions

Figure 2.3 also indicates the rankings of the mean, median and mode. For example, the mean is more influenced than the median and mode by the few extremely high values if there is positive skewness (Fig. 2.3(b)). For this reason the median is preferred to the mean if data are markedly skew. The mode is rarely used since it may not be unique, and for other theoretical reasons. As we have seen, however, the mode is the only measure of location for categorical data.

2.9 GEOMETRIC MEAN, WEIGHTED MEAN AND INDEX NUMBERS

In certain special applications, measures of location other than those discussed so far in this chapter are employed. For example, if we measure a variable for an 'individual' and we have reason to believe that the value of the variable is increasing (or decreasing) at a rate proportional to its previous value then the *geometric mean* (GM) is used.

The geometric mean of the n values $x_1, x_2, \ldots, x_n$ is defined by the formula:

$$GM = \sqrt[n]{x_1 x_2 \ldots x_n} \tag{2.4}$$

Taking logs of (2.4) gives

$$\log(GM) = \frac{1}{n}\sum \log x_i = \frac{1}{n}[\log x_1 + \log x_2 + \ldots + \log x_n],$$

so the geometric mean is the antilog of the mean of the logs of the observations. Note that the geometric mean can only be used if all observations are positive since logs of negative numbers are not defined.

Example 2.11

If I invested £100 on 1 January 1984, at a compound interest rate of 10%, I would have:

$£100(1+0.1) = £110$	on 1 January 1985
$£100(1+0.1^2) = £121$	,, ,, ,, 1986
$£100(1+0.1)^{16} = £459.5$	,, ,, ,, 2000

The amount I would have exactly half-way through the period from 1984 to 2000, i.e. 1 January 1992, is not the mean:

$$\frac{100+459.5}{2} = £279.75$$

but the geometric mean of the values 100 and 459.5, i.e.

$$\sqrt{100 \times 459.5} = £214.4$$

This can be checked by calculating $100(1+0.1)^8$. □

Other examples of the use of the geometric mean are the weights of children during the growing period, and the size of populations over a period of time. We must be careful to check that the assumption of a constant rate of growth is justified.

The *weighted mean* (WM) has an important application in the calculation of *index numbers*, which are of particular interest in business and finance. Instead of simply adding the n values, $x_1, x_2, \ldots, x_n$, we attach weights $w_1, w_2, \ldots, w_n$, and calculate the weighted mean using:

$$WM = \frac{w_1 x_1 + w_2 x_2 + \ldots + w_n x_n}{w_1 + w_2 + \ldots + w_n} = \frac{\sum w_i x_i}{\sum w_i} \tag{2.5}$$

Note that, if the weights are all equal, the weighted mean and the mean are identical.

Example 2.12

Suppose that I wish to set up an index number[†] for three items of food, namely potatoes, carrots and mushrooms. Table 2.6 shows the quantity I consume per week and the price of each item in Year 0, the *base* year. The table also shows the total expenditure and the weights, i.e. the proportions of total expenditure for each item.

Table 2.6 Calculation of the weights for an index number

Item	*Quantity* (kg)	*Price in Year* 0 (p/kg)	*Expenditure in Year* 0 (p)	*Weight*
Potatoes	5	20	100	0.606 $(=\frac{100}{165})$
Carrots	1	40	40	0.242
Mushrooms	0.1	250	25	0.152
			165	1.000

The value of the index number in the base year is usually taken to be 100, and the weights are assumed to be constant in the calculation of future values of the index.

Suppose that one year later (Year 1) the prices of the three items of food are 25, 40 and 200 p/kg respectively. The new value of the index number is calculated as in Table 2.7, in which it is assumed that the quantities consumed have remained constant.

Table 2.7 The index number for Year 1

Item	*Weight* (w)	*Price relative* (x)	*Weight* × *price relative* (wx)
Potatoes	0.606	125 $(=\frac{25}{20}\times 100)$	75.75
Carrots	0.242	100	24.20
Mushrooms	0.152	80	12.16
			112.11

The 'price relative' column is the ratio of the price in Year 1 to the price in Year 0, multiplied by 100. The index number for Year 1 = 112.1.

We see that the index number has risen by 12.1% from the base of 100. □

The calculation of a Laspeyres price index[‡] for the same example would give the same numerical answer. The calculations are set out differently and perhaps more conveniently in one table, rather than two; see Table 2.8.

[†] a fuller title is 'weighted mean of relatives index number'.
[‡] a fuller title is 'Laspeyres weighted aggregate price index with base year quantity weights'.

Example 2.13

Table 2.8 Calculation of a Laspeyres price index

Item	*Price Year* 0 (p/kg)	*Quantity Year* 0 (kg)	*Expenditure Year* 0 (p)	*Price Year* 1 (p/kg)	*Expenditure in Year* 1 (*Year* 0 *quantities*) (p)
Potatoes	20	5	100	25	125
Carrots	40	1	40	40	40
Mushrooms	250	0.1	25	200	20
			165		185

$$\text{Laspeyres price index for Year 1} = \frac{185}{165} \times 100 = 112.1$$

This is the same answer as in Table 2.7. Table 2.8 shows clearly the total expenditure in Years 0 and 1, and that the Laspeyres price index is simply based on the ratio of these expenditures, assuming the quantities are constant. □

For a Paasche's price index[†], the quantities consumed in Year 1 is used (rather than the Year 0 (base year) quantities) and the index is based on the ratio of the expenditure in Year 1 to that of Year 0. An example is shown in Table 2.9, in which it has been assumed that the Year 1 quantities are 4, 1.2 and 0.12 kg respectively.

Example 2.14

Table 2.9 Calculation of a Paasche's price index

Item	*Price Year* 0 (p/kg)	*Quantity Year* 1 (kg)	*Expenditure in Year* 0 (*Year* 1 *quantities*) (p)	*Price Year* 1 (p/kg)	*Expenditure in Year* 1 (*Year* 1 *quantities*) (p)
Potatoes	20	4	80	25	100
Carrots	40	1.2	48	40	48
Mushrooms	250	0.12	30	200	24
			158		172

Paasche's price index for Year 1 $= \frac{172}{158} \times 100 = 108.9$, an increase of 8.9% from Year 0. □

[†] A fuller title is 'Paasche's weighted aggregate price index with given year quantity weights'.

2.10 SUMMARY

A measure of location is an attempt to summarize a set of data with a single value. For symmetrical unimodal data, the mean is preferred to the median and the mode. For markedly skew data, the median is the preferred measure of location. The mode is useful only for categorical data. The geometric mean is used for a variable which is increasing or decreasing with time at a constant rate. A weighted mean is used in the calculation of index numbers, two examples being the Laspeyres price index and the Paasche's price index.

WORKSHEET 2

Section 2.3

1. Use both the coded and uncoded methods of calculating the mean time to complete a task using the data from Worksheet 1, Exercise 6. Try two different values for a, the assumed mean, for the coded method. Compare your answers.

2. An experiment was carried out to test whether the digit 8 occurred randomly in tables of random numbers. Successive sets of 20 single digits (0, 1, 2, 3, 4, 5, 6, 7, 8, 9) were examined and the number of times (out of 20) that the digit 8 occurred was noted for each of 200 sets. The results were as follows:

Numbers of 8s	0	1	2	3	4	5	6 or more
Number of sets	25	45	70	35	15	10	0

   (a) Calculate the mean number of 8s per set.
   (b) If the random numbers are truly random, what would you expect the answer to (a) to be?

3. (a) Suppose a set of n_1 observations has a mean of $\bar{x}_1$, and another set of n_2 observations has a mean of $\bar{x}_2$. Find the mean of the combined set of $(n_1 + n_2)$ observations. Write down a similar formula for the case of k sets consisting of $n_1, n_2, \ldots, n_k$ observations, respectively.
   (b) Using the data from Worksheet 1, Exercise 9, find the mean number of goals for: (i) Division 1, (ii) Division 2, (iii) both Divisions. Do not group the data.

4. The mean number of calls per minute received at a telephone exchange between 9.00 a.m. and 9.10 a.m. is 4.6, and between 9.10 a.m. and 10.00 a.m. is 3.2. What is the mean for the period 9.00 a.m. to 10.00 a.m.?

5. (a) $\sum_{i=1}^{n} x_i$ means $x_1 + x_2 + \ldots + x_n$ (see Section 2.2).

   Similarly $\sum_{i=1}^{n} (x_i - \bar{x}) = (x_1 - \bar{x}) + (x_2 - \bar{x}) + \ldots + (x_n - \bar{x})$, noting the constant term $\bar{x}$ in each bracket (since $\bar{x}$ does not contain the subscript i).

   Show that $\sum_{i=1}^{n} (x_i - \bar{x}) = 0$.

   (b) The mean of 10 values is 4.3. The first 9 values are 3, 4, 7, 8, 10, 11, 4, 3, 12. What is the 10th value?

6. In the formula for the mean of grouped data it is assumed that each value in a group is that of the mid-point. This assumption must introduce some difference between the means obtained using the grouped and ungrouped formula on the same data set, but it is expected that those differences will be small and unimportant.

   Verify this using the height data from Table 1.1, which were grouped in the solution to Worksheet 1, Exercise 2.

▶

Section 2.5

7. The scores of 150 students who took a verbal scholastic aptitude test were as follows:

Score	0–9	10–19	20–29	30–39	40–49	50–59	60–69	70–79	80–89	90–99
Number of students	2	13	9	10	12	18	40	27	16	3

Instead of receiving their scores the students were assigned grades A, B, C, D, E or F. Grade boundaries were fixed so that:

The 10% of students with the highest marks were assigned grade A; the next 15% were assigned grade B; the next 20% were assigned grade C; the next 15% were assigned grade D; the next 10% were assigned grade E; and the 30% with the lowest marks were assigned grade F.

Draw up a table relating mark ranges to grades.

Section 2.8

8. Find the mean, median and mode for the following variables from Table 1.1.
(a) Distance from home (start with Table 1.2)
(b) Number of siblings (start with your solution to Worksheet 1, Exercise 3)
Decide which measure of location is the most appropriate in each case.

9. The table below shows the salaries of those with the qualification MIS who took part in the 1983 salary survey conducted by the Institute of Statisticians. Calculate the mean, median and the mode. What kind of skewness is exhibited by these data?

Salary £(000s)	*Number of members*
5 up to but not including 10	49
10 15	133
15 20	62
20 25	15
25 30	6
30 40	7
40 50	1
50 60	0
60 70	1

Note: The data in this exercise is published with permission from the Institute of Statisticians.

10. The data below show the sizes of trade unions in 1970 in terms of the number of members.

Draw up a table from which a histogram could be drawn, using 'frequency per 1000 members group width' as one heading. (Do *not* draw a histogram.) ►

Number of members			*Number of unions*
	under	100	90
100	and under	500	116
500	" "	1,000	50
1,000	" "	2,500	58
2,500	" "	5,000	50
5,000	" "	10,000	30
10,000	" "	15,000	13
15,000	" "	25,000	21
25,000	" "	50,000	13
50,000	" "	100,000	16
100,000	" "	200,000	23

What is the median number of members per union and why is this the best measure of location in this case?

Section 2.9

11. A baby weighed 3600 g at birth, and 7000 g at six months old. What did it weigh at the age of three months? What assumption have you made?

12. Show that the geometric mean of two positive values x_1 and x_2 is always less than or equal to the arithmetic mean. When are the two means equal?

13. For four cereals, the prices per unit and the number of units produced in two years were as follows:

	Year 0		*Year 1*	
	Price	*Quantity*	*Price*	*Quantity*
Barley	1.4	240	1.2	470
Corn	1.2	3200	1.1	3800
Oats	0.7	1000	0.6	1400
Wheat	2.2	1100	2.3	1500

Calculate
(a) a weighted mean of relatives index number for Year 1;
(b) a Laspeyres weighted aggregate price index with base year quantity weights for Year 1;
(c) a Paasche's weighted aggregate index with given year quantity weights for Year 1.

In each case assume an index number of 100 for Year 0. Comment on the difference between (i) the answers to (a) and (b); (ii) the answers to (b) and (c). ▶

14. A factory uses four raw materials I, II, III and IV to manufacture a component. For every 1 tonne of I used, 2 tonnes of II, 4 tonnes of III and 3 tonnes of IV are used. The prices of the materials in £/tonne in the years 1983 and 1984 are as follows:

	I	*II*	*III*	*IV*
1983	3	4	6	2
1984	7	6	5	8

Using 1983 as the base year, calculate an index number for the total cost of the raw materials used for the manufacture of the component in 1984.

3

Measures of dispersion and skewness

▼

3.1 INTRODUCTION

In Chapter 2 various measures of location were discussed. For example, we saw how to calculate the mean of a set of observations of a variable. In addition to a measure of location, or 'middle value', we often need to calculate a second type of measure, namely a measure of *dispersion.* This measures the variation in the observations about the middle value. Consider the following two data sets, each consisting of three observations:

(i) 11, 12, 13 (ii) 1, 12, 23.

The mean of each set is 12, but there is clearly more dispersion (variation) in the second set.

Just as there are many different ways of defining a measure of location, there are many ways of defining a measure of dispersion. In this chapter we will consider the following measures of dispersion:

standard deviation
variance
inter-quartile range
percentiles and deciles
range

In some cases we need not only measures of location and dispersion but also a third type of measure, such as a measure of skewness. Two measures of skewness will be discussed briefly.

3.2 STANDARD DEVIATION AND VARIANCE OF UNGROUPED DATA

It might seem that a suitable measure of dispersion would simply be the range, in other words the difference between the largest and the smallest of a set of observations. However, such a measure has drawbacks as discussed in Section 3.6.

A second measure of dispersion which might suggest itself is the mean deviation from the mean, $\bar{x}$, in other words:

$$\frac{\sum(x_i - \bar{x})}{n}$$

However, this is always zero! (see Worksheet 2, Exercise 5).

A very important measure of dispersion (which statisticians have shown is theoretically sound in many situations) is the square root of the mean of the squared deviations from the mean, which is called the *standard deviation.*

A formula for s', the standard deviation of n observations $x_1, x_2, \ldots, x_n$, is:

$$s' = \sqrt{\frac{\sum(x_i - \bar{x})^2}{n}} = \sqrt{\frac{\sum x_i^2}{n} - \bar{x}^2} \tag{3.1}$$

In some applications the square of the standard deviation, called the *variance*, is useful. So the variance is $(s')^2$ where:

$$(s')^2 = \frac{\sum(x_i - \bar{x})^2}{n} = \frac{\sum x_i^2}{n} - \bar{x}^2 \tag{3.2}$$

Both standard deviation and variance are measures of dispersion about the mean, $\bar{x}$. Of the two we will make more use of standard deviation in this book.

The result that the alternative formulae for s' in (3.1) are equivalent is proved in the Appendix to this chapter.

Note

The reason for using s' rather than s for standard deviation is because s is reserved for $\sqrt{\frac{\sum(x_i - \bar{x})^2}{n-1}}$; see Section 9.7 and subsequently.

Example 3.1

For the first 5 students in Table 1.1, the 'distances from home' are:

80 3 90 272 80 (km)

$$n = 5, \quad \sum x_i = 80 + 3 + 90 + 272 + 80 = 525 \text{ km}$$

$$\sum x_i^2 = 80^2 + 3^2 + 90^2 + 272^2 + 80^2 = 94{,}893 \text{ km}^2$$

$$(s')^2 = \frac{\sum x_i^2}{n} - \bar{x}^2 = \frac{94{,}893}{5} - \left(\frac{525}{5}\right)^2 = 7{,}953.6 \text{ km}^2$$

$$s' = 89.2 \text{ km} \quad (1 \text{ d.p.})$$

The standard deviation of the distance from home for the 5 students is 89.2 km. We note that the units of standard deviation are the same as the units of the variable, namely km. Note also that, instead of substituting a rounded value for $\bar{x}$, we substituted exact values for Σx_i and n. Doing this will, in general, give a more accurate value for s' (although in this example it makes no difference). □

3.3 STANDARD DEVIATION AND VARIANCE OF GROUPED DATA

If we have so many values that we decide to form a frequency distribution table (see Section 1.3), the formula for the standard deviation, s', is

$$s' = \sqrt{\frac{\sum f_i (x_i - \bar{x})^2}{n}} = \sqrt{\frac{\sum f_i x_i^2}{n} - \bar{x}^2} \tag{3.3}$$

where $n = \Sigma f_i$, and x_i and f_i are the mid-point and frequency of the ith group respectively. In (3.3) the summations are over all the groups.

The variance $(s')^2$ for grouped data is given by

$$(s')^2 = \frac{\sum f_i (x_i - \bar{x})^2}{n} = \frac{\sum f_i x_i^2}{n} - \bar{x}^2 \tag{3.4}$$

As we saw in Section 3.2, the variance is the square of the standard deviation.

That the alternative formulae for s' are equivalent is proved in the Appendix to this chapter.

Example 3.2 for continuous data.

Extending Table 2.1 to include a column headed $\Sigma f_i x_i^2$ we obtain Table 3.1.

$$\bar{x} = \frac{1}{n}\sum f_i x_i \qquad \text{see (2.2)}$$

$$= \frac{1}{70} \times 14{,}137.0$$

$$(s')^2 = \frac{\sum f_i x_i^2}{n} - \bar{x}^2 \qquad \text{see (3.4)}$$

$$= \frac{2{,}855{,}149.50}{70} - \left(\frac{14{,}137}{70}\right)^2 \qquad \text{see note (ii) below}$$

Table 3.1 Standard deviation of continuous grouped data

Weight (*g*)	*Group mid-pt* x_i	*Frequency* f_i	$f_i x_i$	$f_i x_i^2$ (see note (i))
200 upto but excluding 201	200.5	13	2,606.5	522,603.25
201 ,, ,, ,, 202	201.5	27	5,440.5	1,096,260.75
202 ,, ,, ,, 203	202.5	18	3,645.0	738,112.50
203 ,, ,, ,, 204	203.5	10	2,035.0	414,122.50
204 ,, ,, ,, 205	204.5	1	204.5	41,820.25
205 ,, ,, ,, 206	205.5	1	205.5	42,230.25
		$n = \sum f_i = 70$	$\sum f_i x_i = 14{,}137.0$	$\sum f_i x_i^2 = 2{,}855{,}149.50$

$$(s')^2 = 1.16\ \text{g}^2$$

$$s' = 1.08\ \text{g}$$

The standard deviation of the weight of coffee is 1.08 g.

Notes

(i) The $f_i x_i^2$ column may be obtained by multiplying the x_i column by the $f_i x_i$ column.

N.B. $f_i x_i^2$ should not be confused with $(f_i x_i)^2$.

(ii) Do not round at this stage, since two large numbers, which are very nearly equal, are being subtracted. This calculation should be carried out in the memory of the calculator. □

Now that we know how to calculate standard deviation, what does it tell us? We must postpone a full discussion of this until Chapter 6, but we can say the following at this stage with reference to the data in Table 3.1: the standard deviation of 1.1 g (1 d.p.) is a measure of the variation about the mean weight of 202.0 g (see Section 2.3). The more variation there is about the mean, the larger the standard deviation, and vice versa. If there is no variation at all, i.e. the values are all the same, then the standard deviation is zero. It can never be negative.

Just as there were coding methods for obtaining the mean of grouped data, the same idea can be used in the calculation of standard deviation and variance:

$$\left.\begin{aligned} & s' = c\sqrt{\frac{\sum f_i d_i^2}{n} - \overline{d}^2}, \quad (s')^2 = c^2\frac{\sum f_i d_i^2}{n} - c^2\overline{d}^2, \\ \text{where} \quad & d_i = \frac{x_i - c}{a} \quad \text{and} \quad \overline{d} = \frac{1}{n}\sum f_i d_i \end{aligned}\right\} \tag{3.5}$$

The proof that these formulae are equivalent to (3.3) and (3.4), when no coding was used, is given in the Appendix to this chapter.

Example 3.3 for continuous data.

Extending Table 2.2, in which we used $c = 1$ and $a = 201.5$, to include a column headed $f_i d_i^2$, we obtain Table 3.2.

$$\overline{d} = \frac{1}{n}\sum f_i d_i \qquad \text{see (3.5)}$$

$$= \frac{32}{70}$$

$$(s')^2 = c^2\frac{\sum f_i d_i^2}{n} - c^2\overline{d}^2 \qquad \text{see (3.5)}$$

Table 3.2 Standard deviation of continuous grouped data, using coding

Group mid-pt x_i	$d_i = \frac{x_i - 201.5}{1}$	f_i	$f_i d_i$	$f_i d_i^2$
200.5	−1	13	−13	13
201.5	0	27	0	0
202.5	1	18	18	18
203.5	2	10	20	40
204.5	3	1	3	9
205.5	4	1	4	16
		$n = \sum f_i = 70$	$\sum f_i d_i = 32$	$\sum f_i d_i^2 = 96$

$$= 1^2 \times \frac{96}{70} - 1^2 \times \left(\frac{32}{70}\right)^2$$

$$= 1.16 \text{ g}^2$$

$$s' = 1.08 \text{ g}$$

This agrees exactly (as it should) with the answer we obtained without coding.

The advantage of the coding method is more apparent in the calculation of the standard deviation and variance than it is in the calculation of the mean. For example, the numbers in the final column of Table 3.2 are much simpler than those in the final column of Table 3.1, so arithmetical errors are much less likely. Also rounding errors are not a problem. One disadvantage of the coding method is that it cannot be used if the groups do not all have the same width (a point we noted in Section 2.3 in the calculation of the mean). □

Finally in this section, we will give an example of the calculation of the standard deviation and variance of *discrete* data.

Example 3.4 for discrete data.

Extending Table 2.3 to include a column headed $f_i x_i^2$, we obtain Table 3.3.

$$\bar{x} = \frac{1}{n}\sum f_i x_i \qquad \text{see (2.2)}$$

$$= \frac{239}{40}$$

$$(s')^2 = \frac{\sum f_i x_i^2}{n} - \bar{x}^2 \qquad \text{see (3.4)}$$

Table 3.3 Standard deviation of discrete grouped data

A-level count	*Frequency*		
x_i	f_i	f_ix_i	$f_ix_i^2$
2	1	2	4
3	6	18	54
4	6	24	96
5	8	40	200
6	8	48	288
7	2	14	98
8	2	16	128
9	3	27	243
10	0	0	0
11	2	22	242
12	1	12	144
13	0	0	0
14	0	0	0
15	0	0	0
16	1	16	256
	$n = \sum f_i = 40$	$\sum f_ix_i = 239$	$\sum f_ix_i^2 = 1753$

$$= \frac{1753}{40} - \left(\frac{239}{40}\right)^2$$

$$= 8.12$$

$$s' = 2.85$$

The standard deviation of A-level count is 2.9 (1 d.p.). Note that we did *not* use a rounded value for $\bar{x}$. □

For categorical data we cannot calculate a standard deviation (or a variance) since the 'values' of the variable are not numerical.

3.4 INTER-QUARTILE RANGE, PERCENTILES AND DECILES OF GROUPED DATA

'Quantiles' is the general name we give to values which divide up ranked (i.e. ordered) data into subsets, each subset containing the same number of observations. So the median is an example of a quantile because it divides the ranked data into two subsets each containing half the observations. *Quartiles*, *deciles* and *percentiles* are also examples of quantiles, since they divide ranked data in 4, 10 and 100 subsets respectively.

Dealing first with quartiles, there are three quartiles which we will call Q_1, Q_2 and Q_3:

Q_1, the 'lower' quartile, is such that one-quarter of the observations are less than it in value.

Q_2, is such that two-quarters, i.e. one-half, of the observations are less than it in value.

Q_3, the 'upper' quartile, is such that three-quarters of the observations are less than it in value.

We see that Q_2 is simply the median. Q_1 and Q_3 may be obtained from the cumulative frequency polygon in a similar way to that used to obtain the median.

A measure of dispersion which is used for markedly skew data is the *inter-quartile range* which is the difference between the upper and lower quartiles:

$$\text{Inter-quartile range} = Q_3 - Q_1 \qquad (3.6)$$

Example 3.5 Inter-quartile range for continuous data.

For the coffee data, using Fig. 2.1 reproduced as Fig. 3.1.

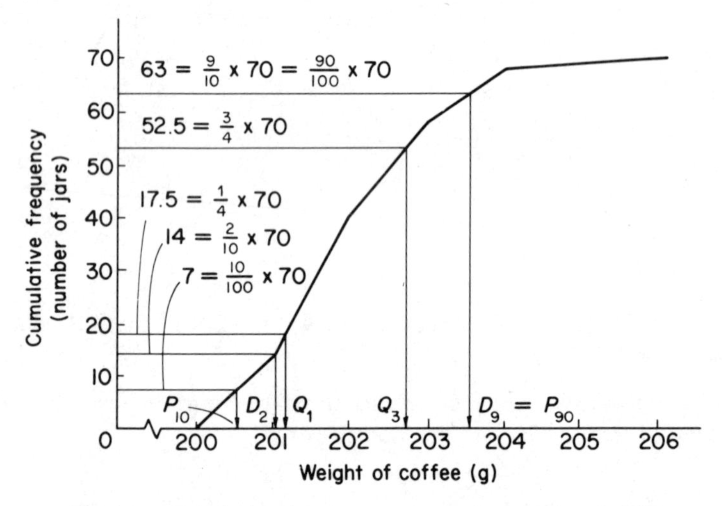

Fig. 3.1 Cumulative frequency polygon for weight of coffee

One-quarter of the total frequency equals $\frac{1}{4} \times 70 = 17.5$.

From Fig. 3.1 we read that the lower quartile, $Q_1 = 201.2$ g approximately.

Three-quarters of the total frequency equals $\frac{3}{4} \times 70 = 52.5$.

From Fig. 3.1 we read that the upper quartile, $Q_3 = 202.7$ g approximately.
So the inter-quartile range $= Q_3 - Q_1 = 202.7 - 201.2 = 1.5$ g.

For more precise values of Q_1 and Q_3 we can use similar triangles (as we did in Section 2.5 to find the median).

It is left as an exercise for the reader to show that

$$Q_1 = 201 + \frac{17.5 - 13}{40 - 13}(202 - 201) = 201.2 \qquad \text{(1 d.p.)}$$

$$Q_3 = 202 + \frac{52.5 - 40}{58 - 40}(203 - 202) = 202.7 \qquad \text{(1 d.p.)}$$

These values agree with those taken directly from Fig. 3.1. □

Deciles, which we will refer to as $D_1, D_2, \ldots, D_9$, divide the values of a ranked set of data into ten subsets, each containing one-tenth of the observations. Similarly, percentiles, which we will refer to as $P_1, P_2, \ldots, P_{99}$, divide the data set into 100 subsets, each containing one-hundredth of the observations. Deciles and percentiles will usually only be needed when we have a very large data set, but we will give a few examples which refer to Fig. 3.1.

Example 3.6

What are the second and ninth deciles, D_2 and D_9, for the data represented in Fig. 3.1?

At a cumulative frequency of $\frac{2}{10} \times 70 = 14, \quad D_2 = 201.0$

At a cumulative frequency of $\frac{9}{10} \times 70 = 63, \quad D_9 = 203.5$ □

Example 3.7

What are the tenth and ninetieth percentiles, P_{10} and P_{90}, for the data represented in Fig. 3.1?

At a cumulative frequency of $\frac{10}{100} \times 70 = 7, \quad P_{10} = 200.5$

At a cumulative frequency of $\frac{90}{100} \times 70 = 63, \quad P_{90} = 203.5$

We note that $D_9 = P_{90}$, since both correspond to 90% of the total frequency. Similarly $D_1 = P_{10}$, and so on.

The 10 *to* 90 *percentile range* is defined as $P_{90} - P_{10}$, and so equals 3 g for the above example. More precise values of the deciles and percentiles could be obtained by using similar triangles and interpolation. □

For *discrete* data, the inter-quartile range, deciles and percentiles can be obtained by inspecting the frequency distribution table, if each group consists of only a single value of the variable, as in Table 2.3. For these data, there are 40 values:

2 3 3 3 3 3 3 4 4 4 4 . . . 6 7 7 8 8 9 9 9 11 11 12 16

Since the 10th and 11th values are both 4, $Q_1 = 4$
Since the 30th and 31st values are both 7, $Q_3 = 7$
Inter-quartile range $= 7 - 4 = 3$

For *categorical* data we cannot obtain quartiles, deciles or percentiles.

3.5 WHICH MEASURE OF DISPERSION TO USE?

Standard deviation should be used as a measure of dispersion for roughly symmetrical, unimodal data. We have already seen (in Section 2.8) that the mean is the preferred measure of location for such data. We also note that the mean occurs in the formulae for the standard deviation, so the two measures 'go together' (although they measure different aspects of a data set). The variance is simply the square of the standard deviation, but variance is useful in its own right in many important applications of statistics, for example in a topic called 'analysis of variance' which is introduced in Chapter 17 of this book.

For markedly skew data the inter-quartile range is useful as a measure of dispersion, since it is less affected by a few extreme values than is the standard deviation. We have already seen (in Section 2.8) that the median is the preferred measure of location for such data. We also note that both the median and inter-quartile range are obtained from the same diagram, namely the cumulative frequency polygon, so these two measures 'go together' (although they measure different aspects of a data set).

An alternative to the inter-quartile range is the percentile range, but each measures dispersion in a different way. Whereas the former, $Q_3 - Q_1$, is the range for the middle 50 % of the ranked observations, the latter, $P_{90} - P_{10}$, is the range for the middle 80 % of the ranked observations.

3.6 RANGE

For a set of observations the range is the (largest value – smallest value). Why do we not use such a simple measure of dispersion? The answer is that it is not used very often for two reasons. Firstly, it is not a very precise measure since it uses only two of the values in a data set, whereas the standard deviation uses all the values (to calculate Σx_i^2, for example). Secondly, the more observations we take, the larger the range is likely to be. Suppose, for example, we select 700 instead of 70 jars of coffee. The chances are that in the larger sample there will be one with a weight smaller than the smallest weight for the smaller sample. Similarly for the largest weight. The range increases as the number of observations increases. The formula for standard deviation, however, takes account of the number of observations, n.

There is one application for the range as a measure of dispersion and this is discussed in Chapter 18, but this applies only when we know that the variable varies in a particular way, namely is 'normally distributed' (see Section 6.6).

3.7 MEASURES OF SKEWNESS

Two measures of skewness which are independent of the units we choose for a given variable are:

$$\frac{\text{mean} - \text{mode}}{\text{standard deviation}} \quad \text{and} \quad \frac{3(\text{mean} - \text{median})}{\text{standard deviation}} \qquad (3.7)$$

These give approximately the same value for unimodal data. For perfectly symmetrical data, both will give a value of zero because then the mean, median and the mode are all equal (see Fig. 2.3(a)). For positively skewed data both measures will be positive and for negatively skewed data both measures will be negative (see Figs 2.3(b) and (c)).

It can be shown that the value of the second measure of skewness above must be in the range -3 to $+3$. As a guide we can conclude that the data show marked skewness if this measure of skewness is outside the range -1 to $+1$.

3.8 SUMMARY

In order to summarize a set of univariate data a measure of dispersion, as well as a measure of location, should be calculated. A measure of skewness is also useful, since the appropriate measures of dispersion and location depend on the shape of the distribution.

Different formulae apply depending on whether data are grouped or ungrouped.

Finally, the range is a measure of dispersion which is generally unreliable, but may be used in special applications. Table 3.4 summarizes the appropriate measures of location and dispersion.

Table 3.4

Type of variable	*Shape of distribution*	*Measure of location*	*Measure of dispersion*
continuous or discrete	roughly symmetrical and unimodal	mean	standard deviation (or variance)
continuous or discrete	markedly skew	median	inter-quartile range (or percentile range)
categorical	–	mode	–

APPENDIX TO CHAPTER 3

Σ notation and some proofs

We have already met Σ notation in Chapters 2 and 3. For example, in Section 2.2, we used

$$\sum_{i=1}^{n} x_i \quad \text{instead of} \quad x_1 + x_2 + \cdots + x_n$$

Σ implies summing, and $i = 1$ to $i = n$ means that we put $i = 1, i = 2, \cdots, i = n$ in turn as the suffix for x and then sum all n terms. We will now look at some other results of using this notation.

1.
$$\sum_{i=1}^{n} k = nk, \text{ where } k \text{ is independent of } i \tag{A3.1}$$

Proof Since k does not contain the suffix i, it does not change as i changes.

$$\therefore \sum_{i=1}^{n} k = k + k + \cdots + k = nk$$

2.
$$\sum_{i=1}^{n} (x_i + y_i) = \sum_{i=1}^{n} x_i + \sum_{i=1}^{n} y_i \tag{A3.2}$$

Proof
$$\begin{aligned}\sum_{i=1}^{n} (x_i + y_i) &= (x_1 + y_1) + (x_2 + y_2) + \cdots + (x_n + y_n) \\ &= (x_1 + x_2 + \cdots + x_n) + (y_1 + y_2 + \cdots + y_n) \\ &= \sum_{i=1}^{n} x_i + \sum_{i=1}^{n} y_i\end{aligned}$$

3.
$$\sum_{i=1}^{n} (x_i + k) = \sum_{i=1}^{n} x_i + nk \tag{A3.3}$$

Proof This result follows from the application of results 2 and 1 above.

4.
$$\sum_{i=1}^{n} kx_i = k \sum_{i=1}^{n} x_i \tag{A3.4}$$

Proof
$$\begin{aligned}\sum_{i=1}^{n} kx_i &= kx_1 + kx_2 + \cdots + kx_n \\ &= k(x_1 + x_2 + \cdots + x_n) \\ &= k \sum_{i=1}^{n} x_i\end{aligned}$$

So we can 'put k in front of Σ', if k is independent of i.

5. The two formulae for the mean of grouped data

$$\bar{x} = \frac{1}{n}\sum f_i x_i \quad \text{and} \quad \bar{x} = \frac{c}{n}\sum f_i d_i + a, \quad \text{where} \quad d_i = \frac{x_i - a}{c} \qquad \text{(A3.5)}$$

are equivalent.

Proof

$$\begin{aligned}
\sum f_i d_i &= \sum f_i \left(\frac{x_i - a}{c}\right) \\
&= \sum \frac{f_i x_i}{c} + \sum \frac{-f_i a}{c} \qquad \text{by (A3.2)} \\
&= \frac{1}{c}\sum f_i x_i - \frac{a}{c}\sum f_i \qquad \text{by (A3.4)} \\
&= \frac{1}{c}\sum f_i x_i - \frac{an}{c} \qquad \text{since } n = \sum f_i \\
\therefore \frac{c}{n}\sum f_i d_i + a &= \frac{c}{n}\left(\frac{1}{c}\sum f_i x_i - \frac{an}{c}\right) + a \\
&= \frac{1}{n}\sum f_i x_i.
\end{aligned}$$

So the right-hand sides of the two formulae for $\bar{x}$ are equal.

6. The two formulae for the standard deviation of ungrouped data

$$s' = \sqrt{\frac{\sum (x_i - \bar{x})^2}{n}} \quad \text{and} \quad s' = \sqrt{\frac{\sum x_i^2}{n} - \bar{x}^2} \qquad \text{(A3.6)}$$

are equivalent.

Proof

$$\begin{aligned}
\sum (x_i - \bar{x})^2 &= \sum (x_i^2 - 2\bar{x}x_i + \bar{x}^2) \\
&= \sum x_i^2 + \sum -2\bar{x}x_i + \sum \bar{x}^2 \qquad \text{by (A3.2)} \\
&= \sum x_i^2 - 2\bar{x}\sum x_i + \bar{x}^2 \sum 1 \qquad \text{by (A3.4)} \\
&= \sum x_i^2 - 2\bar{x}n\bar{x} + \bar{x}^2 n \qquad \text{by (A3.1) and (2.1)} \\
&= \sum x_i^2 - n\bar{x}^2
\end{aligned}$$

Dividing both sides by n and taking the square root, we see that the right-hand sides of the two formulae for s' are equal.

7. The two formulae for the standard deviation of grouped data

$$s' = \sqrt{\frac{\sum f_i (x_i - \bar{x})^2}{n}} \quad \text{and} \quad s' = \sqrt{\frac{\sum f_i x_i^2}{n} - \bar{x}^2} \qquad \text{(A3.7)}$$

are equivalent.

Proof

$$\begin{aligned}\sum f_i(x_i-\bar{x})^2 &= \sum f_i x_i^2 - 2\bar{x}\sum f_i x_i + \bar{x}^2\sum f_i && \text{by (A3.2) and (A3.4)}\\ &= \sum f_i x_i^2 - 2\bar{x}n\bar{x} + \bar{x}^2 n && \text{by (2.2)}\\ &= \sum f_i x_i^2 - n\bar{x}^2\end{aligned}$$

Dividing both sides by n and taking the square root, the right-hand sides of the formulae for s' are equal.

8. The two formulae for the standard deviation of grouped data

$$s' = \sqrt{\frac{\sum f_i x_i^2}{n} - \bar{x}^2} \quad \text{and} \quad s' = c\sqrt{\frac{\sum f_i d_i^2}{n} - \bar{d}^2} \qquad \text{(A3.8)}$$

where $\quad d_i = \dfrac{x_i - a}{c}$ and $\bar{d} = \dfrac{1}{n}\sum f_i d_i$, are equivalent.

Proof We need to show that

$$\frac{\sum f_i x_i^2}{n} - \bar{x}^2 = c^2\frac{\sum f_i d_i^2}{n} - c^2\bar{d}^2$$

Now
$$c^2\frac{\sum f_i d_i^2}{n} = \frac{c^2}{n}\sum f_i\left(\frac{x_i - a}{c}\right)^2$$

$$= \frac{1}{n}\sum f_i(x_i - a)^2 \qquad \text{by (A3.4)}$$

$$= \frac{1}{n}\sum f_i x_i^2 - \frac{2a}{n}\sum f_i x_i + \frac{a^2}{n}\sum f_i$$

by (A3.2) and (A3.4)

$$= \frac{\sum f_i x_i^2}{n} - 2a\bar{x} + a^2 \qquad \text{by (2.2)}$$

Also, since
$$\bar{x} = \frac{c}{n}\sum f_i d_i + a \qquad \text{by (2.3)}$$

$$= c\bar{d} + a \qquad \text{by (3.5),}$$

it follows that $c\bar{d} = \bar{x} - a$.

Hence
$$c^2\frac{\sum f_i d_i^2}{n} - c^2\bar{d}^2 = \frac{\sum f_i x_i^2}{n} - 2a\bar{x} + a^2 - (\bar{x} - a)^2$$

$$= \frac{\sum f_i x_i^2}{n} - \bar{x}^2$$

Section 3.2

1. Eleven bags of sugar each marked 1 kg actually contained the following weights of sugar:

   1.02 1.05 1.08 1.03 1.00 1.06 1.08 1.01 1.04 1.07 1.00

   (a) Without grouping these data, calculate the mean and standard deviation of the weight of sugar, using appropriate formulae.
   (b) Suppose the total weights (bag and sugar) are calculated in kg. Assuming that each packet weighs exactly 10 g, can you guess the new mean and standard deviation? Verify your answers by calculation, using appropriate formulae.
   (c) Suppose the total weights in (b) are calculated in grams (instead of kg). Guess the new mean and standard deviation in grams. Verify your answers by calculation, using appropriate formulae.
   (d) Check your answers to (a), (b) and (c) using the mean and standard deviation facilities on your calculator. (*Warning*: some calculators have *two* buttons for standard deviation.)

2. Use both the coded and uncoded methods of calculating the standard deviation of the time to complete a task using the data from Worksheet 1, Exercise 6. Try two different values for a, the assumed mean, for the coded method. Compare your answers.

3. The mean and standard deviation of some data for the time taken to complete a chemical test have been calculated with the following results:

   number of observations = 25, mean = 18.2 seconds,
   standard deviation = 3.25 seconds.

   A further set of 15 observations ($x_1, x_2, \cdots, x_{15}$), also in seconds, is now available and for these data

   $$\sum_{i=1}^{15} x_i = 279 \quad \text{and} \quad \sum_{i=1}^{15} x_i^2 = 5524$$

   Calculate a mean and standard deviation based on all 40 observations.

   (IOS)

4. The mean and standard deviation of a set of n_1 observations are $\bar{x}_1$ and s'_1 respectively, while the mean and standard deviation of another set of n_2 observations are $\bar{x}_2$ and s'_2. What are the mean and standard deviation of the combined set of $(n_1 + n_2)$ observations?

   Use your result to verify the answer to Question 3.

5. In an experiment in extrasensory perception (ESP) four cards marked A, B, C and D are used. The 'experimenter', unseen by the 'subject', shuffles the cards and selects one. The subject decides which card he thinks has been selected. This procedure is repeated five times for each of a random sample of 50 subjects. The number of times out of five when each subject correctly identifies a selected card is counted. These data are recorded in the following frequency distribution table: ▶

Number of correct decisions	0	1	2	3	4	5
Number of subjects	15	18	8	5	3	1

(a) Calculate the mean and standard deviation of the number of correct decisions per subject.
(b) Calculate the total number of correct decisions made by all subjects.
(c) Calculate the percentage of the total number of decisions made which were correct. What percentage would you have expected if you knew that all subjects were guessing at each selection of a card?

6. The following data are the 'distances travelled' in thousands of kilometres by 60 tyres in simulated wear tests before they reached the critical minimum tread.

Distance (000s km)	*Number of tyres*
16 upto but excluding 24	5
24 ,, ,, ,, 32	10
32 ,, ,, ,, 40	20
40 ,, ,, ,, 48	16
48 ,, ,, ,, 56	9

Find $\bar{x}$ and s', the mean and standard deviation of the distance travelled. Draw a cumulative frequency polygon and estimate what percentage of tyres 'travelled' between the following distances:

(a) 25,000 to 50,000 km
(b) $(\bar{x}-s')$ to $(\bar{x}+s')$ km

Section 3.4

7. A random sample of 1000 surnames is drawn from a local telephone directory. The distribution of the lengths of the names is as shown:

Number of letters in surname	3	4	5	6	7	8	9	10	11	12
Frequency	13	102	186	237	215	113	83	32	13	6

Calculate the sample mean and sample standard deviation. Obtain the upper quartile.
Represent graphically the data in the table.
Give a reason why the sample of names obtained in this way may not be truly representative of the population of Great Britain.

(JMB)

8. An investigation into alcohol consumption in England and Wales was carried out by recording, for a random sample of adults, their alcohol consumption to the nearest unit, in the week preceding the interview. The results, for 933 men and for 1063 women, are shown separately, in the table below.

(A unit is equivalent to half a pint of beer, 1/6 gill of spirits, a glass of wine (4 fl. oz.) or a small glass of fortified wine (2 fl. oz.)).

▶

Consumption during week preceding interview	%	
	Men	*Women*
0 units	24	43
1–5 units	20	34
6–10 units	14	10
11–20 units	16	10
21–35 units	13	2
36–50 units	8	1
51 or more units	5	

Source: Social Trends

Plot the data for men as a histogram.

Showing all your working, obtain values for:

(i) the sample median for men's consumption;
(ii) the sample inter-quartile range for men's consumption;
(iii) the sample mean for women's consumption;
(iv) the sample proportion of women drinking more than 15 units in the previous week.

Suggest reasons why the interview asked for alcohol consumption over the preceding week and not over a longer period. Indicate how the sample should be taken if the results are to be representative of alcohol consumption *throughout the year*.

(JMB)

9. The data in the table are taken from the National Travel Survey of 1978/79. They relate to the distance in miles travelled by the 95.2 thousand people who were going to or from work. The figures are percentages.

Journey distance	*Percentage*
under 1 mile	16
1 but under 3 miles	31
3 " " 10 miles	39
10 " " 15 miles	7
15 " " 30 miles	5
30 miles and over	2

Plot these data as a histogram.

Showing all your working, obtain values for:

(i) the mean distance travelled;
(ii) the median distance travelled;
(iii) the inter-quartile range of distance travelled;
(iv) the proportion of people travelling less than the mean distance.

The same survey interviewed 48.3 thousand people travelling for the purpose of education. The distances they were travelling are given below: ►

Journey distance	*Percentage*
under 1 mile	51
1 but under 3 miles	31
3 " " 10 miles	15
10 " " 15 miles	2
15 " " 30 miles	1
30 miles and over	0

Describe *one* major difference between this distribution and the previous one and suggest a possible explanation for the difference you describe.

(JMB)

10. The table below shows the cumulative distribution of gross weekly earnings in £'s among men aged 21 years and over in a survey period in 1977.

Earnings	< 30	< 35	< 40	< 45	< 50	< 55
% of men	0.5	1.1	2.4	5.5	10.8	18.2
Earnings	< 60	< 65	< 70	< 75	< 80	< 85
% of men	27.0	36.6	45.7	54.6	62.2	69.0
Earnings	< 90	< 95	< 100	< 120	< 150	< 200
% of men	74.2	78.9	82.7	92.1	97.0	99.2

Find the probability that a man chosen at random earns more than £100 per week.

Estimate the median earnings and the semi-interquartile range.

Illustrate these data by drawing a histogram showing earnings in the ranges: 0–40, 40–60, 60–80, 80–100, 100–120 and 120–200 £ per week.

(O&C)

11. The following table shows the frequency distribution of the masses, to the nearest kilogram, of a batch of 150 steel bars.

Mass	10–19	20–29	30–39	40–49	50–59	60–69	70–79	80–89
Frequency	4	18	28	56	25	15	3	1

Working with an origin at the mid-point of the fourth class interval and with units of 10 kilograms, prepare a table for the given data showing all the terms in the summations that have to be calculated to obtain the mean and the standard deviation of the masses of the bars. Calculate these two quantities correct to one decimal place.

Compile a cumulative frequency distribution table and represent this distribution graphically by a cumulative frequency polygon. Using your graph, or otherwise, estimate the median mass of the bars.

Estimate from the cumulative frequency polygon, or otherwise, the smallest and the largest percentiles of the distribution which lie within two standard deviations of the mean.

(JMB) ►

12. The table shows the number of addicts of dangerous drugs on record as receiving treatment, classified by age. The figures are for the United Kingdom for 1980.

Age	*15–19*	*20–24*	*25–29*	*30–34*	*35–49*	*50 and over*
Male	18	221	705	697	238	110
Female	16	175	277	155	88	99

Plot the data for male addicts as a histogram.
Showing all your working, obtain values for:

(i) the mean age of the male addicts;
(ii) the median age of the male addicts;
(iii) the inter-quartile range of the age of the male addicts;
(iv) the proportion of the male addicts aged 40 and over.

Without carrying out any further calculations state *two* major differences between the male and female addicts as shown by these data.

(JMB)

13. The weights of 225 fifteen-year-old children were measured, and the following frequency table was constructed:

Mid-interval weight (lb) $= w$	75	80	85	90	95
Number $= f$	1	3	6	9	13
Mid-interval weight (lb) $= w$	100	105	110	115	120
Number $= f$	17	22	26	30	31
Mid-interval weight (lb) $= w$	125	130	135	140	145
Number $= f$	29	21	11	4	2

$\sum f = 225 \quad \sum fw = 25{,}600 \quad \sum fw^2 = 2{,}957{,}650$

Illustrate these data in a histogram.
Estimate (to 1 decimal place) the mean, median and standard deviation of these weights, using (where necessary) the results at the end of the table.

Frequency distributions which are not symmetrical about their mean are called skew; the parameter S, where

$$S = \frac{\text{mean} - \text{median}}{\text{standard deviation}}$$

is sometimes used as a measure of the degree of skewness. Estimate S for the data given in the table. What can be deduced about the shape of another histogram for which S has the same magnitude as that for the data above, but the opposite sign?

Give one reason why S is a better measure of the degree of skewness than S', where $S' = (\text{mean} - \text{median})$.

(MEI) ►

Section 3.7

14. Find the standard deviation and inter-quartile range for the data shown in Worksheet 2, Exercise 9. Which one measure of dispersion would you quote if asked?

15. Find the standard deviation and inter-quartile range for the following variables from Table 1.1:

    (a) Distance from home (start with your solution to Worksheet 2, Exercise 8(a)).
    (b) Number of siblings (start with your solution to Worksheet 2, Exercise 8(b)).

    Also calculate $\frac{3(\text{mean} - \text{median})}{\text{standard deviation}}$ in each case, and hence decide which measure of dispersion is the more appropriate in each case.

16. Two random samples each consisting of 40 students took a course in statistics. Their examination marks were summarized as follows:

	Number of students	
Mark range	*Sample 1*	*Sample 2*
10–19	1	1
20–29	3	1
30–39	4	3
40–49	7	4
50–59	12	8
60–69	8	12
70–79	4	6
80–89	1	4
90–99	0	1

Treating the variable Examination mark as continuous, so that the group 10–19 really represents marks between 9.5 and 19.5, and so on,

(a) draw a cumulative frequency polygon for each sample on the same graph;
(b) hence, or otherwise, obtain the median, inter-quartile range and percentile range for sample 1. Does sample 2 have a large or smaller median?
(c) find the mean, standard deviation and $\frac{3(\text{mean} - \text{median})}{\text{standard deviation}}$ for sample 1, and comment on the degree of skewness of the data for sample 1.

17. The data below are the yields in kilograms from a random sample of 50 cherry trees in an orchard.

    (a) Summarize these data:

        (i) in a table;
        (ii) graphically;
        (iii) by suitable measure of location and dispersion. ►

21	40	33	29	34	12	59	36	54	46
30	30	21	35	42	38	43	47	33	28
43	38	47	27	57	35	33	64	43	24
52	24	51	40	36	41	57	30	29	35
37	46	23	25	28	32	22	10	37	49

(b) Use a cumulative frequency polygon to estimate the percentage of trees which lie within one standard deviation of the mean.

Appendix to Chapter 3

18. A set of n values $x_1, x_2, \cdots, x_n$ has a mean $\bar{x}$ and standard deviation s'_x.

(a) A new set of n values $y_1, y_2, \cdots, y_n$ is obtained by adding a constant k to each of the x values, so that

$$y_i = x_i + k \quad \text{for} \quad i = 1, 2, \cdots, n$$

Show that $\bar{y} = \bar{x} + k$ and that $s'_y = s'_x$, where $\bar{y}$ and s'_y are the mean and standard deviation of the y values.

Hint: sum the n equations $y_i = x_i + k$ for $i = 1, 2, \cdots, n$,
then show that $y_i - \bar{y} = x_i - \bar{x}$ for $i = 1, 2, \cdots, n$

and use $s'_x = \sqrt{\dfrac{\sum(x_i - \bar{x})^2}{n}}$.

(b) A new set of n values $z_1, z_2, \cdots z_n$ is obtained by multiplying each of the x values by a positive constant l, so that

$$z_i = lx_i \quad \text{for} \quad i = 1, 2, \cdots, n$$

Show that $\bar{z} = l\bar{x}$ and that $s'_z = ls'_x$, where $\bar{z}$ and s'_z are the mean and standard deviation of the z values.

(c) A new set of n values $w_1, w_2, \cdots, w_n$ is obtained by multiplying each of the x values by a positive constant l and adding a constant k, so that

$$w_i = lx_i + k$$

Show that $\bar{w} = l\bar{x} + k$ and that $s'_w = ls'_x$, where $\bar{w}$ and s'_w are the mean and standard deviation of the w values.

(d) The result of (a) implies that if we add a constant to each of a number of values the mean is increased by the value of the constant but the standard deviation is unaltered. What similar statements can be made in relation to (b) and (c)?

19. (a) Given the set (a, b, c, d, e) with mean m and standard deviation s, what are:

(i) the mean and standard deviation of the set

$(a+k, b+k, c+k, d+k, e+k)$;

(ii) the mean and standard deviation of the set

(ak, bk, ck, dk, ek)?

In a test, five candidates obtained marks of 6, 8, 10, 14, 17. Calculate the mean and standard deviation of these marks (your method must be clearly shown). ▶

It is necessary to scale these marks so that they have a mean of 60 and a standard deviation of 20. Calculate the new marks.

(b) Two sets, each of 20 members, have the same standard deviation of 5. The first set has a mean of 17, and the second a mean of 22. Find the standard deviation of the set obtained by combining the given two sets.

(SUJB)

20. A set of n values $x_1, x_2, \cdots, x_n$ are to be scaled into a set of values $w_1, w_2, \cdots, w_n$ using $w_i = lx_i + k$, for $i = 1, 2, \cdots, n$, where l and k are constants.

If the mean and standard deviation of the x's are 48 and 12, respectively, and the mean and standard deviation of the w's are required to be 55 and 15, what values of l and k should be chosen?

For the x's, the lower quartile, median, and upper quartile are 40, 52 and 60 respectively. What are the corresponding statistics for the w's? Show that the inter-quartile range for the w's is independent of the value of k.

21. (a) Find the mean and standard deviation of the first n positive integers.
(b) For $n = 10$, show that the mean is 5.5 and the standard deviation is 2.87.

Find the new mean, standard deviation and variance if:

(i) we add 1 to each value;
(ii) we multiply each value by -1;
(iii) we multiply each value by -1 and then add 1 to each value.

(c) Repeat (b) for $n = 100$.

4

Basic ideas of probability

4.1 INTRODUCTION

There are many ways of approaching the subject of probability, ranging from the highly theoretical to the highly practical. In this book the latter approach is taken, for the most part, in the belief that most readers will wish to have a thorough knowledge of a number of methods of dealing with a range of probability problems.

If we consider the probability that there will be an all-out nuclear war before the end of this century, then we may be thinking of probability as a *measure of uncertainty*, in which case we may wish to assign a value to this measure. Because degrees of optimism about the future vary, different people will assign different values to this probability. We will not consider such *subjective* estimates of probability.

Instead we will consider probability as a measure of uncertainty in connection with *experiments* whose *outcomes* are uncertain, in the sense that one of a number of outcomes may occur when we carry out the experiment on a particular occasion. Two approaches to this concept of probability will be considered. The *relative frequency* approach requires that the experiment is repeated a large number of times. The second approach requires only that we think about what would happen if the experiment were to be carried out, although we must be sure that the outcomes of the experiment are *equally likely* to occur.

4.2 SOME TERMINOLOGY

Definition 4.1

> A *trial* is an action which results in one of several possible outcomes, called the *outcome set, S.*

Example 4.1

Throw a die once. The trial is the action of throwing the die. The outcome set, S, is:

$$S = \{1, 2, 3, 4, 5, 6\}$$ □

Example 4.2

Ask a student to indicate how well he or she likes statistics as a subject. The trial is the action of asking the student and the thought processes which follow. A possible outcome set, S, is:

$$S = \{\text{love it, like it, so-so, dislike it, hate it}\}$$ □

Example 4.3

A customer selects a light-bulb in a shop and the assistant tests it for the customer. Here the trial is selecting and testing the light-bulb. The outcome set, S, is:

$$S = \{\text{bulb lights, bulb does not light}\}$$ □

Example 4.4

A ball is drawn from a bag containing three red, five white and seven blue balls. The trial is selecting the ball. The outcome set, S, is:

$$S = \{\text{red ball, white ball, blue ball}\}$$ □

Example 4.5

A card is selected from a standard pack of well-shuffled playing cards. The trial is selecting the card. The outcome set, S, is:

$$S = \{\text{Ace of spades, King of spades, \ldots, 3 of clubs, 2 of clubs}\}$$ □

Definition 4.2

> An *experiment* consists of a number of trials.

The five examples above (4.1–4.5) can all be enlarged to become examples of experiments rather than single trials.

Example 4.6

(i) Throw a die ten times.
(ii) Ask twenty students to indicate how well they like statistics.
(iii) A customer selects four light bulbs to be tested.
(iv) Three balls are drawn from a bag.
(v) Two cards are selected from a pack. □

Definition 4.3

An *event*† is a subset of the outcome set.

Example 4.7

Throw a die once. The outcome set, S, is:

$$S = \{1, 2, 3, 4, 5, 6\}$$

Suppose event A is 'even number'. Then event

$$A = \{2, 4, 6\}$$

Suppose event B is 'number $\leqslant 4$'. Then event

$$B = \{1, 2, 3, 4\}$$

We can illustrate events A and B in a *Venn diagram* as in Fig. 4.1.

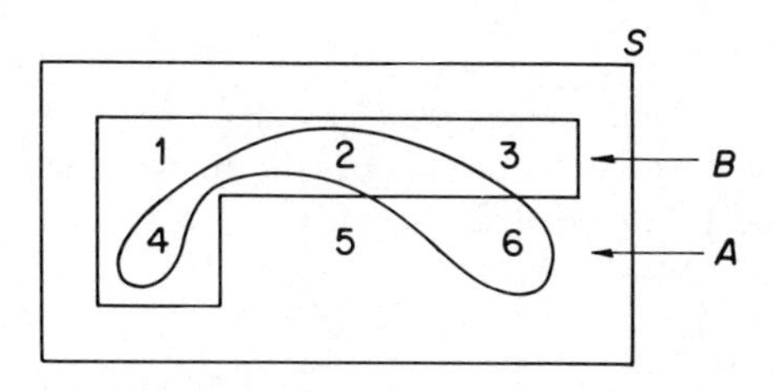

Fig. 4.1 Venn diagram for one throw of a die □

Example 4.8

Draw a card from a well-shuffled pack. The outcome set, S, is:

$$S = \{\text{Ace of spades, King of spades, } \ldots \text{, 3 of clubs, 2 of clubs}\}$$

† Some texts describe the distinct and indivisible outcomes of a trial as 'elementary events'. We will not use this terminology.

Suppose event A is 'spade'. Then event

$$A = \{\text{Ace of spades, King of spades, } \ldots \text{, 2 of spades}\}$$

Suppose event B is '4'. Then event

$$B = \{\text{4 of spades, 4 of hearts, 4 of diamonds, 4 of clubs}\}$$

We can illustrate events A and B in a Venn diagram as in Fig. 4.2.

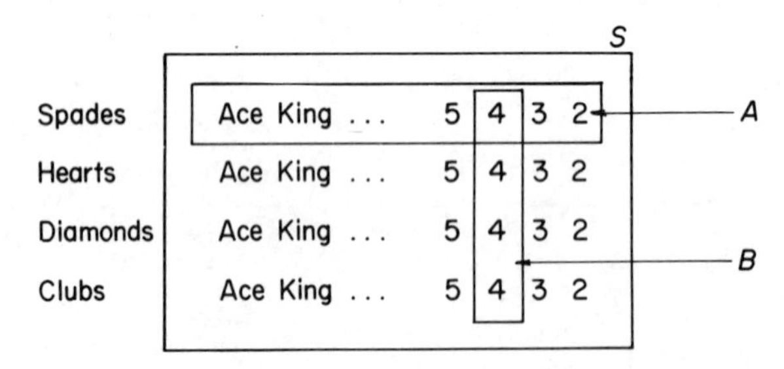

Fig. 4.2 Venn diagram for drawing a card from a pack □

4.3 THE DEFINITION OF PROBABILITY FOR THE CASE OF EQUALLY LIKELY OUTCOMES

With the background of the last section we can now define probability for the case in which each of the outcomes of a trial is equally likely to occur.

Definition 4.4

> If an outcome set S consists of $n(S)$ equally likely outcomes, and $n(E)$ of these correspond to an event E, then
>
> $$P(E) = \frac{n(E)}{n(S)},$$
>
> where $P(E)$ denotes the *probability of event E.*

This definition is sometimes referred to as the 'a priori' definition, because the Latin phrase 'a priori' means 'obtained by deduction without sensory experience'. This implies that, as long as we know the outcomes are equally likely, we can calculate the probabilities of various events *without carrying out an experiment.*

The main application area of the 'equally likely' or 'a priori' definition is in games of chance.

Example 4.9

Consider the outcome set corresponding to one throw of a die, and suppose the event E is 'even number'.

Then $n(S) = 6$, $n(E) = 3$, and $P(E) = \frac{3}{6}$

Here we are assuming we have a 'fair' die, i.e. that the six possible outcomes 1, 2, 3, 4, 5, and 6 are equally likely to occur. □

Example 4.10

Consider the outcome set corresponding to drawing one ball from a bag containing three red, five white and seven blue balls, and suppose the event E is 'blue ball'. Assuming each of the 15 balls is equally likely to be drawn, $n(S) = 15$, $n(E) = 7$, and $P(E) = \frac{7}{15}$. □

Example 4.11

Consider the outcome set corresponding to selecting a card at random from a standard pack of 52 playing cards, and suppose the event E is 'spade'. The phrase 'at random' is another way of indicating that each of the 52 cards is equally likely to be selected.

So $n(S) = 52$, $n(E) = 13$, and $P(E) = \frac{13}{52}$ □

It follows from the 'equally likely' definition of probability that the minimum value for a probability is 0. This occurs when none of the outcomes in the outcome set correspond to some event. It also follows that the maximum value for a probability is 1. This occurs when all of the outcomes in the outcome set correspond to some event.

Example 4.12

For one throw of a fair die

P (number 7) $= 0$

P (number < 7) $= 1$ □

One curious aspect of the 'equally likely' definition is that it is a *circular* definition because equally likely really means equally probable. However, this is not a practical problem since our other definition of probability (the relative frequency definition, to be discussed in Section 4.4) leads to estimates of probability which agree well with values obtained using the 'equally likely' definition.

What approach should we use if the outcomes are not equally likely? Look again at Examples 4.2 and 4.3 (in Section 4.2). In Example 4.2, we cannot say that the probability that a student will indicate 'love it' is $\frac{1}{5}$ because we have no reason to believe that the five choices in the outcome set are equally likely. And in

Example 4.3 we cannot say that the probability that a bulb will not light is $\frac{1}{2}$ because we have no reason to believe that the two outcomes 'bulb lights' and 'bulb does not light' are equally likely.

Another approach is needed to deal with outcomes which are not equally likely to occur, and this will be discussed in the next section.

4.4 THE RELATIVE FREQUENCY DEFINITION OF PROBABILITY

Suppose the person who is asking students how well they like the subject of statistics has carried out similar surveys in the past and has carefully recorded the data in a frequency distribution table (Table 4.1).

Table 4.1 The opinions of 150 students concerning statistics as a subject

Opinion	love it	like it	so-so	dislike it	hate it
Number of students	25	60	40	20	5

From these data we can estimate that the probability of the event 'love it' is $\frac{25}{150}$, that is the frequency of the event in question relative to the total frequency, expressed as a ratio.

The formal relative frequency definition of probability is as follows.

Definition 4.5

If a large number of trials, n, are carried out, and $n(E)$ of these trials result in event E, then an estimate of the probability of event E is

$$P(E) = \frac{n(E)}{n}$$

The larger the value of n the better will be the estimate, the latter settling down to a limit as n increases.

Notice that, in order to obtain our estimate of probability, we must carry out an experiment or survey. For this reason the relative frequency definition is sometimes called the 'a posteriori' definition, after the Latin phrase 'a posteriori' implying estimating a probability *after an experiment.*

Example 4.13

A shop assistant who tests light-bulbs for customers finds that, of 50 tested, 48 bulbs light and 2 bulbs do not light. If E is the event 'bulb lights', then $n = 50$, $n(E) = 48$, and we estimate that $P(\text{bulb lights}) = P(E) = 48/50$.

As with the 'equally likely' definition, the minimum and maximum values of probability using the relative frequency definition are 0 and 1 respectively. If none of the n trials result in the event, the estimate of probability is $0/n$ or 0. If all of the n trials result in the event, the estimate of probability is n/n or 1. □

4.5 PROBABILITY, PROPORTION, PERCENTAGE AND ODDS

In the example of the previous section we estimated $P(\text{bulb lights}) = 48/50$. We could also say that 48/50 is the *proportion* of the light-bulbs which light, or we could say that the *percentage* of light-bulbs which light is $\frac{48}{50} \times 100 = 96\%$, or we could say the *odds* are 48 to 2 (or 24 to 1) that a light-bulb will light.

4.6 PROBABILITIES OF THE INTERSECTION OF EVENTS; THE MULTIPLICATION LAW

Suppose that, instead of considering the probabilities of various events resulting from a single trial, we consider the possible sequences of events resulting from a series or sequence of trials.

Example 4.14

A die is thrown twice. The outcome set can be written as follows:

$$S = \{1 \text{ and } 1, 1 \text{ and } 2, \ldots, 6 \text{ and } 6\}$$

where 1 and 2 means 1 with the first throw and 2 with the second throw, and so on. The 36 outcomes may be displayed graphically as in Fig. 4.3.

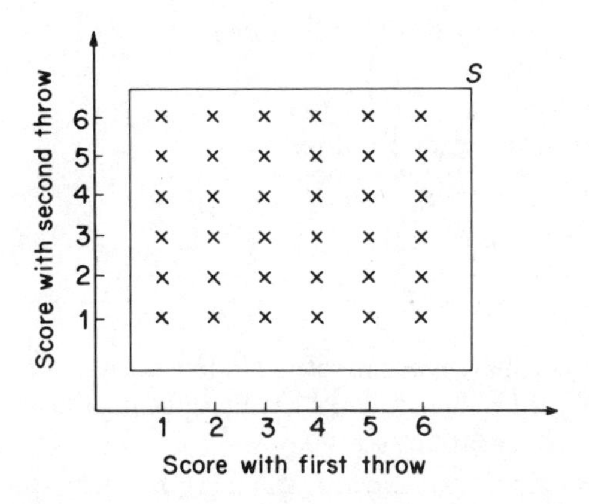

Fig. 4.3 Venn diagram for two throws of a die

Consider the two events '3 with first throw' and '4 with second throw', and suppose we are interested in the probability that both events occur, i.e. the probability that we will throw 'a 3 and then a 4'.

To evaluate this probability we may use the multiplication law of probability:

$$P(E_1 \cap E_2) = P(E_1)P(E_2 | E_1) \tag{4.1}$$

Notes

(i) $P(E_1 \cap E_2)$ means the probability of the intersection of the two events E_1 and E_2, which simply means the probability that both E_1 and E_2 will occur.

(ii) $P(E_2 | E_1)$ means the probability that event E_2 will occur, given that event E_1 has already occurred. $P(E_2 | E_1)$ is called a *conditional probability*. □

Example 4.15

What is P(a 3 and then a 4) if we throw a die twice?

Here E_1 is the event '3 with first throw' and E_2 is the event '4 with second throw'.

$$P(E_1) = \tfrac{1}{6}, \text{ using the 'equally likely' definition}$$

$$\begin{aligned} P(E_2 | E_1) &= P \text{ (4 with second throw given 3 with first throw)} \\ &= \tfrac{1}{6}, \text{ using the 'equally likely' definition} \end{aligned}$$

So $$P(E_1 \cap E_2) = \tfrac{1}{6} \times \tfrac{1}{6} = \tfrac{1}{36}, \quad \text{using (4.1)}$$

The probability of 'a 3 and then a 4' is $\frac{1}{36}$.

Note that the condition 'given 3 with first throw' is redundant. Whatever happens as a result of the first throw will not affect the probabilities for the second throw. In this case we say that the events E_1 and E_2 are *statistically independent*.

More mathematically we say that:

Events E_1 and E_2 are statistically independent if

$$P(E_2) = P(E_2 | E_1) \tag{4.2}$$

and, for such events, the multiplication law is

$$P(E_1 \cap E_2) = P(E_1)P(E_2) \tag{4.3}$$

□

Example 4.16

(With two events which are *not* statistically independent). A card is drawn from a pack and a second card is then drawn without replacing the first card. What is the probability that both cards are aces?

Here E_1 is the event 'first card is an ace', and E_2 is the event 'second card is an ace'. We wish to find $P(E_1 \cap E_2)$. The probability that the second card will be an ace is affected by the first card drawn, since the latter is not replaced. E_1 and E_2 are not statistically independent, so we use (4.1).

$$\begin{aligned} P(E_1 \cap E_2) &= P(E_1)P(E_2 \mid E_1) \\ &= \frac{4}{52} \times P(\text{second card is an ace given first card is an ace}) \\ &= \frac{4}{52} \times \frac{3}{51} \end{aligned}$$

$$\text{So } P(2 \text{ aces}) = \frac{4}{52} \times \frac{3}{51}$$ □

Example 4.17

Consider the previous example but this time for the case in which the first card is replaced and the pack is shuffled before the second card is selected. Then the events 'first card is an ace' and 'second card is an ace' are statistically independent. Using (4.3) we obtain

$$P(2 \text{ aces}) = \frac{4}{52} \times \frac{4}{52}$$ □

The multiplication law for n events (where $n > 2$) is as follows:

$$P(E_1 \cap E_2 \cap E_3 \ldots \cap E_n) = P(E_1)P(E_2 \mid E_1)P(E_3 \mid E_1 \cap E_2) \ldots P(E_n \mid E_1 \cap E_2 \cap \ldots \cap E_{n-1})$$

If all n events are statistically independent, the law becomes:

$$P(E_1 \cap E_2 \cap E_3 \ldots \cap E_n) = P(E_1)P(E_2)P(E_3) \ldots P(E_n)$$

Example 4.18

If four cards are drawn from a pack without replacement

$$P(4 \text{ aces}) = \frac{4}{52} \times \frac{3}{51} \times \frac{2}{50} \times \frac{1}{49}$$ □

Example 4.19

If a die is thrown three times,

$$P(3 \text{ and then } 4 \text{ and then } 5) = \frac{1}{6} \times \frac{1}{6} \times \frac{1}{6}$$ □

4.7 PROBABILITIES OF THE UNION OF EVENTS; THE ADDITION LAW

Suppose we are considering events E_1 and E_2 which can occur as possible results of an experiment. Suppose we wish to know the probability of the occurrence of

$$E_1, \text{ or } E_2, \text{ or both } E_1 \text{ and } E_2$$

We write this as $P(E_1 \cup E_2)$, where $E_1 \cup E_2$ is called the union of events E_1 and E_2. To evaluate this probability we may use the addition law of probability:

$$P(E_1 \cup E_2) = P(E_1) + P(E_2) - P(E_1 \cap E_2) \tag{4.4}$$

Notice that we can use the multiplication law to evaluate $P(E_1 \cap E_2)$, as discussed in the previous section.

Example 4.20

What is the probability of obtaining 'at least one 6' with two throws of a die?

We can describe 'at least one 6' as '6 with first throw or 6 with second throw or two 6s'.

So if E_1 denotes the event '6 with first throw', and E_2 denotes the event '6 with second throw', we need to evaluate $P(E_1 \cup E_2)$.

Using (4.4) and (4.3),

$$P(E_1 \cup E_2) = \frac{1}{6} + \frac{1}{6} - \frac{1}{6} \times \frac{1}{6}$$

$$= \frac{11}{36}$$

The probability of 'at least one 6' is $\frac{11}{36}$. □

Sometimes the events E_1 and E_2 will be such that if one occurs, the other cannot occur. In other words, they cannot both occur. In this case we say that the events E_1 and E_2 are *mutually exclusive* (or *disjoint*). Such events have no common area on a Venn diagram (Fig. 4.4).

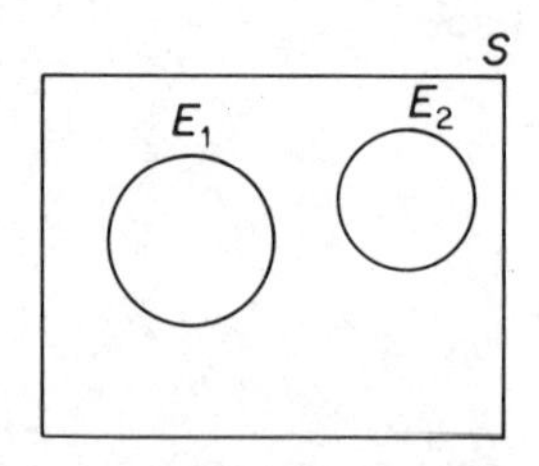

Fig. 4.4 Venn diagram for mutually exclusive events E_1 and E_2

More mathematically we say that:

Events E_1 and E_2 are mutually exclusive if $P(E_1 \cap E_2) = 0$ (4.5)

And, for such events, the addition law is:

$$P(E_1 \cup E_2) = P(E_1) + P(E_2) \quad (4.6)$$

Example 4.21

What is the probability of obtaining a 3 or a 6 when a die is thrown once? The events '3' and '6' cannot both occur with one throw of a die, i.e. they are mutually exclusive events. Hence

$$P(3 \text{ or } 6) = P(3) + P(6) = \frac{1}{6} + \frac{1}{6} = \frac{1}{3}$$ □

Example 4.22

Based on the data in Table 4.1, estimate the probability that a student either loves statistics or hates statistics.

Since the events 'loves statistics' and 'hates statistics' are mutually exclusive, we use (4.6) and obtain:

$$\begin{aligned} P(\text{loves or hates statistics}) &= P(\text{loves statistics}) + P(\text{hates statistics}) \\ &= \frac{25}{150} + \frac{5}{150} \\ &= \frac{30}{150} \end{aligned}$$ □

Example 4.23

What is the probability that the total score from two throws of a die will be 10? There are three events which result in a total of 10, namely 4 and then 6; 5 and then 5; 6 and then 4. All three are mutually exclusive. So, extending the addition law to the case of the three mutually exclusive events, we may write:

$$\begin{aligned} P(\text{total of } 10) &= P(4 \text{ and then } 6 \textit{ or } 5 \text{ and then } 5 \textit{ or } 6 \text{ and then } 4) \\ &= P(4 \text{ and then } 6) + P\ (5 \text{ and then } 5) + P\ (6 \text{ and then } 4) \\ &= \frac{1}{6} \times \frac{1}{6} + \frac{1}{6} \times \frac{1}{6} + \frac{1}{6} \times \frac{1}{6} \\ &= \frac{3}{36} \end{aligned}$$

(using the multiplication law for independent events three times) □

4.8 COMPLEMENTARY EVENTS, A MUTUALLY EXCLUSIVE AND EXHAUSTIVE SET OF EVENTS, AND THE PROBABILITY OF 'AT LEAST ONE'

If *two* events are mutually exclusive and together they exhaust all the possibilities, they are called *complementary* events.

If E denotes one of the events, we denote the complementary event by E' (see Fig. 4.5). E' is usually called 'not E' rather than 'E dashed'.

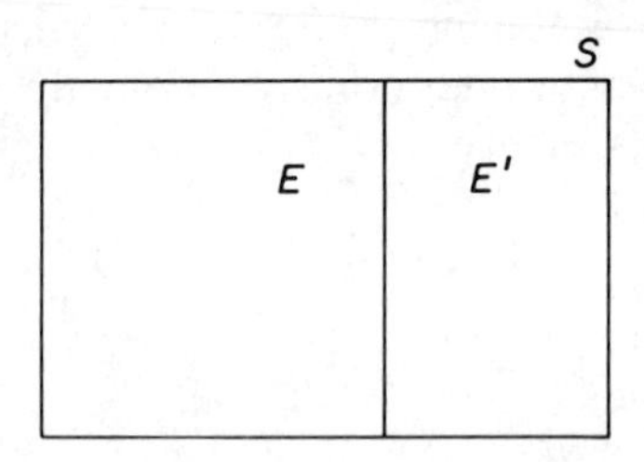

Fig. 4.5 Venn diagram for complementary events

Since E and E' are mutually exclusive, $P(E \cup E') = P(E) + P(E')$ by (4.6). Since E and E' are exhaustive, $P(E \cup E') = 1$. Hence $1 = P(E) + P(E')$, or

$$P(E') = 1 - P(E) \tag{4.7}$$

Example 4.24

What is the probability that a 5 will not be thrown with one throw of a die?

If E represents 'throw a 5', then we require $P(E')$

$$\begin{aligned}\text{So } P(E') &= 1 - P(E)\\ &= 1 - P(\text{throw a 5})\\ &= 1 - \frac{1}{6}\\ &= \frac{5}{6}\end{aligned}$$ □

If the outcomes of an experiment form a set of n mutually exclusive and exhaustive events $E_1, E_2, \ldots, E_n$, then using similar arguments to the case of two events we obtain the result:

$$P(E_1) + P(E_2) + \ldots + P(E_n) = 1 \tag{4.8}$$

Equation (4.8) is useful in checking that we have correctly calculated the probabilities of a set of mutually exclusive and exhaustive events.

Example 4.25

What are the probabilities that a family of three children will contain no girls, one girl, two girls, three girls?

Assume boys and girls are equally likely at each birth. Let E_0, E_1, E_2, E_3 denote the events no girls, one girl, two girls and three girls respectively. Then the various possible outcomes for families with three children may be listed as follows, where GBG for example implies the birth order 'girl then boy then girl'.

BBB	E_0
BBG, BGB, GBB	E_1
BGG, GBG, GGB	E_2
GGG	E_3

So
$$P(E_0) = P(\text{BBB}) = \frac{1}{2} \times \frac{1}{2} \times \frac{1}{2} = \frac{1}{8}, \qquad \text{using (4.3)}$$

$$P(E_1) = P(\text{BBG}) + P(\text{BGB}) + P(\text{GBB}), \qquad \text{using (4.6)}$$

$$= \frac{1}{2} \times \frac{1}{2} \times \frac{1}{2} + \frac{1}{2} \times \frac{1}{2} \times \frac{1}{2} + \frac{1}{2} \times \frac{1}{2} \times \frac{1}{2}, \qquad \text{using (4.3)}$$

$$= \frac{3}{8}$$

Similarly, $P(E_2) = \frac{3}{8}$, $P(E_3) = \frac{1}{8}$ and we verify, as expected from (4.8), that $P(E_0) + P(E_1) + P(E_2) + P(E_3) = 1$. □

Equation (4.8) is also useful when we want to evaluate the probability that 'at least one . . .' will occur.

Example 4.26

A die is thrown five times. What is the probability that at least one 6 will occur? If $P(1)$ stands for the probability that one 6 will occur in five throws and so on,

then $$P(0) + P(1) + P(2) + P(3) + P(4) + P(5) = 1$$

i.e. $$P(0) + P(\text{at least } 1) = 1$$

i.e. $$P(\text{at least } 1) = 1 - P(0)$$

$$= 1 - P(6' \text{ and } 6' \text{ and } 6' \text{ and } 6' \text{ and } 6')$$

$$= 1 - \frac{5}{6} \times \frac{5}{6} \times \frac{5}{6} \times \frac{5}{6} \times \frac{5}{6}$$

$$= 0.598$$ □

4.9 USING BOTH LAWS OF PROBABILITY, TREE DIAGRAMS

There are many probability problems for which we can enumerate all the possible outcomes of an experiment as a set of mutually exclusive and exhaustive events. These outcomes and their corresponding probabilities may then be displayed graphically in what is called a *tree diagram* (which some texts refer to as a probability tree).

Example 4.27

A bag contains 5 red balls and 3 black balls. If 3 balls are drawn *without* replacement, what is the probability that:

(a) no black balls will be selected;
(b) exactly one red ball will be selected;
(c) at least one red ball will be selected?

The first trial, namely selecting the first ball, is represented by branches of a tree, the number of branches depending on the number of possible outcomes of the trial. The events and their probabilities are entered as in Fig. 4.6.

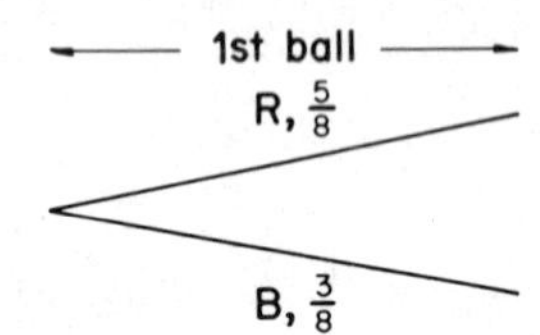

Fig. 4.6 Part of a tree diagram

Subsequent trials are represented by further branches as shown in Fig. 4.7. It is important to note that the probabilities for all trials after the first trial are *conditional* on the outcomes of previous trials. For example, the probability that the second ball is red, given that the first ball is also red is 4/7 (since only 4 reds remain out of 7 balls).

Notice that the probabilities emanating from each point sum to 1 since the corresponding events are mutually exclusive and exhaustive. Figure 4.7 shows that there are 8 possible outcomes to the whole experiment whose probabilities are obtained by multiplying probabilities along the branches as indicated. The total probability of these 8 outcomes is also 1, since they also form a mutually exclusive and exhaustive set.

(a) $$P(\text{no black balls}) = P(\text{RRR}) = \frac{60}{336}$$

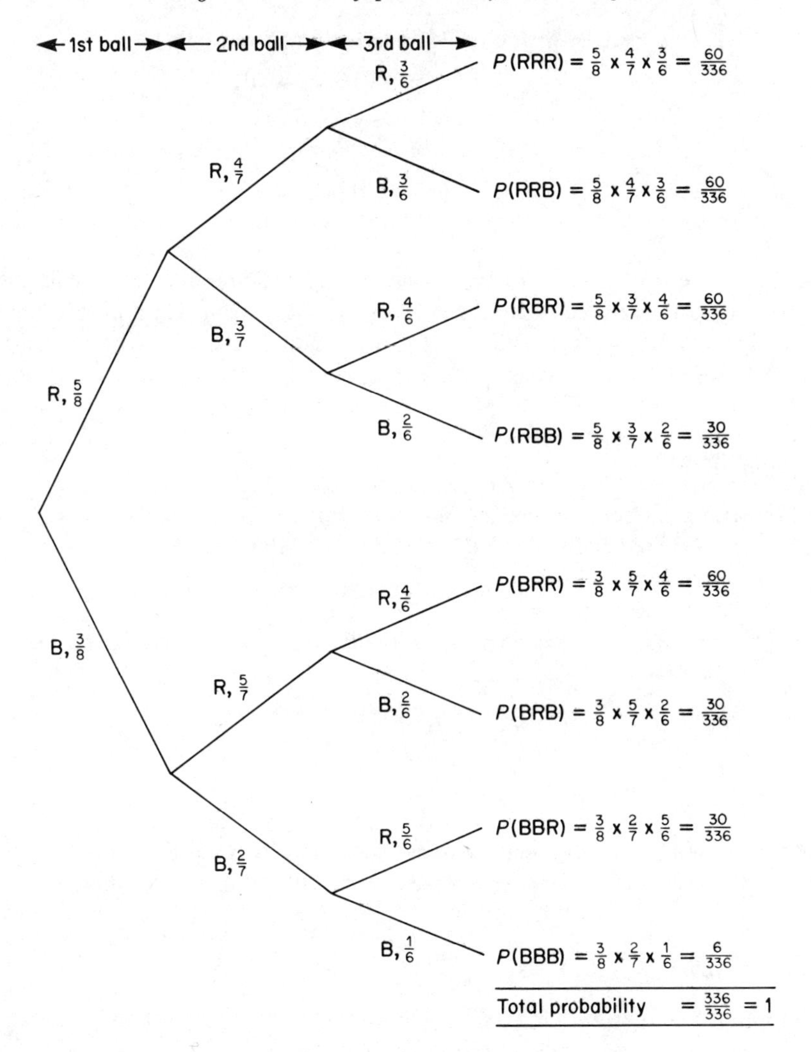

Fig. 4.7 Tree diagram

(b)

$$P(\text{exactly one red ball}) = P(\text{RBB}) + P(\text{BRB}) + P(\text{BBR})$$

$$= \frac{30}{336} + \frac{30}{336} + \frac{30}{336}$$

$$= \frac{90}{336}$$

(c)
$$\begin{aligned}P(\text{at least one red ball}) &= 1 - P(\text{no red balls})\\ &= 1 - P(\text{BBB})\\ &= 1 - \frac{6}{336}\\ &= \frac{330}{336}\end{aligned}$$ □

Essentially we use the multiplication law along the branches to calculate the probabilities of the outcomes of the experiment. Then we use the addition law for mutually exclusive events to calculate the probabilities of events which consist of two or more of the outcomes of the experiment.

Example 4.28

Without using a tree diagram, calculate the probability of exactly one red ball.

$$\begin{aligned}P(\text{exactly one red ball}) &= P(\text{RBB or BRB or BBR})\\ &= P(\text{RBB}) + P(\text{BRB}) + P(\text{BBR}) \qquad \text{by (4.6)}\\ &= P(R)\cdot P(B|R)\cdot P(B|R\cap B) + 2 \text{ similar terms}\\ &= \frac{5}{8}\times\frac{3}{7}\times\frac{2}{6} + 2 \text{ equal terms}\\ &= \frac{90}{336}\end{aligned}$$ □

The tree diagram can also be used to evaluate certain conditional probabilities, which are in the opposite sense to those represented on the tree diagram.

Example 4.29

Using the tree diagram of Fig. 4.7, what are the following probabilities:

(a) the probability that, if exactly two of the three balls drawn were red,
   (i) the first ball was red;
   (ii) the second ball was black?

(b) the probability that, if at least one of the three balls drawn was black,
   (i) the first ball was red
   (ii) the second ball was black?

Solution

(a) Let $R2$ be the event that two of the three balls drawn were red.

$$P(R2) = \frac{60}{336} + \frac{60}{336} + \frac{60}{336} = \frac{180}{336}$$

(i)
$$P(\text{1st red}|R2) = \frac{P(R2 \cap \text{1st red})}{P(R2)} \quad \text{by (4.1)}$$
$$= \frac{P(\text{RRB}) + P(\text{RBR})}{P(R2)}$$
$$= \frac{60/336 + 60/336}{180/336}$$
$$= \frac{2}{3}$$

(ii)
$$P(\text{2nd black}\ |R2) = \frac{P(R2 \cap \text{2nd black})}{P(R2)} \quad \text{by (4.1)}$$
$$= \frac{P(\text{RBR})}{P(R2)}$$
$$= \frac{60/336}{180/336}$$
$$= \frac{1}{3}$$

(b) Let B^* be the event that at least one of the three balls drawn was black.

$$P(B^*) = 1 - P(\text{RRR}) = 1 - \frac{60}{336} = \frac{276}{336}$$

(i)
$$P(\text{1st red}\,|\,B^*) = \frac{P(B^* \cap \text{1st red})}{P(B^*)} \quad \text{by (4.1)}$$
$$= \frac{P(\text{RRB}) + P(\text{RBR}) + P(\text{RBB})}{P(B^*)}$$
$$= \frac{60/336 + 60/336 + 30/336}{276/336}$$
$$= \frac{150}{276}$$

(ii)
$$P(\text{2nd black}\,|\,B^*) = \frac{P(B^* \cap \text{2nd black})}{P(B^*)} \quad \text{by (4.1)}$$
$$= \frac{P(\text{RBR}) + P(\text{RBB}) + P(\text{BBR}) + P(\text{BBB})}{P(B^*)}$$
$$= \frac{60 + 30 + 30 + 6}{276}$$
$$= \frac{126}{276}$$

□

It is always useful when considering probability experiments to have a picture in our minds, and even better, to have one we can look at. The tree diagram provides such a picture, and has the additional advantage of providing an arithmetic check for the total probability.

However, the use of the tree diagram is limited when the number of outcomes (8 in the example above) becomes large. Other methods are useful in certain cases. For example, the binomial distribution (Chapter 6) can be used if the trials are independent and each trial can result in one of only two possible outcomes. In other situations the enumeration methods discussed in the next section are useful.

4.10 PERMUTATIONS AND COMBINATIONS

Consider the letters of the alphabet, A to Z. How many different arrangements are there of these 26 letters, assuming each is used once and only once? The answer is $26 \times 25 \times \ldots \times 2 \times 1 = 4.0329 \times 10^{26}$, since there are 26 ways of choosing the first letter, 25 ways of choosing the second letter, and so on. (We can relate this to a probability experiment of choosing, from a bag containing 26 cards labelled A to Z, one card at a time without replacement). A shorthand for the number $26 \times 25 \times \ldots \times 2 \times 1$ is 26! which we call 'factorial 26' or '26 factorial'.

Definition 4.6

> If n is a positive whole number (integer),
> $n! = 1 \times 2 \times 3 \times \ldots \times n$

Sometimes we need 0! which cannot be defined in the same way, since 0 is not a positive integer.

Definition 4.7

> $0! = 1$

Suppose we now consider taking a group of only four letters from the alphabet. How many different arrangements are there of four letters, assuming no letter is repeated in any arrangement? The answer is $26 \times 25 \times 24 \times 23 = 358{,}800$, since there are 26 ways of choosing the first letter, 25 ways of choosing the second letter, and so on.

The answer can be written in terms of factorials:

$$26 \times 25 \times 24 \times 23 = \frac{26 \times 25 \times 24 \times 23 \times 22 \times \ldots \times 3 \times 2 \times 1}{22 \times \ldots \times 3 \times 2 \times 1} = \frac{26!}{22!} = \frac{26!}{(26-4)!}$$

$$\frac{26!}{(26-4)!} \text{ is also written } {}^{26}P_4$$

Definition 4.8

The number of arrangements of r objects selected from a set of n different objects is:

$$^{n}P_{r}=\frac{n!}{(n-r)!}$$

The letter P is used because we are considering what is called a *permutation* of r objects from n objects.

Some of the $^{n}P_{r}$ arrangements will contain the same r objects. For the example of selecting four letters from the alphabet, the letters A, B, C, D occur in exactly 4! of the arrangements:

A B C D	B A C D	C A B D	D A B C
A B D C	B A D C	C A D B	D A C B
A C B D	B C A D	C B A D	D B A C
A C D B	B C D A	C B D A	D B C A
A D B C	B D A C	C D A B	D C A B
A D C B	B D C A	C D B A	D C B A

(The reason why there are 4! arrangements containing A, B, C, D is because there are 4 ways of choosing the first letter, 3 ways of choosing the second letter, and so on). If we decide to ignore the order in which the 4 letters are arranged there are only

$$\frac{26!}{4!(26-4)!}=14{,}950$$

different *combinations* of 4 letters from 26 letters.

$$\frac{26!}{4!(26-4)!}\text{ is also written } {}^{26}C_{4} \text{ or } \binom{26}{4}.$$

Definition 4.9

The number of ways of selecting r objects from a set of n different objects if we ignore the order of selection is:

$$^{n}C_{r}\text{ or }\binom{n}{r},\text{ where }{}^{n}C_{r}=\binom{n}{r}=\frac{n!}{r!(n-r)!}$$

The letter C is used because we are considering what is called a *combination* of r objects from n objects.

Example 4.30

A gambler selects sets of three football matches from the eleven first-division matches played on a particular Saturday. How many such sets are there?

Since choosing matches 1, 2, 3, say, is the same as choosing 3, 2, 1, we are clearly not interested in the order of selection. We want the number of combinations of '3 from 11', i.e.

$$^{11}\text{C}_3 = \binom{11}{3} = \frac{11!}{3!(11-8)!} = \frac{11 \times 10 \times 9}{3 \times 2 \times 1} = 165 \text{ sets}$$ □

Example 4.31

Suppose that the eleven players in a ladies' cricket team are all equally good at bowling. How may ways are there of choosing the first three bowlers who will be referred to as first bowler, second bowler, third bowler?

Since order is important here, the number of ways is:

$$^{11}\text{P}_3 = \frac{11!}{(11-3)!} = 11 \times 10 \times 9 = 990 \text{ ways.}$$

We will now consider examples in which the ideas of permutations and combinations will be used to determine the probabilities of events. □

Example 4.32

A pack of playing cards consists only of the thirteen spades. If three cards are selected without replacement, what is the probability that ace, king, queen will be selected in any order?

There are $^{13}\text{C}_3$ ways of choosing 3 from 13, ignoring the order of selection. Ace, king, queen is one of these ways.

Therefore $P(\text{ace, king, queen in any order}) = 1/^{13}\text{C}_3 = 0.0035$

Here we are using the 'equally likely' definition of probability, since we are assuming that the $^{13}\text{C}_3$ outcomes are equally likely to occur. □

Example 4.33

For the same pack as in the previous example, what is the probability that ace, king, queen will be selected in that order?

Since order is now important,

$$P(\text{ace, king, queen in that order}) = 1/^{13}\text{P}_3 = 0.0006.$$

What if some of the objects are alike? Suppose there are n objects in total, consisting of n_1 objects of the first type, n_2 objects of the second type, . . . , n_k objects of the kth type.

Then the number of ways of arranging n objects, if n_1 are of the first type, . . . , n_k are of the kth type is:

$$\frac{n!}{n_1!\,n_2!\ldots n_k!}$$ □

Example 4.34

How many ways are there of arranging the letter of the word BALL?

If all the letters had been different there would have been $4! = 24$ ways.
But there are two L's. So $n = 4, n_1 = 1, n_2 = 1, n_3 = 2$, and the number of ways of arranging the letters is:

$$\frac{4!}{1!\,1!\,2!} = 12$$

Two more examples now follow on the use of combinations when some of the objects are alike. □

Example 4.35

A bag contains 5 red balls and 3 black balls. If 3 balls are drawn *without* replacement, what are the probabilities that:

(a) no black balls will be selected;
(b) exactly one red ball will be selected;
(c) at least one red ball will be selected?

(Note that this example was covered in Section 4.9 using tree diagrams.)

Here the order of selection is not important, so we are dealing with combinations. The number of ways of choosing 3 balls from 8 balls is ${}^8C_3 = 56$.

(a) The number of ways of choosing no black balls from 3 balls is ${}^5C_3 = 10$, since this is equivalent to choosing 3 balls from only the 5 red balls.

$$P(\text{no black balls}) = \frac{10}{56}$$

(b) The number of ways of choosing exactly one red ball from 3 balls is ${}^5C_1 \times {}^3C_2 = 15$, since we must choose 1 red from 5 reds and hence 2 blacks from 3 blacks.

$$P(\text{exactly one red ball}) = \frac{15}{56}$$

(c) The number of ways of choosing at least one red ball from 3 balls is:

$${}^5C_1 \times {}^3C_2 + {}^5C_2 \times {}^3C_1 + {}^5C_3 \times {}^3C_0 = 55$$

since at least one red implies:

one red and two black, or
two red and one black, or
three red and no black

$$P(\text{at least one red}) = \frac{55}{56}$$ □

Example 4.36

A committee of three is selected from four Conservative, four Labour, and four Alliance Members of Parliament.

(a) If the committee is chosen at random, what is the probability that each party will be represented?
(b) If one of the four Conservatives is a government minister and insists on being on the committee, what is the probability that the Labour and Alliance parties will be represented if the remaining two committee members are chosen randomly from the other eleven Members of Parliament?

Solution

(a) There are ${}^{12}C_3 = 220$ possible committees each having 3 members. If each party is represented, there must be one from each party.
There are ${}^4C_1 \times {}^4C_1 \times {}^4C_1 = 64$ ways of choosing one from each party.

$$P(\text{each party represented}) = \frac{64}{220}$$

(b) Since one committee member has already been decided, there are ${}^{11}C_2 = 55$ possible committees. If the Labour and Alliance parties are represented, there must be one from each party.
There are ${}^4C_1 \times {}^4C_1 = 16$ ways of choosing one from each of the two parties.

$$P(\text{each party represented} \mid \text{Conservative minister on committee}) = \frac{16}{55}$$

Notice that the answers to (a) and (b) are the same. Why is this? □

4.11 THE LAW OF TOTAL PROBABILITY AND BAYES' FORMULA

Suppose that A_1 and A_2 are two mutually exclusive and exhaustive events, and suppose that B is any other event (see Fig. 4.8).
Then $P(B) = P(B \text{ and } A_1 \text{ or } B \text{ and } A_2)$, since B must occur with A_1 or A_2.

$$= P(B \text{ and } A_1) + P(B \text{ and } A_2),$$

because 'B and A_1' and 'B and A_2' are mutually exclusive.

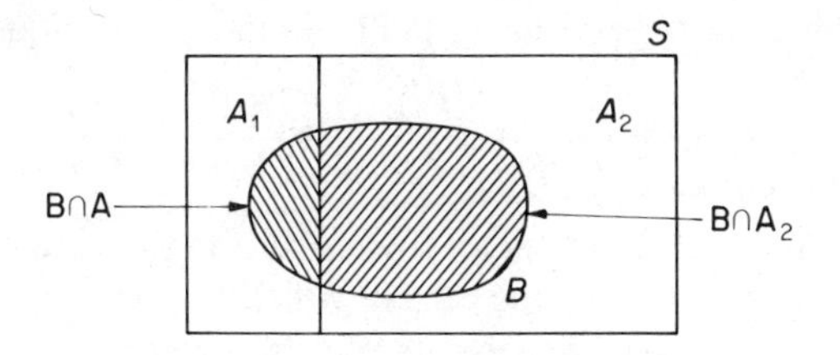

Fig. 4.8 Venn diagram

Therefore $$P(B) = P(B|A_1)P(A_1) + P(B|A_2)P(A_2) \qquad (4.9)$$

using (4.1)

This result is an example of the *law of total probability* and is useful in its own right in probability problems, although its main use here is to derive Bayes' formula. The law can easily be extended to cover the case in which $A_1, A_2, \ldots, A_n$ are a set of n mutually exclusive and exhaustive events:

$$P(B) = P(B|A_1)P(A_1) + P(B|A_2)P(A_2) + \ldots + P(B|A_n)P(A_n) \qquad (4.10)$$

Example 4.37

Suppose 10% of men and 5% of women suffer from some form of colour-blindness. If a theatre audience consists of 60% men and 40% women, what is the probability that a randomly chosen individual will be colour-blind?

Let B be the event 'colour-blind'.

Let A_1, A_2 be the event 'man chosen', 'woman chosen', respectively.

Then we are given $P(B|A_1) = 0.1$, $P(B|A_2) = 0.05$, $P(A_1) = 0.6$; $P(A_2) = 0.4$, and using (4.9),

$$P(\text{colour-blind}) = (0.1 \times 0.6) + (0.05 \times 0.4) = 0.08.$$

Can you see how to tackle this example using a tree diagram? □

Continuing now with the derivation of Bayes' formula, we note that

$P(A_1 \cap B) = P(A_1)P(B|A_1)$ and $P(B \cap A_1) = P(B)P(A_1|B)$ using (4.1).

But since the left-hand sides are equal, so are the right-hand sides.

$$\therefore P(A_1|B) = \frac{P(A_1)P(B|A_1)}{P(B)}$$

and hence

$$P(A_1|B) = \frac{P(B|A_1)P(A_1)}{P(B|A_1)P(A_1) + P(B|A_2)P(A_2)} \qquad (4.11)$$

rearranging the numerator, and using (4.9). Similarly we could obtain:

$$P(A_2|B) = \frac{P(B|A_2)P(A_2)}{P(B|A_1)P(A_1)+P(B|A_2)P(A_2)} \tag{4.12}$$

Equations (4.11) and (4.12) are Bayes' formulae for the case of two mutually exclusive events A_1 and A_2. For the case in which $A_1, A_2 \ldots, A_n$ are a set of n mutually exclusive and exhaustive events, Bayes' formula is:

$$P(A_r|B) = \frac{P(B|A_r)P(A_r)}{P(B|A_1)P(A_1)+P(B|A_2)P(A_2)+\ \ldots\ +P(B|A_n)P(A_n)} \tag{4.13}$$

which is true for $r = 1, 2, \ldots, n$

Notes

1. The condition 'A_r given B' on the left-hand side of (4.13) is reversed in the numerator on the right-hand side.
2. For the case $n = 2$, once we have found $P(A_1|B)$ using (4.11) we can find $P(A_2|B)$ using $P(A_1|B)+P(A_2|B) = 1$ (a result which is obtained by adding (4.11) and (4.12)). In general, however, $P(B|A_1)+P(B|A_2) \neq 1$.

In many probability problems involving the use of Bayes' formula we will need to evaluate a conditional probability of the form $P(A_1|B)$ by substituting known values for probabilities in the right-hand side of Bayes' formula.

Example 4.38

Suppose I throw a die. If I obtain a 1 or a 2, I toss a coin 3 times and note the number of heads. If I obtain a 3, 4, 5, or 6, I toss a coin once and note whether heads or tails is obtained.

If all you are told about the result of my experiment is that I obtained exactly one head, what is the probability that:

(a) I threw a 1 or 2 with the die;
(b) I threw a 3, 4, 5, or 6 with the die?

In (a) and (b) we are being asked to evaluate the following conditional probabilities:

$$P\text{ (I threw a 1 or 2}|\text{I obtained one head)} \quad \text{and}$$
$$P\text{ (I threw 3, 4, 5 or 6}|\text{I obtained one head)}$$

If we let 'I obtained one head' be the event B,

'I threw a 1 or 2' be the event A_1,

and 'I threw 3, 4, 5 or 6' be the event A_2,

we notice that A_1 and A_2 are mutually exclusive and exhaustive, and we also

notice that we need to evaluate $P(A_1|B)$ and $P(A_2|B)$, the left-hand sides of Bayes' formulae (4.11) and (4.12).
Using (4.11),

$$P(A_1|B) = \frac{P(B|A_1)P(A_1)}{P(B|A_1)P(A_1) + P(B|A_2)P(A_2)}$$

Now $P(A_1) = \frac{2}{6}$, $\quad P(A_2) = \frac{4}{6}$

Also
$$\begin{aligned} P(B|A_1) &= P(\text{1 head in 3 tosses of a coin}) \\ &= P(\text{HTT or THT or TTH}) \\ &= P(\text{HTT}) + P(\text{THT}) + P(\text{TTH}) \\ &= \frac{1}{2}\times\frac{1}{2}\times\frac{1}{2}+\frac{1}{2}\times\frac{1}{2}\times\frac{1}{2}+\frac{1}{2}\times\frac{1}{2}\times\frac{1}{2} \\ &= \frac{3}{8} \end{aligned}$$

Also $P(B|A_2) = P(\text{1 head in 1 toss of a coin}) = \frac{1}{2}$

Therefore
$$P(A_1|B) = \frac{3/8 \times 2/6}{3/8 \times 2/6 + 1/2 \times 4/6} = \frac{3}{11}$$

Similarly,
$$P(A_2|B) = \frac{1/2 \times 4/6}{3/8 \times 2/6 + 1/2 \times 4/6} = \frac{8}{11}$$ □

Example 4.39

Of 100 people, 40 are Conservative, 35 are Labour and 25 are Alliance voters. Also, the percentages of Conservatives, Labour and Alliance voters who read *The Times* are 50%, 40% and 80% respectively. If one of the 100 people is observed reading *The Times*, what is the probability that he is:

(a) A Conservative voter;
(b) A Labour voter;
(c) An Alliance voter?

Let event B be 'reads *The Times*', and let events A_1, A_2, A_3 be 'Conservative voter', 'Labour voter' and 'Alliance voter' respectively.

(a)
$$P(A_1|B) = \frac{P(B|A_1)P(A_1)}{P(B|A_1)P(A_1) + P(B|A_2)P(A_2) + P(B|A_3)P(A_3)}$$

$$= \frac{0.5 \times 0.4}{0.5 \times 0.4 + 0.4 \times 0.35 + 0.8 \times 0.25}$$

$$= \frac{10}{27}$$

(b) Similarly,
$$P(A_2|B) = \frac{7}{27}$$

(c) Similarly,
$$P(A_3|B) = \frac{10}{27}$$

Notice that the three probabilities sum to 1. □

Example 4.40

Three bags contain a number of green and yellow marbles as follows:

Bag 1 contains 3 green marbles
Bag 2 contains 2 green and 1 yellow marble
Bag 3 contains 3 yellow marbles

(a) If the probability is $i/6$ for $i = 1, 2$ and 3, that bag i will be chosen and a marble selected from it, what is the probability that:

(i) a green marble will be chosen;
(ii) a yellow marble will be chosen?

(b) If a yellow marble is chosen, what is the probability that it was selected from:

(i) Bag 1;
(ii) Bag 2;
(iii) Bag 3?

We will use G and Y to stand for green and yellow marble chosen, respectively and 1, 2, and 3 to stand for Bag 1, Bag 2 and Bag 3 respectively.

(a)
$$P(G) = P(G|1)P(1) + P(G|2)P(2) + P(G|3)P(3), \quad \text{by (4.10)}$$
$$= 1 \times \frac{1}{6} + \frac{2}{3} \times \frac{2}{6} + 0 \times \frac{3}{6}$$
$$= \frac{7}{18}$$
$$P(Y) = 1 - \frac{7}{18} = \frac{11}{18}, \text{ since } G \text{ and } Y \text{ are complementary events.}$$

(b) (i)
$$P(1|Y) = \frac{P(Y|1)(P(1)}{P(Y|1)P(1) + P(Y|2)P(2) + P(Y|3)P(3)} \quad \text{by (4.13)}$$
$$= \frac{0 \times 1/6}{0 \times 1/6 + 1/3 \times 2/6 + 1 \times 3/6}$$
$$= \frac{0}{11/18}$$
$$= 0$$

(ii) $P(2|Y) = \frac{2}{11}$, similarly.

(iii) $P(3|Y) = \frac{9}{11}$, similarly. □

4.12 SUMMARY

Two ways of defining probability as a measure of uncertainty are the relative frequency and the 'equally likely' definitions.

When considering two or more events, the multiplication and addition laws are useful:

Multiplication law

$$P(E_1 \cap E_2) = P(E_1)P(E_2|E_1) \tag{4.1}$$

or, if E_1 and E_2 are statistically independent,

$$\text{i.e. } P(E_2) = P(E_2|E_1), \tag{4.2}$$

$$\text{then } P(E_1 \cap E_2) = P(E_1)P(E_2) \tag{4.3}$$

Addition law

$$P(E_1 \cup E_2) = P(E_1) + P(E_2) - P(E_1 \cap E_2) \tag{4.4}$$

or, if E_1 and E_2 are mutually exclusive,

$$\text{i.e. } P(E_1 \cap E_2) = 0, \tag{4.5}$$

$$\text{then } P(E_1 \cup E_2) = P(E_1) + P(E_2) \tag{4.6}$$

If two events are mutually exclusive and exhaustive, they are called complementary events, E and E', say, and

$$P(E') = 1 - P(E) \tag{4.7}$$

If n events form a mutually exclusive and exhaustive set,

$$P(E_1) + P(E_2) + \ldots + P(E_n) = 1 \tag{4.8}$$

Tree diagrams are a useful way of displaying the outcomes of small probability experiments. Permutations and combinations help us to enumerate the number of ways in which events can occur.

The law of total probability,

$$P(B) = P(B|A_1)P(A_2) + P(B|A_2)P(A_2) \tag{4.9}$$

and its more general form, (4.10), help us to evaluate the probability of some event B which can only occur with one of a set of mutually exclusive and exhaustive events.

Finally, Bayes' formula,

$$P(A_1|B) = \frac{P(B|A_1)P(A_1)}{P(B|A_1)P(A_1) + P(B|A_2)P(A_2)} \tag{4.11}$$

and (4.12), (4.13) help us to evaluate the probabilities of each of a set of mutually exclusive and exhaustive events, $A_1, A_2 \ldots, A_n$ given that some event B has already occurred.

Section 4.4

1. Either there will be or there will not be an all-out nuclear war before the end of this century, so the probability of each event is $\frac{1}{2}$. Discuss.
2. In the game of 'two up' two coins are tossed. The possible outcomes are 2 heads, 1 head and 1 tail, and 2 tails. Hence the probability of each outcome is $\frac{1}{3}$. Discuss.
3. If I toss a coin 5 times and it comes down heads on each occasion, my estimate of the probability of heads is 1. Discuss.
4. A bag contains only 3 red balls, 5 white balls and 7 blue balls. What is the probability that a ball drawn at random will be:

   (a) red; (b) not red; (c) red or white; (d) neither red nor white; (e) red, white or blue; (f) yellow?

   What do you notice about the answer to (a) and (b), and also the answers to (c) and (d)?
5. One red die, one white die and one blue die are placed in a bag. One die is selected at random and rolled, its colour and the number on its uppermost face is noted.
   (a) Write down all the outcomes in the outcome set.
   (b) Are they all equally likely? What is the probability of each outcome?
   (c) What are the probabilities of the following events:
      (i) red with any number;
      (ii) red with an odd number;
      (iii) any colour with an even number;
      (iv) red with an even number or white with an odd number;
      (v) neither red with an even number, white with an odd number, nor blue with any number?
6. From the data in the following table, which gives the distance from home to Oxford for 40 students, estimate the probability that a randomly selected student lives:
   (a) within 100 km of Oxford;
   (b) between 100 and 300 km from Oxford;
   (c) at least 300 km from Oxford.

Distance from home (km)	*Number of students*
less than 100	20
100 up to but not including 200	8
200 " " " " " 300	8
300 " " " " " 400	0
400 " " " " " 500	4

Also, draw a histogram of these data. Show how the answers to (a), (b) and (c) may be expressed in terms of areas of the histogram. Explain in terms of areas why the answers to (a), (b) and (c) sum to 1.

Section 4.8

7. Two events A and B are such that $P(A) = 0.3$, $P(B) = 0.4$ and $P(A \cup B) = 0.6$. What is:
   (a) $P(A \cap B)$;
   (b) $P(B|A)$;
   (c) $P(A|B)$?

   Are the events A and B (i) statistically independent; (ii) mutually exclusive?

8. Two events A and B are statistically independent. If $P(A) = 0.3$, $P(A \cup B) = 0.5$, what is:
   (a) $P(B)$;
   (b) $P(A \cap B)$;
   (c) $P(B|A)$;
   (d) $P(A|B)$?

9. Two events A and B are mutually exclusive. If $P(A) = 0.3$, $P(A \cup B) = 0.5$, what is:
   (a) $P(B)$;
   (b) $P(A \cap B)$;
   (c) $P(B|A)$;
   (d) $P(A|B)$?

10. Can two events A and B be both mutually exclusive and statistically independent? Assume $P(A) > 0$, $P(B) > 0$.

11. Two dice are thrown and the total score is noted. The events A, B and C are 'a total of 4', 'a total of 9 or more', and 'a total divisible by 5'. Calculate $P(A)$, $P(B)$ and $P(C)$ and decide which pairs of events, if any, are:
    (a) mutually exclusive, (b) statistically independent.

12. One hundred students took examinations in mathematics and statistics with the following results:

		Statistics Pass	*Statistics* Fail
Mathematics	*Pass*	70	10
	Fail	15	5

    Let A be the event 'pass in mathematics' and B be the event 'pass in statistics'.
    (a) Calculate $P(A)$, $P(B)$, $P(A \cap B)$, $P(A \cup B)$, $P(B|A)$, $P(A|B)$.
    (b) Are A and B (i) mutually exclusive, (ii) statistically independent?

13. (a) Show that, for two events A and B,. $P(A' \cup B') = 1 - P(A \cap B)$
    and $P(A' \cap B') = 1 - P(A \cup B)$
    (b) If $P(A) = 0.4$, $P(B) = 0.3$, $P(A \cap B) = 0.2$, calculate:
    (i) $P(A')$
    (ii) $P(B')$

▶

(iii) $P(A' \cup B')$
(iv) $P(A' \cap B')$
(v) $P(B'|A')$
(vi) $P(A'|B')$

14. If A and B are statistically independent, are A' and B' as well?

15. If A and B are mutually exclusive, are A' and B' as well?

16. When a card is drawn at random from a standard pack of 52 playing cards, what is the probability of:

    (a) a red card or a picture card (Ace, King, Queen or Jack);
    (b) a 2 or a 3;
    (c) a numbered card (2, 3, . . . , 10) with a number less than 6;
    (d) a numbered card or a picture card?

    State two events named in this question which are:

    (i) mutually exclusive but not exhaustive;
    (ii) mutually exclusive and exhaustive.

17. Of eight players who enter a tennis tournament, two are left-handed and six are right-handed players. In the first round of the tournament four singles are played. The players are selected randomly, so that the first two players selected play against each other in the first match, and so on.

    (a) What are the probabilities that the two left-handed players will be selected for the:

        (i) first match;
        (ii) second match;
        (iii) third match;
        (iv) fourth match?

    (b) Assuming now that both left-handed players play against a right-handed player in the first round, and assuming that all players are of equal ability, what are the probabilities that:

        (i) both left-handed players go through to the second round;
        (ii) only one left-handed player goes through to the second round?

18. A certain type of switch has a probability p of working when it is turned on. Two such switches are involved in the operation of machine A and two other such switches in the operation of machine B. For machine A both switches must work for the machine to work, but machine B will work provided at least one of the switches works.

    Calculate the probabilities that:

    (i) machine A, (ii) machine B, will *fail* to work. Show in a sketch how these probabilities vary with p.

    Hence show that one machine is, in general, more likely to fail to work than the other machine.

    For what value of p is the difference between the probabilities of failure a maximum?

19. A coin is tossed until a head appears. If p is the probability of heads for each toss, what is the probability that exactly r tosses will be required? ▶

20. Explain what is meant by a pair of independent events.

For a loaded die the probabilities of obtaining each of the possible 'scores' when the die is rolled are as follows:

Score	1	2	3	4	5	6
Probability	0.2	0.2	0.1	0.3	0.1	0.1

The die is rolled twice. Let A denote the event that the same score is obtained each time and B the event that a total score of 10 or more is obtained. Determine whether or not events A or B are independent. Would these events be independent if the die were fair?

(IOS)

21. Two events A and B are such that

$$P(A) = \frac{1}{2}, \quad P(B) = \frac{1}{3}, \quad P(A|B) = \frac{1}{4}$$

Evaluate:

(i) $P(A \cap B)$;
(ii) $P(A \cup B)$;
(iii) $P(A' \cap B')$.

Another event C is such that A and C are independent, and

$$P(A \cap C) = \frac{1}{12}, \quad P(B \cup C) = \frac{1}{2}$$

Show that B and C are mutually exclusive.

(JMB)

22. Assume that, in any family, each child born is equally likely to be a boy or a girl. A family with three children is chosen at random. Find the probability that the oldest child is a girl:

(i) given no further information;
(ii) given that the family contains exactly one girl;
(iii) given that the family contains exactly two girls;
(iv) given that the family contains at least one girl.

(JMB)

Section 4.9

23. Three men take it in turns to fire at a target. Their probabilities of hitting the target are 0.2, 0.3 and 0.4 respectively. What are the probabilities of:

(a) no hits;
(b) one hit;
(c) two hits;
(d) three hits?

Check that your answers sum to 1. Why should they?

24. Suppose that the probability of a boy at each birth is 0.5. Calculate the probabilities that families with three children will have:

(a) 0, (b) 1, (c) 2, (d) 3 boys.

In families with three children of which two are boys, what is the probability that the eldest child is a boy?

►

25. Two friends, Richard and Brian, have played chess against each other many times and estimate that the probabilities of Richard winning a game, Brian winning a game, and a drawn game are 0.3, 0.5, and 0.2 respectively.

    (a) In a two-game tournament, what is the probability of Richard winning (i) neither game, (ii) one game, (iii) both games?
    (b) Suppose two points are awarded to a player who wins a game, while 1 point is awarded to each player in a drawn game. What are the probabilities that:

        (i) Richard will win the tournament by gaining more points than Brian;
        (ii) The tournament will result in a tie on points?

26. Secretarial staff in a company use a particular photocopying machine to produce half-size copies. The machine is temperamental, particularly in warm weather when it works better if switched off between jobs. Normally it is desirable to leave it switched on during working hours to save waiting for the machine to warm up before each job. The following notice is prominently displayed beside the machine:

    Please leave this machine as you found it:
    that is, switched on if you found it on and switched off if you found it off.

    (a) Suppose that all secretarial staff correctly obey the instructions on the machine with probability 4/5 and do exactly the opposite with probability 1/5. On a day when the machine was originally switched on, what is the probability that the fourth member of staff to use the machine finds the machine switched on?
    (b) Suppose now that all secretarial staff have a probability of 0.9 of correctly remembering to leave the machine on when they find it on, but only a probability of 0.7 of remembering to switch it off if they find it off. What is the probability that the fourth member of staff to use the machine finds it switched off, on a day when it is originally switched off?

    (*Note*: You may find it helpful to construct a tree diagram.)

    (IOS)

27. A large store sells spare bulbs for Christmas-tree lights, types A and B. Type A have an average defective rate of one bulb in twenty, while type B have an average defective rate of one bulb in thirty. It may be assumed that bulbs sold are selected independently at random from the productions of the two types.

    (i) Mr Smith buys two type A bulbs. What is the probability that only one of them is satisfactory?
    (ii) Mr Jones buys three type B bulbs. What is the probability that they are all satisfactory?
    (iii) Which of the two customers is the more likely to need further spare bulbs because he has purchased one or more defective ones?
    (iv) One of these two customers is chosen randomly as he leaves the store, by spinning a fair coin. Draw a tree diagram showing the probability that the chosen customer has no defective bulbs, and state this probability.

    (IOS)

28. A box contains ten objects of which 1 is a red ball, 2 are white balls, 3 are red cubes and 4 are white cubes. Three objects are drawn at random from the box, in succession and without replacement. Events B and R are defined as follows: ►

B: Exactly two of the objects drawn are balls;
R: Exactly one of the objects drawn is red.

Show that $P(B) = 7/40$ and calculate $P(R)$, $P(B \cap R)$, $P(B \cup R)$ and $P(B|R)$.
(Cambridge)

29. At a particular sixth-form college 60% of the students are classified as science students, and 40% are classified as arts students. Of the arts students, 65% are female. Of the science students, 20% are female. Because of the widespread catchment area, some students live in halls of residence during the week. Of the male science students, 30% live in halls of residence, and of the male arts students, 30% live in halls of residence. The corresponding figures for female students are 50%. Find the probability that a student chosen at random is:

    (i) a female science student;
    (ii) female;
    (iii) a science student, given that the student is female;
    (iv) male and lives in a hall of residence, given that the student is a science student;
    (v) a science student, given that the student is male and lives in a hall of residence.

    (JMB)

30. For two mutually exclusive events A and B:

    (i) state the value of $P(A \cap B)$,
    (ii) give the expression for $P(A \cup B)$ in terms of $P(A)$ and $P(B)$.

    State the value of $P(A \cup B)$ when A and B are also exhaustive.

    Every packet of a certain cereal contains either a plastic orange or a plastic lemon. For each packet the probability of each type of fruit is 1/2, independently of the contents of other packets. If you collect one of each type of fruit you can obtain a refund on your next packet.

    Suppose that you buy n packets. Consider the events A: 'an orange in each of the n packets' and B: 'a lemon in each of the n packets'. Show that

    $$P(A) = P(B) = (\tfrac{1}{2})^n.$$

    Describe in words the event $A \cup B$. Hence, or otherwise, obtain the probability that you will not qualify for a refund.

    If you wish to have a probability of at least 0.9 of obtaining one or more refunds, determine how many packets you must buy.

    Suppose that the machine which inserts the plastic fruits is not reliable, and that for any packet there is a probability of 0.05 that no fruit is placed in the packet. If both types of fruit are still equally likely in packets which receive a single plastic fruit, calculate the probability that a randomly selected packet contains a plastic orange. Obtain the probability that you will qualify for a refund if you buy three packets.
    (JMB)

31. Write down the possible values of the total score, S, that can be obtained when a pair of fair dice is thrown once, giving in each case the corresponding probability.

    (i) What is the probability that in a single throw of the dice one of the results $S = 7$ or $S = 11$ is obtained? ►

(ii) Suppose that, instead of being thrown only once, the pair of dice is thrown repeatedly until one of the results $S = 4$ or $S = 7$ is obtained (all other scores being disregarded). Show that the probability that $S = 4$ occurs before $S = 7$ is equal to 1/3.

Find the probabilities that, in similar repeated throwings,

(iii) $S = 5$ occurs before $S = 7$,

(iv) $S = 6$ occurs before $S = 7$.

In the game of Craps, the person throwing the dice wins on his first throw if S is either 7 or 11, and loses if S is one of 2, 3 or 12. For any other value, k, of this first throw he must then throw both dice repeatedly and he wins provided $S = k$ occurs before $S = 7$. By drawing a suitable tree diagram, or otherwise, show that his total probability of winning is slightly less than 1/2.

(SMP)

Section 4.10

32. (a) Cards are drawn with replacement from a standard pack until an ace is drawn. What are the probabilities that the following numbers of cards will be required:

(i) 1; (ii) 2; (iii) 3; (iv) n (where $n \geqslant 1$)?

(b) Cards are drawn with replacement from a standard pack until two aces are drawn. What are the probabilities that the following number of cards will be required:

(i) 1; (ii) 2; (iii) 3; (iv) n (where $n \geqslant 2$)?

(c) Cards are drawn with replacement from a standard pack until r aces are drawn. What is the probability that n cards will be required (where $n \geqslant r$ and $r \geqslant 1$)?

33. A coin is tossed n times and r heads are observed. Show that there are $\binom{n}{r}$ ways in which this can occur. If the probability of heads is p for each toss show that the probability for each way in which r heads can occur in n tosses is $p^r(1-p)^{n-r}$. Hence deduce the probability that r heads will occur in n tosses of a coin.

34. The Vice-Chancellor of the University of Melchester has asked a group of his staff to sit on a committee to consider future developments in the light of the levels of funding promised to all universities. The committee consists of twelve people each representing a different department, and of these, four are professors, namely the professors of Physics, Mathematics, Law and Engineering. At their first meeting all members of the committee sit down at random in a row.

(a) In how many distinct ways may they do this?

(b) What is the probability that the professors of Mathematics and Physics are next to each other?

(c) What is the probability that the professors of Mathematics and Physics are next to each other, but that the professors of Law and Engineering are not?

At their second meeting the professor of Law has arranged for the committee to sit round a circular table. Assuming that all members of the committee still sit down at random, calculate the new answers to questions (a), (b) and (c) above.

(AEB (1983)) ►

35. An inspector takes a random sample of n articles from a batch containing $s+d$ articles, of which s are sound and d are defective. Explain why the probability that the sample contains exactly x defective articles, given that $x \leqslant d$ and $n-x \leqslant s$, is

$$\frac{\binom{d}{x}\binom{s}{n-x}}{\binom{d+s}{n}}.$$

Given that a batch consists of 16 articles, 5 of which are defective, show that if the sample contains 8 articles, the probability of at least 3 of these being defective is 0.5.

(Cambridge)

36. (*In this question answers should be given as simplified fractions or as decimals correct to three places.*)

At a certain school a Recreation Committee comprises three representatives from each of four school houses, White House, Red House, Yellow House and Blue House. A sub-committee of four pupils is randomly selected from this committee to organize the school sports day. Show that the number of different sub-committees that can be formed is 495.

Find the probabilities that

(i) no one is selected from White House;
(ii) one pupil is selected from each house;
(iii) one pupil is selected from White House, two from Red House and one from either of the other two houses.

Given that there are five girls on the Recreation Committee, find the probability that just one girl is on the sub-committee.

It is later decided that the Head Prefect, who is a White House representative on the Recreation Committee, must be on the sub-committee. In this case find the probability that each house will be represented on the sub-committee when the other three members are randomly chosen.

(JMB)

37. A committee of 4 people is to be selected from a group consisting of 8 men and 4 women. Determine the number of ways in which the committee may be formed if it is to contain

(a) exactly one woman,
(b) at least one woman.

(London)

38. A group of 5 pupils is selected at random from a class which has 20 girls and 15 boys. The probability distribution of the number of boys in the group is:

Number of boys, x	0	1	2	3	4	5
Probability	0.048	0.224	0.369	0.266	0.084	0.009

Verify the calculation for $x = 1$. ►

Suppose K groups are selected independently, with each group being 'replaced' after selection so that it might be selected again. Find the probabilities that

(i) with $K = 2$, both groups have the same number of boys,
(ii) with $K = 4$, at least one group is all of the same sex.

(OLE)

39. A bag contains N beads of which R are red and the remainder are of other colours. A sample of n beads is taken from the bag at random without replacement. Explain why the probability that the sample contains exactly r red beads is equal to

$$\binom{R}{r}\binom{N-R}{n-r}\Big/\binom{N}{n}$$

A bag contains 2 red, 4 yellow and 6 blue beads. What is the probability that a sample of 6 beads taken from the bag at random without replacement consists of 1 red, 2 yellow and 3 blue beads?

(O&C)

40. A bag contains 8 counters, 4 black and 4 white. The counters are drawn at random one at a time and placed in a row in the order of withdrawal. What is the probability that no counter is next to one of the same colour?

What is the probability that, among the first 5 counters in the row, there are 2 black counters and 3 white?

(O&C)

41. Two cards are dealt consecutively from a well-mixed pack of ordinary playing cards. Show that the probability that both cards are queens is 1/221.

Determine also the probabilities that

(i) at least one of the two cards is a queen, and
(ii) the second card dealt is a queen given that the first is.

Each of 50 women has a pack of well-mixed playing cards, and each deals two cards consecutively from her pack. Find, correct to 3 decimal places, the probability that exactly one woman deals a pair of queens.

(MEI)

42. (a) How many different arrangements are there of the letters in the word S T A T I S T I C S ?
(b) Six married couples meet together for an afternoon's tennis, playing sessions of mixed doubles (one man and one woman play together on each side of the net).
   (i) In how many ways can the first match of the afternoon be arranged if no husband and his wife are allowed to play in the foursome (together or on opposite sides)?
   (ii) If a husband and his wife are not allowed to play together on the same side, but may play on opposite sides, in how many ways can the first mixed double be arranged?

(SUJB)

43. (a) Four cards are to be drawn at random *without replacement* from a pack of nine cards which are numbered from 1 to 9, respectively. Calculate the probabilities that the numbers on the four cards drawn will be such that ►

(i) the highest of them is 8,
(ii) two of them are 1 and 9,
(iii) at least three of them are odd numbers,
(iv) their product is exactly divisible by 9.

(b) A local council has 40 members, of whom 30 are male and 24 are over 60 years old. One member is chosen at random. Let A denote the event that the chosen member is male and let B denote the event that the chosen member is over 60 years old. Given that A and B are independent, determine the number of male members over 60 years old.

(WJEC)

Section 4.11

44. If it rains on one day the probability that it rains on the following day is 2/3, whereas if it is fine on one day the probability that the next day is fine is 3/4. Assuming days are classified only as fine or rainy, show that the probability of a rainy day is 3/7.
   (a) What is the probability that, in a two-day period, there will be one fine day and one rainy day?
   (b) If it is fine on one day, what are the probabilities that the previous day was
      (i) fine,
      (ii) rainy?

45. Three machines of varying efficiency are used to manufacture a component. Machines A1, A2, A3 produce 20%, 30% and 50% respectively of the total production of the components. The percentages of each machine's production which are defective are 8%, 5% and 6% respectively.
   (a) What percentage of the total production will be defective?
   (b) If a component is randomly selected from the total production and is found to be defective, what are the probabilities that it came from
      (i) Machine A1, (ii) Machine A2, (iii) Machine A3?

46. A diagnostic test for a certain disease is such that, if a person has the disease, the test will detect this with probability 0.9; also that, if a person does not have the disease, the test will detect this with probability 0.95.

   Suppose only 1% of the population have the disease in question. If a person is chosen at random from the population and the diagnostic test indicates that the person has the disease, what is the probability that the person actually does have the disease?

   Calculate the same probability for the case in which the person is randomly selected from patients in a hospital in which it is estimated that 80% have the disease.

   Comment on the difference in the two answers, and the value of the test in each case.

47. Three women, A, B and C, share an office with one telephone. Calls for this office arrive at random during working hours in the ratios 3, 2, 1 for A, B, C, respectively. Because of the nature of their work the three women leave the office independently at random times, so that A is absent for 1/5 of her working hours and B and C are each absent for 1/4 of their working hours. On occasions when the telephone rings during working hours find the probabilities that: ▶

(i) no one is in the office to answer the telephone;
(ii) a caller who wishes to speak to C is able to do so;
(iii) a caller cannot speak to the person he wishes to contact;
(iv) a caller who wishes to speak to B is unable to on three consecutive occasions;
(v) three successive calls are all for different people;
(vi) given that two consecutive calls are for the same person, they are both for C.

(JMB)

48. An unbiased cubical die has four faces numbered 1 and two faces numbered 2. Two boxes are numbered 1 and 2; the box numbered 1 contains three red discs and two blue discs, and the box numbered 2 contains two red discs and three blue discs. Given that the die is rolled and three discs are then drawn at random without replacement from the box with the same number as that uppermost on the die, calculate the probabilities that

(i) two red discs and one blue disc are drawn from the box numbered 1;
(ii) two red discs and one blue disc are drawn;
(iii) the discs came from the box numbered 1, given that two of the drawn discs were red and one disc was blue.

(JMB)

49. A box contains 15 tulip bulbs of which 10 will have red flowers and 5 yellow, though it is not possible to determine the colour of the flowers by visual inspection of the bulbs. A package of 5 is to be made up from the 15. If they are selected at random, what is the probability that the package will contain exactly 3 red-flowering bulbs?

One of the 15 bulbs is found to be damaged and is discarded before the selection is made. Find the conditional probability that the package of 5 contains exactly 3 red-flowering bulbs, given that the damaged bulb is red.

Hence or otherwise find the probability that the package of 5 contains exactly 3 red-flowering bulbs when one bulb in the original 15 has been thrown away as damaged, given that all bulbs are equally likely to be damaged.

(*Give all your answers to 3 decimal places.*)

(OLE)

50. (a) If A and B are any two events, explain in words, and briefly justify, the result $P(A|B)P(B) = P(A \cap B)$. Use this result to show that

$$P(A|B) = \frac{P(A)P(B|A)}{P(A)P(B|A) + P(A')P(B|A')}$$

where A' is the complement of A.
Given that $P(B|A) = P(B|A')$, deduce that A and B are independent.

(b) Eustace and his family go on holiday every summer either to Acapulco or to Mudflats-by-Sea. If Eustace receives a pay rise in the autumn, the probability that next summer's holiday will be in Acapulco is 0.7, while the probability that he does not receive a pay rise and that the holiday is in Mudflats is 0.6. Denote by C and D respectively the events that the holiday is in Acapulco, and that Eustace receives a pay rise in the autumn. Given that $P(D) = 0.3$, determine $P(C)$, and show that C and D are not independent. One summer a neighbour of Eustace sees him sun- ►

bathing at Mudflats (under an umbrella). What are the chances that Eustace had a pay rise in the preceding autumn?

(MEI)

51. (a) Three numbers are chosen at random from the set (1, 2, 3, . . ., 10).

    What is the probability that the smallest of the three numbers chosen is even?

    (b) Two boys, A and B, play a game of Beetle; each throws a die in turn and cannot start drawing his Beetle until he obtains a 6.

    The boy A throws first.

    (i) What is the probability that B is the first boy to begin drawing a Beetle, starting on his second throw?

    (ii) What is the probability that B is the first to start drawing a Beetle?

(SUJB)

52. Each of three boxes A, B and C contains four balls. Each of the four balls in A is red. Two of the four balls in B are red and the remaining two balls are white. Three of the fours balls in C are white and the remaining ball is yellow. Two fair coins are tossed together. If two heads are tossed, one ball is drawn at random from box A; if two tails are tossed, one ball is drawn at random from box B; otherwise, one ball is drawn at random from box C.

    (i) Show that the probability of the drawn ball being white is four times that of it being yellow.

    (ii) Given that the drawn ball is red, find the conditional probability that it came from box B.

    The box from which the ball was drawn is then set aside, and a ball is drawn at random from one of the other two boxes, the choice of box being determined by the outcome of one toss of a fair coin.

    (iii) Calculate the probability that the two balls drawn are of the same colour.

(WJEC)

5

Random variables and their probability distributions

▼

5.1 INTRODUCTION

If, as a result of carrying out an experiment (as defined in Chapter 4), an event is assigned a quantitative measure, this measure is called a *random variable*. Random variables may be either discrete or continuous depending on whether the variable may take only certain distinct numerical values in a given range or any numerical value in a given range (see Section 1.2). For example, the number of heads in three tosses of a coin is a discrete random variable, while the height of a randomly selected student is a continuous random variable.

The relationship between the possible values of a random variable and the corresponding probabilities is called the *probability distribution* of the random variable. Just as a random variable may be discrete or continuous, so may a probability distribution.

In this chapter the general properties of random variables and their probability distributions will be discussed.

5.2 DISCRETE RANDOM VARIABLES, PROBABILITY FUNCTION

Let X denote the name of a random variable, and let x denote an observed value of X.

If X is a discrete random variable its probability distribution may be specified in terms of a *probability function*,

$$P(X = x), \text{ which we will simply write as } p(x).$$

Example 5.1

If X is the number of heads in three tosses of a coin,

$$p(x) = \binom{3}{x}\left(\frac{1}{2}\right)^3, \qquad x = 0, 1, 2, 3.$$

The probability function may be expressed in a tabular form or graphically, as in the next example. □

Example 5.2

If X is the number of heads in three tosses of a coin, the probability function can be represented as in Table 5.1 or Fig. 5.1.

Table 5.1 An example of a discrete probability distribution

x	0	1	2	3
$p(x)$	$\frac{1}{8}$	$\frac{3}{8}$	$\frac{3}{8}$	$\frac{1}{8}$

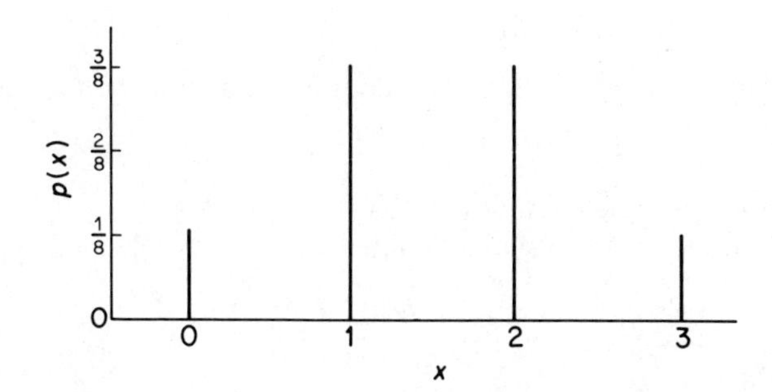

Fig. 5.1 Example of a discrete probability distribution □

The sum of the probabilities of an 'proper' discrete probability distribution is 1, since the events corresponding to the various possible values of X form a mutually exclusive and exhaustive set. So $\Sigma p(x) = 1$ for any discrete probability distribution, where the summation is over all values of x.

Example 5.3

Suppose that X, the score on the uppermost face of a loaded die, has a probability function

$$p(x) = kx \qquad x = 1, 2, 3, 4, 5, 6.$$

Find k, write $p(x)$ in tabular form, and find the probability that $X \geqslant 3$.

Since $$\Sigma p(x) = 1, \quad k + 2k + \cdots + 6k = 1$$

$$\therefore k = \frac{1}{21} \quad \text{and}$$

$$p(x) = \frac{x}{21}, \quad x = 1, 2, 3, 4, 5, 6 \qquad \text{(see Table 5.2).}$$

$$P(X \geqslant 3) = \frac{3}{21} + \frac{4}{21} + \frac{5}{21} + \frac{6}{21} = \frac{6}{7},$$

since the events '3', '4', '5' and '6' are mutually exclusive.

Table 5.2 An example of a discrete probability distribution

x	1	2	3	4	5	6
$p(x)$	$\frac{1}{21}$	$\frac{2}{21}$	$\frac{3}{21}$	$\frac{4}{21}$	$\frac{5}{21}$	$\frac{6}{21}$

□

5.3 EXPECTATION, MEAN AND VARIANCE OF A DISCRETE RANDOM VARIABLE

Let $E(X)$ denote the expectation of a random variable X. If X is discrete, then

$$E(X) = \Sigma x p(x)$$

where the summation is over all possible values of X.

The mean value of X is usually denoted by μ,† and is defined as follows:

$$\mu = E(X)$$

so we may write

$$\mu = E(X) = \Sigma x p(x) \tag{5.1}$$

for a *discrete* random variable X. We note that 'mean' and 'expectation' are synonymous in this context.

† In this section we have used μ and σ to denote the mean and standard deviation of a random variable. Compare the use of $\bar{x}$ and s' in earlier chapters to denote the mean and standard deviation of a sample of observed values.

The definition of expectation, $E(X) = \Sigma x p(x)$ is a special case of a more general definition:

$$E[g(X)] = \Sigma g(x)p(x) \tag{5.2}$$

where $g(X)$ is a function of the random variable X.

The variance of X is usually denoted by σ^2, but Var(X) is also used. The variance of X is defined in terms of expectation:

$$\sigma^2 = \text{Var}(X) = E[(X - E(X))^2]$$
$$= E[(X - \mu)^2] \qquad \text{by (5.1)}$$

which can be simplified to give a more useful result as follows (for a proof see the Appendix to this chapter):

$$\sigma^2 = \text{Var}(X) = E(X^2) - \mu^2 \tag{5.3}$$

where, from (5.2),

$$E(X^2) = \Sigma x^2 p(x) \tag{5.4}$$

The *standard deviation* of X is the square root of the variance and consequently is denoted by σ. □

Example 5.4

If X is the number of heads in three tosses of a coin, what are the mean, variance and standard deviation of X? Extending Table 5.1 to include rows labelled $xp(x)$ and $x^2 p(x)$ gives Table 5.3.

Table 5.3 An example of the calculation of the mean and variance of a discrete random variable

x	*0*	*1*	*2*	*3*	*Total*
$p(x)$	$\frac{1}{8}$	$\frac{3}{8}$	$\frac{3}{8}$	$\frac{1}{8}$	1
$xp(x)$	0	$\frac{3}{8}$	$\frac{6}{8}$	$\frac{3}{8}$	$\frac{12}{8}$
$x^2p(x)$	0	$\frac{3}{8}$	$\frac{12}{8}$	$\frac{9}{8}$	$\frac{24}{8}$

Using (5.1). $\quad \mu = E(X) = \Sigma xp(x) = \dfrac{12}{8} = 1.5$

Using (5.3) and (54), $\quad \sigma^2 = \text{Var}(X) = \dfrac{24}{8} - 1.5^2 = 0.75$

$$\sigma = 0.866$$ □

Example 5.5

In a dice game a player pays a stake of £1 for each throw of a die. The player receives £5 if a 3 is thrown, £2 if a 1 or 6 is thrown, and nothing otherwise. What is the player's mean (expected) profit per throw over a long series of tosses? What is the standard deviation of the profit per throw?

Let X be the profit per throw. The profit is £4 (£5 − £1) with probability $\frac{1}{6}$, since this is the probability of throwing a 3, and so on. Table 5.4 shows the probability distribution of X and rows labelled $xp(x)$ and $x^2p(x)$.

Table 5.4 An example of the calculation of the mean and variance of a discrete random variable

x	*4*	*1*	*−1*	*Total*
$p(x)$	$\frac{1}{6}$	$\frac{2}{6}$	$\frac{3}{6}$	1
$xp(x)$	$\frac{4}{6}$	$\frac{2}{6}$	$-\frac{3}{6}$	$\frac{3}{6}$
$x^2p(x)$	$\frac{16}{6}$	$\frac{2}{6}$	$\frac{3}{6}$	$\frac{21}{6}$

Using (5.1), $$\mu = E(X) = \Sigma xp(x) = \frac{3}{6} = £0.50$$

Using (5.3) and (5.4), $$\sigma^2 = \text{Var}(X) = \frac{21}{6} - 0.5^2 = 3.25$$

$$\sigma = £1.80$$

The player's mean (expected) profit per throw is £0.50, while the standard deviation of the profit per throw is £1.80. □

5.4 PROBABILITY GENERATING FUNCTION FOR A DISCRETE RANDOM VARIABLE

The probability distribution of a discrete random variable may be described in terms of a *probability generating function* (*p.g.f.*) rather than in terms of a probability function. The mean, variance and standard deviation may be obtained directly from the p.g.f.

If $G(t)$ is the p.g.f. of a discrete random variable X which can take values $0, 1, 2, \ldots$, then

$$G(t) = E(t^X),$$

where t is a dummy variable.

$G(t)$ can be expressed as a series in t such that the coefficients of the powers of t are the probabilities of the various possible values of X. The proof follows:

$$\begin{aligned} G(t) &= E(t^X) \\ &= \Sigma t^x p(x) \qquad \text{by (5.2)} \end{aligned}$$

where the summation is over all possible values of X. This is a series in t with probabilities as coefficients; for example, the probability that X takes the value 2 is the coefficient of t^2, and so on.

The following results hold for any p.g.f.:

$$G(1) = 1 \tag{5.5}$$

$$\mu = G'(1) \tag{5.6}$$

$$\sigma^2 = G''(1) + \mu - \mu^2 \tag{5.7}$$

where $G'(1)$ means differentiating $G(t)$ once with respect to t and then putting $t = 1$; $G''(1)$ means differentiating $G(t)$ twice with respect to t and then putting $t = 1$.

Proof of (5.5)

$$\begin{aligned} G(1) &= E(1^X) && \text{from the definition of } G(t) \\ &= E(1) \\ &= \Sigma p(x) && \text{using (5.2)} \\ &= 1 && \text{for a 'proper' discrete distribution} \end{aligned}$$

Proof of (5.6)

$$\begin{aligned} G(t) &= E(t^X) && \text{definition} \\ &= \Sigma t^x p(x) && \text{using (5.2)} \\ G'(t) &= \Sigma x t^{x-1} p(x) \\ G'(1) &= \Sigma x p(x) \\ &= \mu && \text{using (5.1)} \end{aligned}$$

Proof of (5.7)

$$\begin{aligned} G''(t) &= \Sigma x(x-1) t^{x-2} p(x) && \text{differentiating } G'(t) \\ G''(1) &= \Sigma x(x-1) p(x) \\ &= \Sigma x^2 p(x) - \Sigma x p(x) \\ &= E(X^2) - E(X) && \text{using (5.4) and (5.1)} \\ &= \sigma^2 + \mu^2 - \mu && \text{using (5.3)} \\ \therefore \sigma^2 &= G''(1) + \mu - \mu^2 \end{aligned}$$

If we only know the p.g.f. of a discrete random variable, (5.5) is a check that the probabilities sum to 1. (5.6) and (5.7) can be used to find the mean and variance of X.

Example 5.6

A discrete random variable X has a p.g.f.

$$G(t) = \frac{(1+t)^3}{8}$$

What are the mean and variance of X?

$$G'(t) = \frac{3(1+t)^2}{8}$$

$$\therefore \mu = G'(1) = \frac{3(1+1)^2}{8} = 1.5$$

$$G''(t) = \frac{6(1+t)}{8}$$

$$\begin{aligned} \therefore \sigma^2 &= G''(1) + \mu - \mu^2 \\ &= \frac{6(1+1)}{8} + 1.5 - (1.5)^2 \\ &= 0.75 \end{aligned}$$

In fact $(1+t)^3/8$ is the p.g.f. of the number of heads in three tosses of a coin. We can check this by deriving $G(t)$ from the probability function $p(x)$ for the number of heads in three tosses of a coin:

$$\begin{aligned} G(t) &= E(t^X) && \text{definition} \\ &= \Sigma t^x p(x) && \text{using (5.2)} \\ &= t^0 \times \frac{1}{8} + t^1 \times \frac{3}{8} + t^2 \times \frac{3}{8} + t^3 \times \frac{1}{8} && \text{using Table 5.1} \\ &= \frac{1 + 3t + 3t^2 + t^3}{8} \\ &= \frac{(1+t)^3}{8} \end{aligned}$$

□

5.5 CONTINUOUS RANDOM VARIABLES, PROBABILITY DENSITY FUNCTION

If X is a continuous random variable its probability distribution may be specified in terms of a *probability density function (p.d.f.)*

$f_X(x)$, which we will simply write $f(x)$

(Compare with *probability function* for a random variable; see Section 5.2).

It is most important to realize that $f(x)$ is *not* a probability. To find the probability that X lies between two values, x_1 and x_2 say, where $x_1 \leqslant x_2$, we use:

$$P(x_1 \leqslant X \leqslant x_2) = \int_{x_1}^{x_2} f(x)\,dx$$

The probability density function, $f(x)$, may be expressed graphically, as in the next example.

Example 5.7

A continuous random variable, X, has a p.d.f.

$$f(x) = \frac{x}{8}, \qquad 0 \leqslant x \leqslant 4$$

$$= 0, \qquad \text{otherwise}$$

Draw a graph of $f(x)$ against x, and find the probability that $2 \leqslant X \leqslant 3$

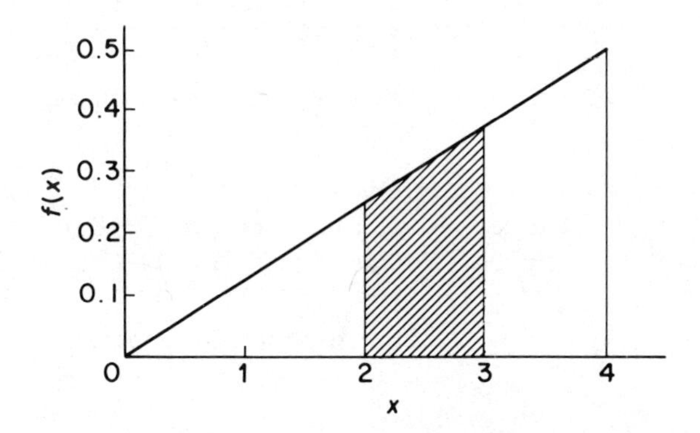

Fig. 5.2 Example of a continuous probability distribution

The graph is shown in Fig. 5.2.

$$P(2 \leqslant X \leqslant 3) = \int_2^3 f(x)\mathrm{d}x$$

$$= \int_2^3 \frac{x}{8}\mathrm{d}x$$

$$= \left[\frac{x^2}{16}\right]_2^3$$

$$= \frac{9}{16} - \frac{4}{16}$$

$$= \frac{5}{16}$$

We notice that this probability is the area between 2 and 3 under the graph of $f(x)$ vs. x in Fig. 5.2.

When we are dealing with continuous random variables, the idea that probabilities may be represented as areas is a very useful one.

Generally, $P(x_1 \leqslant X \leqslant x_2)$, for any continuous probability distribution, is the area under the graph of $f(x)$ vs. x between x_1 and x_2 (Fig. 5.3).

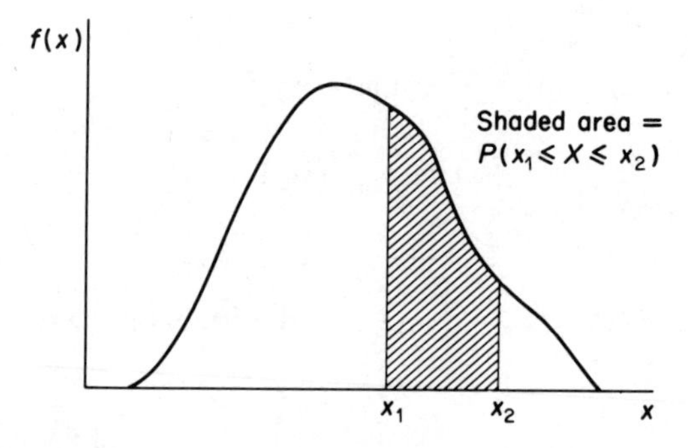

Fig. 5.3 A continuous probability distribution

For a 'proper' continuous probability distribution, the total area under the graph of $f(x)$ vs. x is 1 (this corresponds to the property for discrete distributions that the sum of probabilities is 1). For example 5.7, the total area

$$= \int_0^4 f(x)\mathrm{d}x = \left[\frac{x^2}{16}\right]_0^4 = 1,$$ so that distribution is 'proper'. □

Example 5.8

A continuous random variable, X, has a p.d.f.

$$\begin{aligned} f(x) &= kx, && 0 \leqslant x \leqslant 1 \\ &= k(2-x), && 1 < x \leqslant 2 \\ &= 0, && \text{elsewhere} \end{aligned}$$

Sketch the distribution and find k.

The graph is shown in Fig. 5.4.

When $x = 0, f(x) = 0$; $x = 1, f(x) = k$; $x = 2, f(x) = 0$

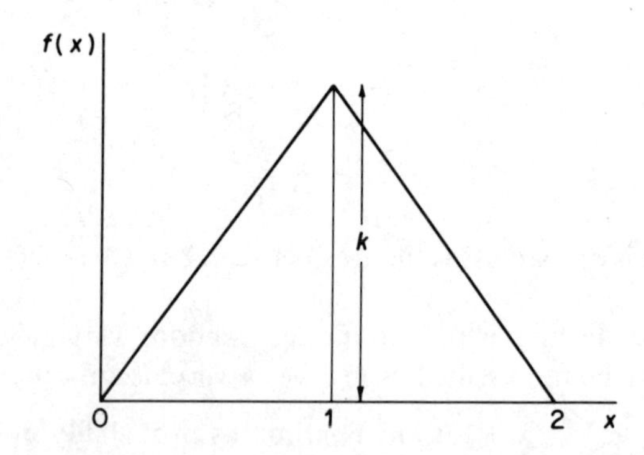

Fig. 5.4 Example of a continuous probability distribution

Since total area = 1, $\frac{1}{2}\times 2\times k = 1$ (area of a triangle)

$$\therefore k = 1$$

Alternatively we could have used:

$$\int_0^2 f(x) = 1, \text{ i.e.}$$

$$\int_0^1 kx\,\mathrm{d}x + \int_1^2 k(2-x)\,\mathrm{d}x = 1$$

$$\left[\frac{kx^2}{2}\right]_0^1 + \left[k\left(2x - \frac{x^2}{2}\right)\right]_1^2 = 1$$

$$\frac{k}{2} + k(4-2-2+\tfrac{1}{2}) = 1$$

$$k = 1 \qquad \square$$

5.6 EXPECTATION, MEAN AND VARIANCE OF A CONTINUOUS RANDOM VARIABLE

The notation $E(X)$ was introduced in Section 5.3 to represent the expectation of a random variable X. If X is continuous, then

$$E(X) = \int x f(x)\,\mathrm{d}x$$

where the integration is over the possible range of values of x.

The mean value of X, μ, was defined as $\mu = E(X)$ in Section 5.3. So we may write

$$\mu = E(X) = \int x f(x)\,\mathrm{d}x \tag{5.8}$$

for a *continuous* random variable X. Again we note that 'mean' is synonymous with 'expectation'.

The definition of expectation, $E(X) = \int x f(x)\mathrm{d}x$, is a special case of a more general definition:

$$E[g(X)] = \int g(x) f(x)\,\mathrm{d}x \tag{5.9}$$

where $g(X)$ is a function of the continuous random variable X.

To find the variance of X, we use (5.3), i.e.

$$\sigma^2 = \mathrm{Var}(X) = E(X^2) - \mu^2$$

where, from (5.9),

$$E(X^2) = \int x^2 f(x)\,\mathrm{d}x \tag{5.10}$$

Example 5.9

If X is a continuous random variable with p.d.f.

$$f(x) = \frac{x}{8}, \qquad 0 \leqslant x \leqslant 4$$
$$= 0, \qquad \text{otherwise,}$$

find the mean and variance of X.

$$\mu = E(X) = \int x f(x) \mathrm{d}x$$
$$= \int_0^4 x \frac{x}{8} \mathrm{d}x$$
$$= \left[\frac{x^3}{24}\right]_0^4$$
$$\mu = \frac{8}{3} = 2.67$$
$$\sigma^2 = \operatorname{Var}(X) = E(X^2) - \mu^2$$
$$= \int_0^4 x^2 \frac{x}{8} \mathrm{d}x - \left(\frac{8}{3}\right)^2$$
$$= \left[\frac{x^4}{32}\right]_0^4 - \left(\frac{8}{3}\right)^2$$
$$\sigma^2 = \frac{8}{9} = 0.89$$

□

Example 5.10

If X is a continuous random variable with p.d.f.

$$f(x) = x, \qquad 0 \leqslant x \leqslant 1$$
$$= 2 - x, \qquad 1 < x \leqslant 2$$
$$= 0, \qquad \text{otherwise,}$$

find the mean and variance of X.

$$\mu = E(X) = \int_0^1 x x \mathrm{d}x + \int_1^2 x(2-x) \mathrm{d}x$$
$$= \left[\frac{x^3}{3}\right]_0^1 + \left[x^2 - \frac{x^3}{3}\right]_1^2$$
$$\mu = 1$$

$$\sigma^2 = E(X^2) - \mu^2$$

$$= \int_0^1 x^2 x\mathrm{d}x + \int_1^2 x^2(2-x)\mathrm{d}x - 1^2$$

$$= \left[\frac{x^4}{4}\right]_0^1 + \left[\frac{2x^3}{3} - \frac{x^4}{4}\right]_1^2 - 1^2$$

$$\sigma^2 = \frac{1}{6}$$

□

5.7 DISTRIBUTION FUNCTION FOR A CONTINUOUS RANDOM VARIABLE

The *distribution function*†, $F(x)$, of a continuous random variable X is defined as follows:

$$F(x) = P(X \leqslant x) = \int_{-\infty}^{x} f(x)\,\mathrm{d}x$$

Since probabilities can be expressed in terms of areas under the graph of $f(x)$ vs. x, we see that

$$F(x_1) = \int_{-\infty}^{x_1} f(x)\,\mathrm{d}x, \quad \text{where } x_1 \text{ is any value of } X$$

$$= \text{area to left of } x_1 \text{ under graph of } f(x) \text{ vs. } x \qquad \text{(Fig. 5.5)}$$

It follows that $F(x_{max}) = 1$, since the total area under the graph is 1. Also $F'(x) = f(x)$, so we can derive $f(x)$ from $F(x)$ and vice versa.

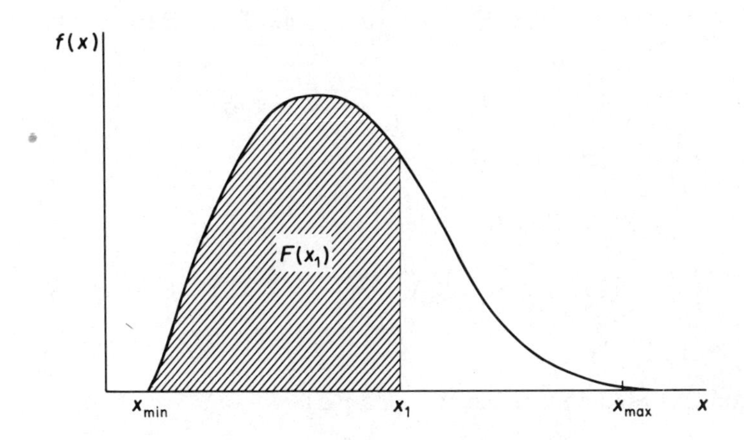

Fig. 5.5 A continuous probability distribution

†also sometimes called the *cumulative* distribution function.

Example 5.11

A continuous random variable, X, has a p.d.f.

$$f(x) = \frac{x}{8}, \qquad 0 \leqslant x \leqslant 4$$
$$= 0, \qquad \text{otherwise}$$

Find the distribution function, $F(x)$.
Express $P(2 \leqslant X \leqslant 3)$ in terms of (i) $f(x)$, (ii) $F(x)$.

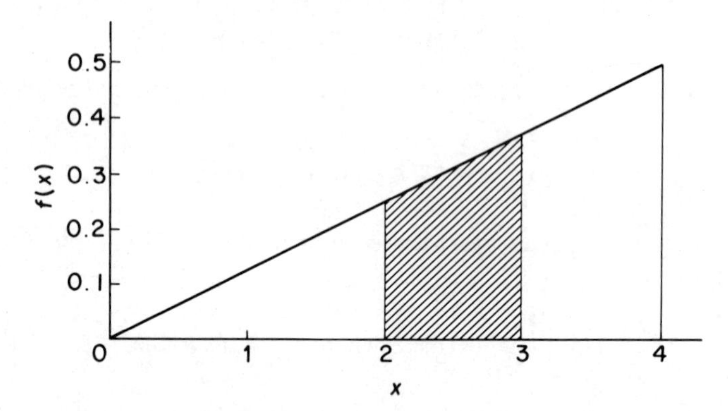

Fig. 5.6 Example of a continuous probability distribution

The p.d.f. is shown in Fig. 5.6.

$$F(x) = \int_{-\infty}^{x} f(x)\,\mathrm{d}x = \int_{0}^{x} \frac{x}{8}\,\mathrm{d}x, \quad \text{since } f(x) = 0 \text{ for } x < 0$$

$$\therefore F(x) = \frac{x^2}{16}, \quad 0 \leqslant x \leqslant 4;$$

Also $F(x) = 0, x < 0$ and $F(x) = 1, x > 1$.

(i) $$P(2 \leqslant X \leqslant 3) = \int_{2}^{3} f(x)\,\mathrm{d}x$$

(ii) $$P(2 \leqslant X \leqslant 3) = P(X \leqslant 3) - (P(X \leqslant 2) = F(3) - F(2)$$ □

Example 5.12

A continuous random variable, X, has a p.d.f.

$$f(x) = x, \qquad 0 \leqslant x \leqslant 1$$
$$= 2 - x, \quad 1 < x \leqslant 2$$
$$= 0, \qquad \text{otherwise}$$

Find the distribution function, $F(x)$, and check that $F(x_{\max}) = 1$.

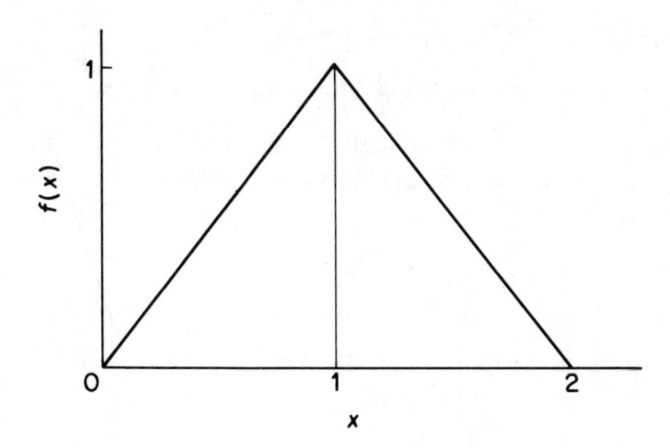

Fig. 5.7 Example of a continuous probability distribution

The p.d.f. is shown in Fig. 5.7.
We must treat the four cases $x < 0$, $0 \leqslant x \leqslant 1$, $1 < x \leqslant 2$, $x > 2$ separately.
When $x < 0$, $\quad F(x) = 0$

When $0 \leqslant x \leqslant 1$,

$$F(x) = \int_0^x x\,dx$$
$$= \frac{x^2}{2}$$

When $1 < x \leqslant 2$,

$$F(x) = P(X \leqslant 1) + P(1 < X \leqslant x)$$
$$= \frac{1}{2} + \int_1^x (2-x)\,dx, \quad \text{from previous result}$$
$$= \frac{1}{2} + \left[2x - \frac{x^2}{2}\right]_1^x$$
$$= \frac{1}{2} + 2x - \frac{x^2}{2} - 2 + \frac{1}{2}$$
$$F(x) = -\frac{x^2}{2} + 2x - 1$$

Since the maximum value of x is 2, $F(2)$ should equal 1

$$F(2) = -\frac{2^2}{2} + 2 \times 2 - 1 = 1, \quad \text{which checks.}$$

When $x > 2$, $\quad F(x) = 1.$ □

5.8 MEDIAN OF A CONTINUOUS RANDOM VARIABLE

The median, m, of a continuous random variable, X, is such that

$$P(X \leqslant m) = \tfrac{1}{2}$$

i.e. $F(m) = \frac{1}{2}$, where $F(x)$ is the distribution function.

Alternatively, we can use $f(x)$, the p.d.f., since $\int_{-\infty}^{m} f(x) = \frac{1}{2}$

The median is the value which divides the curve of $f(x)$ vs. x into two halves.

Similarly, quartiles, deciles and percentiles of continuous random variables can be obtained.

Example 5.13

A continuous random variable, X, has a p.d.f.

$$f(x) = 3e^{-3x}, \qquad x \geqslant 0$$
$$= 0, \qquad \text{elsewhere}$$

Find $F(x)$. Check that $F(x_{\max}) = 1$, and find the median and inter-quartile range of X.

$$F(x) = \int_0^x 3e^{-3x}\,dx, \qquad \text{when } x \geqslant 0$$
$$= [-e^{-3x}]_0^x$$
$$= -e^{-3x} - (-e^0)$$

So $\quad F(x) = 1 - e^{-3x}, \qquad x \geqslant 0$

and $\quad F(x) = 0, \qquad x < 0$

The maximum value of X is ∞.
$F(\infty) = 1$, since $e^{-3x} \to 0$ as $x \to \infty$
$F(m) = \frac{1}{2}$, where m is the median value of X

$$\therefore 1 - e^{-3m} = \frac{1}{2}$$
$$e^{-3m} = \frac{1}{2}$$
$$m = -\frac{1}{3}\log_e\left(\frac{1}{2}\right)$$
$$m = 0.23 \qquad \text{to 2 d.p.}$$

If q_1 is the lower quartile value of X,

$$F(q_1) = \frac{1}{4}$$
$$q_1 = -\frac{1}{3}\log_e\left(\frac{3}{4}\right) = 0.10 \qquad \text{to 2 d.p.}$$

If q_3 is the upper quartile value of X,

$$F(q_3) = \frac{3}{4}$$

$$q_3 = -\frac{1}{3}\log_e\left(\frac{1}{4}\right) = 0.46 \qquad \text{to 2 d.p.}$$

$$\begin{aligned}\text{Inter-quartile range of } X &= q_3 - q_1 \\ &= 0.46 - 0.10 \\ &= 0.36\end{aligned}$$

Figure 5.8 shows the graph of $f(x)$.

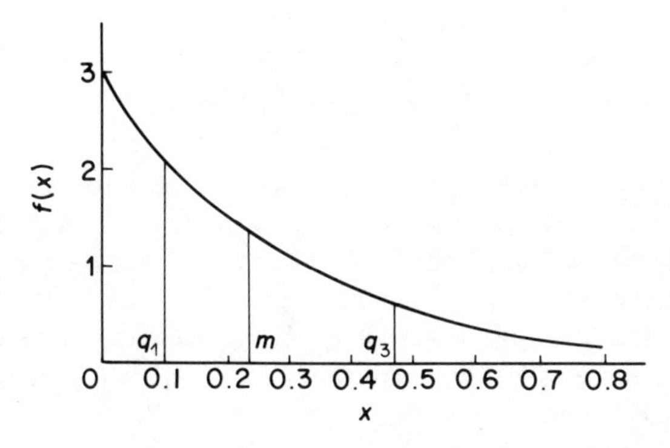

Fig. 5.8 Example of a continuous probability distribution

The vertical lines at $q_1 = 0.10$, $m = 0.23$, $q_3 = 0.46$ divide the graph of $f(x)$ vs. x into four quarters. □

5.9 MOMENT GENERATING FUNCTION FOR A CONTINUOUS RANDOM VARIABLE

The probability distribution of a continuous random variable may be described in terms of a *moment generating function* (*m.g.f.*) rather than in terms of a p.d.f. or distribution function. The mean, variance and standard deviation may be obtained directly from the m.g.f.

If $M(t)$ is the m.g.f. of a continuous random variable X, then

$$M(t) = E(e^{tX})$$

where t is a dummy variable. $M(t)$ can be expressed as a series in t:

$$\begin{aligned}
M(t) &= E(e^{tX}) \\
&= \int e^{tx} f(x)\,dx, \qquad \text{using (5.9)} \\
&= \int \left[1 + \frac{tx}{1!} + \frac{(tx)^2}{2!} + \ldots\right] f(x)\,dx \\
&= \int f(x)\,dx + \frac{t}{1!}\int x f(x)\,dx + \frac{t^2}{2!}\int x^2 f(x)\,dx + \ldots \\
&= 1 + E(X)\frac{t}{1!} + E(X^2)\frac{t^2}{2!} + \ldots \qquad \text{by (5.9)}
\end{aligned}$$

It follows that

$$M(0) = 1 \tag{5.11}$$

$$M'(0) = E(X) = \mu \tag{5.12}$$

$$M''(0) = E(X^2) = \sigma^2 + \mu^2 \tag{5.13}$$

where $M'(0)$ means differentiating $M(t)$ with respect to t and then putting $t = 0$; $M''(0)$ means differentiating $M(t)$ twice with respect to t and then putting $t = 0$. (These are similar to (5.5), (5.6) and (5.7) for the p.g.f. for a discrete random variable).

(5.11) is a check that the total probability is 1. (5.12) and (5.13) can be used to find the mean and variance of X if we know the m.g.f. The proofs of these three results are left as an exercise for the reader.

Example 5.14

A continuous random variable X has a m.g.f.

$$M(t) = \frac{1}{1-t}$$

Does X have a 'proper' distribution? Find the mean and variance of X.

Since $M(0) = \dfrac{1}{1-0} = 1$, distribution of X is 'proper'.

$$\begin{aligned}
M'(t) &= (1-t)^{-2} \\
M'(0) &= (1-0)^{-2} = 1 \\
\therefore \mu &= M'(0) = 1 \qquad \text{from (5.12)} \\
M''(t) &= 2(1-t)^{-3} \\
M''(0) &= 2(1-0)^{-3} = 2
\end{aligned}$$

$$\therefore \sigma^2 = M''(0) - \mu^2 \qquad \text{from (5.13)}$$

$$= 2 - 1^2$$

$$\sigma^2 = 1$$

$$\sigma = 1$$

The mean, variance and standard deviation are all equal to 1. □

Strictly speaking, $M(t)$ is the moment generating function (m.g.f.) *about zero*. The result shown above, namely,

$$M(t) = 1 + E(X)\frac{t}{1!} + E(X^2)\frac{t^2}{2!} + \ldots$$

may also be written $$M(t) = 1 + \mu_1'\frac{t}{1!} + \mu_2'\frac{t^2}{2!} + \ldots \qquad (5.14)$$

where μ_r' is called the rth moment about zero, for $r = 1, 2 \ldots$ Comparing coefficients of powers of t, we see that:

$$\left.\begin{aligned} \mu_1' &= E(X) = \mu \\ \mu_2' &= E(X^2) = \sigma^2 + \mu^2 \end{aligned}\right. \qquad (5.15)$$

We also note from (5.14) that:

$$\left.\begin{aligned} \mu_1' &= M'(0) \\ \mu_2' &= M''(0) \\ &\text{and so on} \end{aligned}\right\} \qquad (5.16)$$

There is another m.g.f., called the moment generating function *about the mean*, which is denoted by $M_\mu(t)$. It is defined as follows:

$$M_\mu(t) = E[e^{t(X-\mu)}] \qquad \text{(compare this with the definition of } M(t)\text{)}$$

$$= \int e^{t(x-\mu)} f(x)\,\mathrm{d}x$$

$$= \int \left[1 + \frac{t(x-\mu)}{1!} + \frac{t^2}{2!}(x-\mu)^2 + \ldots\right] f(x)\,\mathrm{d}x$$

$$= \int f(x)\,\mathrm{d}x + \frac{t}{1!}\int (x-\mu) f(x)\,\mathrm{d}x + \ldots$$

$$= 1 + E(X-\mu)\frac{t}{1!} + E[(X-\mu)^2]\frac{t^2}{2!} + \ldots$$

which may also be written

$$M_\mu(t) = 1 + \mu_1\frac{t}{1!} + \mu_2\frac{t^2}{2!} + \ldots \qquad (5.17)$$

where μ_r is called the rth moment about the mean for $r = 1, 2 \ldots$ Comparing coefficients of powers of t, we see that:

$$\mu_1 = E(X - \mu)$$
$$= \int_{-\infty}^{\infty} (x - \mu) f(x) \, dx$$
$$= \int_{-\infty}^{\infty} x f(x) \, dx - \mu \int_{-\infty}^{\infty} f(x) \, dx$$
$$= \mu - \mu \times 1$$
$$= 0$$

$$\left.\begin{aligned} \therefore \mu_1 &= 0 \\ \text{and } \mu_2 &= E[(X - \mu)^2] = \sigma^2 \end{aligned}\right\} \qquad (5.18)$$

We also note from (5.1) that

$$\left.\begin{aligned} \mu_1 &= M'_\mu(0) \\ \text{and} \quad \mu_2 &= M''_\mu(0) \end{aligned}\right\} \qquad (5.19)$$

The two moment generating functions $M(t)$ and $M_\mu(t)$ are related by the equation:

$$M_\mu(t) = e^{-\mu t} M(t) \qquad (5.20)$$

(where μ is the mean of the distribution)
Proof

$$M_\mu(t) = E[e^{t(X-\mu)}]$$
$$= E[e^{tX} e^{-\mu t}]$$
$$= e^{-\mu t} E(e^{tX})$$

using (A5.2), referred to in Section 5.10, since $e^{-\mu t}$ does not contain X and may therefore be treated as a constant.

$$= e^{-\mu t} M(t)$$

It follows that if we have either m.g.f. we can quickly obtain the other using (5.20).

Example 5.15

A continuous random variable, X, has m.g.f. about zero, $M(t) = \dfrac{1}{1-t}$

Find $M_\mu(t)$ and hence obtain the first two moments about the mean.

$$\mu = M'(0) \qquad \text{from (5.12)}$$
$$= (1-0)^{-2}$$
$$= 1$$

$$\therefore M_\mu(t) = e^{-\mu t} M(t) \qquad \text{from (5.20)}$$

$$= \frac{e^{-t}}{1-t}$$

Also
$$M'_\mu(t) = \frac{-e^{-t}}{1-t} + \frac{e^{-t}}{(1-t)^2}$$

$$\therefore M'_\mu(0) = -1 + 1 = 0$$

$$\therefore \mu_1 = M'_\mu(0) = 0, \qquad \text{as it always is (see below)}$$

Also
$$M''_\mu(t) = \frac{e^{-t}}{1-t} - \frac{2e^{-t}}{(1-t)^2} + \frac{2e^{-t}}{(1-t)^3}$$

$$\therefore M''_\mu(0) = 1 - 2 + 2 = 1$$

$$\therefore \mu_2 = M''_\mu(0) = 1$$

□

Finally in this section we can expand (5.20) to derive equations which relate moments about zero to moments about the mean:

(5.20) is
$$M_\mu(t) = e^{-\mu t} M(t)$$

Using (5.14) and (5.17),

$$1 + \mu_1 \frac{t}{1!} + \mu_2 \frac{t^2}{2!} = \left[1 - \frac{(\mu t)^1}{1!} + \frac{(\mu t)^2}{2!} - \ldots\right] \times \left[1 + \mu'_1 \frac{t}{1!} + \mu'_2 \frac{t^2}{2!} + \ldots\right]$$

Equating coefficients of t, t^2, t^3, we obtain

$$\mu_1 = \mu'_1 - \mu = 0, \qquad \text{by (5.15)}$$

$$\mu_2 = \mu'_2 - (\mu'_1)^2$$

$$\mu_3 = \mu'_3 - 3\mu'_1 \mu'_2 + 2(\mu'_1)^2$$

and so on.

5.10 MEAN AND VARIANCE OF A LINEAR FUNCTION OF A RANDOM VARIABLE

Suppose X is a random variable and a and b are constants. Then $aX + b$ is a linear function of X and is itself a random variable. Let's call it Y, so

$$Y = aX + b$$

How do the mean and variance of Y relate to the mean and variance of X? The answers are:

$$E(Y) = a\,E(X) + b \tag{5.21}$$

$$\text{Var}\,(Y) = a^2\,\text{Var}(X) \tag{5.22}$$

Proof

$$
\begin{aligned}
E(Y) &= E(aX + b)\\
&= aE(X) + b \qquad \text{by (A5.3); see Appendix to this chapter.}\\
\text{Var}\,(Y) &= E[(Y - E(Y))^2] \qquad \text{see definition of variance in Section 5.3}\\
&= E[(aX + b - aE(X) - b)^2], \qquad \text{substituting for } Y \text{ and } E(Y)\\
&= E[a^2(X - E(X))^2]\\
&= a^2\,E[(X - E(X))^2] \qquad \text{by (A5.2)}\\
&= a^2\,\text{Var}\,(X), \qquad \text{see definition of variance in Section 5.3.}
\end{aligned}
$$

Example 5.16

A temperature, measured in °C, is subject to random variation such that the mean reading is 10 °C with a standard deviation of 3 °C. If the temperature had been measured in °F instead, what would the mean and standard deviation be?
If X stands for temperature in °C, and Y stands for temperature in °F,

then
$$Y = 1.8X + 32$$

Using $a = 1.8$, $b = 32$ in results (5.21) and (5.22) we obtain

$$
\begin{aligned}
E(Y) &= E(1.8X + 32)\\
&= 1.8E(X) + 32\\
&= 1.8 \times 10 + 32\\
&= 50\,^\circ\text{F}\\
\text{Var}\,(Y) &= \text{Var}\,(1.8X + 32)\\
&= 1.8^2\,\text{Var}\,(X)\\
&= 1.8^2 \times 3^2\\
&= 5.4^2\\
\sigma_Y &= \sqrt{\text{Var}\,(Y)} = 5.4\,^\circ\text{F}
\end{aligned}
$$

The mean temperature is 50 °F, the standard deviation of the temperature is 5.4 °F. □

5.11 THE PROBABILITY DISTRIBUTION FOR A FUNCTION OF A CONTINUOUS RANDOM VARIABLE

Suppose that X is a continuous random variable with probability density function $f(x)$ and distribution function $F(x)$.

Suppose that Y is a continuous random variable with probability function $g(y)$ and distribution function $G(y)$.

Also suppose that X and Y are related by the equation $Y = h(x)$, where the function h is a strictly increasing or decreasing function. It follows that:

(a) if h is an increasing function,

$$G(y) = F(x) \tag{5.23}$$

(b) if h is a decreasing function

$$G(y) = 1 - F(x) \tag{5.24}$$

Example 5.17

X is a continuous random variable with p.d.f.

$$\begin{aligned} f(x) &= 4x^3, \qquad 0 \leqslant x \leqslant 1 \\ &= 0 \qquad\quad \text{elsewhere} \end{aligned}$$

If Y is a continuous random variable such that $Y = X^2$, find $G(y)$ and $g(y)$, the distribution function and p.d.f. of Y.

Clearly x^2 is a strictly increasing function over the range $x = 0$ to $x = 1$ (see Fig. 5.9).

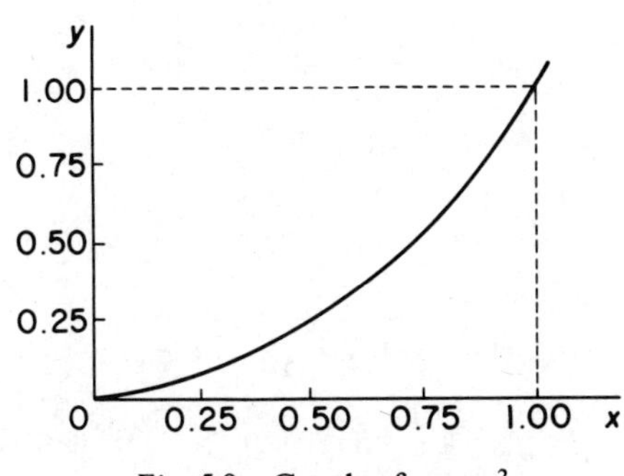

Fig. 5.9 Graph of $y = x^2$

From (5.23),
$$G(y) = F(x)$$
$$= \int_0^x f(x)\,dx, \qquad \text{see Section 5.7}$$
$$= \int_0^x 4x^3\,dx$$
$$= x^4$$

Also since $y = x^2$, $x = y^{\frac{1}{2}}$.
$$\therefore G(y) = (y^{\frac{1}{2}})^4 = y^2$$
and
$$g(y) = G'(y) = 2y$$
Also when $x = 0$, $y = 0$, and when $x = 1$, $y = 1$
$$\therefore g(y) = 2y, \qquad 0 \leqslant y \leqslant 1$$
$$g(y) = 0, \qquad \text{elsewhere}$$
□

Example 5.18

X is a continuous random variable with p.d.f.
$$f(x) = x - 2, \qquad 2 \leqslant x \leqslant 2 + \sqrt{2}$$
$$= 0, \qquad \text{elsewhere}$$

If Y is a continuous random variable such that $y = 12 - x^2$, find $g(y)$, the p.d.f. of Y.

Clearly, as x increases from 2 to $2 + \sqrt{2}$, y decreases from 8 to 0.34.
From (5.24),
$$G(y) = 1 - F(x)$$
$$= 1 - \int_2^x (x-2)\,dx$$
$$= 1 - \left[\frac{x^2}{2} - 2x\right]_2^x$$
$$= -\frac{x^2}{2} + 2x - 1$$

Also, since $y = 12 - x^2$, $x = (12 - y)^{\frac{1}{2}}$
$$\therefore G(y) = -\frac{(12-y)}{2} + 2(12-y)^{\frac{1}{2}} - 1$$
$$\text{and } g(y) = G'(y) = \frac{1}{2} - (12-y)^{-\frac{1}{2}}$$
So
$$g(y) = \frac{1}{2} - (12-y)^{-\frac{1}{2}}, \qquad 0.34 \leqslant y \leqslant 8$$
$$= 0, \qquad \text{elsewhere}$$
□

5.12 SUMMARY

There are two types of random variable, discrete and continuous.

The probability distribution of a *discrete* random variable may be described in terms of a probability function, $p(x)$, or in terms of a probability generating function, $G(t)$.

The mean and variance of the distribution may be obtained using

either $$\mu = E(X) = \Sigma x p(x) \quad (5.1)$$

and $$\sigma^2 = E(X^2) - \mu^2 = \Sigma x^2 p(x) - \mu^2 \quad (5.3) \text{ and } (5.4)$$

or $$\mu = G'(1) \quad (5.6)$$

and $$\sigma^2 = G''(1) - \mu^2 + \mu \quad (5.7)$$

The probability distribution of a *continuous* random variable may be described in terms of a probability distribution function, $f(x)$, or in terms of a moment generating function, $M(t)$. The mean and variance of the distribution may be obtained using

either $$\mu = E(X) = \int x f(x)\,\mathrm{d}x \quad (5.8)$$

and $$\sigma^2 = E(X^2) - \mu^2 = \int x^2 f(x)\,\mathrm{d}x - \mu^2 \quad (5.3) \text{ and } (5.10)$$

or $$\mu = M'(0) \quad (5.12)$$

and $$\sigma^2 = M''(0) - \mu^2 \quad (5.13)$$

For a continuous random variable, the distribution function, $F(x) = P(X \leqslant x)$, can be used to find m, the median of the distribution, by solving $F(m) = \frac{1}{2}$.

The mean and variance of a linear function of a random variable, X, can be found by using

$$E(aX + b) = aE(X) + b \quad (5.21)$$

and $$\text{Var}(aX + b) = a^2 \text{Var}(X) \quad (5.22)$$

The p.d.f. of a strictly increasing or decreasing function of a continuous random variable, X, can be found using

$$G(y) = F(x) \quad (5.23)$$

if $Y = h(X)$ is a strictly increasing function of x,

or $$G(y) = 1 - F(x) \quad (5.24)$$

if $Y = h(X)$ is a strictly decreasing function of x.

APPENDIX TO CHAPTER 5

Expectation algebra – some useful results

1. If a is a constant, $$E(a) = a \tag{A5.1}$$

Proof (assuming X is a discrete random variable)

$$\begin{aligned} E(a) &= \Sigma a p(x) && \text{by (5.2)} \\ &= a \Sigma p(x) && \text{since } a \text{ does not vary as } x \text{ varies} \\ &= a && \text{since the total probability is 1} \end{aligned}$$

A similar proof follows if X is a continuous random variable.

2. If a is a constant, $$E(aX) = aE(X) \tag{A5.2}$$

Proof (assuming X is a discrete random variable)

$$\begin{aligned} E(aX) &= \Sigma a x p(x) && \text{by (5.2)} \\ &= a \Sigma x p(x) && \text{since } a \text{ does not vary as } x \text{ varies} \\ &= aE(X) && \text{by (5.1)} \end{aligned}$$

A similar proof follows if X is a continuous random variable.

3. If a and b are constants, $$E(aX + b) = a\,E(X) + b \tag{A5.3}$$

Proof (assuming X is a discrete random variable)

$$\begin{aligned} E(aX + b) &= \Sigma\,(ax + b)p(x) && \text{by (5.2)} \\ &= \Sigma\,axp(x) + \Sigma\,bp(x) && \text{splitting into two summations} \\ &= E(aX) + E(b) && \text{by (5.2)} \\ &= aE(x) + b && \text{by (A5.1) and (A5.2) above.} \end{aligned}$$

A similar proof follows if X is a continuous random variable.

4. $$E[(X - \mu)^2] = E(X^2) - \mu^2 \tag{A5.4}$$

Proof (assuming X is a discrete random variable)

$$\begin{aligned} E[(X - \mu)^2] &= E[X^2 - 2\mu X + \mu^2] \\ &= \Sigma(x^2 - 2\mu x + \mu^2)\,p(x) && \text{by (5.2)} \\ &= \Sigma x^2 p(x) + \Sigma - 2\mu x p(x) + \Sigma \mu^2 p(x) \\ &= E(X^2) - 2\mu\Sigma x p(x) + \mu^2 \Sigma p(x) \\ & && \text{since } \mu \text{ does not change as } x \text{ changes} \\ &= E(X^2) - 2\mu^2 + \mu^2 && \text{by (5.1)} \\ &= E(X^2) - \mu^2 \end{aligned}$$

WORKSHEET 5

Section 5.1

1. In each of the following state the name of the random variable and whether it is discrete or continuous. Also state the set or range of possible values which the variable may take:
   (a) The number of draws in eight matches selected each week for a football pool.
   (b) The time taken by army recruits to run a mile at the beginning of their initial training period.
   (c) The number of counts per minute over a long period recorded by a Geiger counter in a test of a weak source of radioactivity.
   (d) The number of germinating seeds in packets of 100 seeds.
   (e) The height of ten-year-old school boys in the UK.
   (f) The amount of money gained by a gambler who always bets on the number 6 when a die is thrown (the gambler's stake is £1 per bet, the gambler receives £4 if a 6 is thrown but nothing otherwise).
   (g) The number of correct answers in a multiple-choice test taken by a large number of candidates. The test consists of 50 questions, each having 5 possible answers only one of which is correct. Assume each candidate guesses the answer to each question.
   (h) The number of cracked eggs found in cartons of six eggs in a large batch of eggs.
   (i) The weight of new-born lambs born on one farm.
   (j) The number of rainy days per week in the first week of July in the UK.

Section 5.2

2. A sample of 4 people is selected at random from a set of 20 people, composed of 10 married couples. Explaining your arguments carefully, find the distribution of the number of married couples in the sample.

(OLE)

Section 5.3

3. For Exercise 1(f) above, what is the probability distribution of the gambler's gain per bet? Find the mean, variance and standard deviation of the gambler's gain per bet. What amount of money would the gambler need to receive when a 6 is thrown in order that his (or her) mean gain is zero (making the game 'fair')?

4. In a raffle 10,000 tickets are sold for £1 each. With a first prize of £1000, second prize of £500 and ten other prizes of £100, what is the expected gain of a person who buys a ticket? Assume that a ticket can win more than one prize.

5. A discrete random variable X has a probability function:

x	1	2	3
$p(x)$	$\frac{1}{2}$	$\frac{1}{3}$	$\frac{1}{6}$

Find $E(X)$, $E(X^2)$ and the mean and variance of X.

►

6. A survey of 200 cars entering a car park revealed the following numbers of persons travelling per car:

Number per car	1	2	3	4	5
Number of cars	110	40	30	15	5

Write down your estimate of the probability distribution of the number of persons travelling per car. Use expectation to estimate the mean and variance of the distribution.

7. Let X denote the number of heads in four tosses of a coin. Express the probability distribution of X in a table and graphically. Find the mean and variance of X.

8. The probability function of a discrete random variable, X, is:

$$p(x) = k, \quad x = 1, 2, 3, \ldots, n.$$

Find k and the mean and variance of X. (This is an example of a discrete uniform distribution since the probability is constant. See Section 6.5 for the *continuous* uniform distribution.)

9. There are n balls in a bag, r of which are red. Two balls are drawn out with replacement. What is the probability distribution of the number of red balls drawn out? What are the mean and variance of the number of red balls drawn out?

10. Three alternative designs of an electronic component are considered:

    Design A includes 3 microchips, all of which must function to enable the component to function.
    Design B includes 6 microchips arranged in two sets of three. Provided at least one set of three function, the component will function.
    Design C includes 6 microchips arranged in three pairs. Provided at least one microchip of each pair functions, the component will function.

    If the probability of a microchip functioning is p, find the probability of the component functioning for each of the three alternative designs.
    Assuming that components which do not function are scrapped, find, in terms of p, the expected number of functioning components obtained from 6000 microchips for each of the three designs. Show that this expected number is greater for design A than for design B for all $p > 0$.

    Evaluate the expected number of functioning components for design A and for design C at $p = 0$, $p = 0.5$, and $p = 1.0$.

    Hence, or otherwise, sketch graphs on the same axes of the expected number of functioning components against p for design A and design C.

    State briefly the main point illustrated by the graph relevant to choice of design.

    (AEB (1984))

11. (i) E_1 and E_2 are two independent events such that:

$$P(E_1) = p_1, \qquad P(E_2) = p_2.$$

Describe in words each of the events whose probabilities are:

(a) $p_1 p_2$;
(b) $(1 - p_1)p_2$;

(c) $1-(1-p_1)(1-p_2)$;
(d) $p_1+p_2-2p_1p_2$.

(ii) A box A contains 9 red balls and one white ball. A box B contains 8 red balls and one white ball. A ball is to be taken at random from box A and put into box B, and then a ball is to be taken at random from box B. Find the probability that this ball from box B will be white.

Of the two balls drawn, one from A and one from B, let X denote the number that are white. Find the probability distribution of X.

Find the mean of X.

Find also, to 2 decimal places, the variance of X.

(London)

12. n people are asked a question in a random order, and exactly two of the n know the answer. If $n > 5$, what is the probability that the first four of those asked do not know the answer?

Show that the probability that the rth person asked is the first to know the answer is

$$\frac{2(n-r)}{n(n-1)} \quad \text{if } 1 < r < n.$$

Show that this expression for the probability also holds when $r = 1$ or $r = n$.

Verify that the sum of these probabilities over all possible values of r is 1, and show that the expected number of people to be asked before the correct answer is obtained is $\frac{1}{3}(n+1)$.

(OLE)

13. In the village of Jieyuhao families are strictly limited to two children. The probability distribution of the number of children of any given individual is as follows:

Number of children	0	1	2
Probability	$\frac{1}{6}$	$\frac{1}{2}$	$\frac{1}{3}$

Find the mean and variance of the number of children.

Assuming that individuals are independent, find the probability distribution of the number of grandchildren of a given individual. What is the expected number of grandchildren? Comment on the assumption of independence.

(Give answers as fractions, not decimals).

(OLE)

14. Two players, A and B, play a game in which they throw a fair six-sided die alternatively, A having the first throw. The first player to throw a 6 is the winner. What is the probability that A is the winner?

Given that A wins eventually, write down the conditional probability that A wins on his nth throw, and hence determine the expected number of throws before A wins.

What is the expected duration of the game?

(OLE)

15. The random variable, X, has a discrete distribution, and ▶

$$P(X=0)=\lambda,\ P(X=n)=\lambda\frac{k(k+1)\ldots(k+n-1)}{n!}a^n \text{ for } n=1,2,\ldots$$
with $0<a<1$.

Show that $\lambda=(1-a)^k$ and that the mean of X is $ka/(1-a)$. What is the variance of X?

(OLE)

16. A fisherman casts his line into a river with probability p for each cast of getting a bite. If X is the number of casts up to and including the first bite, what is the probability function for X, $p(x)$? Show that $\Sigma p(x)=1$, where the summation is over all possible values of X, and find the mean and variance of X.

17. A monkey has a bag containing 4 apples, 3 bananas and 2 coconuts. It selects the fruits at random and eats them until it comes upon one fruit of a type which it has eaten before, whereupon it throws away that fruit and also the bag and its remaining contents. What is the probability distribution of the number of fruits eaten? How many of the fruits can it expect to eat?

(OLE)

18. A radio network is launching a new music game (based on one actually broadcast in France). Each contestant is given the title of a song and a list of 7 words, exactly 3 of which occur in the lyric of the song. The contestant is asked to choose the 3 correct words (the choice being made before listening to the song). Assuming that a particular contestant makes this choice completely by guesswork, find the probabilities that he gets 0, 1, 2, 3 words correct.

    Prizes of £1, £3 and £r are to be awarded for 1, 2 and 3 words correct respectively. Find, in terms of r, the expected prize paid to a contestant choosing completely by guesswork. Hence determine the greatest integer value of r for which the expected prize is less than £3.

(MEI)

Section 5.4

19. Find the probability generating function for X in Exercise 16. Use it to verify:
    (a) that the total probability is 1;
    (b) the mean value of X;
    (c) the variance of X.

20. Find the probability generating function for X in Exercise 5. Use it to verify:
    (a) that the total probability is 1;
    (b) the mean value of X;
    (c) the variance of X.

21. Show that the probability generating function for X in Exercise 7 is $\left(\frac{1+t}{2}\right)^4$. Use it to verify the values for the mean and variance of X.

22. The probability generating function of a discrete random variable, X, is $(q+pt)^n$, where $q=1-p$. Show that the mean and variance of X are np and npq respectively. ►

23. The variable R has probability generating function given by $G(t) = ke^{t^2+t}$, where k is a constant. Find k and obtain:
    (i) $P(R = 0)$, $P(R = 1)$, $P(R = 2)$, $P(R = 3)$ and $P(R = 4)$;
    (ii) the mean of R;
    (iii) the variance of R;
    (iv) the smallest value r_0 of R for which $P(R > r_0) < 0.4$.

(Cambridge)

Section 5.6

24. A continuous random variable, X, has a probability density function (p.d.f.)

$$f(x) = k(x - x^3), \quad 0 \leqslant x \leqslant 1$$
$$= 0, \quad \text{elsewhere}$$

Find k and the mean and variance of X. Find the value of X for which $f(x)$ is a maximum and sketch the curve of $f(x)$ vs. x.

25. A continuous random variable, X, has a p.d.f.

$$f(x) = \frac{1}{2}e^{-x}, \quad x \geqslant 0$$
$$= \frac{1}{2}e^{x}, \quad x < 0$$

Show that $\int_{-\infty}^{\infty} f(x)\,dx = 1$. Sketch $f(x)$ vs. x. Find the mean and variance of X and find $P(|X| < \frac{1}{2})$.

26. A woman is employed to clean a hen-house. The number of minutes that she takes is a random variable X with probability density function

$$f(x) = \alpha^2 x e^{-\alpha x}, \quad x > 0,$$

where $\alpha = 0.04$. Show that the probability that the work is completed in x minutes or less is given by

$$\Pr(X \leqslant x) = 1 - (1 + \alpha x)e^{-\alpha x}$$

Also calculate the mean time she takes to complete the work.

The cleaner catches either an early bus or a late bus to get to work and as a result starts work at 10 a.m. or 10.20 a.m. each morning. Calculate the probabilities that on any one day she finishes work by 11 a.m. if

(a) she catches the early bus,
(b) she catches the early bus with probability 0.4.

(IOS)

27. The continuous random variable X has probability density function given by

$$f(x) = \begin{cases} k(1 + x^2) & \text{for } -1 \leqslant x \leqslant 1, \\ 0 & \text{otherwise,} \end{cases}$$

▶

where k is a constant. Find the value of k, and determine $E(X)$ and $\mathrm{Var}(X)$.
A is the event $X > \frac{1}{2}$, B is the event $X > \frac{3}{4}$.
Find (i) $P(B)$, (ii) $P(B|A)$.

(Cambridge)

28. Based upon substantial experience with a large number of water pumps of a particular type, the manufacturer draws up the following table to indicate how many of the batch of 100 pumps may be expected still to be operating after n hours.

Running time (hours)	0	100	200	300	400	500	600	700
Estimated no. of pumps still working	100	95	85	60	30	12	3	0

Use this table to calculate the probabilities that a pump will last:

(i) more than 200 hours,
(ii) less than 400 hours,
(iii) between 200 and 400 hours.

A particular pump has been running for 400 hours. Write down the probabilities that it will fail within the next 100 hours and within the next 200 hours. Deduce the probability that the machine will break down after more than 100 but less than 200 hours further running. Find the mean number of hours it can be expected to continue running.

A firm has a machine which is cooled by water from three of the pumps operating independently. The machine will continue to operate as long as at least two of the pumps are working. Each of the three pumps has already been running for 400 hours. Find the probability that the machine will break down within the next 100 hours. Find also the mean time that the machine can be expected to operate.

(JMB)

29. The continuous random variable X has probability density function $(x-a)(2a-x)$ for $a \leqslant x \leqslant 2a$, and zero elsewhere. Show that $a^3 = 6$. Obtain the mean and variance of X.

(OLE)

30. A random variable X has probability density function

$$f(x) = \begin{cases} kx^2(1-x) & 0 \leqslant x \leqslant 1 \\ 0 & \text{otherwise} \end{cases}$$

Find k and sketch the curve $y = f(x)$.

Find μ, the mean of X; find also the mode of X. Obtain the mean deviation of X, defined by $E(|X-\mu|)$.

(O&C)

31. The guarantee by a retailer ensures that a faulty television tube is replaced within the first year of purchase. The probability distribution of the life of the tube can be modelled by

$$f(x) = \frac{A}{x^4} \qquad \text{for } 1 \leqslant x \leqslant \infty,$$

$$f(x) = 0 \qquad \text{for } x < 1.$$

where x is the lifetime in years.

▶

Find:

(i) the value of A;
(ii) the mean and variance of x;
(iii) the probability that the tube lasts longer than five years.

Show that, if a tube has lasted for n years, the probability that it will fail during the next year is

$$1-\left(\frac{n}{n+1}\right)^3.$$

(SUJB)

Section 5.7

32. Find the (cumulative) distribution function for a continuous random variable, X, if the p.d.f. of X is:

(i) $$f(x)=4(x-x^3), \quad 0\leqslant x\leqslant 1$$
$$=0 \quad \text{elsewhere}$$

(ii) $$f(x)=\frac{1}{9}(3+2x-x^2), \quad 0\leqslant x\leqslant 3$$
$$=0 \quad \text{elsewhere}$$

Verify in each case that $F(x_{MAX})=1$

33. A continuous random variable, X, has a distribution function

$$F(x)=kx^2\left(1-\frac{x}{3}\right), \quad 0\leqslant x\leqslant 2$$

Find the value of k and the p.d.f. of X.

34. The probability that an electrical component lasts longer than t hours is e^{-kt}, where k is a positive constant. State the distribution function for the lifetime of a component.

Find the probability density function for the lifetime of a component.

This mean lifetime is in fact $1/k$ hours.

If the mean lifetime of a component is 2000 hours, find the probability of a component lasting at least 4000 hours.

If the manufacturer wants to ensure that, on average, less than one in a thousand components will fail after less than 5 hours use, find, to one significant figure, the lowest mean lifetime he can allow his components to have.

(JMB)

Section 5.8

35. A continuous random variable, X, has a p.d.f.

$$f(x)=\frac{(3-x)(1+x)}{9}, \quad 0\leqslant x\leqslant 3$$
$$=0 \quad \text{elsewhere}$$

▶

Find the mean and variance of X. Also find $P(X > 2)$ and show that m, the median value of X, satisfies the equation

$$2m^3 - 6m^2 - 18m + 27 = 0$$

Verify that the median value of X is 1.21.

36. A continuous random variable, X, has a p.d.f.

$$\begin{aligned} f(x) &= k(2-x), \quad && 0 \leqslant x \leqslant 2 \\ &= 0 && \text{elsewhere} \end{aligned}$$

Find k. Also find the median and inter-quartile range of X.

37. A continuous variable Z has the probability density function

$$f(z) = \begin{cases} k \sin \tfrac{1}{2}z, & 0 \leqslant z < \tfrac{2}{3}\pi \\ k \sin(z - \tfrac{1}{3}\pi), & \tfrac{2}{3}\pi \leqslant z \leqslant \tfrac{4}{3}\pi \\ 0 & \text{elsewhere} \end{cases}$$

Find the value of k and sketch the graph of $f(z)$.

Obtain the distribution function of Z and sketch its graph.

Show that the median of Z is approximately 2.365.

(Cambridge)

38. A continuous random variable X has probability density function f given by

$$\begin{aligned} f(x) &= 1 - \tfrac{1}{2}x, \quad && 0 \leqslant x \leqslant 2, \\ f(x) &= 0, && \text{otherwise.} \end{aligned}$$

Find the exact values of the mean and the median of X.

(JMB)

39. The random variable X has the probability density function given by

$$f(x) = \begin{cases} kx & 0 \leqslant x \leqslant 1 \\ k & 1 \leqslant x \leqslant 3 \\ \tfrac{1}{3}k(6-x) & 3 \leqslant x \leqslant 6 \\ 0 & \text{otherwise} \end{cases}$$

Sketch the function and calculate the value of k. Find the median and the mean of this distribution.

(O & C)

40. In the study of atmospheric pollution, the concentration X of the pollutant is a random variable. Units of concentration can be chosen so that the value of X always lies between 1 and 10. In certain circumstances, an appropriate mathematical model of the probability that $X \leqslant x$ is

$$c\{\sqrt{(\ln 10)} - \sqrt{(\ln 10 - \ln x)}\}, \qquad (1 \leqslant x \leqslant 10)$$

where c is a constant. Explain how the value of c can be determined by considering the probability that $X \leqslant 10$. Hence show that $c = 0.659$ correct to 3 decimal places. ▶

Find, correct to 2 decimal places, the probabilities that

(a) $X \leqslant 8$,
(b) $2 \leqslant X \leqslant 8$.

Show that the median concentration is about 5.6.

(MEI)

41. A continuous random variable X has a probability density function given by:

$$f(x) = Ax(2-x)(2+x) \qquad \text{if } 0 \leqslant x \leqslant 2,$$
$$f(x) = 0 \qquad \text{elsewhere}$$

(i) Find the value of A.
(ii) Find the probability that a value of X taken at random will exceed 1.
(iii) show that, if the median value is m, then m satisfies the quartic equation $m^4 - 8m^2 + 8 = 0$.

By using the substitution $m^2 = t$, or otherwise, find the value of m.

If the lower quartile is q, which quartic equation would you need to solve in order to find the value of q? (Obtain the equation, but do not attempt to solve it).

(iv) Find the mean (expectation) of X.
(v) Find the variance of X.

(SUJB)

42. The continuous random variable X has density function $\dfrac{\theta}{x}$ $(1 \leqslant x \leqslant 2)$ and zero elsewhere.

Find the value of the constant θ, and the mean and variance of X.

What are the median and the quartiles of the distribution?

Show that the probability that X is less than the mean is

$$-\frac{\ln(\ln 2)}{\ln 2}$$

(OLE)

Section 5.9

43. A continuous random variable, X, has a p.d.f.

$$f(x) = 2e^{-2x}, \qquad x \geqslant 0$$
$$= 0 \qquad \text{elsewhere}$$

Find $M(t)$, the moment generating function (about zero) of X, and use it to find the mean and variance of X.

44. Find the moment generating function (about zero) of a continuous random variable, X, with p.d.f.

$$f(x) = xe^{-x}, \qquad x \geqslant 0$$
$$= 0 \qquad \text{elsewhere}$$

Hence find the mean and variance of X.

45. The moment generating function (about zero) of a continuous random variable, X, is

$$M(t) = \frac{\lambda}{\lambda - t}$$

Find the mean and variance of X. Also find the m.g.f. about the mean, $M_{\mu}(t)$, and hence obtain μ_1 and μ_2. Are you surprised at the value obtained for μ_1? Is it a coincidence that the values you obtain for the variance and μ_2 are the same?

46. Show that, if μ_r' is the rth moment about the origin ($r = 2, 3, 4, \ldots$) of a probability distribution of which μ_1' is the mean, then the third moment μ_3 about the mean is given by

$$\mu_3 = \mu_3' - 3\mu_2'\mu_1' + 2(\mu_1')^3$$

The continuous random variable X takes all positive values with probability density function $(1-\theta+\theta x)e^{-x}$. Find its moment generating function.

Show that $\mu_r' = (1+\theta)(r!)$
and find μ_3 as a polynominal in θ.

(O&C)

Section 5.10

47. The mean and variance of a continuous random variable, X, are 120 and 100 respectively. Find the mean and variance of:

(i) $X - 10$
(ii) $X - 120$
(iii) $10X$
(iv) $10(X - 12)$
(v) $\dfrac{X - 120}{10}$

48. When a market gardener takes n cuttings from a shrub and plants them, the number X that will root successfully is a discrete random variable with

$$P(X = r) = \frac{2r}{n(n+1)}, \qquad r = 1, 2, \ldots, n.$$

The total cost in pence to the gardener of taking and planting n cuttings is equal to $20 + 0.8n^2$. Cuttings that root successfully are sold by the gardener for 60 pence each.

(a) Show that $E(X) = (2n+1)/3$.
(b) For $n = 20$, calculate

   (i) the probability that the gardener will make a loss,
   (ii) the gardener's expected profit.

(c) Determine the value of n which will maximize the gardener's expected profit and evaluate this maximum expected profit.

(WJEC) ►

49. A couple decide to continue to have children until they have both a boy and a girl in their family and then stop. The probabilities of any particular child being a boy or a girl are p and q respectively. You may assume that there are no multiple births (twins, triplets, and so on). If such a couple eventually succeed in their intention and have $2+x$ children, what is the probability that x has the value r ($r \geqslant 0$)?

    Find the generator for these probabilities, assuming that x may be *any* integer $\geqslant 0$. Hence show that in the case $p = q = 1/2$ the expected number of children will be three. For this case find also the variance of the family size.

    (SMP)

50. A small newsagent sells a particular monthly magazine. Provided that he has sufficient copies, the number of copies that he can sell in a given month has the probability distribution shown in the table. Each month he buys a supply and sells them at £1.20 per copy, making a profit of 50p on each magazine sold. Any remaining unsold at the end of the month are wasted and valueless.

Number of magazines	10	11	12	13	14
Probability	0.15	0.25	0.30	0.20	0.10

Find:

(i) the mean number that he could sell if he had sufficient copies;
(ii) the probability that he could sell more than 12 if he had sufficient copies;
(iii) the mean monthly profit if he always bought 11 copies from the wholesaler;
(iv) the mean monthly profit if he always bought 13 copies from the wholesaler;
(v) the number he should buy if he is to maximize his mean monthly profit.

Suppose now that the profit margin is x pence per copy, the original price still being 70p. Find, in terms of x, the mean monthly profit if he were to buy regularly 10 magazines and the mean monthly profit if he were to buy regularly 11 magazines. Hence find the values of x for which it is most profitable for him to buy only 10 magazines a month.

(JMB)

Section 5.11

51. A variable X has a p.d.f.

$$f(x) = \frac{6x(x+1)}{5}, \qquad 0 \leqslant x \leqslant 1$$
$$= 0, \qquad \text{otherwise}$$

Find $G(y)$ and $g(y)$, the distribution function and p.d.f. of Y, when:

(i) $Y = X + 1$
(ii) $Y = 1 - X$
(iii) $Y = X^2$
(iv) $Y = 1 - X^2$
(v) $Y = e^x$
(vi) $Y = e^{-x}$

Remember to state the range of possible values of Y in each case. ►

52. A machine produces square plates. The lengths of the sides of the squares are uniformly distributed between 9.9 cm and 10.1 cm. The area of a square plate is denoted by $Y\,\text{cm}^2$.
Fine the probability density function of Y.

(JMB)

53. The maximum length to which a string of natural length a meters can be stretched before it snaps is $a(1+X^2)$ where X is a continuous random variable whose probability density function is

$$f(x) = 4x, \qquad 0.25 \leqslant x \leqslant 0.75$$
$$f(x) = 0, \qquad \text{otherwise.}$$

(i) Calculate the probability that a string can be stretched to $1\frac{1}{2}$ times its natural length without snapping.
(ii) Find the value of $E(X^2)$ and hence find μ, the mean maximum stretched length of strings of natural length 1 metre.
(iii) Find the probability density function of $Y = 1 + X^2$, the maximum stretched length of a string of natural length of 1 metre, and use it to verify the value of μ you obtained in (ii).

(WJEC)

54. The continuous random variable X has probability density function f, where

$$f(x) = cx(2-x), \qquad 0 \leqslant x \leqslant 2,$$
$$f(x) = 0, \qquad \text{otherwise.}$$

(i) Find the value of c.
(ii) Show that the variance of X is 0.2.
(iii) Determine the cumulative distribution function of X.
(iv) Given that

$$y = \frac{x}{4-x},$$

determine the range R corresponding to values of x ranging from 0 to 2. Show that the probability density function g of

$$Y = \frac{X}{4-X}$$

is given by

$$g(y) = \frac{ky(1-y)}{(1+y)^4}, \qquad \text{for } y \text{ in } R,$$
$$g(y) = 0, \qquad \text{otherwise,}$$

and write down the value of the constant k.

(WJEC)

6

Some standard discrete and continuous probability distributions

6.1 INTRODUCTION

There are a number of standard probability distributions which are 'models' for how various random variables behave. In this chapter three discrete distributions† (binomial, Poisson and geometric) and three continuous distributions‡ (rectangular (uniform), normal and exponential) will be introduced and some of their properties derived.

A random variable is distributed according to a particular distribution if certain conditions hold. The conditions which apply in the case of each of the six distributions named above will be stated, which should help the reader to recognize situations when one of the six is the correct 'model'.

6.2 BINOMIAL DISTRIBUTION

Consider the discrete random variable, X, where X denotes the number of heads in ten tosses of a fair coin. We will show below that X has a binomial distribution, but first we will note that the following conditions apply to tossing a fair coin ten times:

†You should read Sections 5.1 to 5.4 before reading Sections 6.2 to 6.4 inclusive.
‡You should read Sections 5.1, 5.5 to 5.9 before reading Sections 6.5 to 6.7 inclusive.

(i) There are a fixed number of trials (i.e. tosses), namely 10. (We are considering how the number of heads will vary in repetitions of exactly the same experiment.)
(ii) There are only *two* possible outcomes to each trial (i.e. toss), namely heads and tails. (Compare throwing a die when there are six possible outcomes.)
(iii) The probability of heads is the same for each trial (i.e. toss), namely 0.5 (because we are considering repetitions of exactly the same trial).
(iv) The trials (i.e. tosses) are independent, in the sense that the outcome of each toss does not depend on the outcome of any previous toss.

We can see that X, the number of heads in ten tosses of a coin, is a discrete random variable which can take values 0, 1, 2, . . . , 10. What is its probability function, $p(x)$?

$$\begin{aligned} p(x) &= P[x \text{ heads and } (10-x) \text{ tails}], \qquad x = 0, 1, 2, \ldots, 10 \\ &= 0, \qquad \text{otherwise} \end{aligned}$$

The probability of x heads followed by $(10-x)$ tails is $0.5^x(1-0.5)^{10-x}$, using the multiplication law for independent events (Section 4.6). But x heads and $(10-x)$ tails can occur in ${}^{10}C_x = \binom{10}{x}$ mutually exclusive and exhaustive ways (Section 4.10), and the probability of each is $0.5^x(1-0.5)^{10-x}$. And so *adding* these probabilities (Section 4.8),

$$\begin{aligned} p(x) &= \binom{10}{x} 0.5^x(1-0.5)^{10-x}, \qquad x = 0, 1, 2, \ldots, 10 \\ &= 0, \qquad \text{otherwise} \end{aligned}$$

Generalizing on this example, suppose we have an experiment which satisfies the following four conditions:

(i) there are a fixed number of trials, n;
(ii) there are two possible outcomes to each trial, which we will call 'success' and 'failure';
(iii) the probability of 'success' in each trial is p (so the probability of 'failure' in each trial is $(1-p)$);
(iv) the trials are independent.

Then X, the number of successes in n trials, is a discrete random variable with probability function

$$\left.\begin{aligned} p(x) &= \binom{n}{x} p^x(1-p)^{n-x}, \qquad x = 0, 1, 2, \ldots, n \\ &= 0, \qquad \text{otherwise} \end{aligned}\right\} \qquad (6.1)^\dagger$$

† The term $\binom{n}{x} p^x(1-p)^{n-x}$ is the general term in the expansion of $[(1-p)+p]^n$, or $[q+p]^n$ if q is used instead of $(1-p)$. Hence the sum of all the binomial probabilities is $[(1-p)+p]^n = 1^n = 1$.

(The reader is asked to justify this formula using arguments similar to those of the coin tossing example above.)

This is the probability function of the *binomial distribution* with parameters n and p, which we refer to as the $B(n, p)$ distribution for short. If we can assign values to n and p in a particular case (as in the coin-tossing experiment in which $n = 10$ and $p = 0.5$), then we can calculate the probabilities for the various values of X using either the probability function or tables (see below). An experiment in which the four conditions for a binomial distribution hold is called a 'binomial experiment' and the individual trials in a binomial experiment are called 'Bernoulli trials'.

Some other examples of binomial experiments and distributions are now listed. The reader should check in each case that the four conditions hold:

(a) Five cards are drawn from a well-shuffled pack, each card being replaced and the pack shuffled before the next card is drawn. The number of spades will have a $B(5, 0.25)$ distribution.
(b) In a multiple-choice test there are 50 questions. Each question has five possible answers, only one of which is correct. If candidates guess the answer to each question, the number of correct answers per candidate will have a $B(50, 0.2)$ distribution.
(c) If cracked eggs are distributed randomly among a large consignment of eggs, and 10% of the eggs are cracked, the number cracked per box (of six eggs) will have a $B(6, 0.1)$ distribution.
(d) If boys and girls are equally likely at each birth, the number of girls in families with three children will have a $B(3, 0.5)$ distribution.
(e) If 90% of all seeds germinate, the number of germinating seeds in packets of 100 seeds will have a $B(100, 0.9)$ distribution.

We will now derive the mean and variance of the binomial distribution from the probability function (6.1), using the methods of Section 5.3. Then we will derive its probability generating function (p.g.f.) and verify the values for the mean and variance, using the methods of Section 5.4.

The mean, μ

$$\mu = E(X) = \sum xp(x) \qquad \text{from (5.1)}$$

$$= \sum_{x=0}^{n} x\binom{n}{x}p^x(1-p)^{n-x} \qquad \text{from (6.1)}$$

$$= 0 + \binom{n}{1}p(1-p)^{n-1} + 2\binom{n}{2}p^2(1-p)^{n-2} + \dots$$

$$+ n\binom{n}{n}p^n(1-p)^{n-n}$$

$$= np(1-p)^{n-1}\left[1+\binom{n-1}{1}\left(\frac{p}{1-p}\right)+\ldots+\left(\frac{p}{1-p}\right)^{n-1}\right]$$

$$= np(1-p)^{n-1}\left[1+\left(\frac{p}{1-p}\right)\right]^{n-1}$$

recognizing the binomial series in the square brackets

$$= np(1-p)^{n-1}\left[\frac{1}{1-p}\right]^{n-1}$$

$$\therefore \mu = np$$

The mean value of X, the number of successes in a binomial experiment, is np.

The variance, σ^2

The usual way of finding the variance of a probability distribution is to find $E(X^2)$ and then use $\sigma^2 = E(X^2) - \mu^2$ (see (5.3) and (5.4)). For the binomial distribution however, it is better to find $E[X(X-1)]$ first.

$$E[X(X-1)] = \sum x(x-1)p(x) \qquad \text{from (5.2)}$$

$$= \sum_{x=0}^{n} x(x-1)\binom{n}{x}p^x(1-p)^{n-x} \qquad \text{from (6.1)}$$

$$= 0+0+\sum_{x=2}^{n} x(x-1)\binom{n}{x}p^x(1-p)^{n-x}$$

$$= 2\times 1\binom{n}{2}p^2(1-p)^{n-2}+\ldots+n(n-1)\binom{n}{n}p^n(1-p)^{n-n}$$

$$= n(n-1)p^2(1-p)^{n-2}\left[1+(n-2)\left(\frac{p}{1-p}\right)+\ldots+\left(\frac{p}{1-p}\right)^{n-2}\right]$$

$$= n(n-1)p^2(1-p)^{n-2}\left[1+\frac{p}{1-p}\right]^{n-2}$$

$$= n(n-1)p^2(1-p)^{n-2}\left[\frac{1}{1-p}\right]^{n-2}$$

$$= n(n-1)p^2$$

Now
$$E[X(X-1)] = \sum x(x-1)p(x)$$
$$= \sum x^2 p(x) - \sum xp(x)$$
$$= E(X^2) - E(X)$$
$$\therefore E(X^2) = E[X(X-1)] + E(X) \qquad (6.2)$$

and this result is true for any distribution.
For the binomial,

$$E(X^2) = n(n-1)p^2 + np,$$

using results proved above in this section.

$$\therefore \sigma^2 = E(X^2) - \mu^2 \qquad \text{from (5.3)}$$
$$= n(n-1)p^2 + np - (np)^2 \qquad \text{for the binomial}$$
$$\therefore \sigma^2 = np(1-p)$$

The variance of X, the number of successes in a binomial experiment, is $np(1-p)$. Hence the standard deviation of X is $\sqrt{np(1-p)}$.

The p.g.f. $G(t)$

$$G(t) = E(t^X), \qquad \text{from Section 5.4}$$
$$= \sum_{x=0}^{n} t^x \binom{n}{x} p^x (1-p)^{n-x} \qquad \text{from (5.2) and (6.1)}$$
$$= \sum_{x=0}^{n} \binom{n}{x} (pt)^x (1-p)^{n-x}$$
$$= (1-p)^n + \binom{n}{1}(pt)^1(1-p)^{n-1} + \ldots + \binom{n}{n}(pt)^n(1-p)^0$$
$$= (1-p)^n \left[1 + \binom{n}{1}\left(\frac{pt}{1-p}\right) + \ldots + \binom{n}{n}\left(\frac{pt}{1-p}\right)^n\right]$$
$$= (1-p)^n \left[1 + \left(\frac{pt}{1-p}\right)\right]^n$$
$$= (1-p)^n \left(\frac{1-p+pt}{1-p}\right)^n$$
$$\therefore G(t) = (1-p+pt)^n \qquad (6.3)$$

Using (5.5), (5.6) and (5.7) as they apply to this p.g.f.,

$$G(1) = (1-p+p)^n = 1,$$

which is really a check that the sum of the binomial probabilities is 1.

$$G'(t) = np(1-p+pt)^{n-1}$$

$$\therefore \mu = G'(1) = np(1-p+p)^{n-1} = np$$

$$G''(t) = n(n-1)p^2(1-p+pt)^{n-2}$$

$$\begin{aligned} \therefore \sigma^2 &= G''(1) + \mu - \mu^2 \\ &= n(n-1)p^2 + np - (np)^2 \\ &= np(1-p) \end{aligned}$$

So we have verified the values of μ and σ^2 obtained earlier in this section when the probability function was used.

The calculation of probabilities for a particular binomial distribution may be performed using (1) the probability function (6.1); (2) tables of cumulative binomial probabilities; (3) approximate methods (see Chapter 7) when the other two methods either are very time-consuming or cannot be used.

Example 6.1

Calculating probabilities for the $B(10, 0.5)$ distribution. The probability function for the $B(10, 0.5)$ distribution is:

$$\begin{aligned} p(x) &= \binom{10}{x}(0.5)^x(1-0.5)^{10-x}, \qquad x = 0, 1, 2, \ldots, 10 \\ &= 0 \qquad \text{otherwise.} \end{aligned}$$

For example, when $x = 2$,

$$\begin{aligned} p(2) &= \binom{10}{2}(0.5)^2(1-0.5)^{10-2} \\ &= \frac{10!}{2!\,8!}(0.5)^2\,(0.5)^8 \\ &= 0.044 \qquad \text{(3 d.p.)} \end{aligned}$$

Similarly,

$$\begin{aligned} p(0) &= \binom{10}{0}(0.5)^0(1-0.5)^{10-0} \\ &= 1 \times 1 \times 0.5^{10} \\ &= 0.001 \qquad \text{(3 d.p.)} \end{aligned}$$

From such calculations, Table 6.1 can be formed.

Table 6.1 Probabilities for the $B(10, 0.5)$ distribution

x	0	1	2	3	4	5	6	7	8	9	10
$P(x)$	0.001	0.010	0.044	0.117	0.205	0.246	0.205	0.117	0.044	0.010	0.001

These probabilities may also be obtained using tables of cumulative binomial probabilities; see Table C.1 (Appendix C). These give $P(X \leqslant r)$ for a range of different binomial distributions. Selecting the column of probabilities for the $B(10, 0.5)$ distribution we use the result that:

$$p(r) = P(X \leqslant r) - P(X \leqslant r-1) \qquad \text{for } r = 1, 2, \ldots, n,$$

noting that (i) $P(X \leqslant n) = 1$ (certain event)

and (ii) $p(0) = P(X \leqslant 0)$ (since the number of successes cannot be negative)

So for $r = 2$,

$$\begin{aligned} p(2) &= P(X \leqslant 2) - P(X \leqslant 1) \\ &= 0.055 - 0.011, \qquad \text{for our example} \\ &= 0.044, \qquad \text{which checks with the value in Table 6.1} \end{aligned}$$

And for $r = 0$,

$$\begin{aligned} p(0) &= P(X \leqslant 0) \\ &= 0.001, \qquad \text{which checks with the value in Table 6.1.} \end{aligned}$$

The reader should check the other values in Table 6.1 using tables of cumulative binomial probabilities (Table C.1). □

The $B(10, 0.5)$ distribution may also be represented graphically as in Fig. 6.1.

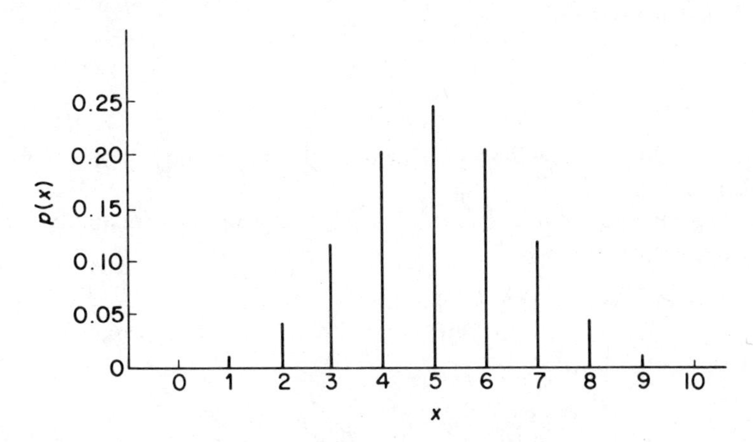

Fig. 6.1 The $B(10, 0.5)$ distribution

We note that there is symmetry in Fig. 6.1. This is because the value of p is 0.5, i.e. in the middle of the probability range 0 to 1. Binomial distributions with p values less than 0.5 have positive skewness, whereas negative skewness is exhibited if the p value is greater than 0.5.

The mean and variance of the $B(10, 0.5)$ distribution are $np = 10 \times 0.5 = 5$ and $np(1-p) = 2.5$ respectively. (Looking at Fig. 6.1 it should come as no surprise that the mean is 5.)

Note that the tables of cumulative probabilities give only a limited number of values of n and p. In particular, no value of p greater than 0.5 is listed in these tables. This is not so restrictive as it first appears since we may always refer to as our 'success' that outcome of a trial which has a probability of $\leqslant 0.5$.

Example 6.2

Suppose that 90% of people are right-handed. What is the probability that at most 15 of a random sample of 20 people are right-handed?

Since the probability of being left-handed is 0.1 (and is < 0.5), we will think of being left-handed as being a 'success'.

$$
\begin{aligned}
&P(\leqslant 15 \text{ out of 20 are right-handed}) \\
&= P(\geqslant 5 \text{ out of 20 are left-handed}), \quad \text{logically equivalent} \\
&= 1 - P(\leqslant 4 \text{ out of 20 are left-handed}), \\
&\quad \text{since Table C.1 gives probabilities of the '}\leqslant\text{' kind.} \\
&= 1 - 0.957, \quad \text{from Table C.1 for the } B(20, 0.1) \text{ distribution.} \\
&= 0.043
\end{aligned}
$$

There are cases in which the tables of cumulative binomial probabilities cannot be used, and when it would be tedious to calculate the required probability using the probability function. □

Example 6.3

For the $B(120, 0.2)$ distribution, calculate the probability that 25 or more successes occur.

The highest value of n in Table C.1 is 50, and using the probability function would involve calculating

$$
\begin{aligned}
P(X \geqslant 25) &= 1 - P(X \leqslant 24) \\
&= 1 - P(X=0) - P(X=1) - \ldots - P(X=24) \\
&= 1 - p(0) - p(1) - \ldots - p(24),
\end{aligned}
$$

where $$p(x) = \binom{120}{x}(0.2)^x(1-0.2)^{120-x}, \qquad x = 0, 1, 2, \ldots, 120$$

Luckily there are ways of calculating approximate probabilities in such cases; see Chapter 7. □

6.3 POISSON DISTRIBUTION

Our second discrete probability distribution concerns random events in time or space. The word 'random' here implies that there is a constant probability that an event will occur in each unit of time or space. For example, suppose a geologist keeps a record of the number of earthquakes which occur in a particular geographical region over a period of years. If earthquakes occur randomly then the probability that one will occur on any given day is constant. Suppose this is the case and that the mean number of earthquakes recorded per year is 5, then X, the number of earthquakes per year, will have a Poisson distribution with probability function:

$$p(x) = \frac{e^{-5}5^x}{x!}, \qquad x = 0, 1, 2, \ldots, \infty$$

$$= 0 \qquad \text{otherwise}$$

In general, if a is the mean number of random events per unit time (or space), then X, the number of random events per unit time (or space), will have a Poisson distribution with probability function:

$$\left.\begin{aligned} p(x) &= \frac{e^{-a}a^x}{x!}, \qquad x = 0, 1, 2, \ldots, \infty \\ &= 0 \qquad \text{otherwise} \end{aligned}\right\} \tag{6.4}$$

where a is the parameter of the distribution. Below we will verify that a is the mean of the Poisson distribution, and show that a is also the variance of this distribution. A shorthand way of writing a Poisson distribution with parameter a is Poi(a).

Some other examples of events which might be randomly distributed in time or space, and may therefore be 'modelled' by a Poisson distribution are:

(i) The number of meteorites which have fallen per 100 hectares of a desert (2-dimensional space).
(ii) The number of misprints per page of a book (2-dimensional space).
(iii) The number of defects per car coming off the production line in a factory (3-dimensional space).
(iv) The number of snags per 100 metres of finished wool in a spinning process from the raw wool (1-dimensional space).
(v) The number of postilions struck by lightning per year in the 18th and 19th centuries (time).

We will now derive the mean and variance of the Poisson distribution from the probability function (6.4), using the methods of Section 5.3. Then we will derive its probability generating function (p.g.f.) and verify the values for the mean and variance, using the methods of Section 5.4.

The mean, μ

$$\mu = E(x) = \sum xp(x) \qquad \text{from (5.1)}$$
$$= \sum_{x=0}^{\infty} x\frac{e^{-a}a^x}{x!} \qquad \text{from (6.4)}$$
$$= 0 + \sum_{x=1}^{\infty} x\frac{e^{-a}a^x}{x!}$$
$$= e^{-a}\left(\frac{a}{1!} + \frac{2a^2}{2!} + \frac{3a^3}{3!} + \ldots\right)$$
$$= ae^{-a}\left(1 + \frac{a}{1!} + \frac{a^2}{2!} + \ldots\right)$$
$$= ae^{-a}e^{a}$$
$$\therefore \mu = a$$

The mean value of X, the number of random events per unit time or space, is a (the parameter of the distribution).

The variance, σ^2

As with the binomial distribution it is convenient to find $E[X(X-1)]$ initially:

$$E[X(X-1)] = \sum x(x-1)p(x) \qquad \text{from (5.2)}$$
$$= \sum_{x=0}^{\infty} x(x-1)\frac{e^{-a}a^x}{x!} \qquad \text{from (6.4)}$$
$$= 0 + 0 + \sum_{x=2}^{\infty} x(x-1)\frac{e^{-a}a^x}{x!}$$
$$= e^{-a}\left(\frac{a^2}{0!} + \frac{a^3}{1!} + \ldots\right)$$
$$= a^2e^{-a}\left(1 + \frac{a}{1!} + \ldots\right)$$
$$= a^2e^{-a}e^{a}$$
$$= a^2$$

From (6.2), which is true for any distribution,

$$E(X^2) = E[X(X-1)] + E(X)$$

$$= a^2 + a \qquad \text{for the Poisson distribution}$$

$$\therefore \sigma^2 = E(X^2) - \mu^2 \qquad \text{from (5.3)}$$

$$= a^2 + a - a^2$$

$$\therefore \sigma^2 = a$$

The variance of X, the number of random events per unit time or space, is a. Hence the standard deviation of X is $\sqrt{a}$.

We note that the mean and variance of the Poisson distribution are equal. One way of deciding whether observed data are consistent with a Poisson distribution is to see if the mean and variance are approximately equal. Such 'goodness-of-fit' tests are discussed in detail in Chapter 13.

The p.g.f. $G(t)$

$$G(t) = E(t^X) \qquad \text{from Section 5.4}$$

$$= \sum_{x=0}^{\infty} t^x \frac{e^{-a}a^x}{x!} \qquad \text{from (5.2) and (6.4)}$$

$$= \sum_{x=0}^{\infty} \frac{e^{-a}(at)^x}{x!}$$

$$= e^{-a}\left[1 + \frac{(at)^1}{1!} + \frac{(at)^2}{2!} + \dots\right]$$

$$= e^{-a}e^{at}$$

$$G(t) = e^{a(t-1)} \qquad (6.5)$$

Using (5.5), (5.6) and (5.7) as they apply to this p.g.f., $G(1) = e^{a(1-1)} = e^0 = 1$, which is really a check that the sum of the Poisson probabilities is 1.

$$G'(t) = ae^{a(t-1)}$$

$$\therefore \mu = G'(1) = ae^0 = a$$

$$G''(t) = a^2e^{a(t-1)}$$

$$\therefore \sigma^2 = G''(1) + \mu - \mu^2$$

$$= a^2 + a - a^2$$

$$= a$$

So we have verified the values of μ and σ^2 obtained earlier in this section when the probability function was used.

The calculation of probabilities for a particular Poisson distribution may be performed using.

1. the probability function (Result 6.4);
2. tables of cumulative Poisson probabilities;
3. special graph paper;
4. approximate methods (see Chapter 7) when the other three methods are either time-consuming or cannot be used.

Example 6.4

Calculating probabilities for the Poi(5) distribution. The probability function for the Poi(5) distribution is $p(x) = \dfrac{e^{-5}5^x}{x!}$, $x = 0, 1, 2, \ldots$

For example, when $x = 2$,

$$p(2) = \frac{e^{-5}5^2}{2!}$$
$$= 0.084$$

Similarly,
$$p(0) = \frac{e^{-5}5^0}{0!}$$
$$= 0.007$$

From such calculations Table 6.2 can be formed:

Table 6.2 Probabilities for the Poi(5) distribution

x	0	1	2	3	4	5	6	7	8	9	10
$p(x)$	0.007	0.034	0.084	0.140	0.175	0.175	0.146	0.104	0.065	0.036	0.018

These probabilities sum to 0.984 (rather than 1), since values of X above 10 are possible, although their probabilities are diminishingly small. The probabilities in Table 6.2 may also be obtained using tables of cumulative Poisson probabilities; see Table C.2. These give $P(X \leqslant r)$ for a range of different Poisson distributions. Selecting the column of probabilities for the Poi(5) distribution we use the result that

$$p(r) = P(X \leqslant r) - P(X \leqslant r-1) \qquad r = 1, 2, \ldots.$$

noting that $p(0) = P(X \leqslant 0)$ (since the number of random events occurring per unit time cannot be negative).
So, for $r = 2$,

$$p(2) = P(X \leqslant 2) - P(X \leqslant 1)$$
$$= 0.125 - 0.040, \quad \text{for our example}$$
$$= 0.085, \quad \text{which is close to the value in Table 6.2.}$$

And for $r = 0$,

$$\begin{aligned} p(0) &= P(x \leqslant 0) \\ &= 0.007, \quad \text{which is the same as the value in Table 6.2.} \end{aligned}$$

The reader should check the other values in Table 6.2 using tables of cumulative Poisson probabilities (Table C.2).

The third method of obtaining Poisson probabilities is to use Poisson probability graph paper[†] (see Fig. 6.2).

For a Poisson distribution with a particular value of a we locate this value on the horizontal axis. We can then read $P(X \geqslant c)$ for various values of c. For example, for the Poi(5) distribution, and $c = 2$, we find:

$$P(X \geqslant 2) = 0.959$$

To find $P(X = 2)$, we use the result that

$$\begin{aligned} P(X = 2) &= P(X \geqslant 2) - P(X \geqslant 3) \\ &= 0.959 - 0.875 \\ &= 0.084, \quad \text{which agrees with the value in Table 6.2.} \end{aligned}$$

Similarly,

$$\begin{aligned} P(X = 0) &= P(X \geqslant 0) - P(X \geqslant 1) \\ &= 1 - 0.993, \\ &= 0.007, \quad \text{which agrees with the value in Table 6.2.} \end{aligned}$$

In general, we can use:

$$P(X = c) = P(X \geqslant c) - P(X \geqslant c + 1) \qquad \text{for } c = 0, 1, 2, \ldots$$

and $\quad P(X \geqslant 0) = 1$ (certain event) □

The Poi(5) distribution may also be represented graphically as in Fig. 6.3.

We note that Fig. 6.3 shows some positive skewness. Poisson distributions with values of a less than 5 will be more skew than Fig. 6.3, whereas less skewness is exhibited for values of a larger than 5.

The mean and variance of the Poi(5) distribution are $a = 5$ and $a = 5$ respectively.

There are cases when the three methods of obtaining Poisson probabilities are either tedious or impossible to use.

Example 6.5

For the Poi(30) distribution, calculate $p(\geqslant 40) = P(X \geqslant 40)$.

[†]e.g. Chartwell, Graph Data Ref. 5591B

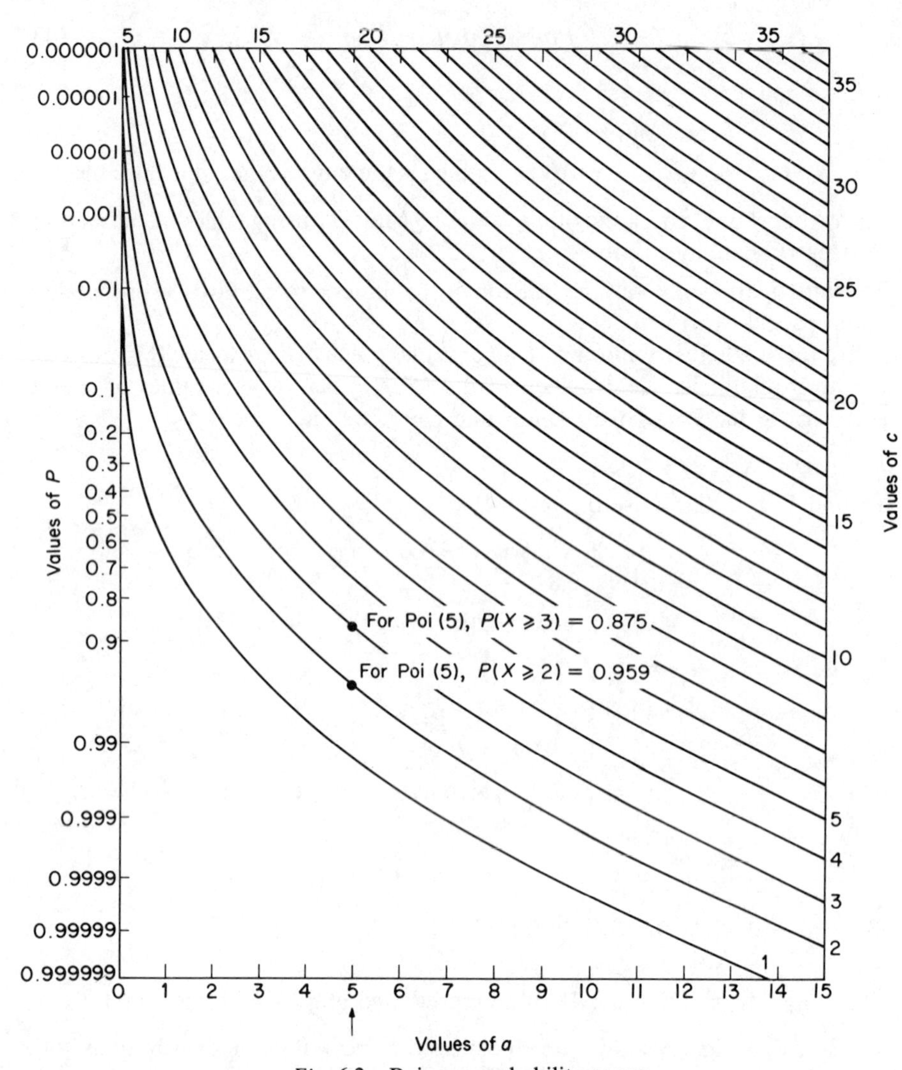

Fig. 6.2 Poisson probability paper

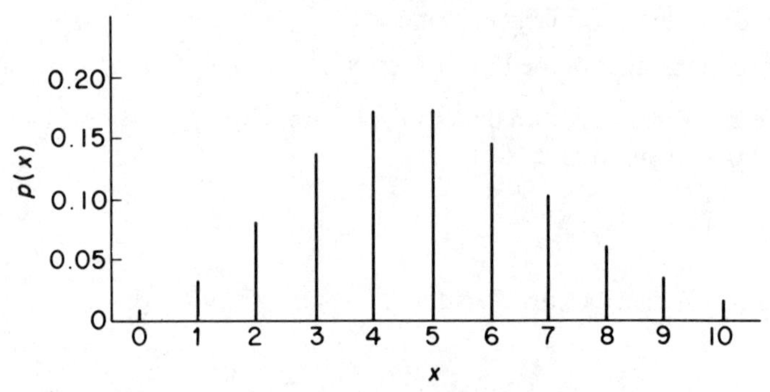

Fig. 6.3 The Poi(5) distribution

It would be tedious to calculate $p(40)+p(41)+\ldots$ using the probability function. Neither Table C.2 nor Fig. 6.2 cover values of a as high as 30.

Luckily there are ways of calculating approximate probabilities in such cases; see Chapter 7.

6.4 GEOMETRIC DISTRIBUTION

The third (and last) discrete probability distribution we will consider is the geometric distribution. This has some similarities with, but also some important differences from, the binomial distribution. The 'geometric' also concerns independent trials, each of which can result in one of only two outcomes called 'success' and 'failure'. The probability of success in each trial, is also the same for each trial. These are the similarities with the 'binomial'. The differences are that the variable, X, is

'the number of trials up to and including the first success'

So the number of successes is always 1, while the number of trials varies from one 'geometric experiment' to another. (Recall that, in the binomial, the number of successes varied from one binomial experiment to another, while the number of trials was always n). In a geometric experiment, therefore, we are thinking of a series of trials each resulting in failure, except for the last trial which results in a success.

Example 6.6

Let X denote the number of throws of a die up to and including the first 3. Then there must be $X-1$ throws resulting in $3'$ (= not 3) followed by a 3. Hence the probability function is:

$$p(x) = \left(\frac{5}{6}\right)^{x-1}\left(\frac{1}{6}\right), \qquad x = 1, 2, \ldots.$$
$$= 0 \qquad \text{otherwise}$$

using the multiplication law for independent events. Note that there must be at least one throw in order that the 3 can be thrown. □

In general, if p is the probability of success in each trial, the probability function for the geometric distribution is:

$$\left.\begin{aligned} p(x) &= (1-p)^{x-1}p; \qquad && x = 1, 2, \ldots, \infty \\ &= 0 && \text{otherwise} \end{aligned}\right\} \tag{6.6}$$

where p is the parameter of the distribution.

Some other examples of variables which might have a geometric distribution are:

(i) the number of driving tests taken by a candidate before he/she passes the test (but is p constant for each test?);
(ii) the number of cards drawn from a pack (with replacement) before an ace is drawn;
(iii) the number of times a fisherman casts a line into a river before he catches a fish;
(iv) the number of days a philatelist has to wait for a first-day cover which he/she has posted to him/herself.

We will now derive the mean and variance of the geometric distribution from the probability function (6.6), using the methods of Section 5.3. Then we will derive its probability generating function (p.g.f.) and verify the values for the mean and variance, using the methods of Section 5.4.

The mean, μ

$$\begin{aligned}
\mu = E(X) &= \sum xp(x) \qquad \text{from (5.2)} \\
&= \sum_{x=1}^{\infty} x(1-p)^{x-1} p \qquad \text{from (6.6)} \\
&= p + 2(1-p)p + 3(1-p)^2 p + \ldots \\
&= p[1 + 2(1-p) + 3(1-p)^2 + \ldots] \\
&= p[1-(1-p)]^{-2} \qquad \text{recognizing the series in square brackets.} \\
&= pp^{-2}
\end{aligned}$$

$$\therefore \mu = \frac{1}{p}$$

The mean value of X, the number of trials up to and including the first success, is $1/p$.

The variance, σ^2

$$\begin{aligned}
E(X^2) &= \sum x^2 p(x) \qquad \text{from (5.2)} \\
&= \sum_{x=1}^{\infty} x^2 (1-p)^{x-1} p \qquad \text{from (6.6)} \\
&= 1^2(1-p)^0 p + 2^2(1-p)^1 p + \ldots \\
&= p[1 + 2^2(1-p) + 3^2(1-p)^2 + \ldots] \\
&= pS, \quad \text{say, where } S \text{ is the series in brackets.}
\end{aligned}$$

Now letting $q = 1 - p$,

$$S = 1 + 2^2 q + 3^2 q^2 + \dots$$

$$\int_0^q S\,dq = q + 2q^2 + 3q^3 + \dots$$

$$= q(1 + 2q + 3q^2 + \dots)$$

$$= q(1-q)^{-2}$$

Differentiating,

$$S = (1-q)^{-2} + 2q(1-q)^{-3}$$

$$= p^{-2} + 2(1-p)p^{-3}$$

$$\therefore E(X)^2 = pS$$

$$= p^{-1} + 2(1-p)p^{-2}$$

$$= \frac{1}{p} + \frac{2(1-p)}{p^2}$$

$$= \frac{2-p}{p^2}$$

$$\therefore \sigma^2 = E(X^2) - \mu^2 \qquad \text{from (5.3)}$$

$$= \frac{2-p}{p^2} - \left(\frac{1}{p}\right)^2$$

$$\therefore \sigma^2 = \frac{1-p}{p^2}$$

The variance of X, the number of trials up to and including the first success, is $\frac{1-p}{p^2}$

Hence the standard deviation of X is $\frac{\sqrt{1-p}}{p}$.

The p.g.f., $G(t)$

$$G(t) = E(t^X) \qquad \text{from Section 6.4}$$

$$= \sum_{x=1}^{\infty} t^x (1-p)^{x-1} p \qquad \text{from (5.2) and (6.6)}$$

$$= pt \sum_{x=1}^{\infty} [t(1-p)]^{x-1}$$

$$G(t) = \frac{pt}{1 - t(1-p)} \tag{6.7}$$

(recognizing the geometric series)

Using (5.5), (5.6) and (5.7) as they apply to this p.g.f.,

$$G(1) = \frac{p}{1-(1-p)} = 1,$$

which is really a check that the sum of the geometric probabilities is 1.

$$G'(t) = \frac{p}{1-t(1-p)} + \frac{pt(1-p)}{[1-t(1-p)]^2}$$

$$\mu = G'(1)$$

$$= \frac{p}{1-(1-p)} + \frac{p(1-p)}{[1-(1-p)]^2}$$

$$= 1 + \frac{(1-p)}{p}$$

$$= \frac{1}{p}$$

$$G''(t) = \frac{p(1-p)}{[1-t(1-p)]^2} + \frac{p(1-p)}{[1-t(1-p)]^2} + \frac{2pt(1-p)^2}{[1-t(1-p)]^3}$$

$$\sigma^2 = G''(1) + \mu - \mu^2$$

$$= \frac{p(1-p)}{p^2} + \frac{p(1-p)}{p^2} + \frac{2p(1-p)^2}{p^3} + \frac{1}{p} - \left(\frac{1}{p}\right)^2$$

$$= \frac{2p(1-p) + 2(1-p)^2 + p - 1}{p^2}$$

$$= \frac{1-p}{p^2}$$

So we have verified the values of μ and σ^2 obtained using the probability function.

The calculation of probabilities for the geometric distribution may be performed using the probability function (6.6) and our knowledge of the geometric series. □

Example 6.7

Calculating probabilities for the geometric distribution with $p = 1/6$.

Let X denote the number of throws of a die up to and including the first 3. We have seen that the probability function for the distribution of X is

$$p(x) = \left(\frac{5}{6}\right)^{x-1} \frac{1}{6}, \qquad x = 1, 2, \ldots\ldots$$

$$\text{so } p(3) = \left(\frac{5}{6}\right)^2 \left(\frac{1}{6}\right) = 0.116, \text{ and so on.}$$

Table 6.3 shows the probabilities for values of X from 1 to 12.

Table 6.3 Probabilities for the geometric distribution with $p = 1/6$

x	1	2	3	4	5	6	7	8	9	10	11	12
$p(x)$	0.167	0.139	0.116	0.096	0.080	0.067	0.056	0.047	0.039	0.032	0.027	0.022

The sum of the probabilities in Table 6.3 is 0.888 (rather than 1) because values of X above 12 are possible although their probabilities are small. The distribution has a long 'tail' (see Fig. 6.4).

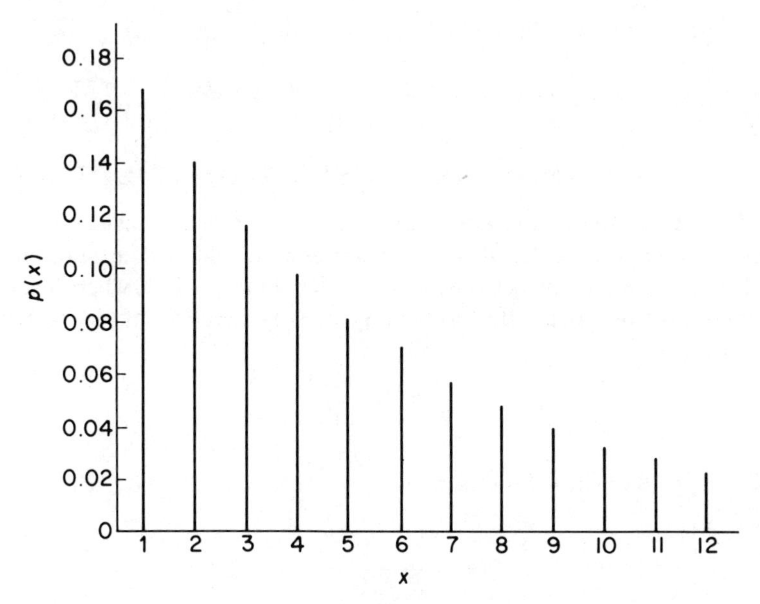

Fig. 6.4 A geometric distribution with $p = \frac{1}{6}$

Figure 6.4 shows extreme positive skewness. All geometric distributions show this type of skewness.

The mean and variance of the above distribution are

$$\frac{1}{p} = 6 \quad \text{and} \quad \frac{1-p}{p^2} = 30, \quad \text{respectively}$$

The mean number of throws up to and including the first 3 is 6. □

Cumulative probabilities for the geometric distribution may be obtained using the sum of a geometric series:

$$\begin{aligned} p(\leqslant r) &= P(X \leqslant r) \qquad \text{where } r = 1, 2, \ldots \\ &= \sum_{x=1}^{r} (1-p)^{x-1} p \\ &= p[1 + (1-p) + (1-p)^2 + \ldots + (1-p)^{r-1}] \\ &= \frac{p[1-(1-p)^r]}{1-(1-p)} \\ &= 1-(1-p)^r \end{aligned}$$

□

Example 6.8

For the geometric distribution with $p = 1/6$ calculate $p(\leqslant 12)$.

$$p(\leqslant 12) = 1 - \left(1 - \frac{1}{6}\right)^{12} = 0.8878$$

6.5 RECTANGULAR (UNIFORM) DISTRIBUTION

The first of the three continuous distributions we shall consider is also the simplest. In the rectangular distribution it is assumed that a continuous random variable, X, is equally likely to lie anywhere in the range a to b, where a and b are constants ($a < b$). Hence the probability density function of the rectangular distribution is

$$\begin{aligned} f(x) &= k, \qquad a \leqslant x \leqslant b \\ &= 0, \qquad \text{otherwise} \end{aligned}$$

Since $\int_a^b f(x)\mathrm{d}x = 1$, it follows that $k = \dfrac{1}{b-a}$ and so the p.d.f. is

$$\left.\begin{aligned} f(x) &= \frac{1}{b-a}, \qquad a \leqslant x \leqslant b \\ &= 0 \qquad\qquad \text{otherwise} \end{aligned}\right\} \tag{6.8}$$

where a and b are the parameters of the distribution. This function is shown graphically in Fig. 6.5.

It is easy to calculate probabilities for this distribution:

$$\begin{aligned} P(x_1 \leqslant X \leqslant x_2) &= \int_{x_1}^{x_2} f(x)\mathrm{d}x \qquad \text{for any continuous distribution} \\ &= \int_{x_1}^{x_2} \frac{1}{b-a} \qquad \text{for the rectangular distribution} \\ &= \frac{x_2 - x_1}{b-a} \end{aligned}$$

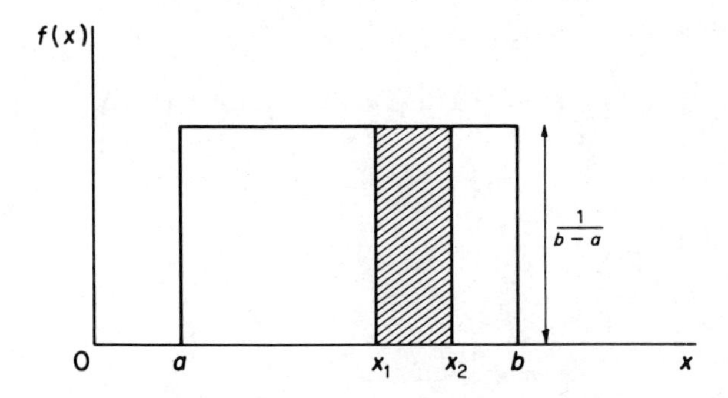

Fig. 6.5 The rectangular distribution

This is also the area of the shaded rectangle in Fig. 6.5 between x_1 and x_2. □

Example 6.9

Suppose that the heights of students are quoted to the nearest centimetre. What is the probability that X, the 'error' in the quoted height, is less than $+2$ mm, where:

$$\text{error} = \text{actual height} - \text{stated height?}$$

Assuming that 'error' is equally likely to be in the range -5 mm to $+5$ mm, the p.d.f. of X is

$$f(x) = \frac{1}{10}, \qquad -5 \leqslant x \leqslant 5$$

$$= 0 \qquad \text{elsewhere.}$$

So

$$P(X \leqslant 2) = \int_{-5}^{2} \frac{\mathrm{d}x}{10} = \frac{2-(-5)}{10} = 0.7$$

□

We will now derive the mean and variance of the rectangular distribution from the probability density function (6.7), using the methods of Section 5.6. Then we will derive its (cumulative) distribution function and median, using the methods of Sections 5.7 and 5.8. Finally, the moment generating function will be found.

The mean, μ

$$\mu = E(X) = \int x f(x)\mathrm{d}x \qquad \text{from (5.8)}$$

$$= \int_a^b \frac{x}{b-a}\mathrm{d}x \qquad \text{from (6.8)}$$

$$= \frac{b^2/2 - a^2/2}{b-a}$$

$$\therefore \mu = \frac{a+b}{2}$$

which should come as no surprise when we consider Fig. 6.5.

The variance, σ^2

$$E(X^2) = \int x^2 f(x)\mathrm{d}x \qquad \text{from (5.10)}$$

$$= \int_a^b \frac{x^2}{b-a}\mathrm{d}x \qquad \text{from (6.8)}$$

$$= \frac{b^3/3 - a^3/3}{b-a}$$

$$= \frac{b^2 + ab + a^2}{3}$$

$$\therefore \sigma^2 = E(X^2) - \mu^2 \qquad \text{from (5.3)}$$

$$= \frac{b^2 + ab + a^2}{3} - \left(\frac{a+b}{2}\right)^2$$

$$= \frac{4b^2 + 4ab + 4a^2 - 3a^2 - 6ab - 3b^2}{12}$$

$$\therefore \sigma^2 = \frac{(b-a)^2}{12}$$

The mean and variance of the rectangular distribution are

$$\frac{a+b}{2} \quad \text{and} \quad \frac{(b-a)^2}{12} \quad \text{respectively.}$$

The standard deviation is $\dfrac{(b-a)}{\sqrt{12}}$.

The distribution function, $F(x)$

$$F(x) = \int_{-\infty}^{x} f(x)\mathrm{d}x \qquad \text{from Section 5.7}$$

$$= \int_a^x \frac{1}{b-a}\mathrm{d}x \qquad \text{from (6.8)}$$

$$\therefore F(x) = \frac{x-a}{b-a}, \qquad a \leqslant x \leqslant b$$

Also $\qquad F(x) = 0, \quad x < a \qquad \text{and} \qquad F(x) = 1, \quad x > b$

The median, m

$$F(m) = \frac{1}{2} \qquad \text{from Section 5.8}$$

$$\therefore \frac{m-a}{b-a} = \frac{1}{2}$$

$$m = \frac{a+b}{2},$$ so the mean and median are equal, as we expect for this symmetrical distribution.

The m.g.f., $M(t)$

$$M(t) = E(e^{tX}) \qquad \text{from Section 5.9}$$

$$= \int_a^b \frac{e^{tx}}{b-a}\,\mathrm{d}x \qquad \text{for the rectangular distribution}$$

$$= \frac{1}{b-a}\left[\frac{e^{tx}}{t}\right]_a^b$$

$$= \frac{e^{bt}-e^{at}}{(b-a)t} \qquad (6.9)$$

One way of verifying the values of the mean and variance is to use (5.12) and (5.13), but we will not use this approach as it leads to messy algebra in this case!

6.6 NORMAL DISTRIBUTION

This is the second continuous distribution we shall consider. Of all probability distributions, the 'normal' is the most important for both practical and theoretical reasons. The normal distribution is an appropriate 'model' whenever we have a continuous variable whose value depends on the effect of a number of factors, each factor exerting small positive or negative influences. For example, if X denotes the height of adult female humans in the UK, then X is likely to be normally distributed. The reason is that X is affected by heredity, diet, exercise, bone structure, metabolism and so on. From the theoretical viewpoint also, the normal distribution is of fundamental importance, as we shall see in Chapters 7, 9, 10 and 12.

Most readers are probably familiar with the 'bell-shape' of the normal distribution (see Fig. 6.6), which is a graphical representation of the probability density function for the distribution,

$$f(x) = \frac{1}{\sigma\sqrt{2\pi}} \exp\left\{-\frac{1}{2}\left(\frac{x-\mu}{\sigma}\right)^2\right\}, \qquad -\infty < x < \infty \qquad (6.10)$$

where μ and σ^2 are the parameters of the distribution.

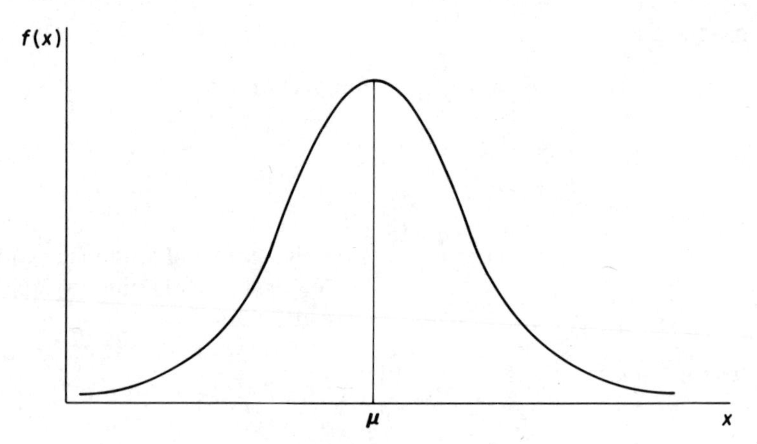

Fig. 6.6 The normal distribution

We used μ and σ^2 in Chapter 5 to denote the mean and variance of any distribution. We have just used μ and σ^2 in (6.10), so we need to verify that the parameters in (6.10) really are the mean and variance, respectively. We will do this by rewriting (6.10) as:

$$f(x) = \frac{1}{b\sqrt{2\pi}} \exp\left\{-\frac{1}{2}\left(\frac{x-a}{b}\right)^2\right\}, \qquad -\infty < x < \infty$$

and show that the mean and variance are a and b^2, respectively, using the methods of Section 5.6.

The mean μ,

$$\mu = E(X) = \int x f(x)\mathrm{d}x \qquad \text{from (5.8)}$$

$$= \int_{-\infty}^{\infty} \frac{x}{b\sqrt{2\pi}} \exp\left\{-\frac{1}{2}\left(\frac{x-a}{b}\right)^2\right\}\mathrm{d}x$$

we will substitute $y = \dfrac{x-a}{b}$ in this expression.

$$\mu = \int_{-\infty}^{\infty} \frac{a+by}{b\sqrt{2\pi}} e^{-\frac{1}{2}y^2} b\,\mathrm{d}y$$

$$= a\int_{-\infty}^{\infty} \frac{1}{\sqrt{2\pi}} e^{-\frac{1}{2}y^2}\mathrm{d}y + \frac{b}{\sqrt{2\pi}}\int_{-\infty}^{\infty} y e^{-\frac{1}{2}y^2}\mathrm{d}y$$

$$= a + \frac{b}{\sqrt{2\pi}}\left[-e^{-\frac{1}{2}y^2}\right]_{-\infty}^{\infty}$$

$$= a + \frac{b}{\sqrt{2\pi}}(0-0)$$

$$\therefore \mu = a$$

We have verified that a is the mean, having quoted without proof that

$$\int_{-\infty}^{\infty} \frac{1}{\sqrt{2\pi}} e^{-\frac{1}{2}y^2} \mathrm{d}y = 1.$$

The variance, σ^2

$$\sigma^2 = E[(X-\mu)^2] \qquad \text{from Section 5.3}$$
$$= \int_{-\infty}^{\infty} (x-a)^2 \frac{1}{b\sqrt{2\pi}} \exp\left\{-\frac{1}{2}\left(\frac{x-a}{b}\right)^2\right\} \mathrm{d}x$$

we will substitute $y = \dfrac{x-a}{b}$ again.

$$\sigma^2 = \int_{-\infty}^{\infty} \frac{b^2 y^2}{b\sqrt{2\pi}} e^{-\frac{1}{2}y^2} b\,\mathrm{d}y$$
$$= \frac{b^2}{\sqrt{2\pi}} \int_{-\infty}^{\infty} y y e^{-\frac{1}{2}y^2} \mathrm{d}y$$
$$= \frac{b^2}{\sqrt{2\pi}} \left[-y e^{-\frac{1}{2}y^2}\right]_{-\infty}^{\infty} + \frac{b^2}{\sqrt{2\pi}} \int_{-\infty}^{\infty} e^{-\frac{1}{2}y^2} \mathrm{d}y \qquad \text{by parts}$$
$$= 0 - 0 + b^2$$

$$\therefore \sigma^2 = b^2, \quad \text{quoting again that} \int_{-\infty}^{\infty} \frac{1}{\sqrt{2\pi}} e^{-\frac{1}{2}y^2} \mathrm{d}y = 1.$$

We have verified that a is the mean and b^2 is the variance.

The distribution function, $F(x)$

$$F(x) = \int_{-\infty}^{x} f(x)\mathrm{d}x \qquad \text{from Section 5.7}$$
$$= \int_{-\infty}^{x} \frac{1}{\sigma\sqrt{2\pi}} \exp\left\{-\frac{1}{2}\left(\frac{x-\mu}{\sigma}\right)^2\right\} \mathrm{d}x$$

Unfortunately this integral cannot be evaluated, but tables exist which enable us to obtain this function for any values of μ, σ and x; see below.

The median, m

Clearly $m = \mu$, because of the symmetry of Fig. 6.6.

The m.g.f., $M(t)$

It can be shown that

$$M(t) = \exp\left\{\mu t + \frac{\sigma^2 t^2}{2}\right\} \tag{6.11}$$

for the normal distribution. We could use this to confirm once more that μ and σ^2 are the mean and variance of the normal distribution using (5.12) and (5.13). This is left as an exercise for the reader.

In order to obtain probabilities for the normal distribution we may use specially prepared tables or special graph paper. We will consider the tables first. We first transform a normally distributed variable X, having a mean μ and variance σ^2, into a new variable Z, using that transformation $Z = (X - \mu)/\sigma$. It can be shown that Z has what is called a standardized normal distribution, that is one with a mean of 0 and variance of 1. We refer to the distribution of X as an $N(\mu, \sigma^2)$ distribution,† so that Z has an $N(0, 1)$ distribution.

The probability density function of Z is:

$$\phi(z) = \frac{1}{\sqrt{2\pi}} e^{-\frac{1}{2}z^2}, \qquad -\infty < z < \infty \tag{6.12}$$

The notation $\phi(z)$, rather than $f(z)$, is specially reserved for the p.d.f. of this very important distribution. The (cumulative) distribution function of Z is referred to as $\Phi(z)$. Referring to Section 5.7, we see that

$$\Phi(z) = P(Z \leqslant z) = \int_{-\infty}^{z} \phi(z)\mathrm{d}z = \int_{-\infty}^{z} \frac{1}{\sqrt{2\pi}} e^{-\frac{1}{2}z^2}\mathrm{d}z \tag{6.13}$$

$\Phi(z)$ is the area to the left of z under the curve of $\phi(z)$, as shown in Fig. 6.7.

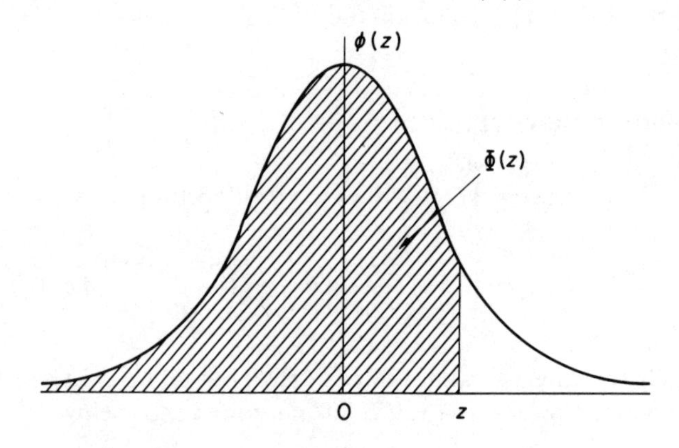

Fig. 6.7 The standardized normal distribution

† $X \sim N(\mu, \sigma^2)$ means X has a normal distribution with mean μ and variance σ^2, so the symbol $\sim$ means 'is distributed as'.

We will now show how to use Table C.3(a), which gives $\Phi(z)$ for various values of z.

Example 6.10

Suppose that X, the height in cm of an adult female human, has an $N(160, 10^2)$ distribution. Calculate:

(i) $P(X \leqslant 170)$
(ii) $P(X > 170)$
(iii) $P(170 < X \leqslant 175)$
(iv) $P(X \leqslant 140)$
(v) $P(140 < X \leqslant 170)$
(vi) $P(X > 175)$

Also calculate the percentages of adult females in the height ranges stated above, and the expected number of females in a random sample of 1000 for the same height ranges.

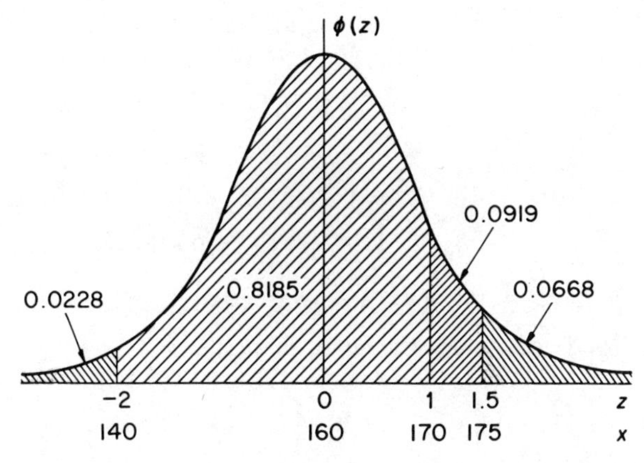

Fig. 6.8 Some areas under the $N(0, 1)$ curve, and the corresponding values of X for Example 6.10

The mean and variance of X are $\mu = 160$ and $\sigma^2 = 10^2$, respectively. So $\sigma = 10$.

Using $Z = \dfrac{X - 160}{10}$

(i)
$$\begin{aligned} P(X \leqslant 170) &= P\left(\frac{X-160}{10} \leqslant \frac{170-160}{10}\right) \\ &= P\left(Z \leqslant \frac{170-160}{10}\right) \\ &= P(Z \leqslant 1) \\ &= \Phi(1) \\ &= 0.8413 \qquad \text{from Table C.3(a)} \end{aligned}$$

This is the area to the left of $X = 170$ $(Z = 1)$ in Fig. 6.8.

(ii)
$$\begin{aligned} P(X > 170) &= 1 - P(X \leqslant 170) \\ &= 1 - 0.8413 \\ &= 0.1587 \end{aligned}$$

This is the area to the right of $X = 170$ $(Z = 1)$ in Fig. 6.8.

(iii)
$$\begin{aligned} P(170 < X \leqslant 175) &= P(X \leqslant 175) - P(X \leqslant 170) \\ &= \Phi\left(\frac{175 - 160}{10}\right) - 0.8413 \qquad \text{from (i)} \\ &= \Phi(1.5) - 0.8413 \\ &= 0.9332 - 0.8413 \qquad \text{from Table C.3(a)} \\ &= 0.0919 \end{aligned}$$

This is the area between $X = 170$ $(Z = 1)$ and $X = 175$ $(Z = 1.5)$ in Fig. 6.8.

(iv)
$$\begin{aligned} P(X \leqslant 140) &= \Phi\left(\frac{140 - 160}{10}\right) \\ &= \Phi(-2) \\ &= 1 - \Phi(2) \qquad \text{by symmetry} \\ &= 1 - 0.9772 \qquad \text{from Table C.3(a)} \\ &= 0.0228 \qquad \text{see Fig. 6.8.} \end{aligned}$$

(v)
$$P(140 < X \leqslant 170) = 1 - P(X \leqslant 140) - P(X > 170),$$

since the total area under the curve of $\phi(z)$ is 1

$$\begin{aligned} &= 1 - 0.0228 - 0.1587 \\ &= 0.8185 \qquad \text{see Fig. 6.8} \end{aligned}$$

(vi)
$$\begin{aligned} P(X > 175) &= 1 - P(X \leqslant 175) \\ &= 1 - \Phi\left(\frac{175 - 160}{10}\right) \\ &= 1 - \Phi(1.5) \\ &= 1 - 0.9332 \qquad \text{from Table C.3(a)} \\ &= 0.0668 \qquad \text{see Fig. 6.8.} \end{aligned}$$

The percentages corresponding to the answers to (i)–(vi) are obtained by multiplying the probabilities by 100 (see Section 4.5). The percentages are 84.13%, 15.87%, 9.19%, 2.28%, 81.85% and 6.68%.

We can obtain the expected numbers in a random sample of 1000 by applying these percentages, e.g. 84.13% of 1000 is $84.13/100 \times 1000 = 841$ (nearest integer).

The calculation is identical to multiplying the probability by the sample size, e.g. $0.8413 \times 1000 = 841$ (nearest integer). The other expected numbers are (ii) 159, (iii) 92, (iv) 23, (v) 819, (vi) 67, to the nearest integers. □

Example 6.11

Show that 95% of the area under any normal distribution curve lie within 1.96 standard deviations of the mean.

Let the mean and variance of a continuous and normally distributed random variable, X, be μ and σ^2.

$$
\begin{aligned}
&P(\mu - 1.96\sigma < X \leqslant \mu + 1.96\sigma)\\
&\quad = P(X \leqslant \mu + 1.96\sigma) - P(X \leqslant \mu - 1.96\sigma) \qquad \text{see Fig. 6.9}\\
&\quad = \Phi\left(\frac{\mu + 1.96\sigma - \mu}{\sigma}\right) - \Phi\left(\frac{\mu - 1.96\sigma - \mu}{\sigma}\right)\\
&\quad = \Phi(1.96) - \Phi(-1.96)\\
&\quad = \Phi(1.96) - (1 - \Phi(1.96))\\
&\quad = 0.975 - (1 - 0.975) \qquad \text{from Table C.3(a)}\\
&\quad = 0.95
\end{aligned}
$$

By symmetry, and since the total area is 1, it follows that the tail areas in Fig. 6.9 are each equal to 0.025.

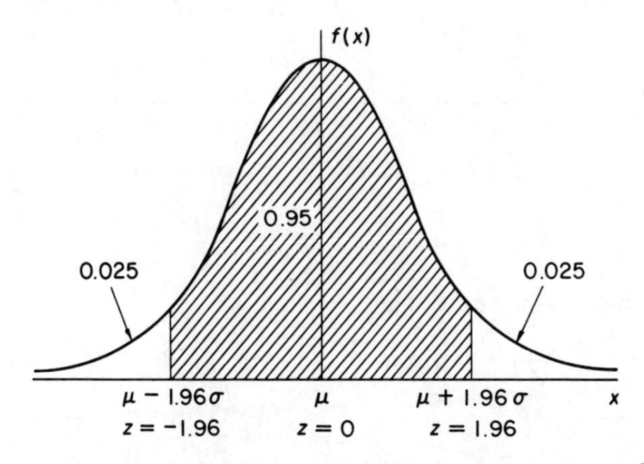

Fig. 6.9 A normal distribution with mean μ and variance σ^2 □

Example 6.12

The length in cm of a component made in a factory has an $N(10, 0.2^2)$ distribution. If the length is greater than 10.35 cm the component must be re-machined and then it can be sold. If the length is less than 9.75 cm the component has to be scrapped as defective. Each component costs 80 p to make, while re-machining (if necessary) costs a further 20 p. Non-defective components can be sold for 150 p each. If a random sample of 1000 components is selected:

(i) How many would you expect to:
  (a) require re-machining?
  (b) be defective?
  (c) be neither defective nor need re-machining?

(ii) What is the expected total profit for the sample of 1000?

(iii) What is the expected mean profit:
  (a) per component made?
  (b) per component sold?

(iv) If a component is randomly selected from those which are to be sold, what is the probability that it is one which has been re-machined?

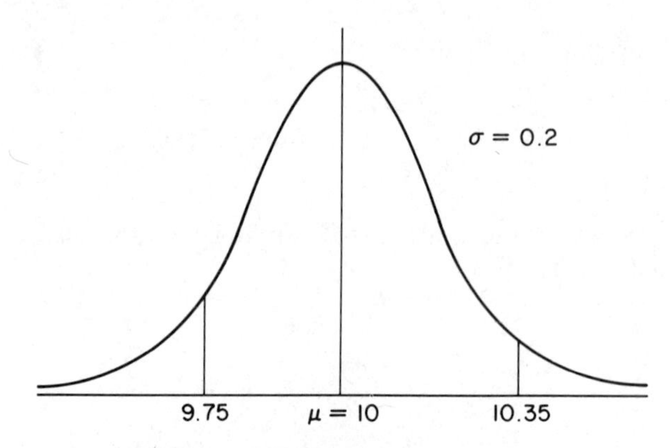

Fig. 6.10 An $N(10, 0.2^2)$ distribution

The distribution is shown in Fig. 6.10.

(i) (a)

$$\begin{aligned} P(X > 10.35) &= 1 - P(X \leqslant 10.35) \\ &= 1 - \Phi\left(\frac{10.35 - 10}{0.2}\right) \\ &= 1 - \Phi(1.75) \\ &= 1 - 0.9599 \qquad \text{from Table C.3(a)} \\ &= 0.0401 \end{aligned}$$

Expect $0.0401 \times 1000 = 40$ components to need re-machining.

(b)
$$P(X \leqslant 9.75) = \Phi\left(\frac{9.75-10}{0.2}\right)$$
$$= \Phi(-1.25)$$
$$= 1-\Phi(1.25) \quad \text{by symmetry}$$
$$= 1-0.8944 \quad \text{from Table C.3(a)}$$
$$= 0.1056$$

Expect $0.1056 \times 1000 = 106$ components to be scrapped.

(c) Expect $1000\ (1-0.0401-0.1056) = 854$ components to be neither defective nor need re-machining.

(ii) Expected total profit:

$$= (854 \times 150) + (40 \times 150) - (1000 \times 80) - (40 \times 20)$$
$$= 53{,}300 \text{ p}$$

(iii) (a) mean profit per component made $= \dfrac{53{,}300}{1000}$

$$= 53.3 \text{ p}$$

(b) mean profit per component sold $= \dfrac{53{,}300}{854+40}$

$$= 59.6 \text{ p}$$

(iv) Expect $854+40 = 894$ to be sold, of which 40 re-machined.

$$\text{Required probability} = \frac{40}{894} = 0.0447$$

In some cases we may require the inverse use of Table C.3(a). The two examples which follow illustrate this. □

Example 6.13

The height of adult males is normally distributed with a mean of 172 cm and standard deviation 8 cm. If 99% of adult males exceed a certain height, what is this height?

Figure 6.11 shows that the required height, X^*, is such that the area to the left of X^* is 0.01.

Therefore
$$\Phi\left(\frac{X^*-172}{8}\right) = 0.01$$

Hence
$$\Phi\left(\frac{172-X^*}{8}\right) = 0.99 \quad \text{by symmetry}$$

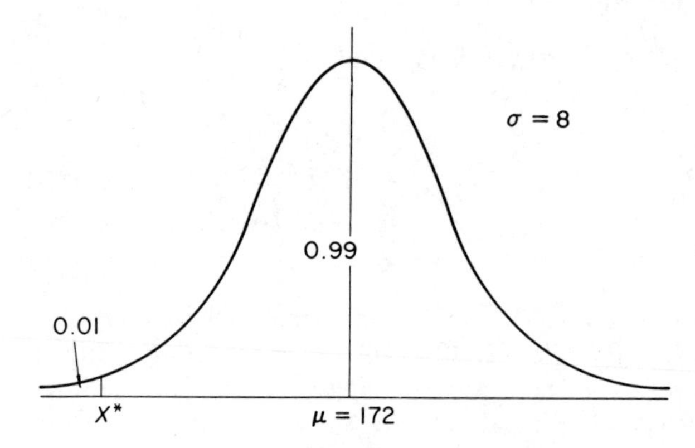

Fig. 6.11 An $N(172, 8^2)$ distribution

therefore $$\frac{172 - X^*}{8} = 2.33 \qquad \text{from Table C.3(a)}$$

$$X^* = 172 - 8 \times 2.33 = 153.4 \text{ cm}$$

99% of adult males exceed a height of 153.4 cm.
(In this example Table C.3(b) could also have been used.) □

Example 6.14

The annual rainfall in a town is normally distributed with a mean of 65 cm. In 15% of years the rainfall is more than 85 cm. What is the standard deviation of annual rainfall?

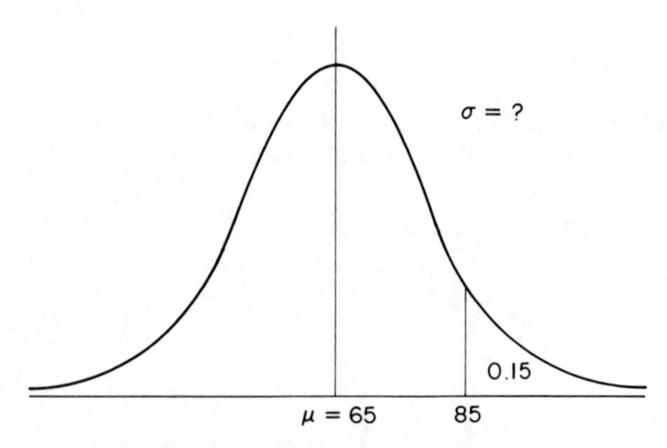

Fig. 6.12 An $N(65, \sigma^2)$ distribution

From Fig. 6.12 we can deduce that

$$\Phi\left(\frac{85-65}{\sigma}\right) = 0.85$$

$$\therefore \frac{85-65}{\sigma} = 1.04, \quad \text{to 2 d.p. from Table C.3(a).}$$

$$\therefore \sigma = \frac{85-65}{1.04} = 19 \text{ cm} \quad (2 \text{ s.f.})$$ □

Probabilities for the normal distribution may also be obtained using special graph paper called arithmetic probability paper† (see Fig. 6.13).

The axes of the graph are distribution function, $\Phi(z)$ (bottom horizontal axis) and the variable of interest (vertical axis). We use the fact that, for a normally distributed variable, X,

$$P(X \leqslant \mu) = 0.5 \text{ or } 50\%$$

and $$P(X \leqslant \mu + 1.96\sigma) = 0.975 \quad \text{or} \quad 97.5\%$$

The reader should verify these statements. A normal distribution may then be represented by a straight line on the graph by joining the points $(50, \mu)$ and $(97.5, \mu + 1.96\sigma)$.

Example 6.15

Represent the $N(160, 10^2)$ distribution by a straight line on arithmetic probability paper. Use the line to verify the answers obtained in Example 6.10.

Since $\mu = 160$, $\sigma = 10$, $\mu + 1.96\sigma = 179.6$, we draw the line through (50, 160) and (97.5, 179.6) (see Fig. 6.13).

We can then read $P(X \leqslant 140) = 2.3\%$ or 0.023

$P(X \leqslant 170) = 84.3\%$ or 0.843

$P(X \leqslant 175) = 93.3\%$ or 0.933

These agree well with the values of 0.0228, 0.8413 and 0.9332 obtained using tables in the earlier example. □

6.7 EXPONENTIAL DISTRIBUTION

The exponential distribution is the third (and last) continuous probability distribution we will consider. It has useful applications in the reliability of components which may fail suddenly and in the theory of queuing. It also has an important connection with the (discrete) Poisson distribution as follows.

† for example; Chartwell, Graph Data Ref 5571

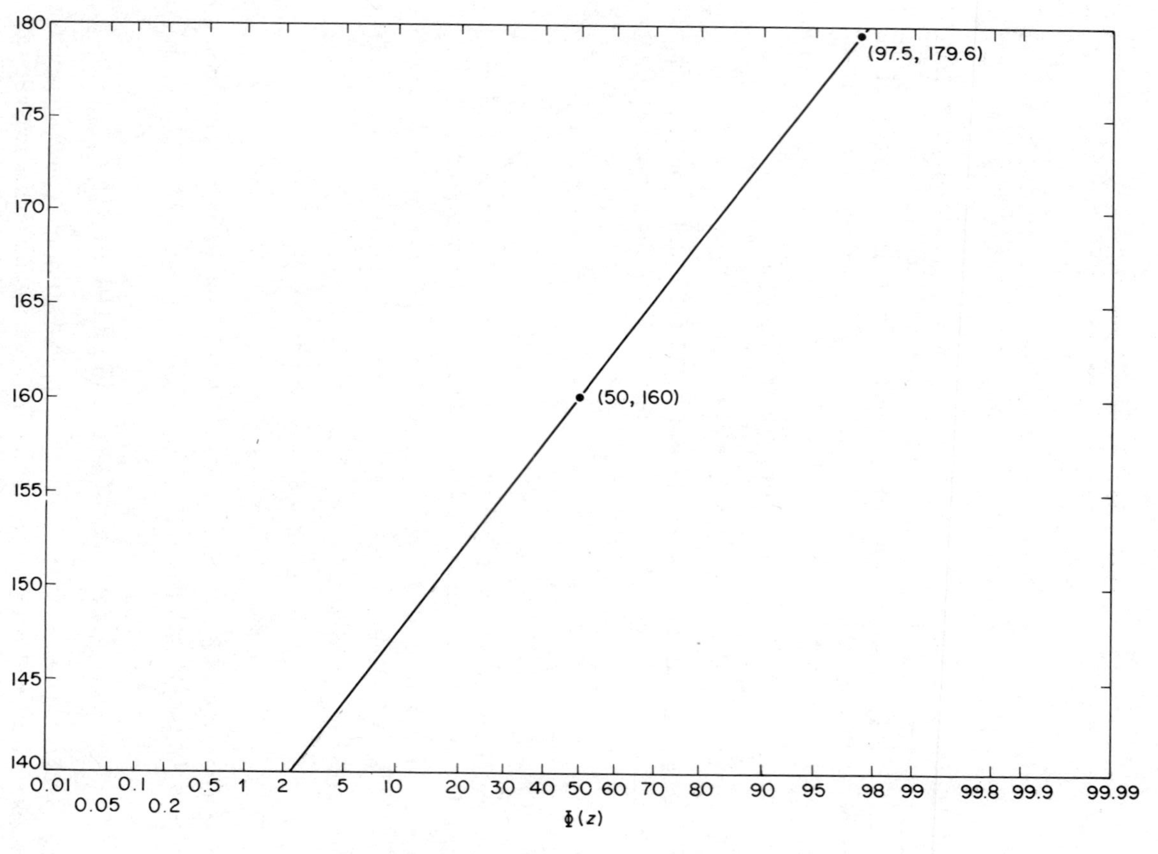

Fig. 6.13 An $N(160, 10^2)$ distribution represented by a straight line

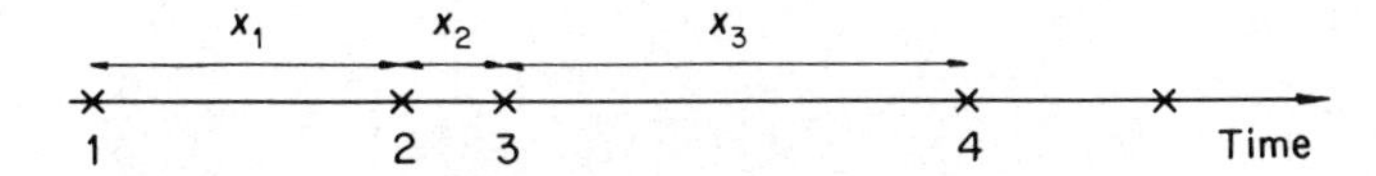

Fig. 6.14 The times between successive random events

Suppose events occur randomly in time. Then we know (from Section 6.3) that the number of events per unit time has a Poisson distribution. It can be shown that the time between successive random events has an exponential distribution. Figure 6.14 shows a possible sequence of random events where x_1, x_2, x_3, and so on, are possible values for the times between events 1 and 2, 2 and 3, 3 and 4, and so on.

The probability density function for the exponential distribution of a continuous random variable, X, is

$$\left.\begin{aligned} f(x) &= \lambda e^{-\lambda x}, \qquad x \geqslant 0, \quad \lambda > 0 \\ &= 0 \qquad \text{otherwise} \end{aligned}\right\} \tag{6.14}$$

where λ is the parameter of the distribution. Figure 6.15 shows the graph of $f(x)$.

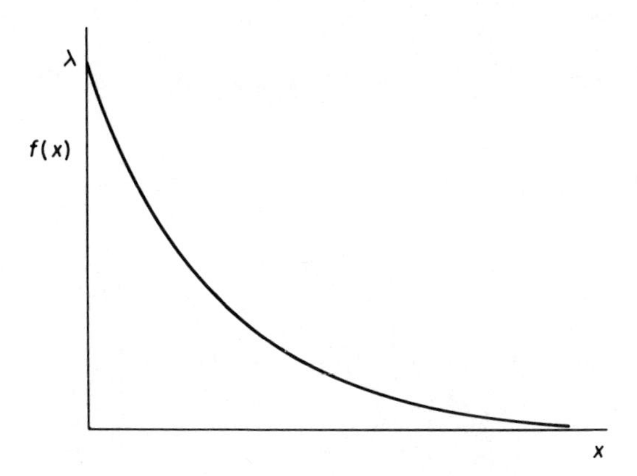

Fig. 6.15 The exponential distribution

Clearly the distribution is positively skew. When $x = 0, f(x) = \lambda$ and when $x \to \infty, f(x) \to 0$.

We will now derive the mean and variance of the exponential distribution from the probability density function (6.14) using the methods of Section 5.6. Then we will derive its (cumulative) distribution function and median, using the methods of Sections 5.7 and 5.8. Finally the moment generating function will be found so that we can verify the values for the mean and variance, using the methods of Section 5.9.

The mean, μ

$$\mu = E(X) = \int x f(x)\,\mathrm{d}x \qquad \text{from (5.8)}$$

$$= \int_0^\infty x\lambda e^{-\lambda x}\,\mathrm{d}x \qquad \text{from (6.14)}$$

$$= \Big[-xe^{-\lambda x}\Big]_0^\infty + \int_0^\infty e^{-\lambda x}\,\mathrm{d}x \qquad \text{by parts}$$

$$= 0 - 0 + \left[\frac{-e^{\lambda x}}{\lambda}\right]_0^\infty$$

$$= -\left(0 - \frac{1}{\lambda}\right)$$

$$\therefore\ \mu = \frac{1}{\lambda}$$

The variance, σ^2

$$E(X^2) = \int x^2 f(x)\,\mathrm{d}x \qquad \text{from (5.10)}$$

$$= \int_0^\infty x^2\lambda e^{-\lambda x}\,\mathrm{d}x \qquad \text{from (6.14)}$$

$$= \Big[-x^2 e^{-\lambda x}\Big]_0^\infty + 2\int_0^\infty xe^{-\lambda x}\,\mathrm{d}x \qquad \text{by parts}$$

$$= 0 - 0 + \frac{2}{\lambda}E(X) \qquad \text{from above}$$

$$= \frac{2}{\lambda^2}$$

$$\therefore\ \sigma^2 = E(X^2) - \mu^2 \qquad \text{from (5.3)}$$

$$= \frac{2}{\lambda^2} - \left(\frac{1}{\lambda}\right)^2$$

$$\sigma^2 = \frac{1}{\lambda^2}$$

The mean and variance of the exponential distribution are $1/\lambda$ and $1/\lambda^2$ respectively. The standard deviation is $1/\lambda$.

The distribution function, $F(x)$

$$F(x) = \int_{-\infty}^{x} f(x)\,\mathrm{d}x \qquad \text{from Section 5.7}$$

$$= \int_{0}^{x} \lambda e^{-\lambda x}\,\mathrm{d}x \qquad \text{from (6.14)}$$

$$= \Big[-e^{-\lambda x}\Big]_0^x$$

$$\therefore\; F(x) = 1 - e^{-\lambda x}, \qquad x \geqslant 0$$

$$= 0 \qquad \text{otherwise}$$

The median, m

$$F(m) = \frac{1}{2} \qquad \text{from Section 5.8}$$

$$1 - e^{-\lambda m} = \frac{1}{2}$$

$$-\lambda m = \log_e\left(\frac{1}{2}\right) = -\log_e 2$$

$$m = \frac{\log_e 2}{\lambda}$$

$$m = \frac{0.6931}{\lambda}$$

So the median is less than the mean, $1/\lambda$, as we would expect for this positively skew distribution (see Section 2.8).

The m.g.f., $M(t)$

$$M(t) = E(e^{tX}) \qquad \text{from Section 5.9}$$

$$= \int_0^\infty e^{tx}\lambda e^{-\lambda x}\,\mathrm{d}x \qquad \text{from (5.9) and (6.14)}$$

$$= \lambda \int_0^\infty e^{(t-\lambda)x}\,\mathrm{d}x$$

$$= \lambda\left[\frac{e^{(t-\lambda)x}}{t-\lambda}\right]_0^\infty$$

$$= \lambda\left[0 - \frac{1}{t-\lambda}\right], \quad \text{if} \quad t < \lambda$$

$$= \frac{\lambda}{\lambda - t}$$

$$M(t) = \left(1 - \frac{t}{\lambda}\right)^{-1} \tag{6.15}$$

We can verify the values for the mean and variance using (5.12) and (5.13).

$$M'(t) = \frac{1}{\lambda}\left(1 - \frac{t}{\lambda}\right)^{-2}$$

$$\therefore \mu = M'(0) = \frac{1}{\lambda}(1-0)^{-2} = \frac{1}{\lambda}$$

$$M''(t) = \frac{2}{\lambda^2}\left(1 - \frac{t}{\lambda}\right)^{-3}$$

$$\sigma^2 = M''(0) - \mu^2$$

$$= \frac{2}{\lambda^2} - \left(\frac{1}{\lambda}\right)^2$$

$$= \frac{1}{\lambda^2}$$

In order to obtain probabilities for the exponential distribution, we may use the distribution function, $F(x) = 1 - e^{-\lambda x}$, since

$$P(x_1 < X \leqslant x_2) = \text{area to left of } x_1 - \text{area to left of } x_2 \quad \text{(see Fig. 6.16)}$$
$$= F(x_2) - F(x_1)$$
$$= 1 - e^{-\lambda x_2} - (1 - e^{-\lambda x_1})$$
$$= e^{-\lambda x_1} - e^{-\lambda x_2}$$

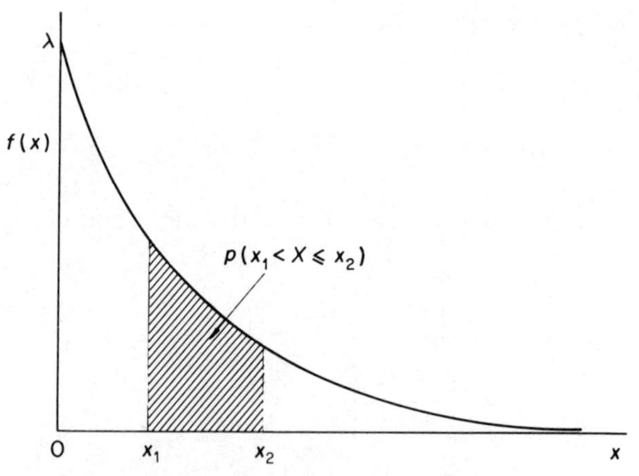

Fig. 6.16 Probabilities for the exponential distribution

Example 6.16

The mean time between successive telephone calls arriving randomly at a switchboard is 30 seconds.

(a) What is the probability that the time between successive telephone calls will be:

(i) less than 15 seconds;
(ii) between 15 and 30 seconds;
(iii) between 30 and 60 seconds;
(iv) more than 60 seconds?

(b) What is the median time between successive telephone calls?

Since calls are arriving randomly with a mean time between calls of 30 seconds, X, the time between calls, has an exponential distribution with a parameter λ such that $1/\lambda = 30$ seconds

therefore $$\lambda = \frac{1}{30} \text{ calls/sec.}$$

(Note that λ is the mean number of calls per second, whereas $1/\lambda$ is the mean time, in seconds, between successive calls.)

(a) (i) $P(X \leqslant 15) = F(15) = 1 - e^{-15/30} = 0.3935$

(ii) $P(15 < X \leqslant 30) = F(30) - F(15) = e^{-15/30} - e^{-30/30} = 0.2387$

(iii) $P(30 < X \leqslant 60) = F(60) - F(30) = e^{-30/30} - e^{-60/30} = 0.2325$

(iv) $P(X > 60) = 1 - F(60) = 1 - (1 - e^{-60/30}) = 0.1353$

(b) $m = \dfrac{0.6931}{\lambda} = 20.8$ seconds □

One of the intriguing properties of the exponential distribution is the 'no memory' property. Suppose that X, the lifetime of a component which can fail suddenly, has an exponential distribution with parameter λ. What is the probability that, having survived for a time x_1, it will fail in the next small unit interval of time dx?

$$P(\text{fail in d}x|\text{survived time } x_1) = \frac{P\text{ (survived time } x_1 \text{ and fails in d}x)}{P(\text{survived time } x_1)}$$

$$= \frac{P\text{ (fails in d}x)}{P(X > x_1)} \quad \text{since 'survived time } x_1\text{' is redundant}$$

$$= \frac{f(x_1)\text{d}x}{1 - F(x_1)}$$

$$= \frac{\lambda e^{-\lambda x_1}\text{d}x}{e^{-\lambda x_1}}$$

$$= \lambda \text{d}x$$

Table 6.4 The properties of three discrete distributions

Distribution	*Probability function*	*Parameters*	*Mean*	*Variance*	*p.g.f.*
binomial, $B(n, p)$	$p(x)=\binom{n}{x}p^x(1-p)^{n-x},\ x=0,1,\ldots,n$	n, p	np	$np(1-p)$	$(1-p+pt)^n$
Poisson, Poi(a)	$p(x)=\dfrac{e^{-a}a^x}{x!},\ x=0,1,\ldots,\infty$	a	a	a	$e^{a(t-1)}$
geometric	$p(x)=(1-p)^{x-1}p,\ x=1,2,\ldots,\infty$	p	$\dfrac{1}{p}$	$\dfrac{1-p}{p^2}$	$\dfrac{pt}{1-t(1-p)}$

Table 6.5 The properties of three continuous distributions and the standardized normal distribution

Distribution	*Probability density function*	*Parameters*	*Mean*	*Variance*	*m.g.f.*
rectangular	$f(x)=\dfrac{1}{b-a},\ a\leqslant x\leqslant b$	a, b	$\dfrac{a+b}{2}$	$\dfrac{(b-a)^2}{12}$	$\dfrac{e^{bt}-e^{at}}{(b-a)t}$
normal, $N(\mu, \sigma^2)$	$f(x)=\dfrac{1}{\sigma\sqrt{2\pi}}\exp\left\{-\dfrac{1}{2}\left(\dfrac{x-\mu}{\sigma}\right)^2\right\},$ $-\infty<x<\infty$	μ, σ	μ	σ^2	$\exp\left\{\mu t+\dfrac{\sigma^2 t}{2}\right\}$
standardized normal, $N(0, 1)$	$\phi(z)=\dfrac{1}{\sqrt{2\pi}}e^{-\frac{1}{2}z^2},\ -\infty<x<\infty$	—	0	1	$e^{\frac{1}{2}t^2}$
exponential	$f(x)=\lambda e^{-\lambda x},\ x\geqslant 0$	λ	$\dfrac{1}{\lambda}$	$\dfrac{1}{\lambda^2}$	$\left(1-\dfrac{t}{\lambda}\right)^{-1}$

Since λdx does not contain x_1, it is independent of x_1. In other words, the probability that the component will fail in the next instant does not depend on how long the component has been working. The component has 'no memory'. Contrast this conclusion with what you know about the probability that an electric light-bulb or a person will 'fail' as they get older. Clearly we are not talking about random events with light-bulbs and people.

6.8 SUMMARY

Three standard discrete and three standard continuous distributions and their main properties and applications were derived or described. Tables 6.4 and 6.5 summarize the main results (including those of the standardized 'normal').

WORKSHEET 6

Section 6.2

1. A binomial experiment consisting of n trials can be thought of as consisting of n Bernoulli trials. A Bernoulli variable takes the value 1 with probability p if the outcome of a trial is a 'success', and 0 if the outcome is a 'failure'. Obtain the probability function for the Bernoulli distribution. Also find its mean, variance, and probability generating function.

2. Suppose that, of all the undergraduates who begin a degree course in higher education, 80 % eventually get a degree. If a random sample of 5 is selected, what is the probability that: (a) none will graduate; (b) all 5 will graduate; (c) at least one will graduate.

3. Suppose that, in a certain country, there are approximately 1080 male children for every 1000 female children. Use this information to calculate the probabilities that families with 4 children will have: (a) no girls; (b) one girl; (c) two girls; (d) three girls; (e) four girls. What are the mean and variance of the number of:

   (i) girls; (ii) boys in families with four children?

4. In a multiple-choice test consisting of 20 questions there are 5 possible answers to each question, only one of which is correct. If a candidate always guesses the answers:

   (a) How many correct answers would you expect the candidate to obtain?
   (b) How many marks would you expect the candidate to be awarded in the test if 4 marks are given for a correct answer and 1 mark is deducted for an incorrect answer?
   (c) What is the probability that the candidate will obtain

      (i) 40 % or more of the correct answers
      (ii) 40 % or more of the maximum possible marks for the test?

5. For a certain strain of a flower the probability that a seed produces a pink flower is 1/4. How many seeds should be sown so that the probability of obtaining at least one pink flower is at least 99 %?

6. Watt Youhavin and Sons are importers of fine wines. Recently they held a wine-tasting evening to publicize a new consignment of French wines. They invited a variety of guests, who could all be classified as either expert or novice at wine identification. Of these guests 20 % were known to be experts. If an expert is given a glass of wine he will identify if correctly with probability 0.9, whereas the novice has a probability of only 0.4 of a correct identification.

   Assuming that these probabilities remain constant throughout the evening and that the results of successive attempts at identification are independent:

   (a) What is the probability that a novice will identify 3 wines correctly out of 5 tried?
   (b) A randomly chosen guest is observed to get 4 wines correct out of 5 tried. What is the probability of this event? And further, what is the probability that this guest is an expert?
   (c) The guest in part (b) tries a sixth wine. What is the probability that it is identified correctly?

(AEB (1983)) ►

7. Packets of crisps are filled by a machine. Over a long period it has been found that $100p\%$ of the packets are underweight. A sample of n packets is selected at random. Write down an expression, in terms of p, n, r, for the probability that exactly r packets are underweight.

   Prove that the mean number of underweight packets is np.

   Given that $p = 0.03$ and $n = 9$, find, to 3 decimal places in each case,

   (a) the probability that none of the packets in the sample is underweight;
   (b) the probability that more than one of the packets in the sample is underweight.

   Find also the expected number of underweight packets in the sample.

   (London)

8. State clearly conditions under which it is appropriate to assume that a random variable has a binomial distribution. Explain briefly any use which you made, or could have made, of the binomial distribution in one of your projects.

   A door-to-door canvasser tries to persuade people to have a certain type of double-glazing installed. The probability that his canvassing at a house is successful is 0.05. Using the table of cumulative binomial probabilities provided, or otherwise, find, to 3 decimal places, the probability that he will have at least 2 successes out of the first 10 houses he canvasses.

   Find the number of houses he should canvass per day in order to average 3 successes per day.

   Calculate the least number of houses that he must canvass in order that the probability of his getting at least one success exceeds 0.99.

   (London)

9. A chain-store is supplied with 60% of its light-bulbs from manufacturer A and the remainder from manufacturer B. It is known that, of the light bulbs from A, 6% are defective and 3% are wrongly labelled, these faults occurring independently, while, of those from B, 9% are defective but all are correctly labelled.

   (i) What is the probability that, in a sample of 15 bulbs taken from the batch from A, all are found to be working and correctly labelled?
   (ii) What is the probability that a sample of 15 taken at random from the combined stock of bulbs contains exactly two bulbs that are not working?
   (iii) What is the probability that a defective bulb in the store's stock comes from manufacturer A?

   (O&C)

10. (a) Quote and derive formulae for the mean and standard deviation of a binomial distribution.

    If the mean of a binomial distribution is 24 and the variance is 14.4, find the values of n and p.

    (b) It is known that in a large consignment of battery-driven model cars, 10% are defective. Fifty cars are delivered to a shopkeeper, and he examines five of them taken at random. Find the chance that there will be at least two defective cars in this sample of five.

    If the shopkeeper takes these cars one at a time and examines them without replacing them, what is the probability that the third car will be first to be satisfactory?

    (SUJB) ►

Section 6.3

11. A rolling machine in a paper mill produces on average one flaw every 500 feet of paper. Assuming the number of flaws per unit length of paper has a Poisson distribution, find the probability that:

    (a) a 500 foot roll has (i) no flaws; (ii) at least one flaw
    (b) a 2000 foot roll has (i) no flaws; (ii) at most three flaws
    (c) out of two 1000 foot rolls,

    (i) neither has a flaw;
    (ii) both have at least one flaw;
    (iii) at least one has at least one flaw.

12. A hire-car firm has two cars which it hires out day by day. The number of demands has a Poisson distribution with a mean of 1.5 cars per day. Calculate the probability that on a given day:

    (i) neither car is hired out;
    (ii) demand exceeds supply.

    On how many days a week (7 days) would you expect exactly one car will be hired out?

13. The number of insects per plant on the leaves of a certain type of plant has a Poisson distribution with mean a, except that leaves on which there are no insects are not observed. Show that the probability function for the distribution of the observed number of insects per plant is:

$$p(x) = \frac{e^{-a}a^x}{(1-e^{-a})x!}; \qquad x = 1, 2, \ldots$$
$$= 0 \qquad \text{otherwise}$$

    Hence find the mean and variance of the distribution.

14. Because of faults in an apparatus used to supply a gas, the apparatus requires attention on average 1.2 times a day. In a working week of five days, what is the probability that the apparatus will require no attention, assuming faults occur randomly? On how many days per year would you expect the apparatus to require no attention? (Assume that there are 52 weeks per year each with 5 working days.)

15. The discrete random variable X has probability distribution defined by:

$$P(X = x) = \begin{cases} \dfrac{e^{-\mu}\mu^x}{x!} & \text{for } x = 0, 1, 2, \ldots, \mu > 0 \\ 0 & \text{otherwise} \end{cases}$$

    Derive the mean and variance of X.

    Sixty fish caught in a pond were examined and the number of parasites on each fish counted. The results were as follows:

Number of parasites on fish	0	1	2	3	4	5	6	7	8
Number of fish	6	9	10	4	4	2	3	1	4

Number of parasites on fish	9	10	11	12	13	14	15	16	17
Number of fish	2	4	4	2	3	0	1	0	1

Calculate the arithmetic mean and the standard deviation of the number of parasites on a fish, for this sample.

If the fish had all been hatched in the same season, the number of parasites on a fish would be expected to follow a Poisson distribution. Without further calculation state whether or not you think it likely that all the fish were hatched in the same season. Justify your answer.

If the number of parasites per fish did follow a Poisson distribution with mean 5.35, what is the probability of finding less than 4 parasites on a fish?

Does this result support your conclusion about whether or not the fish were hatched in the same season?

(AEB (1984))

16. Given that the random variable X has the Poisson distribution with mean a, show that:

(i) $\sum_{r=0}^{n} rP(X=r) = aP(X \leqslant n-1)$;

(ii) $\sum_{r=0}^{n} r^2P(X=r) = a^2P(X \leqslant n-2) + aP(X \leqslant n-1)$

A garage receives delivery of new cars at the beginning of each month and accepts as many new cars as is necessary to bring its stock of new cars to 10. The monthly demand for new cars at the garage has a Poisson distribution with mean 8. Find, to two significant figures in each case, the mean and the variance of the number of new cars sold per month by the garage.

(JMB)

17. Samples of a production batch of a vaccine are taken at random and the number x of live viruses in each sample is counted. The results for 20 samples each of $3\ \text{cm}^3$ of vaccine are as follows:

x	0	1	2	3	more than 3
Frequency	10	6	3	1	0

Assuming that the distribution of X is Poisson with mean equal to that of the sample, find the probability that a $3\ \text{cm}^3$ sample taken at random from the batch contains 4 live viruses.

The batch is to be rejected if the probability that it contains more than 4 live viruses in a randomly chosen $3\ \text{cm}^3$ sample exceeds 0.001. Should the batch be accepted or rejected?

Comment on the assumptions that have been made.

(O & C) ►

Section 6.4

18. Using the probability function only, check that the sum of the probabilities for the binomial, Poisson and geometric distributions are each equal to 1.

19. A fisherman casts his line into a river with probability p for each cast of getting a bite. If X denotes the number of casts up to but not including the first bite, show that the probability function for X is:

$$\begin{aligned} p(x) &= (1-p)^x p, \qquad x = 0, 1, \ldots \\ &= 0 \qquad \text{otherwise} \end{aligned}$$

   (a) Show that $\sum_{x=0}^{\infty} p(x) = 1$, and find the mean and variance of X
   (b) Also find the p.g.f. of this distribution and verify the values for the mean and variance.
   (c) If the fisherman can make 100 casts in an afternoon, and $p = 0.05$, what is the probability that the fisherman gets:
      (i) his first bite with his last cast of the afternoon;
      (ii) no bites all afternoon;
      (iii) at least two bites during the afternoon?

20. A bag contains 2 hard and 8 soft sweets. Assume it is not possible to distinguish between the two types of sweet while they are still in the bag, but it is possible immediately they are removed from the bag. A child selects a sweet at random but replaces it if it is not the child's favourite type of sweet. What is the expected number of selections before the favourite is selected if the favourite is (i) the hard type, (ii) the soft type of sweet?

21. Two ordinary dice are thrown repeatedly. What are the mean and variance of the number of throws before a total of 7 is obtained?

22. The constant probability of success in a series of independent random trials is $1/k$. Given that the first success is achieved at the Xth trial, show that $E(X) = k$ and $\text{Var}(X) = k(k-1)$.

   Two dice are thrown simultaneously until two sixes are showing. Find the probability that the process will terminate before the expected number of throws is reached.
   (Cambridge)

Section 6.5

23. When asked how old they are, people often give their 'age at last birthday', i.e. in whole number of years. If X denotes the error in months in the stated age, so that

$$X = \text{actual age} - \text{age at last birthday},$$

   what distribution will X have? What are the mean and variance of the distribution? What is $P(1 \leqslant X \leqslant 3)$?

24. (a) A continuous random variable X is uniformly distributed in the interval $2 \leqslant x \leqslant 6$ and can take no other values.

▶

(i) Write down the probability density function for X.
(ii) Find the mean of X and the standard deviation of X.
(iii) Obtain the distribution function for X.

(b) Petrol is delivered to a filling station every Monday morning before it opens for business, and the tanks are filled. The weekly demand Y, in thousands of gallons, has the continuous probability density function given by:

$$f(x) = kx(7-2x), \qquad 1 < x < 3,$$
$$f(x) = 0, \qquad \text{otherwise},$$

where k is a constant.

(i) Find the value of k.
(ii) Hence find the mean number of gallons sold per week, assuming that the filling station tanks hold a total of 3500 gallons.
(iii) During a time of structural change at the filling station, one tank is out of use. This reduces the total capacity to 2500 gallons. Calculate the probability that in a week, chosen at random during the period of structural change, the filling station will run out of petrol.

(JMB)

25. The random variable X has a rectangular distribution whose probability density function $f(x)$ is given by:

$$f(x) = \begin{cases} \frac{1}{6} & \text{for } -2 \leqslant x \leqslant 4 \\ 0 & \text{otherwise} \end{cases}$$

Sketch the probability density function of X and hence, or otherwise, find the probabilities that:

(i) $X \leqslant 2$;
(ii) $|X| \leqslant 2$;
(iii) $|X| \leqslant x$ for $0 \leqslant x \leqslant 2$;
(iv) $|X| \leqslant x$ for $2 \leqslant x \leqslant 4$;
(v) $|X| \leqslant x$ for $x > 4$.

Hence obtain, and sketch, the probability density function of $|X|$. Thus determine the mean of $|X|$.

(MEI)

Section 6.6

26. Obtain $M(t)$, the moment generating function of the standardized normal distribution, using $M(t) = E(e^{tZ})$, and hence verify its mean and variance.

27. (a) Using tables, find what proportion of the area under any normal distribution curve lies within: (i) one; (ii) two; (iii) three standard deviations of the mean.

(b) How many standard deviations from the mean correspond to areas of (i) 95%, (ii) 90%?

28. A distribution has a moment generating function $M(t) = \exp\left\{at + \frac{b^2t^2}{2}\right\}$

    What are its mean and variance? What type of distribution is it?

29. A railway company keeps a record of how late trains are when they arrive at their final destination:

    (a) Suppose they are equally likely to arrive anywhere between 5 minutes early and 35 minutes late. What is the probability that a randomly selected train will:

    (i) be late; (ii) be more than 20 minutes late?

    Draw a sketch of the distribution of the number of minutes late.

    (b) Suppose, instead, that the number of minutes late is normally distribution with mean 12 minutes and standard deviation 10 minutes. What is the probability that a randomly selected train will:

    (i) be late; (ii) be more than 20 minutes late?

    Draw a sketch of the distribution of the number of minutes late, and compare it with your sketch for (a) above.

30. The lengths of mass-produced components are normally distributed with mean 15 cm and standard deviation 0.03 cm. Each cost 8 p to produce. A component shorter than 14.95 cm has to be scrapped, whereas a component larger than 15.05 cm can be trimmed at a further cost of 3 p. Components having lengths between these limits are satisfactory and cost the manufacturer nothing extra.

    (a) Of 1000 components made, how many of them would you expect to:

    (i) be scrapped;
    (ii) be acceptable after trimming;
    (iii) be immediately acceptable?

    Hence calculate the expected cost of an acceptable component.

    (b) What is the probability that an acceptable component had needed machining?

31. A machine which automatically packs plastic granules into bags is known to operate with a standard deviation of 1 kg. Assuming a normal distribution, to what mean (target) weight should the machine be set so that 95 % of bags weigh over 20 kg. In this case what weight would be exceeded by 1 bag in 1000 on average?

32. (a) $2\frac{1}{2}$ % of light-bulbs have lifetimes greater than 1235 hours and $2\frac{1}{2}$ % have lifetimes less than 765 hours. Assuming lifetime is normally distributed, what are the mean and standard deviation of this distribution (to the nearest hour)? What is the probability that a randomly selected light-bulb will last longer than 1250 hours?

    (b) Suppose that an improved light-bulb has a mean lifetime 10 % higher than before, but with the same standard deviation. What is the probability that:

    (i) a randomly selected improved light-bulb will last longer than 1250 hours?
    (ii) all four of a set of four randomly selected improved light-bulbs will last longer than 1250 hours?

33. Using the result, $M_\mu(t) = e^{-\mu t}M(t)$ (from Chapter 5), find $M_\mu(t)$, the moment generating function about the mean, for the normal distribution. Hence verify that $\mu_1 = 0$, $\mu_2 = \sigma^2$, $\mu_3 = 0$, $\mu_4 = 3\sigma^4$.

▶

34. A machine produces bolts whose lengths are normally distributed with mean 8.54 cm and standard deviation 0.05 cm. The diameters of the bolts vary independently of their lengths and are also normally distributed with mean 1.57 cm and standard deviation 0.01 cm. The specifications require that a bolt should have a length of at least 8.45 cm and a diameter between 1.55 and 1.60 cm. Find the probability that a bolt chosen at random will meet:

    (i) both specifications;
    (ii) at least one of the specifications;
    (iii) exactly one of the specifications.

    The bolts are inspected by gauges which automatically reject those which do not meet the above specifications. Find the probability that a bolt chosen at random from those passed by the gauges will have a length less than 8.54 cm.

    (IOS)

35. The heights of two-year seedlings of *Tammia sinensis* from a certain nursery have a normal distribution with mean 30 cm and standard deviation 6 cm. The seedlings are graded according to height: size P, over 38 cm; size Q, between 24 cm and 38 cm; size R, under 24 cm. Seedlings of size P are sold for £25 per hundred; size Q, £20 per hundred; size R, £10 per hundred.

    (i) Find the proportion of the seedlings in each size.
    (ii) Find the height exceeded by two seedlings in a hundred on average.
    (iii) Find the expected selling price of 1000 seedlings.
    (iv) A man buys 1000 seedlings selected at random from the nursery but rejecting any of size R. How much would he expect to pay for his 1000 seedlings?

    (Cambridge)

36. The random variable Z is normally distributed with mean 0 and variance 1. The information below gives (correct to 5 decimal places) the probabilities that $Z \leqslant z$ for various values of z.

z	1.03	1.04	1.05	1.06
$P(Z \leqslant z)$	0.84849	0.85083	0.85314	0.85543

    Using this information, and linear interpolation (where necessary), find, correct to 4 decimal places, the probabilities that (i) $Z \geqslant -1.04$; (ii) $|Z| \leqslant 1.05$; (iii) $0 \leqslant Z \leqslant 1.034$; (iv) $-1.057 \leqslant Z \leqslant -1.034$.

    Find, correct to 4 decimal places, the value of z such that:

$$P(Z \leqslant z) = 0.14740.$$

    (MEI)

37. A machine produces components, the lengths of which are normally distributed with mean 102.30 mm and standard deviation 1.20 mm.

    (i) If four components are picked at random, calculate the probability that at least one will have a length of more than 102.70 mm.
    (ii) All components which are less than 100.50 mm in length are rejected. What proportion is this?

    Draw a cumulative frequency graph of the lengths of all components at the time of ▶

making and hence determine the median length of the components which are accepted. (Only those less than 100.50 mm in length are rejected.)

(SUJB)

38. A random variable X has a normal distribution with mean μ and standard deviation σ. What percentage of the population is such that:

(i) $X < \mu - \sigma$; (ii) $X < \mu + \sigma$?

Two hundred samples of liquid were analysed for the presence of an impurity, which was measured in parts per million correct to the nearest whole number. The results were as follows:

p.p.m. (X)	50–59	60–69	70–79	80–89	90–99	100–109	110–119	120–129
Frequency	1	6	23	57	51	42	16	4

(iii) Using an assumed mean of 84.5 (or otherwise) calculate the mean and standard deviation of X.

(iv) Draw a percentage cumulative frequency curve and, on the assumption that the distribution is normal, use the results of (i) and (ii) to estimate the mean and standard deviation of X. Comment on your results.

(SUJB)

39. A random variable X has a normal distribution with mean, μ, and variance, σ^2. Find the value of X for which $f(x)$ is a maximum. What are the mode and median of this normal distribution?

Section 6.7

40. Consider light-bulbs produced with a lifetime, X (in hours), having a distribution with probability density function:

$$f(x) = \frac{1}{1000} e^{-x/1000}, \qquad x > 0$$
$$= 0 \qquad \text{otherwise}$$

(a) What is the mean lifetime of these light-bulbs?
(b) What is the the probability that a randomly selected bulb will have a lifetime of at least 1200 hours?
(c) If 5 light-bulbs are randomly selected, what is the probability that (i) none, (ii) at least 3, will have a lifetime of at least 1200 hours?

41. The queue for a telephone kiosk is such that:

(i) 'callers' arrive randomly at a rate of 1 per 5 minutes;
(ii) when it is their turn to use the telephone the time a caller spends in the kiosk is exponentially distributed with a mean of 4 minutes.

(a) State the probability function for X, the number of arrivals at the kiosk per minute. Also state the mean and variance of the distribution of X.
(b) State the probability density function for T, the time in minutes a caller spends in the kiosk. Also state the mean and variance of the distribution of T.
(c) What is the probability that:
(i) exactly 1 caller arrives in one minute?
(ii) the time a caller spends in the kiosk is less than 2 minutes?

42. Suppose the accident rate in a large factory is 0.03 accidents per day. Assuming accidents occur randomly what is:

   (i) the mean time between successive accidents?
   (ii) the median time between successive accidents?

43. A garage which carries out MOT tests finds that the time to carry out the test is exponentially distributed with a mean of 20 minutes.

   (a) What is the probability that a randomly selected vehicle will take between 15 and 25 minutes to test?
   (b) What percentage of vehicles take less than 10 minutes to test?

44. (a) Observations by traffic engineers at a crossroads, where there had been a number of accidents, showed that of vehicles approaching from a particular direction 45% turned left, 35% turned right and 20% went straight on. Assuming that the choice made for each car is independent, what is the probability that, of the next five cars arriving, exactly three will turn left?
   (b) An audience research survey among married couples revealed that the probability of a woman watching a particular programme was 0.16, but the probability of a man watching it was 0.23. However, if a woman did watch it, the probability that her husband would also watch was 0.62. If a man watched the programme, what is the probability that his wife also watched?
   (c) The arrival time of customers who enter a village store during 60-minute period is a random variable X with probability density function:

$$f(x) = \begin{cases} \frac{1}{60}, & 0 < x < 60 \\ 0\ , & \text{otherwise,} \end{cases}$$

   where X is measured in minutes from the start of the period.

   What is the probability that a customer who arrives during the period has an arrival time between 45 and 60?

   The time, in minutes, taken to serve a particular customer is a random variable Y with probability density function:

$$g(y) = \begin{cases} 0.4e^{-0.4y}, & y > 0 \\ 0\quad , & \text{otherwise.} \end{cases}$$

   What is the probability that a customer requires more than 3 minutes to be served?

   If in a particular 60 minute period only one customer arrives, find the probability that she will arrive in the last fifteen minutes and will require more than 3 minutes to be served.

(AEB (1984)

45. Describe one practical situation (other than that used later in this question) for which the exponential distribution might be an appropriate model.

   During business hours the time in minutes between the arrival of two successive telephone calls to a particular firm may be taken to have an exponential distribution with mean 1.2 minutes.

   Find:

   (i) the median time between the arrival of two successive telephone calls; ▶

(ii) the probability that the time between two successive telephone calls is greater than three minutes;

(iii) the probability that, of four successive intervals between telephone calls, exactly two are greater than three minutes.

One hundred successive time intervals between calls are to be recorded exactly. Find the expected number of intervals that would have durations of 0–30 seconds, 30–60 seconds, 60–90 seconds, and 90–180 seconds, respectively.

The firm finds that in practice this particular model is not very good when applied to telephone calls throughout the whole of the working day. Give one possible practical reason for this and suggest a possible modification to the model that might improve it.

(JMB)

46. Give two examples, from your project work if possible, where the Poisson distribution is appropriate, explaining how the necessary conditions for its use are satisfied.

People entering a museum are counted by the warden on duty at the door. The number of people entering in any time interval of length t minutes has a Poisson distribution with mean $\frac{1}{2}t$. The warden leaves the front door unattended for 5 minutes. Calculate, to 3 decimal places, the probability that:

(a) no one,
(b) three or more people,

will enter the museum during this time.

Find, to 3 significant figures, the length of time, in seconds, for which the warden could leave the front door unattended for there to be a probability of 0.95 that no one will enter during that time.

(London)

47. The continuous random variable X has the exponential distribution whose probability density function is given by:

$$f(x) = \lambda e^{-\lambda x}, \qquad x \geqslant 0$$
$$f(x) = 0, \qquad \text{otherwise,}$$

where λ is a positive constant. Obtain expressions, in terms of λ, for

(a) the mean $E(X)$ of the distribution;
(b) $F(x)$, the (cumulative) distribution function.

Television sets are hired out by a rental company. The time in months, X, between major repairs has the above exponential distribution with $\lambda = 0.05$. Find, to 3 significant figures, the probability that a television set hired out by the company will not require a major repair for at least a two-year period.

Find also the median value of X.

The company agreed to replace any set for which the time between major repairs is less than M months. Given that the company does not want to have to replace more than one set in 5, find M.

(London)

7

Approximations to the binomial and Poisson distributions

▼

7.1 INTRODUCTION

In Section 6.2 we calculated binomial probabilities either using the probability function (6.1) or a table of cumulative binomial probabilities. When the use of the formula becomes tedious, or when the values of n and p are not available in the tables, we may use other methods as follows to obtain approximate probabilities:

1. If n is large and p is small, we may use the Poisson distribution as an approximation to the binomial by putting $a = np$. In practice, quite good approximations may be obtained if $p < 0.1$, with no condition on the value of n.
2. If n is large and p is neither small nor large, we may use the normal distribution as an approximation to the binomial by putting $\mu = np$ and $\sigma = \sqrt{np(1-p)}$. In practice, quite good approximations may be obtained if $np > 5$ and $n(1-p) > 5$.

Similarly in Section 6.3 we calculated Poisson probabilities either using the probability function (6.4), a table of cumulative Poisson probabilities, or special graph paper. When the use of the formula becomes tedious, or when the values of a are not available in the table or on the graph paper, we may use the following method to obtain approximate Poisson probabilities:

3. If a is large, we may use the normal distribution as an approximation to the Poisson by putting $\mu = a$, $\sigma = \sqrt{a}$. In practice, quite good approximations may be obtained if $a \geqslant 10$.

The three approximations will be discussed below in Sections 7.2, 7.3 and 7.4 respectively.

7.2 POISSON APPROXIMATION TO THE BINOMIAL DISTRIBUTION

It can be shown that the binomial probability function: $p(x) = \binom{n}{x}p^x(1-p)^{n-x}$ becomes the Poisson probability function $p(x) = e^{-a}a^x/x!$, if we let p tend to zero, n tend to infinity, and replace np (the mean for the binomial) by a (the mean for the Poisson).

One way of thinking why this should be so is to consider an interval of time in which random events occur as being divided into a very large number of small equal intervals (Fig. 7.1). Because the intervals are small, the probability that an event will occur in any given small interval is small and the probability that more than one event will occur in the small interval is negligible. Also, because the events occur randomly, the probability is the same for each small interval of time, and what happens in any one time interval does not affect any later interval (independence). So we have the conditions for a binomial distribution, where 'success' is 'event occurs' and 'failure' is 'event does not occur' in a small time interval.

Fig. 7.1 Random events in time

We may use this 'Poisson approximation to the binomial' idea to calculate approximate binomial probabilities if n is large and p is small. In fact the less stringent condition $p < 0.1$ gives quite good approximations.

Example 7.1

Suppose 8% of people are left-handed. What is the probability that 9 or more of a random sample of 50 people are left-handed?

This is a 'binomial' problem if we assume the four conditions for the binomial hold (Section 6.2). If X denotes the number of left-handed people, we could calculate:

$$P(X \geqslant 9) = p(9) + p(10) + \cdots + p(50),$$

or

$$= 1 - p(0) - p(1) - \cdots - p(8),$$

using $p(x) = \binom{50}{x} 0.08^x 0.92^{50-x}$, but this calculation is tedious. Also, cumulative binomial tables for $n = 50, p = 0.08$ may not be available. However, since $p < 0.1$, we may use the Poisson distribution with parameter

$$a = np = 50 \times 0.08 = 4$$

$$\begin{aligned} P(X \geqslant 9) &= 1 - P(X \leqslant 8) \\ &= 1 - 0.979, \quad \text{using Table C.2} \\ &= 0.021 \end{aligned}$$

For those who want reassurance, the *exact* answer (using the binomial formula) is 0.017 to 3 decimal places. The agreement is good. The fact that the probabilities are small indicates that we are working in the 'tails' of the distributions. Approximations are usually better if we are dealing with probabilities in the middle of the possible range of 0 to 1, as the following example will illustrate. □

Example 7.2

Suppose 8 % of people are left-handed. What is the probability that 4 or more of a random sample of 50 are left-handed?

Using the same Poisson distribution as in Example 7.1, since n and p are the same,

$$\begin{aligned} P(X \geqslant 4) &= 1 - P(X \leqslant 3) \\ &= 1 - 0.433, \quad \text{using Table C.2} \\ &= 0.567 \end{aligned}$$

The *exact* answer is 0.575 to 3 decimal places; better agreement in this time (in percentage terms). □

Example 7.3

This example will show that n does not have to be large, as long as $p < 0.1$, to obtain quite good approximations. Suppose 8 % of people are left-handed. What is the probability that 2 or more of a random sample of 25 are left-handed?

Using $a = np = 25 \times 0.08 = 2$,

$$\begin{aligned} P(X \geqslant 2) &= 1 - P(X \leqslant 1) \\ &= 1 - 0.406, \quad \text{using Table C.2} \\ &= 0.594 \end{aligned}$$

The *exact* answer is 0.605, which is good agreement. □

7.3 NORMAL APPROXIMATION TO THE BINOMIAL DISTRIBUTION

It is perhaps surprising that we can sometimes use the normal distribution, which is *continuous*, as an approximation for the binomial distribution, which is *discrete*.

But we see from Fig. 6.1 how the shape of the $B(10, 0.5)$ distribution is similar to the shape of the normal distribution. If we consider a binomial distribution with n much larger than 10, but p still equal to 0.5, we can imagine that, as the vertical lines come closer together to accommodate the large number of values on the horizontal scale, the shape of the normal distribution will become even clearer (see Fig. 7.2).

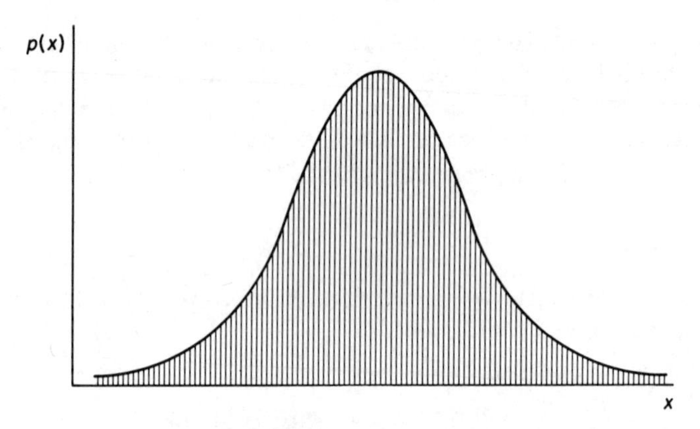

Fig. 7.2 A $B(n, 0.5)$ distribution, where n is large

In practice approximate binomial probabilities may be obtained for a range of values of p (not only 0.5) and for quite small values of n, so that if

$$np > 5 \qquad \text{and} \qquad n(1-p) > 5$$

quite good approximations may be obtained by replacing np (the mean for the binomial) by μ (the mean for the normal), and $\sqrt{np(1-p)}$ (the standard deviation for the binomial) by σ (the standard deviation for the normal). In addition we should apply a 'continuity correction' to allow for the fact that we are using a continuous distribution as an approximation for a discrete distribution. The following examples illustrate the method.

Example 7.4

Suppose it is known that 70% of female students who take statistics at A-level eventually get married. What is the probability that 24 or less of a random sample of 50 such students will eventually get married?

Assuming the conditions for a binomial distribution hold, we have a $B(50, 0.7)$ distribution. We could calculate

$$P(X \leqslant 24) = p(0) + p(1) + \cdots + p(24),$$

using
$$p(x) = \binom{50}{x} 0.7^x \, 0.3^{50-x},$$

but this calculation is tedious. Also, cumulative binomial tables for $n = 50$, $p = 0.7$ or for $n = 50$, $p = 0.3$ may not be available. However, since np and $n(1-p)$ are both greater than 5 in this example (being 35 and 15 respectively), we may use the normal approximation method, i.e. a normal distribution with

$$\mu = np = 35$$

and

$$\sigma = \sqrt{np(1-p)} = 3.240$$

Figure 7.3 shows this distribution.

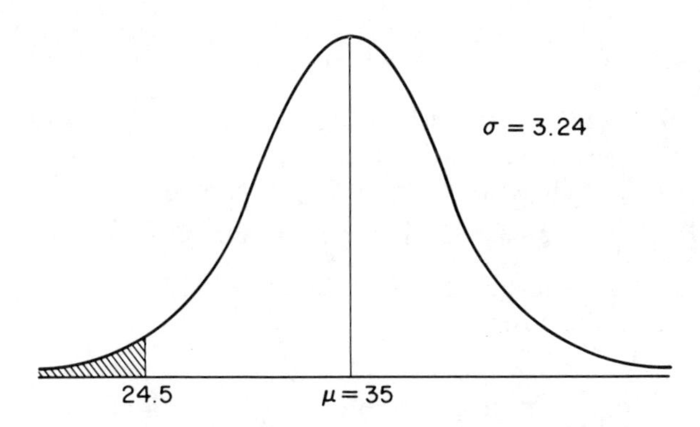

Fig. 7.3 The $N(35, 3.24^2)$ distribution

We now apply the continuity correction as follows:

$\leqslant 24$ on a discrete scale is equivalent to $\leqslant 24.5$ on a continuous scale

The reason for this can be seen by reference to Fig. 7.4 where we see that all values on the discrete scale to the left of the dotted line through 24.5 can be denoted by '$\leqslant 24$', while all values on the continuous scale to the left of the dotted line through 24.5 can be denoted by '$\leqslant 24.5$'.†

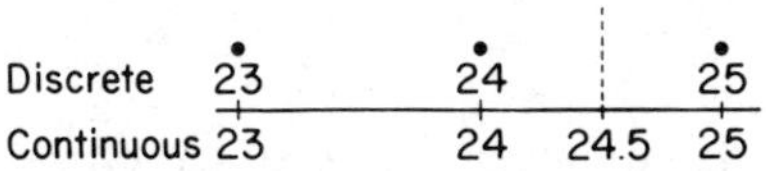

Discrete 23 24 25
Continuous 23 24 24.5 25

Fig. 7.4 Continuity correction example

The required probability is given by the shaded area in Fig. 7.3.

$$P(X \leqslant 24.5) = P\left(Z \leqslant \frac{24.5 - 35}{3.240}\right)$$

$$= P(Z \leqslant -3.24)$$

† Similarly, $\geqslant 24$ on a discrete scale is equivalent to $\geqslant 23.5$ on a continuous scale, while $= 24$ on a discrete scale is equivalent to 23.5 to 24.5 on a continuous scale.

$$= 1 - \Phi(3.24) \qquad \text{by symmetry}$$
$$= 1 - 0.9994 \qquad \text{from Table C.3(a)}$$
$$= 0.0006$$

The exact answer (using the binomial formula) is 0.0009. The fact that the probabilities are small indicates that we are working in the 'tails' of the distributions. Approximations are usually better if we are dealing with probabilities in the middle of the probability range of 0–1, as the following example will illustrate. □

Example 7.5

Suppose it is known that 70% of female students who take statistics at A-level eventually get married. What is the probability that 35 or more of a random sample of 50 such students will eventually get married?

The normal distribution used in the previous example may be used here:

$\geqslant 35$ on a discrete scale is equivalent to
$\geqslant 34.5$ on a continuous scale.

$$P(X \geqslant 34.5) = P\left(Z \geqslant \frac{34.5 - 35}{3.240}\right)$$
$$= P(Z \geqslant -0.15)$$
$$= P(Z \leqslant 0.15) \qquad \text{by symmetry}$$
$$= \Phi(0.15)$$
$$= 0.5596$$

The exact answer (using the binomial formula) is 0.5692, which is good agreement. □

Example 7.6

This example will show that quite good approximation may be obtained even if n is not large (as long as we are not dealing with extremes of probability). What is the probability that 10 or more of a random sample of 15 will eventually get married (using data from last example)?

This time
$$\mu = 15 \times 0.7 = 10.5$$
$$\sigma = \sqrt{15 \times 0.7 \times 0.3} = 1.775$$
$$P(X \geqslant 9.5) = P\left(Z \geqslant \frac{9.5 - 10.5}{1.775}\right)$$

$$= P(Z \geqslant -0.56)$$
$$= P(Z \leqslant 0.56) \qquad \text{by symmetry}$$
$$= \Phi(0.56)$$
$$= 0.7123$$

The exact answer is 0.7216; good agreement.

The final example in this section deals with the problem of finding the value of n in a binomial experiment to satisfy certain probability requirements. □

Example 7.7

Suppose that it is known that one marriage in four ends in divorce within five years. What is the smallest number of randomly selected marriages to ensure that:

(i) the probability of finding at least one which will end in divorce is at least 0.999?
(ii) the probability of finding at most 10 which will end in divorce is less than 0.99?

(i) $p = 1/4$, and assuming $n > 20$ for the moment in order that $np > 5$ and $n(1-p) > 5$ we can use the normal approximation to the binominal with $\mu = np = n/4$, $\sigma = \sqrt{np(1-p)} = \sqrt{3n}/4$ see (ii) below

However, if $X \sim B(n, \frac{1}{4})$,

$$P(X \geqslant 1) = 1 - P(X = 0)$$
$$= 1 - p(0), \quad \text{where} \quad p(x) = \binom{n}{x}\left(\frac{1}{4}\right)^x\left(1-\frac{1}{4}\right)^{n-x}$$
$$= 1 - \left(\frac{3}{4}\right)^n$$

So we require $1 - (\frac{3}{4})^n \geqslant 0.999$,

$$\text{i.e.} \quad \left(\frac{3}{4}\right)^n \leqslant 0.001$$
$$\therefore n \log_e\left(\frac{3}{4}\right) \leqslant \log_e 0.001$$
$$-0.2877n \leqslant -6.9078$$
$$n \geqslant 24.01$$

(note change from $\leqslant$ to $\geqslant$ when we divide by a negative number).

Since n must be an integer, we need a sample of 25 or more.

(ii) This time more than one binomial probability is involved, so we will use the normal approximation:

$$P(X \leqslant 10) = P\left[Z \leqslant \frac{10.5 - n/4}{\sqrt{3n}/4}\right] < 0.99$$

The value of Z corresponding to $\Phi(z) = 0.99$ is 2.33, from Table C.3(b),

so
$$\frac{10.5 - n/4}{\sqrt{3n}/4} < 2.33$$

$$\therefore\ 42 - n < 4.0357\sqrt{n}$$

Squaring, $$n^2 - 100.2867\ n + 1764 < 0$$
The roots of $n^2 - 100.2867\ n + 1764 = 0$ are 22.75 and 77.54.

The root $n = 77.54$ satisfies $42 - n = -4.0357\sqrt{n}$ and so is not the root we want. The required root is $n = 22.75$, but should we round up to 23 or down to 22? $n = 22$ does not quite satisfy $42 - n < 4.0357\sqrt{n}$, but $n = 23$ does. The answer is $n = 23$. □

7.4 NORMAL APPROXIMATION TO THE POISSON DISTRIBUTION

As in the previous section we will discuss the use of a continuous distribution as an approximation to a discrete distribution. We saw in Fig. 6.3 that the shape of the Poisson distribution with parameter $a = 5$ is similar to that of the normal distribution. For larger values of the parameter a the Poisson distribution becomes even more normal in shape, and we get distributions similar to Fig. 7.2.

In practice, if $a \geqslant 10$, approximate Poisson probabilities may be obtained using a normal distribution by replacing a (the mean of the Poisson) by μ (the mean of the normal) and $\sqrt{a}$ (the standard deviation of the Poisson) by σ (the standard deviation of the normal). As with the 'normal approximation to the binomial' we should apply a 'continuity correction' to allow for the fact that we are using a continuous distribution as an approximation for a discrete distribution. The following examples illustrate the method.

Example 7.8

Telephone calls arrive randomly at a switchboard at a rate of 1 per minute on average. What is the probability that 40 or more calls will be received in a period of 30 minutes?

The number of calls received per 30 minutes will have a Poi(30) distribution. We could calculate:

$$\begin{aligned} P(X \geqslant 40) &= 1 - P(X \leqslant 39) \\ &= 1 - p(0) - p(1) - \ldots - p(39) \end{aligned}$$

using $$p(x) = \frac{e^{-30}30^x}{x!}$$

but this calculation is tedious.

Also, cumulative Poisson tables (or special graph paper) for $a = 30$ may not be available. However, since the value of a is greater than 10, we may use the normal approximation method, i.e. a normal distribution with

$$\mu = a = 30$$

and $$\sigma = \sqrt{a} = 5.477$$

Figure 7.5 shows this distribution.

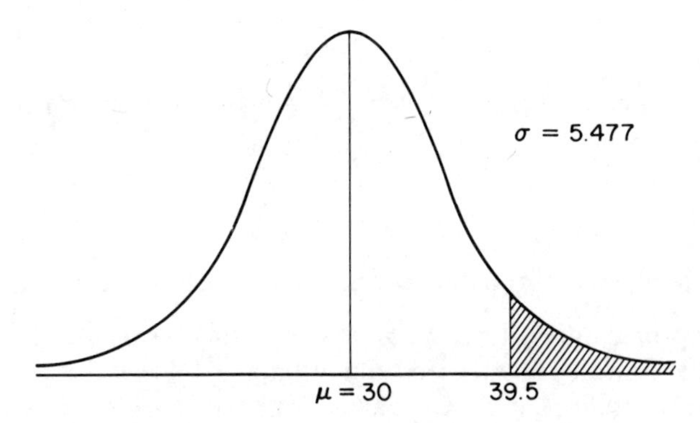

Fig. 7.5 The $(30, 5.477^2)$ distribution

We now apply the continuity correction:
$\geqslant 40$ on a discrete scale is equivalent to
$\geqslant 39.5$ on a continuous scale.

$$\begin{aligned} P(X \geqslant 39.5) &= P\left(Z \geqslant \frac{39.5 - 30}{5.477}\right) \\ &= P(Z \geqslant 1.73) \\ &= 1 - \Phi(1.73) \\ &= 1 - 0.9582 \\ &= 0.0418 \end{aligned}$$

The exact answer is (using the Poisson formula) 0.0463. The agreement is quite good, but once again we are working in the tails of the distributions in both cases. Approximations are usually better if we are working away from the tails, as the following example will illustrate. □

Example 7.9

Telephone calls arrive randomly at a switchboard at a rate of 1 per minute on average. What is the probability that less than 30 will be received in a period of 30 minutes?

The same normal distribution as in the previous example holds, and < 30 on a discrete scale is equivalent to $\leqslant 29.5$ on a continuous scale.

$$\begin{aligned} P(X \leqslant 29.5) &= P\left(Z \leqslant \frac{29.5-30}{5.477}\right) \\ &= P(Z \leqslant -0.09) \\ &= 1-\Phi(0.09) \\ &= 1-0.5359 \\ &= 0.4641 \end{aligned}$$

The exact answer (using the Poisson formula) is 0.4757, which is in good agreement. □

7.5 SUMMARY

Approximate binomial probabilities may be obtained by calculating:

1. Poisson probabilities, if n is large and p is small, by putting $a = np$. As a reasonable working rule the one condition, $p < 0.1$ may be used.
2. Normal probabilities, if n is large and p is neither small nor large, by putting:

$$\mu = np, \qquad \sigma = \sqrt{np(1-p)}$$

and applying a continuity correction. As a reasonable working rule the conditions $np > 5$ and $n(1-p) > 5$ may be used.

Approximate Poisson probabilities may be obtained by calculating normal probabilities if $a \geqslant 10$, by putting $\mu = a$, $\sigma = \sqrt{a}$, and applying a continuity correction.

Section 7.2

1. The average number of defectives in batches of 20 items is 1.6. Find the probability that a batch will contain:

   (a) 3 or more defectives
   (b) 3 defectives

   using (i) an exact method and (ii) an approximate method. Compare the answers given by the two methods in each case.

2. Assuming that on average one car driver in 1000 is colour-blind, and that such drivers pay a surcharge on their car insurance premiums, calculate the approximate probability that at least two of a random sample of 5000 car drivers will have to pay the surcharge.

3. Due to predators, disease and weather, only 5 in 1000 frogs' eggs on average will grow to mature frogs. If a clutch of frog spawn contains 800 eggs, what is the approximate probability that the number of frogs eventually produced is less than 5? What assumptions have you made?

4. The probability that an adult human is a carrier of a particular virus is 0.008. A procedure is devised whereby n adults will be randomly sampled in each town in a given area. If no carrier is found the town will be declared 'clear'. What is the approximate probability that a town will be declared clear if (a) $n = 500$, (b) $n = 1000$?

5. State the conditions under which a binomial distribution may be approximated by a Poisson distribution, and give a reason why this approximation may be useful in practice.

   In a certain population the probability that any sufferer from a certain common complaint is allergic to the drug used in the treatment is 0.001. The occurrences of the allergy in different sufferers may be taken to be independent events. By use of the appropriate table, or otherwise, determine for a random sample of 3000 sufferers;

   (i) the probability that exactly two will be allergic to the drug;
   (ii) the probability that more than two will be allergic to the drug.

   Each sufferer who is allergic has a probability of 0.07 of developing a chronic condition following the administration of the drug. Given that two sufferers develop the allergy from the administration of the drug to the sample of 3000, show that the probability that exactly one develops the chronic condition is 0.1302. Determine also the probability that both sufferers develop the chronic condition. Hence, or otherwise, derive the mean of the random variable which describes how many of these two sufferers develop the chronic condition.

   (JMB)

6. An airline is accepting reservations for the seats on a particular flight of a 100-seater aircraft. It is known from past experience that 5% of the people who reserve seats do not in fact turn up, and so the airline has a policy of allowing 103 people to book seats on the flight. Let P denote the probability that every person turning up for the flight gets a seat.

►

(i) Write down the probability that *more* than 100 people turn up, and hence find an *exact* expression for P.
(ii) Use a suitable Poisson approximation to find a value for P.

(SMP)

Section 7.3

7. A vaccine is to be tested on groups of n individuals. If the probability of success for the vaccine is p, calculate the probability that the vaccine will be successful on:
   (a) 8 or more individuals if $n = 10$, $p = 0.8$.
   (b) more than 415 individuals if $n = 500$, $p = 0.8$.
   (c) fewer than 10 individuals if $n = 100$, $p = 0.1$

   Approximate probabilities only are required for (b) and (c).

8. A geneticist believes that the proportion of males in the population who have a certain blood disorder is 0.3. In a random sample of 5000, what is the approximate probability that more than 1600 will be afflicted by the disorder?

9. A haulage firm has 60 lorries. Because of the need for regular service the probability that a randomly chosen lorry is available for business on any given day is 0.9. Find the probability that 50 or more lorries are available for business on a given day.

10. A new method of treating a disease in children is such that 70% of children treated by the method are able to lead full normal lives. Show that, for a random sample of 10 children chosen from those with the disease, the probability that the method will be successful in at least 8 cases is 0.3828.

    Find the approximate probability that at least 10 but fewer than 19 of a random sample of 20 children will be able to lead full normal lives.

11. It is known that 20% of plants produced by a certain species of corn seed will be infertile. In a random sample of 100 such plants, what is the approximate probability that more than 25 will be infertile?

12. In a certain town 40% of households have a freezer. If 600 households are chosen randomly, what is the approximate probability that at least 50% of them will have a freezer?

13. In a series of 27 Bernoulli trials (see Worksheet 6, Exercise 1) the constant probability of success is 1/3. Calculate the probability of exactly 9 successes using (a) an exact method, (b) an approximate method. Compare your answers. Which is correct?

14. For a marksman firing shots at a target, the distance X cm by which a shot deviates from the centre of the target is a random variable having probability density function

$$f(x) = 2xe^{-x^2}, \qquad x > 0.$$

   (i) Verify that $f(x)$ is a valid probability density function and show that the mean of X is $\sqrt{\pi}/2$.
   (ii) Find the probability that a shot will deviate from the centre of the target by less than the mean.

▶

(iii) If the marksman fires 50 independent shots at the target, find an approximate value for the probability that 25 or more will deviate from the centre by less than the mean.

$$\left[\text{You may use the fact that } \int_0^\infty e^{-x^2}\,dx = \frac{\sqrt{\pi}}{2}\right]$$

(IOS)

15. On the surface of halfpenny postage stamps are either one or two phosphor bands. Ninety per cent of halfpenny stamps have two bands and the rest have one band. Of those having one band, 95 per cent have the band in the centre of the stamp and the remainder have the band on the left-hand edge of the stamp.
    (i) Determine the probability that in a random sample of ten halfpenny stamps there are exactly eight having two phosphor bands.
    (ii) Determine, using a normal approximation, the probability that in a random sample of 100 halfpenny stamps there are between five and fifteen stamps (inclusive) having one phosphor band.
    (iii) Determine, using a Poisson approximation, the probability that in a random sample of 100 halfpenny stamps there are less than three stamps which have only a single band, this band being on the left-hand edge of the stamp.

    (Any expressions evaluated should be clearly exhibited, and answers should be given correct to three significant figures).

    (Cambridge)

16. (a) State, with reasons, which of the distributions you have studied would provide the best model for each of the following random variables. Whenever your answer is a Poisson distribution or a binomial distribution you should estimate the parameters from the data given.
        (i) The number of sixes thrown in 5 minutes playing time of a game which involves four players throwing a single die in turn for a large number of turns.
        (ii) The lifetime, in hours, of 100-watt electric light-bulbs of a particular make.
        (iii) The number of telephone calls received at an exchange in intervals of 6 seconds, when the average number of calls received per minute is 48.
    (b) The random variable X is distributed $B(60, 1/5)$. Using a suitable approximation and giving reasons for your choice, calculate the probability that $9 \leqslant X \leqslant 16$.
    (c) The random variable Y is distributed $B(1200, 1/200)$. Obtain the probability that a random observation of Y will give a value greater than 7.

    (JMB)

17. In breeding a particular type of sweet pea there is a probability p that any offspring will be pink. Let R be the number of pink sweet peas in a random sample of 20 offspring. Write down an expression for $P(R \leqslant 4)$. One theory gives the value of p as 1/4. Using this value of p find, from tables, the value of $P(R \leqslant 4)$.

    The sample size is increased to 200 offspring. Taking $p = 1/4$, obtain, specifying the approximating distribution that you are using, the probability that the sample contains 40 or fewer offspring.

    Find, without the use of a continuity correction, the smallest size of the sample for which the probability of 20% or less of the plants being pink is less than 0.01. ►

Another theory suggests that the true probability is 1/6, not 1/4. In a random sample of 20, exactly 4 were pink. By comparing $P(R = 4$ when $p = 1/4)$ with $P(R = 4$ when $p = 1/6)$, state for which of these two theories this result gives the greater support.

(JMB)

18. A continuous random variable X has probability density function given by

$$f(x) = k(1-x), \qquad 0 \leqslant x \leqslant 1,$$
$$f(x) = 0, \qquad \text{elsewhere,}$$

where k is a constant.

(i) Find the value of k.
(ii) Determine the mean of X.
(iii) Obtain the distribution function of X.

The length X of a component produced by a machine is a random variable having the above probability density function. Show that the probability that the length of a randomly chosen component will be less than the mean is 5/9.

If 180 components are produced by independent operations of the machine, calculate, by using the normal approximation to the binomial, the probability that at most 105 of the components will have length less than the mean.

(JMB)

19. State the conditions under which the Poisson distribution may be used as an approximation to the binomial distribution. Give an example, from your project work if possible, of the use of this approximation.

Independently for each call into the telephone exchange of a large organization, there is a probability of 0.002 that the call will be connected to a wrong extension. Find, to 3 significant figures, the probability that, on a given day, exactly one of the first five incoming calls will be wrongly connected.

Use a Poisson approximation to find, to 3 decimal places, the probability that, on a day when there are 1000 incoming calls, at least three of them are wrongly connected during that day.

Use a suitable approximate distribution to determine, to 3 decimal places, the probability that, of 10,000 incoming calls to this exchange, at most 25 are wrongly connected.

(London)

20. A fair six-sided die is thrown n times. Give the distribution of the number of times a six appears uppermost, and a distribution which can be used to approximate this distribution for large n. How large should n be in order to be 99% certain that a six is uppermost at least 50 times?

(OLE)

21. A test consists of multiple-choice questions. In each question there are five choices, one of which is correct. A candidate has probability p of knowing the correct answer to any question. If he does not know the answer, he makes a random guess from among the five choices given.

(i) For a question selected at random, find the probability of the candidate selecting the correct answer.

▶

(ii) If he selects the correct answer, find the conditional probability that he has guessed.
(iii) Find his expected score if there are 40 questions in the test and one mark is given for a correct answer, zero for an incorrect answer.
(iv) Use a normal approximation to estimate the probability of a candidate with $p = \frac{1}{2}$ scoring more than 27 marks in the test.
(v) In order to make the expected score of an entirely ignorant candidate zero, the examiner decides to alter the marking scheme to give n marks for a correct solution and -1 mark for an incorrect solution. Establish what value he should choose for n.

(SMP)

22. Two fair dice are thrown 72 times. Calculate to 3 decimal places the probability p that the total score is 12 on two or fewer of the 72 throws. Also obtain two approximations to p using:

(a) the normal approximation to the binomial, and
(b) the Poisson approximation to the binomial.

Comment on the different accuracies of the two approximations.

(MEI)

23. At a particular stage of the turkey breeding season, one-half of the turkey chicks hatched are female. A breeder wishes to know the number n of eggs which should be hatched to obtain, with a probability of at least 99.9%, at least 20 females. Use the normal approximation to the binomial to show that

$$n - 39 > 3.09n^{\frac{1}{2}}.$$

By squaring both sides of this inequality, and solving a quadratic equation, obtain the required value of n.

Obtain from the other root of the quadratic equation the value of n for which, with a probability of at least 99.9%, no more than 20 females will be hatched.

(MEI)

24. A multiple-choice examination consists of 20 questions, for each of which the candidate is required to tick as correct one of three possible answers. Exactly one answer to each question is correct. A correct answer gets 1 mark and a wrong answer gets 0 marks. Consider a candidate who has complete ignorance about every question and therefore ticks at random. What is the probability that he gets a particular answer correct? Calculate the mean and variance of the number of questions he answers correctly.

The examiners wish to ensure that not more than 1% of completely ignorant candidates pass the examination. Use the normal approximation to the binomial, working throughout to 3 decimal places, to establish the pass mark that meets this requirement.

(MEI)

25. The probability that a clover has four leaves is 0.002.

(i) Calculate, to three significant figures, the probability that a random sample of five clovers will include exactly one having four leaves. ►

(ii) Find the smallest number of clovers that should be sampled if there is to be a probability of at least 0.9 that the sample will include one or more having four leaves.
(iii) Use one of the tables provided to find an approximate value for the probability that a random sample of 1000 clovers will include at least two having four leaves.
(iv) Each of 50 persons takes a random sample of 1000 clovers. Find an approximate value for the probability that exactly 25 of the persons will each have at least two clovers having four leaves.

(WJEC)

Section 7.4

26. A shopkeeper's sales of an item amount to 40 per month on average. He replenishes his stock once a month. To what number should he make up his stock to reduce the chance of running out of stock to less than 1 in 100? Assume that sales occur randomly.

27. During a certain period a telephone switchboard receives calls at random instants of time at a mean rate of 3 per minute.
    (a) State the distribution which you consider to be appropriate for the number of calls arriving at the switchboard during a given time interval, and assuming this distribution to hold, find the probability of three or more calls arriving during an interval of 30 seconds.
    (b) Use a suitable approximation to estimate:
        (i) the probability of 100 or more calls arriving during an interval of 30 minutes;
        (ii) the value k such that there is probability 0.01 that the number of calls arriving in an interval of 30 minutes exceeds k.

(IOS)

28. Define the Poisson distribution and derive its mean and variance.

    The number of particles emitted from a radioactive source in t seconds has a Poisson distribution with mean $t/20$. Find, correct to two decimal places, the probabilities that in a period of one minute

    (i) no particle,
    (ii) at least three particles

    will be emitted.

    Use the normal approximation to find the probability, correct to two decimal places, that at least 200 particles will be emitted in one hour.

(JMB)

29. During part of the day, cars arrive from a certain direction at a set of traffic lights at random times at an average rate of two per minute. The traffic lights operate on a cycle during which, for traffic from this direction, they are red for a period of two minutes. Find the probability that at least four cars arrive at the lights from this direction during such a period.

    Due to malfunction the traffic lights become stuck on red for 20 minutes. Estimate the probability that 50 or more cars from this direction are held at the lights.

(O & C) ►

30. (a) The random variable X has a Poisson distribution and is such that $P(X = 2) = 3P(X = 4)$. Find, correct to three decimal places, the values of (i) $P(X = 0)$; (ii) $P(X \leqslant 4)$.

(b) The number of characters that are mistyped by a copytypist in any assignment has a Poisson distribution, the average number of mistyped characters per page being 0.8. In an assignment of 80 pages calculate, to three decimal places,

(i) the probability that the first page will contain exactly two mistyped characters;
(ii) the probability that the first mistyped character will appear on the third page;
(iii) an approximate value for the probability that the total number of mistyped characters in the 80 pages will be at most 50.

(WJEC)

8

Linear functions of random variables, and joint distributions

▼

8.1 INTRODUCTION

In Section 5.10 we discussed the mean and variance of a linear function of a random variable, i.e. the mean and variance of $aX + b$. Here X denoted the random variable and a and b were constants. In this chapter we will consider first the mean and variance of $aX + bY$, a linear function of two random variables, X and Y. At the same time the concept of *independent* random variables will be introduced.

Next we will consider the distribution of the sum and difference of two independent and normally distributed random variables. Then the distribution of the sum of independent Poisson variables and the sum of identically distributed geometric variables will be discussed.

Finally there is a brief section on the joint distribution of two discrete random variables, which includes the concepts of marginal and conditional distributions.

8.2 THE MEAN AND VARIANCE OF $aX + bY$

Let $aX + bY$ be a linear function of two random variables X and Y, where a and b are constants. For example, if X denotes the weight of an egg and Y denotes the weight of an egg-box capable of holding six eggs, then $6X + Y$ denotes the total weight of an egg box containing six eggs.

What are the mean and variance of $aX+bY$?
The mean of $aX+bY$ is $E(aX+bY)$ where

$$E(aX+bY) = E(aX)+E(bY) \qquad \text{see note below}$$
$$\therefore E(aX+bY) = aE(X)+bE(Y) \tag{8.1}$$

using (A5.2).

Note
This is an application of a useful result which is quoted without proof, namely

The expectation of a sum equals the sum of the expectations.

To find the variance of $aX+bY$, we will let $W = aX+bY$. Then the variance of $aX+bY$ is also the variance of W, where

$$\begin{aligned}
\text{Var}(W) &= E[(W-E(W))^2] \qquad \text{using the definition of variance given in Section 5.3} \\
&= E[\{aX+bY-aE(X)-bE(Y)\}^2] \qquad \text{from (8.1)} \\
&= E[\{a(X-E(X))+b(Y-E(Y))\}^2] \\
&= E[a^2(X-E(X))^2+2ab(X-E(X))(Y-E(Y))+b^2(Y-E(Y))^2] \\
&= E[a^2(X-E(X))^2]+E[2ab(X-E(X))(Y-E(Y))] \\
&\quad +E[b^2(Y-E(Y))^2]
\end{aligned}$$

$$\begin{aligned}
\therefore \text{Var}(W) = \text{Var}(aX+bY) &= a^2\,\text{Var}(X)+b^2\,\text{Var}(Y) \\
&\quad +2abE[(X-E(X))(Y-E(Y))]
\end{aligned}$$

We now define the *covariance* of X and Y, which we denote by Cov (X, Y), as follows:

$$\text{Cov}(X, Y) = E[(X-E(X))(Y-E(Y))]$$

The expression above for Var (W) then becomes:

$$\text{Var}(aX+bY) = a^2\,\text{Var}(X)+b^2\,\text{Var}(Y)+2ab\,\text{Cov}(X, Y) \tag{8.2}$$

Cov (X, Y) can be expressed in a simpler form:

$$\begin{aligned}
\text{Cov}(X, Y) &= E[(X-E(X))(Y-E(Y))] \\
&= E[XY-E(Y)X-E(X)Y+E(X)E(Y)] \\
&= E(XY)+E(-E(Y)X)+E(-E(X)Y)+E(E(X)E(Y)) \\
&= E(XY)-E(Y)E(X)-E(X)E(Y)+E(X)E(Y),
\end{aligned}$$

using (A5.2) since $E(X)$ and $E(Y)$ do not vary as X and Y vary and hence may be treated as constants

$$\therefore \text{Cov}(X, Y) = E(XY)-E(X)E(Y) \tag{8.3}$$

When X and Y are *independent* (which can be taken to mean that $P(X \leqslant x$ and $Y \leqslant y) = P(X \leqslant x)P(Y \leqslant y))$ they have a covariance of zero.† It follows from (8.2) and (8.3) that if two variables X and Y are independent,

$$\left.\begin{aligned} \operatorname{Var}(aX + bY) &= a^2 \operatorname{Var}(X) + b^2 \operatorname{Var}(Y) \\ \text{and} \quad E(XY) &= E(X)E(Y) \end{aligned}\right\} \tag{8.4}$$

Notice that (8.1), $E(aX + bY) = aE(X) + bE(Y)$, applies whether or not X and Y are independent. Also notice that (8.1) and (8.4) could easily be extended to the case of more than two independent variables.

Example 8.1

In a game with dice, the usual six-sided die and an eight-sided die (with faces numbered 1 to 8) are rolled. The number of points scored is obtained by doubling the number uppermost on the six-sided die and trebling the number uppermost on the eight-sided die. What are the mean and variance of the number of points scored?

Let X and Y denote the numbers uppermost on the six- and eight-sided dice respectively. We require $E(2X + 3Y)$ and $\operatorname{Var}(2X + 3Y)$. From (8.1) and (8.4), since it is reasonable to assume independence here,

$$E(2X + 3Y) = 2E(X) + 3E(Y)$$

$$\operatorname{Var}(2X + 3Y) = 2^2 \operatorname{Var}(X) + 3^2 \operatorname{Var}(Y)$$

Using the results of Section 5.3, we can easily show that

$$E(X) = \frac{7}{2}, \qquad \operatorname{Var}(X) = \frac{35}{12}, \qquad E(Y) = \frac{9}{2}, \qquad \operatorname{Var}(Y) = \frac{21}{4}$$

$$\therefore E(2X + 3Y) = 2 \times \frac{7}{2} + 3 \times \frac{9}{2} = 20.5$$

$$\operatorname{Var}(2X + 3Y) = 4 \times \frac{35}{12} + 9 \times \frac{21}{4} = 58.9$$

□

Example 8.2

In another dice game, two six-sided dice and three eight-sided dice are thrown at the same time. What is the mean and variance of the total of the five numbers shown uppermost?

† The converse is *not* necessarily true. Two variables may have zero covariance and yet be dependent.

If X_1 denotes the number shown uppermost on the first six-sided die and so on, and W denotes the total of the five numbers, then:

$$W = X_1 + X_2 + Y_1 + Y_2 + Y_3$$

and we require $E(W)$ and $\mathrm{Var}(W)$

$$\begin{aligned} E(W) &= E(X_1) + E(X_2) + E(Y_1) + E(Y_2) + E(Y_3), \\ &\quad \text{using (8.1) for the five variables} \\ &= \frac{7}{2} + \frac{7}{2} + \frac{9}{2} + \frac{9}{2} + \frac{9}{2} \\ &= 20.5 \end{aligned}$$

Also, assuming independence and using (8.4) for the five variables,

$$\begin{aligned} \mathrm{Var}(W) &= \mathrm{Var}(X_1) + \mathrm{Var}(X_2) + \mathrm{Var}(Y_1) + \mathrm{Var}(Y_2) + \mathrm{Var}(Y_3) \\ &= \frac{35}{12} + \frac{35}{12} + \frac{21}{4} + \frac{21}{4} + \frac{21}{4} \\ &= 21.6 \end{aligned}$$

You may like to consider why the mean score is the same for the two examples just quoted, but the variance is much lower in the second example. □

8.3 THE DISTRIBUTION OF A LINEAR FUNCTION OF INDEPENDENT NORMALLY DISTRIBUTED VARIABLES

In the previous section no mention was made of the type of distribution exhibited by the random variables X and Y. If, however, we know that $X \sim N(\mu_X, \sigma_X^2)$ and $Y \sim N(\mu_Y, \sigma_Y^2)$, and we know also that X and Y are independent, then $aX + bY$ is also normally distributed. It follows from the previous section that the mean and variance of this normal distribution are $(a\mu_X + b\mu_Y)$ and $(a^2\sigma_X^2 + b^2\sigma_Y^2)$ respectively. This result can be extended to more than two independent normally distributed variables.

A proof that $X + Y$ has a normal distribution with mean $(\mu_X + \mu_Y)$ and variance $(\sigma_X^2 + \sigma_Y^2)$ now follows.

Proof Let $M_X(t)$, $M_Y(t)$, $M(t)$ be the m.g.f.s of X, Y and $X + Y$ respectively.

Then

$$\begin{aligned} M(t) &= E(e^{t(X+Y)}) \qquad \text{see Section 5.9} \\ &= E(e^{tX}e^{tY}) \\ &= E(e^{tX})E(e^{tY}) \end{aligned}$$

since if X and Y are independent, so are e^{tX} and e^{tY}, so (8.4) applies

$$= M_X(t)M_Y(t)$$

$$= \exp\left\{\mu_X t + \frac{\sigma_X^2 t^2}{2}\right\}\exp\left\{\mu_Y t + \frac{\sigma_Y^2 t^2}{2}\right\} \qquad \text{from (6.11)}$$

$$= \exp\left\{(\mu_X + \mu_Y)t + \frac{(\sigma_X^2 + \sigma_Y^2)t^2}{2}\right\}$$

The right-hand side is the m.g.f. of a normally distributed variable with mean $(\mu_X + \mu_Y)$ and variance $(\sigma_X^2 + \sigma_Y^2)$; see (6.11).

Example 8.3

Suppose X denotes the weight (g) of the contents of a can of beans, and Y denotes the weight (g) of the empty can. Assuming X and Y are independent, and that

$$X \sim N(250, 3) \qquad \text{and} \qquad Y \sim N(20, 0.6)$$

(a) What is the distribution of the total weight of can and contents?
(b) What is the probability that the total weight of can and contents exceeds 272 g?
(c) What is the probability that the mean weight of 10 cans and their contents exceeds 272 g?

(a) Let $W = X + Y$ denote the total weight of can and contents,

then $$W \sim N(250 + 20, 3 + 0.6)$$

i.e. $$W \sim N(270, 3.6)$$

(b) $$P(W > 272) = P\left(Z > \frac{272 - 270}{\sqrt{3.6}}\right)$$

$$= P(Z > 1.05)$$

$$= 1 - 0.8531 \qquad \text{from Table C.3(a)}$$

$$= 0.1469$$

(c) For 10 cans and their contents, let T denote the total weight, so that

$$T = W_1 + W_2 + \ldots + W_{10}$$

$$E(T) = E(W_1) + E(W_2) + \ldots + E(W_{10})$$

$$= 270 + 270 + \ldots + 270$$

$$= 2700 \text{ g}$$

$$\text{Var}(T) = \text{Var}(W_1) + \text{Var}(W_2) + \ldots + \text{Var}(W_{10})$$

$$= 3.6 + 3.6 + \ldots + 3.6$$

$$= 36 \text{ g}^2$$

T will also be normally distributed, being the sum of 10 independent normally distributed variables.

$$\begin{aligned}
&P\ (\text{Mean weight of 10 cans and contents exceeds 272 g})\\
&= P\ (\text{Total weight of 10 cans and contents exceeds 2720 g})\\
&= P(T > 2720)\\
&= P\left(Z > \frac{2720 - 2700}{\sqrt{36}}\right)\\
&= P(Z > 3.33)\\
&= 1 - 0.9996 \qquad \text{from Table C.3(a)}\\
&= 0.0004
\end{aligned}$$

□

One result which may come as a surprise to the reader is that, if $X \sim N(\mu_X, \sigma_X^2)$ and $Y \sim N(\mu_Y, \sigma_Y^2)$, and X and Y are independent, then $X - Y$ has a normal distribution with mean $(\mu_X - \mu_Y)$ and variance $(\sigma_X^2 + \sigma_Y^2)$. The surprise is that the variances of $X + Y$ and $X - Y$ are equal, but this follows from (8.4) by putting $a = b = 1$ and $a = 1$, $b = -1$ in turn.

Example 8.4

If the distribution of the weight (kg) of adult male humans is $N(65, 9)$ and the distribution for females is $N(60, 7)$, what is the probability that a randomly selected female will weigh more than a randomly selected male?

If X and Y denote male and female weights, we require $P(U < 0)$, where $U = X - Y$.

Since $U \sim N(65 - 60, 9 + 7)$

i.e. $U \sim N(5, 16)$,

$$\begin{aligned}
P(U < 0) &= P\left(Z < \frac{0-5}{\sqrt{16}}\right)\\
&= P(Z < -1.25)\\
&= 1 - P(Z < 1.25) \qquad \text{by symmetry}\\
&= 1 - 0.8944 \qquad \text{from Table C.3(a)}\\
&= 0.1056
\end{aligned}$$

□

8.4 THE DISTRIBUTION OF THE SUM OF INDEPENDENT POISSON VARIABLES

The sum of two independent Poisson variables is a Poisson variable with a parameter equal to the sum of the two parameters. By this we mean that if X_1 has a Poisson distribution with parameter a_1, and X_2 has a Poisson distribution with parameter a_2, and if X_1 and X_2 are independent, then $X_1 + X_2$ has a Poisson distribution with parameter $(a_1 + a_2)$.

Proof Let $G_1(t)$, $G_2(t)$, $G(t)$ be the p.g.f.s of X_1, X_2 and $X_1 + X_2$.

Then
$$\begin{aligned} G(t) &= E(t^{X_1 + X_2}) && \text{see Section 5.4} \\ &= E(t^{X_1} t^{X_2}) \\ &= E(t^{X_1})E(t^{X_2}) \end{aligned}$$

since if X_1 and X_2 are independent, so are t^{X_1} and t^{X_2}, so (8.4) applies

$$\begin{aligned} &= G_1(t)G_2(t) && \text{see Section 5.4} \\ &= e^{a_1(t-1)}e^{a_2(t-1)} && \text{from (6.5)} \\ &= e^{(a_1 + a_2)(t-1)} \end{aligned}$$

The right-hand side is the p.g.f. of a Poisson variable with parameter $(a_1 + a_2)$; see (6.5).

Example 8.5

Accidents in a factory are known to occur randomly, but the rate at which they occur varies depending on the day of the week. The rates of occurrence are as follows:

Day	Mon.	Tues.	Wed.	Thurs.	Fri.
Mean number of accidents per day	0.6	0.4	0.3	0.3	1.0

What is the probability that:

(a) No accident occur on a given Monday?
(b) Exactly one accident occurs in the first three days of a given week?
(c) There are at least two accidents in a given week?

Since we are dealing with random events in time, the Poisson distribution applies.

(a) The number of accidents per Monday has a Poi (0.6) distribution:

$$P(\text{No accidents}) = e^{-0.6} = 0.5488$$

(b) Here we are dealing with a Poisson distribution with parameter $0.6 + 0.4 + 0.3 = 1.3$

$$P(\text{One accident}) = \frac{e^{-1.3}1.3}{1!} = 0.3543$$

(c) Here we are dealing with a Poisson distribution with parameter $0.6+0.4+0.3+0.3+1.0=2.6$

$$
\begin{aligned}
&= P(\text{at least 2 accidents})\\
&= 1-P(\text{no accidents})-P(\text{one accident})\\
&= 1-e^{-2.6}-\frac{e^{-2.6}2.6}{1!}\\
&= 0.7326
\end{aligned}
$$

□

8.5 THE DISTRIBUTION OF THE SUM OF INDEPENDENT AND IDENTICALLY DISTRIBUTED GEOMETRIC VARIABLES

For the geometric distribution (Section 6.4), the discrete random variable is the number of trials up to and including the first success, with a constant probability, p, of a success in each trial. Suppose we are interested in the discrete random variable, Y_k, where Y_k is the number of trials up to and including the kth success. Then $Y_k = X_1 + X_2 + \ldots + X_k$ where X_1 is the number of trials up to and including the first success, X_2 is the number of trials following the first success up to and including the second success, and so on. $X_1, X_2, \ldots, X_k$ are independent and identically distributed geometric variables and we are interested in the distribution of their sum, Y_k.

Suppose $G_1(t), G_2(t), \ldots, G_k(t)$ are the p.g.f.s of $X_1, X_2, \ldots, X_k$ and let $G(t)$ be the p.g.f. of Y_k. Then, using the same reasoning as in Section 8.4, we obtain:

$$
\begin{aligned}
G(t) &= G_1(t)G_2(t)\ldots G(t)\\
&= \left[\frac{pt}{1-t(1-p)}\right]^k
\end{aligned}
$$

from (6.7), and since the X's are identically distributed.

This is the p.g.f. of what is called the *negative binomial distribution*.

We can find the mean and variance of this distribution using (5.6) and (5.7). We can also find the probability function of this distribution by expanding the p.g.f. as a series in t (see Section 5.4).

Mean and variance

$$G(t) = p^k t^k[1-qt]^{-k} \qquad \text{from above, where } q = 1-p$$

$$G'(t) = kp^k t^{k-1}[1-qt]^{-k} + kqp^k t^k[1-qt]^{-k-1}$$

$$
\begin{aligned}
\therefore \mu = G'(1) &= kp^k p^{-k} + kqp^k p^{-k-1}\\
&= k + \frac{kq}{p}
\end{aligned}
$$

$$\therefore \mu = \frac{k}{p}$$

Differentiating $G'(t)$ and putting $t = 1$ gives

$$G''(1) = k(k-1) + 2k^2\frac{q}{p} + k(k+1)\frac{q^2}{p^2}$$

$$\therefore \sigma^2 = G''(1) + \mu - \mu^2 \qquad \text{from (5.7)}$$

$$\sigma^2 = \frac{k(1-p)}{p^2}$$

The values for μ and σ^2 could have been obtained from (8.1) and (8.4) and the fact that the mean and variance of the geometric distribution are $1/p$ and $(1-p)/p^2$ (see Section 6.4).

Probability function, $p(x)$

$$G(t) = p^k t^k (1-qt)^{-k}$$
$$= p^k t^k \left[1 + kqt + \frac{k(k+1)^2 q^2 t^2}{2!} + \ldots\right]$$

The probability function, $p(x)$, is the coefficient of t^x in this series, and so

$$p(x) = p^k \text{ (coefficient of } t^{x-k} \text{ in expansion of } (1-qt)^{-k})$$
$$= p^k \frac{k(k+1)\ldots(k+x-k-1)q^{x-k}}{(x-k)!}$$
$$= \frac{k(k+1)\ldots(x-1)}{(x-k)!} p^k q^{x-k}$$
$$= \frac{(x-1)!}{(k-1)!\,(x-k)!} p^k q^{x-k}$$
$$\therefore p(x) = \binom{x-1}{k-1} p^k (1-p)^{x-k}$$

This holds for $x = k, k+1, \ldots,$ since we need at least k trials in order to have k successes.

Example 8.6

A die is tossed until the fourth 3 is thrown. Find:
(a) the mean and variance of the number of throws required;
(b) the probability that exactly 24 throws will be required.

(a) $\quad p = \frac{1}{6}, \quad k = 4, \quad \text{mean} = \frac{k}{p} = 24, \quad \text{variance} = \frac{k(1-p)}{p^2} = 120$

(b) $$p(24) = \binom{23}{3}\left(\frac{1}{6}\right)^4\left(\frac{5}{6}\right)^{24-4} = 0.0356$$ □

8.6 JOINT, CONDITIONAL AND MARGINAL DISTRIBUTIONS

Suppose that a discrete random variable X can take values $x_1, x_2, \ldots, x_n$ with probabilities $p_1, p_2, \ldots, p_n$, and that another discrete random variable Y can take values $y_1, y_2, \ldots, y_m$ with probabilities $q_1, q_2, \ldots, q_m$. Suppose further that the *joint* distribution of X and Y is specified as follows:

$$P(X = x_i \text{ and } Y = y_j) = r_{ij}, \qquad i = 1, 2, \ldots, n; \quad j = 1, 2, \ldots, m$$

Notice that we have *not* assumed that X and Y are independent, so that r_{ij} is *not* in general equal to $p_i q_j$

We can set out these probabilities in table form as in Table 8.1.

Table 8.1 The joint distribution of X and Y

	Values of Y				
Values of X	y_1	y_2	...	y_m	*Totals*
x_1	r_{11}	r_{12}	...	r_{1m}	p_1
x_2	r_{21}	r_{22}	...	r_{2m}	p_2
⋮	⋮	⋮		⋮	⋮
x_n	r_{n1}	r_{n2}	...	r_{nm}	p_n
Totals	q_1	q_2	...	q_m	1

Notice how the probabilities in each row may be added to form what is called the *marginal* distribution of X. Similarly, the *marginal* distribution of Y is formed by adding the probabilities in each column. Naturally, if we add all the r's, or p's, or q's together, we get a grand total of 1.

Suppose we are interested in the probability that X will take a particular value given that Y must take some value, for example $P(X = x_1 | Y = y_2)$. Now

$$P(X = x_1 | Y = y_2) = \frac{P(X = x_1 \text{ and } Y = y_2)}{P(Y = y_2)} \qquad \text{from (4.1),}$$

$$= \frac{r_{12}}{q_2} \qquad \text{from Table 8.1}$$

This is, of course, a *conditional* probability.

The *conditional* distribution of X given that $Y = y_2$ is formed similarly, and is shown in Table 8.2.

Table 8.2 The probability distribution of X, given $Y = y_2$

Value of X	x_1	x_2	...	x_n
Conditional probability	$\frac{r_{12}}{q_2}$	$\frac{r_{22}}{q_2}$	...	$\frac{r_{n2}}{q_2}$

Example 8.7

A bag contains 3 red balls, 2 blue balls and 1 white ball. Two balls are drawn *with* replacement. Let X denote the number of red balls drawn and let Y denote the number of blue balls drawn.

(a) Determine the joint distribution of X and Y.
(b) Determine the marginal distributions of X and Y and find their mean and variance.
(c) Determine the conditional distribution of X, given that $Y = 1$, and find its mean and variance.
(d) Show that X and Y are *not* independent.
(e) Determine the probability distribution of XY and hence determine the covariance of X and Y.

Table 8.3 The joint distribution of the number of red and blue balls

	Number of blue (Y)			
Number of red (X)	0	1	2	*Totals*
0	$\frac{1}{36}$	$\frac{4}{36}$	$\frac{4}{36}$	$\frac{9}{36}$
1	$\frac{6}{36}$	$\frac{12}{36}$	0	$\frac{18}{36}$
2	$\frac{9}{36}$	0	0	$\frac{9}{36}$
Totals	$\frac{16}{36}$	$\frac{16}{36}$	$\frac{4}{36}$	1

(a) The nine joint probabilities in Table 8.3 may be obtained using the ideas of Chapter 4 (for example by drawing a tree diagram).
(b) The marginal distribution of X may be read directly from Table 8.3.

Value of X	0	1	2
Probability	$\frac{9}{36}$	$\frac{18}{36}$	$\frac{9}{36}$

In fact $X \sim B(2, \frac{1}{2})$, so $E(X) = 1$ and $\text{Var}(X) = \frac{1}{2}$
The marginal distribution of Y, also directly from Table 8.3, is:

Value of Y	0	1	2
Probability	$\frac{16}{36}$	$\frac{16}{36}$	$\frac{4}{36}$

In fact $Y \sim B(2, \frac{1}{3})$, so $E(Y) = \frac{2}{3}$ and $\mathrm{Var}(Y) = \frac{4}{9}$

(c) The conditional distribution of X, given $Y = 1$, is formed in the same way as Table 8.2:

Value of X	0	1
Conditional probability	$\frac{1}{4}$	$\frac{3}{4}$

$$E(X \mid Y = 1) = 0 \times \frac{1}{4} + 1 \times \frac{3}{4} = \frac{3}{4}$$

$$E(X^2 \mid Y = 1) = 0^2 \times \frac{1}{4} + 1^2 \times \frac{3}{4} = \frac{3}{4}$$

$$\mathrm{Var}(X \mid Y = 1) = \frac{3}{4} - \left(\frac{3}{4}\right)^2 = \frac{3}{16}$$

The mean and variance of X, given $Y = 1$, are $\frac{3}{4}$ and $\frac{3}{16}$ respectively.

(d) X and Y are independent if

$$P(X = x_i \text{ and } Y = y_j) = P(X = x_i)P(Y = y_j),$$

for all i and j

For example, $P(X = 0 \text{ and } Y = 0) = \dfrac{1}{36}$ from Table 8.3

But $P(X = 0)P(Y = 0) = \dfrac{9}{36} \times \dfrac{16}{36} = \dfrac{1}{9}$ from (b) above.

Since $\frac{1}{36} \neq \frac{1}{9}$, X and Y are not independent.

(e) XY can only take values 0 and 1 as shown below:

$$P(XY = 0) = P(X = 0 \text{ or } Y = 0 \text{ or both})$$

$$= \frac{1 + 4 + 4 + 6 + 9}{36} \quad \text{from Table 8.3}$$

$$= \frac{2}{3}$$

$$P(XY = 1) = P(X = 1 \text{ and } Y = 1)$$

$$= \frac{1}{3} \quad \text{from Table 8.3}$$

$$P(XY = 2) = 0 \quad \text{and} \quad P(XY = 4) = 0$$

The probability distribution of XY is:

Value of XY	0	1
Probability	$\frac{2}{3}$	$\frac{1}{3}$

From this we calculate that $E(XY) = \frac{1}{3}$ and

$$\begin{aligned} \text{Cov}(X, Y) &= E(XY) - E(X)E(Y) && \text{from (8.3)} \\ &= \frac{1}{3} - 1 \times \frac{2}{3} && \text{from (b) above} \\ &= -\frac{1}{3} \end{aligned}$$

□

8.7 SUMMARY

1. If X and Y are random variables, and a and b are constants,

$$E(aX + bY) = aE(X) + bE(Y)$$

$$\text{Var}(aX + bY) = a^2 \text{Var}(X) + b^2 \text{Var}(Y) + 2ab\,\text{Cov}(X, Y)$$

where $$\text{Cov}(X, Y) = E(XY) - E(X)E(Y)$$

2. If, in addition, X and Y are independent, then $\text{Cov}(X, Y) = 0$

and $$\text{Var}(aX + bY) = a^2 \text{Var}(X) + b^2 \text{Var}(Y)$$

3. If X and Y are independent and are also normally distributed then $aX + bY$ is also normally distributed.
4. The sum of a number of independent Poisson variables is also a Poisson variable.
5. The sum of a number of identically distributed geometric variables has a negative binomial distribution.
6. Joint, marginal and conditional distributions were discussed briefly for the discrete variable case only.

Section 8.2

1. A hockey team has a constant probability of 0.7 of winning a home match and a constant probability of 0.5 of winning an away match. The team enters competitions in which it has to play 3 home and 3 away matches. If X and Y denote the number of home and away matches won by the team, respectively, calculate $E(X)$, $\text{Var}(X)$, $E(Y)$ and $\text{Var}(Y)$. What are the mean and variance of the total number of matches won by the team in such competitions? What is the probability that the team wins more home matches than away in a given competition?

2. A random variable X takes values 0, 1 and 2 with probabilities 0.4, 0.4 and 0.2 respectively. The independent random variable Y takes values 1, 2 and 3 with probabilities 0.3, 0.4 and 0.3. If $U = X + Y$ and $V = X - Y$, calculate the mean and variance of U and V. Find the mean value of UV and hence the covariance of U and V.

3. The probability of there being X unusable matches in a full box of Surelite matches is given by

$$P(X = 0) = 8k,\ P(X = 1) = 5k,\ P(X = 2) = P(X = 3) = k,\ P(X \geqslant 4) = 0.$$

Determine the constant k and the expectation and variance of X.

Two full boxes of Surelite matches are chosen at random and the total number Y of unusable matches is determined. Calculate $P(Y > 4)$, and state the values of the expectation and variance of Y.

(Cambridge)

4. The random variable X is the score shown when a fair six-sided die is thrown. Derive the mean and variance of X.

The random variables N_A, N_B, N_C are the scores shown when three fair six-sided dice A, B, C are thrown independently. Derive the mean and variance of $N_A + N_B - N_C$.

Derive also the mean and variance of $|N_A - N_B|$.

(O & C)

5. Let X denote the number of heads obtained in two tosses of a fair coin, and let Y denote a random variable which is independent of X and has a Poisson distribution with mean 1.

(i) Find, in terms of e, the probability that both X and Y will have the value zero.
(ii) Find, in terms of e, the probability that X and Y will have the same value.

Given that $Z = XY$,

(iii) express $P(Z = 0)$ in terms of e;
(iv) evaluate the mean and the variance of Z.

(WJEC)

6. The number X of radios sold per week by a certain shop has a binomial distribution with $n = 10$ and $p = 0.3$. Independently, the number Y of televisions sold per week by the same shop may be 0, 1 or 2, with:

$$P(Y = 0) = 0.3, \quad P(Y = 1) = 0.5, \quad P(Y = 2) = 0.2$$

►

The profit made on each radio sold is £2 and on each television is £20. Find the mean and the variance of the profit per week from the sale of radios and televisions.

(WJEC)

Section 8.3

7. Tests on a large batch of light-bulbs indicate that $2\frac{1}{2}\%$ last longer than 1235 hours and $2\frac{1}{2}\%$ last less than 765 hours. Assuming that the lifetime of the light bulbs is normally distributed:

   (a) what are the mean and standard deviation of this distribution?
   (b) what is the probability that the mean lifetime of four bulbs is greater than 1100 hours?
   (c) what is the probability that each of four light-bulbs will have a lifetime greater than 1100 hours?

8. Chocolates are produced by two machines, A and B. For machine A the weights of the chocolates are normally distributed with mean 10.15 g and standard deviation 0.9 g, while the weights of the chocolates from machine B are normally distributed with mean 10.05 g and standard deviation 1.0 g. Twelve chocolates from machine A and eight chocolates from machine B are packed into boxes:

   (a) What is the probability that the total weight of chocolates in a box is less than 200 g?
   (b) What is the probability that the total weight of chocolates in five boxes is less than 1 kg?

9. The random variable X has a normal distribution with parameters μ and σ^2. Derive the mean and variance of X.

   $$\left(\text{You may assume that } \frac{1}{\sqrt{2\pi}}\int_{-\infty}^{\infty} e^{-t^2/2}\,\mathrm{d}t = 1.\right)$$

   In a multi-storey office block there is a passenger-operated lift to take people to the floor of their choice. Inside the lift there is a notice saying: 'This lift can carry at most 6 passengers or 520 kilograms'. It may be assumed that the masses of the people arriving to use this lift are a random sample from a normal distribution with mean 76.2 kg and standard deviation 6.4 kg.

   (a) What is the probability that the mass of any one passenger will lie between 70 and 80 kg?
   (b) What is the probability that if there are six people in the lift their total mass exceeds 520 kg?
   (c) A new and larger lift, capable of carrying 1000 kg, is to be installed in this office block. If the total mass of the maximum number of passengers may only exceed 1000 kg with probability at most 0.01, verify that the warning notice in the lift should read: 'This lift can carry at most 12 passengers or 1000 kilograms'.

   (AEB (1983))

10. The internal diameters of circular tubes from a certain manufacturer are distributed normally with mean 30 mm. It is observed that 97.5 per cent of the tubes have internal diameters greater than 29.02 mm. Show that the proportion of tubes with internal diameters greater than 31 mm is 0.0228, correct to four decimal places. ▶

The manufacturer also produces circular rods, the diameters of which are normally distributed with standard deviation 1.50 mm. Given that 97.5 per cent of the rods have diameters lss than 29.94 mm, find the proportion that have diameters less than 28.50 mm.

If a randomly chosen tube has internal diameter X mm and a randomly chosen rod has diameter Y mm, state the probability distribution of $X - Y$. Hence find the probability that a randomly chosen rod will fit inside a randomly chosen tube.

(JMB)

11. (a) A random variable X is distributed $N(\mu_x, \sigma_x^2)$, a random variable Y is distributed $N(\mu_y, \sigma_y^2)$ and X and Y are independent. Specify the distributions of:

    (i) $X - Y$;
    (ii) $3X$;
    (iii) $X_1 + X_2 + X_3$;

    where X_1, X_2, X_3 are independent random variables each having the same distribution as X.

    (b) The distribution of breaking loads for strands of rope is normal with mean 25 units and standard deviation 2.5 units. A rope is made up of 64 independent strands and may be assumed to have a breaking load which is the sum of the breaking loads of all the strands in it.

    Find the probability that such a rope will support a weight of 1630 units. The manufacturers wish to quote a breaking load for such ropes that will be safe for 99% of the ropes. Calculate the breaking load that should be quoted.

(JMB)

12. The random variables $X_1, X_2, \ldots, X_n$ are each distributed $N(\mu, \sigma^2)$ and are independent. Specify the distributions of:

    (i) $X_1 + X_2 + \ldots + X_n$;
    (ii) $2X_1$;
    (iii) $X_1 - X_2$.

    A fence is to be erected along a straight side of a garden which at present is unfenced. The fence will consist of panels whose widths are independently and normally distributed, each with mean 2 m and standard deviation 0.01 m. The panels will be supported by posts, with one post in between each pair of panels and one at each end of the fence. The thicknesses of these posts are independently and normally distributed, each with mean 0.1 m and standard deviation 0.01 m. The dimensions of the panels and posts are independent of each other, and panels and posts fit exactly side by side with no overlap or gap.

    Obtain the distribution of the random variable which describes the length of a fence which is to be erected with four panels. Calculate the probability that this fence will have a length of between 8.47 m and 8.54 m. This fence is to be erected along a side of length 9 m. Specify the distribution of the unfenced length.

(JMB)

13. Show that the moment generating function (m.g.f.) of a random variable X having a standard normal distribution is $\exp(t^2/2)$.

    Deduce the value of $E(X^4)$, and also the m.g.f. of Y, where Y has a normal distribution with mean 0 and variance σ^2. Y_1, Y_2, Y_3 and Y_4 are independent random ▶

variables, each having the same distribution as Y, and k is a constant. By finding the m.g.f. of $Z = k(Y_1 + Y_2 + Y_3 + Y_4)$, deduce a value of k such that Z is standard normal.

(OLE)

14. A liquid product is manufactured in batches, each weighing 1 tonne. The percentage of acid in a batch is normally distributed with mean 4.0 and standard deviation 0.8.

    (a) Find the probability that a randomly chosen batch contains more than 5% of acid.
    (b) In the long run, one batch in 25 contains less than x% of acid. Find x.
    (c) Two batches are chosen at random. Find the probability that the percentages of acid they contain differ by more than $2\sqrt{2}$.
    (d) N batches chosen at random are thoroughly mixed together. Find the smallest value of N which will ensre that the probability of this mixture containing more then 5% of acid is less than 0.001.

    (O&C)

15. The thickness of some mass-produced hardboard sheets has a normal probability distribution with mean μ and standard deviation σ. It is found that 96% of these sheets will pass through a gauge of width 8 mm while only 1.7% will pass through a gauge of width 7 mm. Find μ and σ.

    A sheet is accepted if it passes through the first gauge but not through the second gauge. Find the probability that two sheets chosen at random are both accepted.

    When four sheets chosen at random are put together, the probability that their combined thickness lies between 29 mm and 31 mm is $1 - \alpha$. Find α.

    (O&C)

16. At one stage in the manufacture of an article, a cylindrical rod with a circular cross-section has to fit into a circular socket. Quality control measurements show that the distribution of rod diameters is normal with mean 5.01 cm and standard deviation 0.03 cm, while that of socket diameters is independently normal with mean 5.11 cm and standard deviation 0.04 cm. If components are selected at random for assembly, what proportion of rods will not fit?

    Rods and sockets are randomly paired for delivery to customers. Batches for delivery are made up of n such pairs. What is the largest value of n for which the probability that all the rods in the batch fit into their respective sockets is greater than 0.9? Given that $n = 30$, find the probability that not more than one rod will fail to fit into its socket.

    (MEI)

17. (a) A bag contains ten coins which appear to be identical. It is known that nine of the coins are fair but the tenth is so biased as to turn up heads with probability 2/3. One coin is selected at random and the following procedure adopted to decide if it is biased. Toss the coin four times. If three or four heads are obtained then the coin is judged to be biased; otherwise it is judged to be fair. Calculate the probability of:

    (i) obtaining fewer than three heads when the coin is biased;
    (ii) obtaining three or four heads when the coin is fair;
    (iii) making a wrong decision.

    (b) The length of a straight line is measured using a ruler. The error in any measurement has a normal distribution with mean zero and standard deviation 0.04 cm. Calculate the probability that the error is greater than 0.1 cm. If the line ▶

were measured four times and an average taken what is the probability that this value has an error greater than 0.05 cm?

(SUJB)

Section 8.4

18. Derive the moment generating function of the Poisson distribution and use it to show that the sum of two independent Poisson variables is also a Poisson variable.

    Customers enter a shop at random at the average rate of 180 persons per hour. What is the probability that during a one-minute interval fewer than three persons will enter?

    Find the time interval such that there is a 50% probability that no one will enter the shop during that interval. Give your answers to two decimal places.

    (IOS)

19. (a) The number of fish caught per hour by a particular angler has a Poisson distribution and on average he catches 1.5 fish per hour. Find the probability that he catches exactly 3 fish in a particular hour.

    The angler's son catches an average of one fish per hour. Assuming that the son's catches per hour also have a Poisson distribution and that the two of them fish at the same time, but sufficiently far apart for their results to be considered independent, find the probabilities that:

    (i) in a given hour, they catch a total of at least two fish;
    (ii) in a given period of three hours, neither father nor son catch any fish.

    (b) State the conditions under which a binomial distribution may be approximated by a Poisson distribution.

    Stating clearly any assumptions that you make, calculate the probability that, of 500 people chosen at random, three or fewer have a birthday on New Year's Day.

    (JMB)

20. A servicing engineer finds that the number of jobs he completes in a working session has a Poisson distribution with mean 4. If sessions are independent, what is the distribution of the number of jobs he completes in n sessions, and how may this be approximated when n is large?

    If he has 100 jobs to do, how many sessions should he allow in order to be 95% sure that he will be able to complete them all?

    (OLE)

21. Two independent random variables X_1 and X_2 have Poisson distributions with means λ_1 and λ_2 respectively. What is the mean and variance of Z where $Z = X_1 + X_2$?

    The probability distributions of the number of daily absences of male and female employees in a factory can be approximated by a Poisson model with mean 0.8 for males and 0.6 for females.

    Assuming that the sum of two independent Poisson variables is a Poisson variable, calculate, correct to three decimal places, the probability that:

    (i) two men and two women are absent on the same day;
    (ii) four employees are absent on any one day;

(iii) during a working week of five days there are no more than two employees absent.

Use the normal approximation to estimate the probability of there being more than twelve absences in any one week.

(SUJB)

Section 8.4

22. Use probability generating functions to show that the sum of n independent Bernoulli variables has a binomial distribution (refer to Worksheet 6, Exercise 1 and its solution).

23. A series of N binomial experiments is carried out, each experiment consisting of n trials with probability of success p for each trial. If $X_1, X_2, \ldots . X_N$ are the number of successes in the N experiments, use probability generating functions to show that:

$$Y = X_1 + X_2 + \ldots + X_N \text{ has a binomial distribution.}$$

24. Find the moment generating function of the distribution of the sum of n independent and exponentially distributed variables, each with parameter λ. Hence find the mean and variance of this distribution (which is called the *gamma* distribution).

25. A coin that falls heads with probability p, or tails with probability $q = 1 - p$, is thrown in a sequence of independent tosses. Show that the probability that tails appears for the first time on the nth toss is qp^{n-1} $(n = 1, 2, 3, \ldots)$.

    Let X and Y be independent random variables, each having the above probability distribution. Find the distribution of $X + Y$.

    Show that, if $1 \leqslant r \leqslant m - 1$, the conditional probability that $X = r$, given that $X + Y = m$, is $1/(m-1)$.

    What does this imply about the position of the first tail in a sequence of tosses, if you know that the second tail appears on the eighth toss?

(OLE)

26. The random variable X takes the values $1, 2, \ldots, n$ with probabilities $p_1, p_2, \ldots, p_n$ respectively. Write down an expression for $g(t)$, the probability generating function of X, and show that $g(1) = 1$.

    The random variable Y is independent of X and takes the values $1, 2, \ldots, n$ with probabilities $q_1, q_2, \ldots, q_n$ respectively, and its probability generating function is $h(t)$. Show that the probability generating function of $X + Y$ is $g(t)h(t)$, and find a similar expression for the probability generating function of $X - Y$.

    An ordinary die is thrown r times. Show that the probability generating function for the total score is:

$$\left\{\frac{t(1-t^6)}{6(1-t)}\right\}^r$$

    Hence show that the probability of the total score being $r + 6$ is:

$$\left[\binom{r+5}{6} - r\right]\frac{1}{6^r}$$

(O&C) ►

27. Three balls a, b, c are placed at random in three boxes A, B, C (with one ball to each box). The variable x is defined by:

$$\begin{cases} x=1 & \text{if ball } a \text{ is in box } A. \\ x=0 & \text{otherwise} \end{cases}$$

Find the probability generator for x.
The variable y is defined in the table:

	Box A	Box B	Box C	Value of y
Contents	a	b	c	1
	c	a	b	0
	b	c	a	0
	a	c	b	0
	b	a	c	1
	c	b	a	1

Deduce the probability generator of y.
The variable s is defined by $s = x + y + 1$. Given that x, y are independent, show that s has generator:

$$\frac{1}{6}(t^3 + 3t^2 + 2t)$$

and deduce the mean and variance of s.

(SMP)

Section 8.6

28. Repeat Example 8.7 in Section 8.6 but this time assuming that the two balls are drawn *without* replacement.

29. The table gives the joint probability distribution of two random variables X and Y; entries in the table represent $P(X = x \text{ and } Y = y)$.

y \ x	-1	0	1
1	$\frac{4}{24}$	$\frac{1}{24}$	0
2	$\frac{3}{24}$	$\frac{3}{24}$	$\frac{1}{24}$
3	$\frac{1}{24}$	$\frac{3}{24}$	$\frac{2}{24}$
4	0	$\frac{2}{24}$	A

(a) What is the value of A?
(b) What is the probability that $X = 0$?
(c) Find $E(X)$ and $E(Y)$

▶

(d) Find $E(XY)$.
(e) Find $E(X+Y)$.
(f) What is the conditional distribution of X given $Y = 1$?
(Give answers as fractions, not decimals.)

(OLE)

30. A factory makes fruit gums in the three flavours 'orange', 'lemon', and 'strawberry'. The proportions of gums that are orange-flavoured, lemon-flavoured, and strawberry-flavoured are 0.4, 0.2 and 0.4, respectively. In a random sample of three of these fruit gums, let X denote the number that are orange-flavoured, and let Y denote the number that are lemon-flavoured.

 (i) Name the distributions of X and Y, and write down their means and variances.
 (ii) The following table displays the joint probability distribution of X and Y. Verify that the entry $P(X = 1, Y = 2) = 0.048$ is correct, and calculate the five missing entries in the table.

	x			
y	*0*	*1*	*2*	*3*
0	0.064	0.192		
1	0.096		0.096	0
2		0.048	0	0
3		0	0	0

 (iii) Calculate the probability that a random sample of three gums will include more strawberry-flavoured gums than the combined number of orange-flavoured and lemon-flavoured gums.

(WJEC)

9

Samples, populations and point estimation

▼

9.1 INTRODUCTION

In many practical statistical investigations we are concerned with drawing conclusions from a limited amount of data. For example, suppose we wish to determine the mean and variance of the weight of 15-year-old children in a particular geographical region. If we were to weigh all these children we could calculate the mean and variance using the methods of Chapters 2 and 3. If however, 40,000 children were involved we might well decide that we could not afford the time or cost of weighing all of them. In that case we might decide to select only a *sample* from the whole *population* of weights, and use the *sample data* (the weights of the children selected in the sample) to *estimate* the mean and variance of the weight of all 40,000 children. The mean and variance are examples of what are called the *parameters* of (the distribution of) a population.

If we quoted a single value for our estimate of the mean weight of all 40,000 children, this would be called a *point estimate.* Another point estimate would be our estimate of the variance of the weight of all 40,000 children. In this chapter we shall deal only with point (= single value) estimation. In the next chapter we will go on to discuss *interval estimation.*

9.2 SAMPLES AND POPULATIONS

The terms population and sample are defined in statistics as follows:

A *population* is the set of measurements which are the subject of an investigation.

A *sample* is a subset of a population.

If the investigation involves only one variable the population is called univariate. In this chapter we will discuss only univariate populations.

The questions which always arise in a sample survey (i.e. an investigation which involves taking a sample of measurements from a defined population) are:

1. How should we select our sample, i.e. what *sampling method* shall we use?
2. How many measurements shall we take in our sample, i.e. what *sample size* shall we use?

The choice of sample size is left to the next chapter since it may be linked with the idea of interval estimation. As for sampling methods, we will restrict ourselves to one, namely random sampling, for two reasons. The first reason is that random sampling generally leads to estimates which are unbiased (see Section 9.5). The second reason is that the mathematical theory concerning estimators obtained by random sampling is well developed (Sections 9.4–9.10, for example).

9.3 RANDOM SAMPLING

A *random sample* is defined as one which is selected so that each measurement in the population has the same chance (probability) of being selected. For example, if we are interested in the population consisting of the weights of 40,000 children, then a random sample is one in which each weight has a 1 in 40,000 chance of being selected.

Suppose we decide to take a random sample of size 100. How shall we select such a sample from the population of 40,000? Since 40,000 has 5 digits we could assign each child a different 5-digit number from 00001 to 40,000, and then use random numbers to select 100 of these. Random numbers are such that each is equally likely to be selected. They are available in tables (see Table C.13), on some hand calculators and on most computers. Suppose our random number tables consist of 50 rows and 50 columns of single digits. We can start by selecting any row and any column – suppose we choose row 13, column 29, and reading left to right find the following:

0 1 4 3 7 0 7 8 7 0 3 1 3 6 1 0 1 . . .

We form 5-digit numbers, ignoring any values above 40,000, and obtain:

01437, 07870, 31361 and so on, until we have 100 5-digit numbers. Each number corresponds to a particular child and their 100 weights constitute the *sample data* for the investigation.

9.4 PROPERTIES OF POINT ESTIMATORS

Let θ denote a parameter of a population (e.g. the population mean) and let $\hat{\theta}$ denote a point estimator of θ based on sample data (e.g. the sample mean). How shall we decide whether the particular way we use the sample data to calculate $\hat{\theta}$ gives 'good' estimates of θ? The answer is that we try to ensure that $\hat{\theta}$ has three desirable properties, namely that:

1. $\hat{\theta}$ is an *unbiased* estimator of θ;
2. $\hat{\theta}$ is a *consistent* estimator of θ;
3. $\hat{\theta}$ is an *efficient* estimator of θ.

The three terms listed above are defined as follows:

1. $\hat{\theta}$ is an unbiased estimator of θ if $E(\hat{\theta}) = \theta$.
2. $\hat{\theta}$ is a consistent estimator of θ if $E(\hat{\theta}) \rightarrow \theta$ and Var $(\hat{\theta}) \rightarrow 0$ as $n \rightarrow \infty$.† However if we already know that $\hat{\theta}$ is an unbiased estimator, then we require only the one condition that Var $(\hat{\theta}) \rightarrow 0$ as $n \rightarrow \infty$.
3. $\hat{\theta}$ is an efficient estimator of θ if Var $(\hat{\theta})$ is small. Since 'small' is a relative term we can say that, of two possible estimators $\hat{\theta}_1$ and $\hat{\theta}_2$, $\hat{\theta}_1$ is relatively more efficient than $\hat{\theta}_2$ if Var $(\hat{\theta}_1) <$ Var $(\hat{\theta}_2)$. We can also say that the efficiency of $\hat{\theta}_2$ relative to $\hat{\theta}_1$ is Var $(\hat{\theta}_1)$/Var $(\hat{\theta}_2)$, a ratio which will always be less than 1 if Var $(\hat{\theta}_1) <$ Var $(\hat{\theta}_2)$. Of course we must compare $\hat{\theta}_1$ and $\hat{\theta}_2$ for the same sample size n.

In order to illustrate the ideas of bias and efficiency, consider Figs 9.1 and 9.2. Suppose the actual value of the parameter θ is represented as a point on a linear scale for the variable which is the subject of a sample survey (e.g. θ is the mean weight of 40,000 children). The actual value of θ will be unknown to the investigator because he has taken only a sample of the population. Nevertheless, the investigator wants to ensure that the method used to calculate $\hat{\theta}$ from the sample data leads to estimates which, on average, are neither too high and hence positively biased (see crosses in Fig. 9.1) nor too low and hence negatively biased (see circles in Fig. 9.1). The investigator also wants to ensure that the method used to calculate $\hat{\theta}$ leads to estimates which have a small variation about the actual value of θ and hence small variance (see diamonds in Fig. 9.2) rather than a large variation about the actual value of θ and hence large variance (see squares in Fig. 9.2).

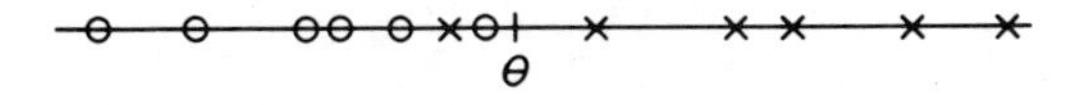

Fig. 9.1 Positive and negative bias

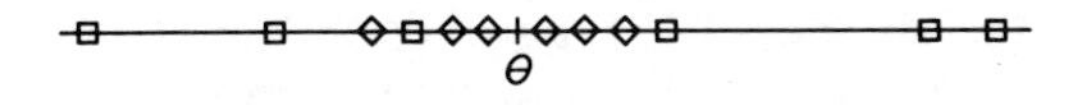

Fig. 9.2 Small and large variance

†The term 'standard error' is used to denote the square root of the variance, when applied to estimators.

Since θ is unknown it might at first seem impossible to guarantee bias or efficiency, but there are theoretical ways of testing a particular estimator for these properties. The property of consistency includes the property of unbiasedness, along with the idea that very large samples lead to estimators with very small variances (one would naturally expect that very large samples will lead to very good estimators).

9.5 SAMPLING DISTRIBUTION OF THE SAMPLE MEAN

This section and the next are concerned with the estimation of the parameter μ, the mean of a population. This section is concerned with how the sample mean $\bar{X}$ varies in repeated sampling, and the next section is concerned with deciding whether $\bar{X}$, as an estimator of μ, has the three desirable properties discussed in Section 9.4.

Let $X_1, X_2, \ldots, X_n$ denote a random sample of size n taken from the population of a variable X with (unknown) mean μ and variance σ^2. If we consider how X_1 would vary in repeated sampling of samples of size n, we see that X_1 is a random variable also with a mean of μ and variance σ^2. The same applies to $X_2, X_3, \ldots, X_n$. The method of random sampling ensures that the n random variables X_1 to X_n are all independent of one another. We can also see that the sample mean, namely:

$$\bar{X} = \frac{X_1 + X_2 + \ldots + X_n}{n}$$

is also a random variable. The distribution of $\bar{X}$ is called the *sampling distribution* of the sample mean. We can think of this sampling distribution as being generated when we repeatedly take samples of size n and each time we calculate the value of $\bar{X}$.

How does this distribution relate to the distribution of the 'parent' population from which the samples are taken? The answer is in three parts:

(1) $$E(\bar{X}) = E(X) = \mu \tag{9.1}$$

(2) $$\text{Var}(\bar{X}) = \frac{\text{Var}(X)}{n} = \frac{\sigma^2}{n} \tag{9.2}$$

Notice that the variance of $\bar{X}$ is less than the variance of X for all $n > 1$. The *standard error* (s.e.) of $\bar{X}$ is the square root of the variance of $\bar{X}$:

$$\text{s.e.}(\bar{X}) = \frac{\sigma}{\sqrt{n}}$$

(3) (a) $\overline{X}$ is approximately normally distributed if n is large ($\geqslant 30$, say[†]) whatever the distribution of the original variable X (see Fig. 9.3). This result is called the *central limit theorem* (C.L.T.) and is very important! (9.3)

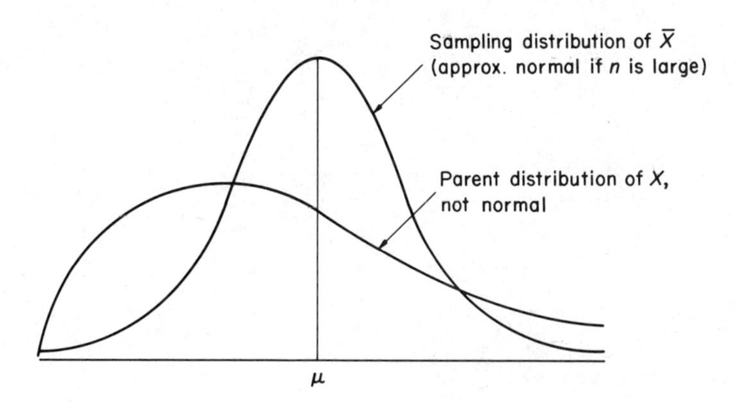

Fig. 9.3 Distribution of $\overline{X}$, non-normally distributed 'parent'

(b) $\overline{X}$ is normally distributed for all n, if X itself is normally distributed (see Fig. 9.4). (9.4)

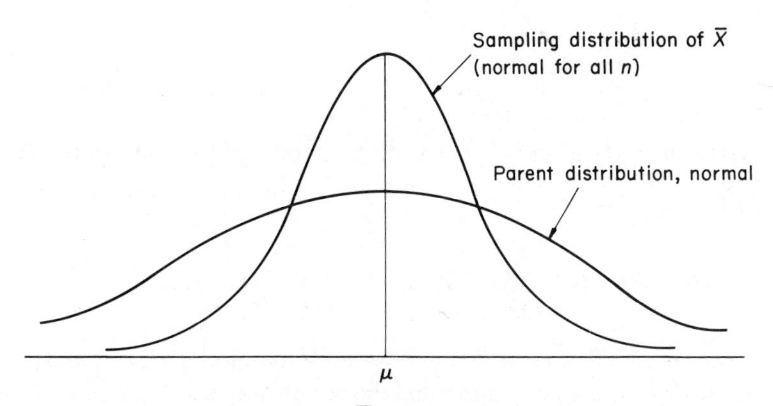

Fig. 9.4 Distribution of $\overline{X}$, normally distributed 'parent'

[†] If the population distribution is not very non-normal, much smaller values of n will ensure that $\overline{X}$ is approximately normally distributed.

Proof of (9.1)

$$E(\overline{X}) = E\left[\frac{X_1 + X_2 + \ldots + X_n}{n}\right]$$

$$= E\left[\frac{X_1}{n} + \frac{X_2}{n} + \ldots + \frac{X_n}{n}\right]$$

$$= \frac{1}{n}E(X_1) + \frac{1}{n}E(X_2) + \ldots + \frac{1}{n}E(X_n) \qquad \text{from (8.1)}$$

$$= \frac{\mu}{n} + \frac{\mu}{n} + \ldots + \frac{\mu}{n}$$

$$= \mu$$

Proof of 9.2

$$\text{Var}(\overline{X}) = \text{Var}\left[\frac{X_1 + X_2 + \ldots + X_n}{n}\right]$$

$$= \text{Var}\left[\frac{X_1}{n} + \frac{X_2}{n} + \ldots + \frac{X_n}{n}\right]$$

$$= \frac{1}{n^2}\text{Var}(X_1) + \frac{1}{n^2}\text{Var}(X_2) + \ldots + \frac{1}{n^2}\text{Var}(X_n)$$

$$= \frac{\sigma^2}{n^2} + \frac{\sigma^2}{n^2} + \ldots + \frac{\sigma^2}{n^2}$$

$$= \frac{\sigma^2}{n}$$

No proof will be given here of (9.3), while the proof of (9.4) is similar to that given in Section 8.3.

9.6 POINT ESTIMATION OF THE MEAN OF A NORMAL DISTRIBUTION

In this section we will look at $\overline{X}$ as a possible estimator of μ. Here $\overline{X}$ and μ are the sample mean for a sample of size n and the population mean respectively, and we will assume that the (parent) population is normally distributed.

Does $\overline{X}$ have the three desirable properties stated in Section 9.4?

1. $\overline{X}$ is an unbiased estimator of μ if $E(\overline{X}) = \mu$.
   We proved that this was the case in the previous section, so $\overline{X}$ is an unbiased estimator of μ.
2. $\overline{X}$ is a consistent estimator of μ if it is unbiased, which we have already

shown, and if Var $(\bar{X}) \to 0$ as $n \to \infty$. From the previous section we know that Var$(\bar{X}) = \sigma^2/n$, which does $\to 0$ as $n \to \infty$, so $\bar{X}$ is a consistent estimator of μ.

3. $\bar{X}$ is an efficient estimator of μ if Var $(\bar{X})$ is small. It can be shown that no other unbiased estimator of μ has smaller variance than $\bar{X}$, for samples taken from a normally distributed population. The proof is beyond the scope of this book (see Hoel, P. G., *Introduction to Mathematical Statistics*, Wiley, 1985).

Example 9.1

A random sample of five values are taken from a population with distribution $N(\mu, \sigma^2)$. The values are 1.06, 1.03, 1.04, 1.05, 1.04. Estimate μ.

We will use $\bar{X}$ as our *estimator*, and we let $\bar{x}$ (lower case) stand for the calculated value of $\bar{X}$.

We say that $\bar{x}$ is the *estimate* of μ, where

$$\bar{x} = \frac{\sum x_i}{n}$$

(Notice that we used exactly the same formula in Section 2.2, see (2.1))
For our sample data,

$$\bar{x} = \frac{1.06 + 1.03 + 1.04 + 1.05 + 1.04}{5}$$

$$= 1.044$$

Our estimate of μ is 1.044.

Note
If the sample data are in the form of a frequency distribution, the formula for $\bar{x}$, equivalent to (2.1), is (2.2), i.e.,

$$\bar{x} = \frac{1}{n}\sum f_i x_i \qquad \text{where } n = \sum f_i$$ □

The following example compares the relative merits of three possible estimators of μ for samples of size 3 taken from a population with distribution $N(\mu, \sigma^2)$.

Example 9.2

Compare the estimators $\hat{\mu}_1$, $\hat{\mu}_2$ and $\hat{\mu}_3$, where

$$\hat{\mu}_1 = \frac{X_1 + X_2 + X_3}{3}, \quad \hat{\mu}_2 = \frac{X_1}{6} + \frac{X_2}{3} + \frac{X_3}{2}, \quad \hat{\mu}_3 = \frac{X_1 - 2X_2 + 3X_3}{6},$$

and X_1, X_2, X_3 denote a sample of size 3 from $N(\mu, \sigma^2)$.

We see that $\hat{\mu}_1$ is the sample mean and so it is unbiased.

Also
$$\text{Var}(\hat{\mu}_1) = \frac{\sigma^2}{3} \qquad \text{from (9.2)}$$

$$\begin{aligned} E(\hat{\mu}_2) &= \frac{1}{6}E(X_1) + \frac{1}{3}E(X_2) + \frac{1}{2}E(X_3) \qquad \text{from (8.1)} \\ &= \frac{\mu}{6} + \frac{\mu}{3} + \frac{\mu}{2} \\ &= \mu \end{aligned}$$

Hence $\hat{\mu}_2$ is also an unbiased estimator.

$$\begin{aligned} \text{Var}(\hat{\mu}_2) &= \frac{1}{36}\text{Var}(X_1) + \frac{1}{9}\text{Var}(X_2) + \frac{1}{4}\text{Var}(X_3) \qquad \text{from (8.2)} \\ &= \frac{\sigma^2}{36} + \frac{\sigma^2}{9} + \frac{\sigma^2}{4} \\ &= \frac{14\sigma^2}{36} \end{aligned}$$

$$E(\hat{\mu}_3) = \frac{\mu}{6} - \frac{2\mu}{6} + \frac{3\mu}{6} = \frac{\mu}{3}, \text{ and hence } \hat{\mu}_3 \text{ is a biased estimator.}$$

$$\text{Var}(\hat{\mu}_3) = \frac{\sigma^2}{36} + \frac{4\sigma^2}{36} + \frac{9\sigma^2}{36} = \frac{14\sigma^2}{36}$$

So $\hat{\mu}_1$ is the best of the three estimators, since it has the smallest variance and is unbiased. $\hat{\mu}_2$ is better than $\hat{\mu}_3$. □

9.7 POINT ESTIMATION OF THE VARIANCE OF A NORMAL DISTRIBUTION

If $X_1, X_2, \ldots, X_n$ denotes a random sample of size n taken from a normally distributed population with unknown mean μ and variance σ^2, what estimator shall we use to estimate σ^2?

The definition of σ^2 is $E[(X-\mu)^2]$ (see Section 5.3). Since we don't know μ we could replace it by the estimator $\bar{X}$. Also, using the idea that 'expectation' and 'mean' are synonymous, we might decide that

$$\frac{\sum(X_i - \bar{X})^2}{n}, \text{ where the summation is from } i = 1 \text{ to } i = n$$

is a 'reasonable' estimator of σ^2. But is this estimator unbiased, consistent and efficient? We will show below that it is a biased estimator of σ^2, but that an unbiased estimator can be found by replacing n by $n-1$.

Proof

$$\sum(X_i-\overline{X})^2 = \sum X_i^2 - 2\overline{X}\sum X_i + n\overline{X}^2$$

$$= \sum X_i^2 - 2\frac{\sum X_i}{n}\sum X_i + n\left(\frac{\sum X_i}{n}\right)^2$$

$$= \sum X_i^2 - \frac{(\sum X_i)^2}{n}$$

$$\therefore\ E\left[\frac{\sum(X_i-\overline{X})^2}{n}\right] = E\left[\frac{\sum X_i^2}{n}\right] - E(\overline{X}^2), \quad \text{since } \overline{X} = \frac{\sum X_i}{n}$$

But $X_i \sim N(\mu, \sigma^2)$ for all $i = 1$ to $i = n$

$$\therefore \sigma^2 = E(X_i^2) - \mu^2$$

$$\therefore E(X_i^2) = \sigma^2 + \mu^2 \qquad \text{for all } i = 1 \text{ to } i = n$$

$$\therefore E\left(\frac{\sum X_i^2}{n}\right) = \frac{1}{n}E(\sum X_i^2) = \frac{n(\sigma^2+\mu^2)}{n} = \sigma^2 + \mu^2$$

Also $\overline{X} \sim N(\mu, \sigma^2/n)$

$$\therefore \frac{\sigma^2}{n} = E(\overline{X}^2) - \mu^2$$

$$\therefore E(\overline{X}^2) = \frac{\sigma^2}{n} + \mu^2$$

Hence

$$E\left[\frac{\sum(X_i-\overline{X})^2}{n}\right] = \sigma^2 + \mu^2 - \frac{\sigma^2}{n} - \mu^2 = \frac{(n-1)}{n}\sigma^2$$

Since $\frac{(n-1)}{n}\sigma^2 \neq \sigma^2$, $\frac{\sum(X_i-\overline{X})^2}{n}$ is a biased estimator of σ^2.

However, it is easy to show (using the same method just outlined) that

$$S^2 = \frac{\sum(X_i-\overline{X})^2}{n-1} \tag{9.5}$$

is an unbiased estimator of σ^2.

We quote without proof that Var $(S^2) = 2\sigma^4/(n-1)$, which $\to 0$ as $n \to \infty$, and so S^2 is a consistent estimator of σ^2. The efficiency of S^2 is less than for the biased estimator, the efficiency of S^2 relative to the biased estimator being $(n-1)/n$, but this ratio is close to 1 for large n, and S^2 is preferred because it is an unbiased estimator of the population variance σ^2.

Example 9.3

A random sample of five values are taken from a population with distribution $N(\mu, \sigma^2)$. The values are 1.06, 1.03, 1.04, 1.05, 1.04.

Estimate σ^2.

We will use S^2 as our *estimator*, and let s^2 (lower case) stand for the calculated value of S^2.

We say that s^2 is the *estimate* of S^2, where

$$s^2 = \frac{\sum(x_i - \bar{x})^2}{n-1}$$

An alternate form of this, which is often used, is

$$s^2 = \frac{\sum x_i^2 - n\bar{x}^2}{n-1} \qquad (9.6)$$

For our sample data,

$$\sum x_i = 1.06 + 1.03 + \ldots + 1.04 = 5.22$$

$$\sum x_i^2 = 1.06^2 + 1.03^2 + \ldots + 1.04^2 = 5.4502$$

$$\bar{x} = \frac{5.22}{5} = 1.044$$

$$s^2 = \frac{5.4502 - 5 \times 1.044^2}{5-1}$$

$$= 0.00013$$

Our estimate of σ^2 is 0.00013.

Notes

(i) In using (9.6) we should be careful not to round prematurely. For example, if 5.4502 is rounded to 5.45, the answer 0.00008 is obtained.

(ii) Most calculators have the facility to calculate s, but the symbol σ_{n-1} is often used instead of s.

(iii) A common question is 'Why was n used in the denominator in Chapter 3 (see (3.2)) while $n-1$ is used in this chapter?' The answer is that $(s')^2$ is the variance of a set of sample values, whereas s^2 is the estimate of the population variance σ^2. From now on in this book, we will use only s^2!

(iv) If the sample data are in the form of a frequency distribution, the formula for s^2, equivalent to (9.5) is

$$s^2 = \frac{\sum f_i x_i^2 - n\bar{x}^2}{n-1} \qquad (9.7)$$

where $$\bar{x} = \frac{1}{n}\sum f_i x_i \qquad \text{from (2.2)}$$ □

9.8 POINT ESTIMATION OF THE BINOMIAL PARAMETER, p

Suppose we carry out a binomial experiment in which p, the probability of success in a single trial, is unknown (for example toss a drawing-pin n times and count the number of times it lands 'point up'). Let X denote the number of successes in the n trials of the experiment. Then X is a discrete random variable (in the sense that it will vary from one repetition of the experiment to another) with a $B(n, p)$ distribution. The sample data from one such experiment is the number of successes we actually observe in the n trials. What estimator shall we use for p? Using the relative frequency ideas of Section 4.4, X/n would seem to be a reasonable estimator of p, as long as n is large. But is X/n (i) unbiased, (ii) consistent, and (iii) efficient?

(i) X/n is an unbiased estimator of p if $E\left(\frac{X}{n}\right) = p$.

Since $X \sim B(n, p)$, $E(X) = np$ (Section 6.2)

$\therefore E\left(\frac{X}{n}\right) = \frac{1}{n}E(X) = \frac{np}{n} = p$, hence X/n is an unbiased estimator of p.

(ii) X/n is consistent if, in addition to being unbiased,

$\text{Var}(X/n) \to 0$ as $n \to \infty$

Since $X \sim B(n, p)$, $\text{Var}(X) = np(1-p)$ (Section 6.2)

$$\therefore \text{Var}\left(\frac{X}{n}\right) = \frac{1}{n^2}\text{Var}(X) = \frac{np(1-p)}{n^2} = \frac{p(1-p)}{n} \qquad \text{(see (5.22))}$$

and this $\to 0$ as $n \to \infty$, and hence X/n is a consistent estimator of σ^2.

(iii) It can be shown that no other unbiased estimator has a smaller variance than X/n (no proof will be given here).

We conclude that X/n is the estimator of p we should use. (9.8)

Example 9.4

A drawing-pin is tossed 100 times. It lands 'point-up' 64 times. Estimate p, the probability of 'point-up' in a single toss, and the variance of the estimate. We can confirm that we have a binomial experiment by checking the four conditions (Section 6.2).

We will use X/n as our *estimator*, and let x/n stand for the calculated value of X/n.

We say that x/n is our *estimate* of p. (9.9)

For our sample data, $x = 64$, $n = 100$. Our estimate of p is 0.64. The variance of this estimate of p is $p(1-p)/n$. Our estimate of p is x/n, so our estimate of

$$\frac{p(1-p)}{n} \quad \text{is} \quad \frac{\frac{x}{n}\left(1-\frac{x}{n}\right)}{n}$$

For the sample data this equals $\dfrac{0.64 \times 0.36}{100} = 0.0023.$

The estimated variance of our estimate of p is 0.0023. □

9.9 POINT ESTIMATION OF THE COMMON VARIANCE OF TWO NORMAL DISTRIBUTIONS, DATA FROM TWO SAMPLES

Consider two normal distributions with different means, μ_1 and μ_2, but the same variance, σ^2. Suppose we take a sample of size n_1 from the $N(\mu_1, \sigma^2)$ distribution and a sample of size n_2 from the $N(\mu_2, \sigma^2)$ distribution. How should we combine the data from the two samples to give one (point) estimator of the common variance, σ^2? Let S_1^2 and S_2^2 denote the unbiased estimators of σ^2 using the data from the first and second samples respectively. Also let S^2 denote a linear combination of S_1^2 and S_2^2, e.g. $S^2 = \lambda_1 S_1^2 + \lambda_2 S_2^2$. We would like S^2 to be an unbiased, consistent and efficient estimator of σ^2.

S^2 is an unbiased estimator of σ^2 if $E(S^2) = \sigma^2$.

Now
$$E(S^2) = \lambda_1 E(S_1^2) + \lambda_2 E(S_2^2)$$
$$= \lambda_1 \sigma^2 + \lambda_2 \sigma^2,$$

since S_1^2 and S_2^2 are both unbiased estimators of σ^2.

$$= \sigma^2 \quad \text{if } \lambda_1 + \lambda_2 = 1.$$

So, replacing λ_1 by λ and λ_2 by $1-\lambda$ we can deduce that $S^2 = \lambda S_1^2 + (1-\lambda) S_2^2$ is unbiased, and we have reduced the problem to one of finding the 'best' value of λ, where 'best' means that value which minimizes the variance of S. We can then check for consistency

$$\begin{aligned} \text{Var}(S^2) &= \text{Var}[\lambda S_1^2 + (1-\lambda) S_2^2] \\ &= \lambda^2 \text{Var}(S_1^2) + (1-\lambda)^2 \text{Var}(S_2^2) \\ &= \frac{\lambda^2 2\sigma^4}{n_1 - 1} + \frac{(1-\lambda)^2 2\sigma^4}{n_2 - 1} \end{aligned}$$

To find the minimum value of Var (S^2) we differentiate with respect to λ, and set the derivative equal to zero (the reader should check that the second derivative is positive, indicating a minimum rather than a maximum).

$$\frac{2\lambda 2\sigma^4}{n_1 - 1} - \frac{2(1-\lambda)2\sigma^4}{n_2 - 1} = 0$$
$$\lambda(n_2 - 1) - (1-\lambda)(n_1 - 1) = 0$$
$$\lambda = \frac{n_1 - 1}{n_1 + n_2 - 2}$$

and
$$1 - \lambda = \frac{n_2 - 1}{n_1 + n_2 - 2}$$

The most efficient unbiased estimator of σ^2 (which combines S_1^2 and S_2^2 linearly) is:

$$S^2 = \frac{(n_1-1)S_1^2 + (n_2-1)S_2^2}{n_1+n_2-2} \tag{9.10}$$

Finally, as a check of consistency,

$$\begin{aligned}
\text{Var}(S^2) &= \left(\frac{n_1-1}{n_1+n_2-2}\right)^2 \text{Var}(S_1^2) + \left(\frac{n_2-1}{n_1+n_2-2}\right)^2 \text{Var}(S_2^2) \\
&= \frac{(n_1-1)^2\, 2\sigma^4}{(n_1+n_2-2)^2\,(n_1-1)} + \frac{(n_2-1)^2\, 2\sigma^4}{(n_1+n_2-2)^2\,(n_2-1)} \\
&= \frac{2\sigma^4}{n_1+n_2-2}, \qquad \text{which} \to 0 \quad \text{as} \quad (n_1+n_2) \to \infty
\end{aligned}$$

So S^2 is a consistent estimator of σ^2 (since we have already shown it to be unbiased).

Example 9.5

A random sample of size five is taken from an $N(\mu_1, \sigma^2)$ distribution. A second random sample of size seven is taken from an $N(\mu_2, \sigma^2)$ distribution. The following statistics were calculated:

$$\bar{x}_1 = 6.8, \quad s_1^2 = 2.7, \quad n_1 = 5$$

$$\bar{x}_2 = 8.0, \quad s_2^2 = 1.3, \quad n_2 = 7$$

Estimate the common variance, σ^2.
We will use S^2, as defined in (9.10), as our estimator of σ^2. Let s^2 stand for the calculated value of S^2, where

$$s^2 = \frac{(n_1-1)s_1^2 + (n_2-1)s_2^2}{n_1+n_2-2} \tag{9.11}$$

From our sample data,

$$s^2 = \frac{(5-1)\times 2.7 + (7-1)\times 1.3}{5+7-2} = 1.86 \quad (2 \text{ d.p.})$$

Our estimate of σ^2 is 1.86. □

Suppose we now consider the special case in which we also know that the distributions have the same mean, so $\mu_1 = \mu_2 = \mu$, say. In other words, the distributions are identical. In this case we should treat the two samples (of sizes n_1 and n_2) as one sample of size $(n_1 + n_2) = n$, say, and use results (2.2) and (9.6) to estimate μ and σ^2 respectively.

Example 9.6

Two random samples of sizes five and seven are taken from an $N(\mu, \sigma^2)$ distribution. The following test statistics were calculated:

$$\bar{x}_1 = 6.8, \quad s_1^2 = 2.7, \quad n_1 = 5$$

$$\bar{x}_2 = 8.0, \quad s_2^2 = 1.3, \quad n_2 = 7$$

Estimate μ and σ^2.
Using (2.2) for all $(n_1 + n_2)$ values,

$$\bar{x} = \frac{\sum x_i}{n} = \frac{n_1 \bar{x}_1 + n_2 \bar{x}_2}{n_1 + n_2}$$

since $n_1 \bar{x}_1$ is the sum of the values in the first sample, and $n_2 \bar{x}_2$ is the sum of the values in the second sample.

So $$\bar{x} = \frac{5 \times 6.8 + 7 \times 8.0}{5 + 7} = 7.50, \quad \text{our estimate of } \mu$$

In order to apply (9.6) to all the data we require Σx_i^2 where the summation is over all 12 values. We can obtain this by applying (9.6) separately to each sample. Thus

$$\sum x_i^2 = (n_1 - 1) s_1^2 + n_1 \bar{x}_1^2 + (n_2 - 1) s_2^2 + n_2 \bar{x}_2^2$$

$$= 4 \times 2.7 + 5 \times 6.8^2 + 6 \times 1.3 + 7 \times 8.0^2 \qquad \text{for our data}$$

$$= 697.8$$

$$\therefore s^2 = \frac{697.8 - 12 \times 7.50^2}{12 - 1}, \qquad \text{using (9.6) for all 12 values}$$

$$= 2.07 \text{ (2 d.p.)}$$

Our estimates of μ and σ^2 are 7.50 and 2.07.

Note
It would have been much easier if we had started with the 12 individual values and used (2.2) and (9.6) once each! □

9.10 POINT ESTIMATION OF THE BINOMIAL PARAMETER, p, DATA FROM TWO BINOMIAL EXPERIMENTS

Suppose we carry out two binomial experiments, one with n_1 trials the other with n_2 trials. If p is the unknown probability of success in each trial, and if X_1 and X_2 denote the number of successes in each experiment, what estimator shall we use for p?

The answer is the 'obvious' one, namely

$$\frac{X_1 + X_2}{n_1 + n_2}$$

since this is the ratio of the total number of successes to the total number of trials. It is easy to show that this estimator is unbiased and consistent (we quote without proof that it has minimum variance):

$$E\left[\frac{X_1+X_2}{n_1+n_2}\right]=\frac{1}{n_1+n_2}E(X_1)+\frac{1}{n_1+n_2}E(X_2)$$

$$=\frac{n_1 p}{n_1+n_2}+\frac{n_2 p}{n_1+n_2}$$

$$=p$$

and hence the estimator is unbiased.

$$\operatorname{Var}\left[\frac{X_1+X_2}{n_1+n_2}\right]=\frac{1}{(n_1+n_2)^2}\operatorname{Var}(X_1)+\frac{1}{(n_1+n_2)^2}\operatorname{Var}(X_2)$$

$$=\frac{n_1 p(1-p)}{(n_1+n_2)^2}+\frac{n_2 p(1-p)}{(n_1+n_2)^2}$$

$$=\frac{p(1-p)}{n_1+n_2}$$

which $\to 0$ as $n_1+n_2\to\infty$, and hence the estimator is consistent.

Example 9.7

A biased coin was tossed 50 times, resulting in 30 heads. A further 100 tosses resulted in 75 heads.

Estimate the probability of heads in a single toss.

We will use $(X_1+X_2)/(n_1+n_2)$ as our *estimator*, and we will let $(x_1+x_2)/(n_1+n_2)$ stand for the calculated value of $(X_1+X_2)/(n_1+n_2)$.
We say that $(x_1+x_2)/(n_1+n_2)$ is our *estimate* of p.
For our sample data, $x_1=30$, $n_1=50$, $x_2=75$, $n_2=100$

$$\frac{x_1+x_2}{n_1+n_2}=\frac{105}{150}=0.7$$

Our estimate of p is 0.7. □

9.11 SUMMARY

Samples are taken from populations to save time and money. Sample data can provide point (single-value) estimators of population parameters which, if random sampling is used, can have the desirable properties of being unbiased, consistent and efficient.

If samples of size n are taken from any distribution with mean μ and variance σ^2, the sample mean, $\bar{X}$, has a distribution with mean μ and variance σ^2/n. If n is large the distribution of $\bar{X}$ is approximately normal (central limit theorem).

Table 9.1 summarizes the formulae which should be used to calculate point estimates of various parameters.

Table 9.1 Point estimates of the parameters of the normal and binomial distributions

Distribution	*Parameter*	*Estimate*
normal	μ	$\bar{x} = \frac{\sum x_i}{n}$ or $\frac{1}{n}\sum f_i x_i$
normal	σ^2	$s^2 = \frac{\sum x_i^2 - n\bar{x}^2}{n-1}$ or $s = \frac{\sum f_i x_i^2 - n\bar{x}^2}{n-1}$
binomial	p	$\frac{x}{n}$

If two normal distributions have different means but the same variance, and a sample is taken from each, then

$$s^2 = \frac{(n_1 - 1)s_1^2 + (n_2 - 1)s_2^2}{n_1 + n_2 - 2}$$

may be used to estimate the common variance. In the special case when the means are also equal, the two samples should be combined and treated as one sample.

If two binomial experiments are carried out in order to estimate a parameter p, they should be treated as one combined experiment.

Section 9.3

1. (a) Explain what is meant by a *random sample*. State two advantages of *random* samples as opposed to other samples.

   In order to obtain data for a project it is required to select a random sample of 25 pupils from a school of 500 pupils. Explain carefully how you would use the random number tables to obtain the sample. Illustrate your explanation by starting at the first digit in the sixth row of the table and give the first six numbers which you obtain.

   (b) A student investigated the number of vehicles passing a particular point on a road during one-minute intervals. He used a Poisson distribution as a model. Comment on the precautions he would need to take in choosing the observation point, giving your reasons.

   (c) Suggest distributions which might serve as suitable models in each of the following experiments, giving where possible appropriate values for the associated parameters:

   (i) recording the length of time in seconds between vehicles as they pass an observation point;
   (ii) recording the number of sixes occurring when five unbiased dice are thrown simultaneously;
   (iii) recording the number of times a particular digit occurs in a page of random number tables.

(JMB)

Section 9.4

2. The operational lifetimes of certain electronic components are found to be normally distributed with mean 5200 hours and standard deviation 400 hours. Calculate, to three significant figures:

   (i) the proportion of such components having lifetimes between 4500 and 5800 hours;
   (ii) the 67th percentile of this distribution;
   (iii) the probability that a random sample of 25 components will have a mean lifetime in excess of 5000 hours.

(JMB)

Section 9.5

3. Let $X_1, X_2, \ldots, X_n$ denote a random sample of size n from a distribution with mean μ and variance σ^2. An estimator $\hat{\mu}$ of μ is to be obtained using

$$\hat{\mu} = a_1 X_1 + a_2 X_2 + \ldots + a_n X_n$$

where $a_1, a_2, \ldots, a_n$ are constants. What equation must the a's satisfy to ensure that $\hat{\mu}$ is an unbiased estimator of μ? If all the a's are equal, what is the value of each? In this case find Var $(\hat{\mu})$, and show that $\hat{\mu}$ is a consistent estimator. ▶

4. The number of vehicles passing a quiet country house is thought to have a Poisson distribution with constant parameter a, where a is the mean number of vehicles passing per hour. Let $X_1, X_2, \ldots, X_n$ denote the number of cars passing in n consecutive periods each lasting one hour. Show that $\bar{X} = \Sigma X_i/n$ is an unbiased estimator of a and find its variance.

5. Assume that a fisherman has a constant probability p of catching a fish with each cast. Let $X_1, X_2, \ldots, X_n$ denote the number of casts required to catch a fish on n separate occasions. Show that $\bar{X} = (\Sigma X_i)/n$ is an unbiased estimator of $1/p$, and finds its variance. Is $\bar{X}$ a consistent estimator of $1/p$?

6. Let $X_1, X_2, \ldots, X_n$ denote the times between successive random events, measured on n separate occasions. Assume the time between successive random events has an exponential distribution with parameter λ, show that $\bar{X} = (\Sigma X_i)/n$ is an unbiased estimator of $1/\lambda$. Find the variance of $\bar{X}$ and show that it is a consistent estimator.

7. Consider the rectangular (uniform) distribution with probability density function:

$$f(x) = \frac{1}{\theta}, \qquad 0 \leqslant x \leqslant \theta$$
$$= 0, \qquad \text{elsewhere}$$

Let $X_1, X_2, \ldots, X_n$ denote a random sample from this distribution. Show that $\bar{X}$ is not an unbiased estimator of θ. Deduce the value of k such that $k\bar{X}$ is an unbiased estimator of θ, and find the variance of this estimator.

8. A continuous random variable, X, has a probability density function:

$$f(x) = \frac{1}{\theta}, \qquad 0 \leqslant x \leqslant \theta$$
$$= 0, \qquad \text{elsewhere}$$

Let another random variable, Y, be the larger of a random sample of size two taken from the distribution of X. By considering the distribution function of Y,

$$\text{i.e.} \qquad G(y) = P(Y \leqslant y)$$

show that the probability density function of Y is

$$g(y) = \frac{2y}{\theta^2}, \qquad 0 \leqslant y \leqslant \theta$$
$$= 0, \qquad \text{elsewhere}$$

Show that $3Y/2$ is an unbiased estimator of θ, and find its variance. If $\bar{X}$ is the mean of a random sample of size two, show that $2\bar{X}$ is also an unbiased estimator of θ and find its variance. Of the two estimators, $3Y/2$ and $2\bar{X}$, which do you prefer? Give a reason.

9. In a football pool it was thought that the probabilities of an away win, a home win and a draw were $\theta, 2\theta, 1 - 3\theta$, respectively. The number of points awarded for an away win, a home win and a draw were 2, 1 and 3, respectively. Find the mean and variance of the number of points awarded in terms of θ.

$\overline{X}$, the mean number of points awarded in n matches, is to be used as an estimator of θ. Show that $(3-\overline{X})/5$ is unbiased estimator of θ. In 60 matches a total of 100 points are awarded. What is your estimate of θ?

10. Let X denote a single random observation from an exponential distribution with probability density function

$$f(x) = \frac{1}{\theta} e^{-x/\theta}, \qquad x \geqslant 0$$

Show that X is an unbiased estimator of θ, and show that $X^2/2$ is an unbiased estimator of θ^2.

11. The random variable X has probability density function given by

$$f(x) = \begin{cases} \dfrac{1}{\theta} & 0 \leqslant x \leqslant \theta, \quad \text{where } \theta > 0 \\ 0 & \text{otherwise} \end{cases}$$

Derive the mean and variance of X.
A random sample $X_1, X_2, \ldots, X_n$ is taken from a population with the above distribution. The estimator T is defined by

$$T = k \sum_{i=1}^{n} X_i, \qquad \text{where } k \text{ is a constant.}$$

Find the value of k such that T is an unbiased estimator of θ, and further show that T is then a consistent estimator of θ.

The values of a random sample of size six from a population with the above distribution are 1.2, 0.9, 2.8, 0.8, 1.3, 1.1. In spite of the above properties, show that with these sample values T yields an unreasonable estimate of θ.

(AEB (1983))

12. The random variable X has probability density function

$$f(x) = \begin{cases} \dfrac{1}{2\theta} & 0 < x < 2\theta, \text{ where } \theta \text{ is a positive constant} \\ 0 & \text{otherwise} \end{cases}$$

Show, by integration, that X has mean θ and variance $\theta^2/3$.

What are the mean and variance of $\overline{X}$, the mean of a random sample of size n from the distribution of X? Show that $\overline{X}$ is a consistent estimator of θ.

If a random sample of four observations is taken from the distribution (of X), the largest observation y will be a random variable Y with probability density function

$$g(y) = \begin{cases} \dfrac{y^3}{4\theta^4} & 0 < y < 2\theta \\ 0 & \text{otherwise} \end{cases}$$

Find $E(Y)$ and show that $T = \frac{5}{8}\overline{Y}$ is a consistent estimator of θ, where $\overline{Y}$ is the mean of a random sample of size m from the distribution. Which of the estimators $\overline{X}$ and T do you prefer and why?

(AEB (1984)) ►

13. Describe briefly the characteristics of the normal distribution, and explain how this distribution was used in one of your projects.

    Research workers hope to diagnose, by means of a screening test, a certain disease in children. The test consists of measuring the concentration level of a certain chemical in the urine. Extensive study has shown that the distribution of the quantity, in mg dl^{-1}, of this chemical present is $N(20, 9)$, that is normally distributed with mean 20 and variance 9, for children not having the disease. For children having the disease the corresponding distribution of the chemical present is $N(30, 16)$. It is decided that any child whose level of the chemical is greater than 23 mg dl^{-1} will be sent for further tests, while all others will be classified as not having the disease. Find, to 3 decimal places, the probability that:

    (a) a child having the disease will be classified as not having it;
    (b) a child not having the disease will be sent for further tests;
    (c) a random sample of 9 children not having the disease will have a mean concentration level of the chemical in excess of 22 mg dl^{-1}.

    (London)

14. A coin, which falls heads with probability p on any toss, is tossed until a total of two heads appear. Show that the probability that n tosses are required is $(n-1)p^2(1-p)^{n-2}$ $(n \geqslant 2)$.

    It is required to estimate p on the basis of the observed value n. Show that the estimator $\hat{p}_1 = 1/(n-1)$ is unbiased, and deduce that the estimator $\hat{p}_2 = 2/n$ has a positive bias. Show also that $\hat{p}_1$ has variance $p^2(p-1-\ln p)/(1-p)$.

    (OLE)

15. A number x is chosen at random uniformly in the interval $(0, a)$. Write down the probability density function of x. Find the values of $E(x)$ and $V(x)$. Deduce the values of $E(\bar{x})$ and $V(\bar{x})$, where $\bar{x}$ is the mean of n numbers $x_1, x_2, \ldots, x_n$ chosen independently and at random from $(0, a)$.

    Let y be the largest number of the set $x_1, x_2, \ldots, x_n$. Show by considering $P(y \leqslant t)$ or otherwise that for $0 \leqslant t \leqslant a$, y has probability density function $na^{-n}t^{n-1}$. Hence find $E(y)$ and $V(y)$.

    The numbers $x_1, x_2, \ldots, x_n$ are actually produced by somebody else's computer program and I have no idea what value of a has been used.

    Explain why either $2\bar{x}$ or $((n+1)/n)y$ would both be sensible guesses for the value of a.

    Calculate the variance of each of these quantities. Hence state which guess you would expect to be better and why.

    (SMP)

16. A cubical die has two each of its faces numbered 1, 2, and 3, respectively, and is such that the probabilities of obtaining these scores in a single throw are 0.1, 0.8, and 0.1, respectively.

    (i) If X is the score obtained in one throw of die, determine the mean and the variance of X.
    (ii) Let M denote the median of the three scores obtained in three independent throws of the die. Show the $P(M = 1) = 0.028$. Evaluate $P(M = 2)$ and ►

$P(M = 3)$, and hence determine the mean and the variance of the sampling distribution of M.

(iii) Let $\overline{X}$ denote the mean of the three scores obtained in three independent throws of the die. Write down the values of the mean and the variance of $\overline{X}$, and verify that the variance of M is 84% of the variance of $\overline{X}$.

(WJEC)

17. A random variable X can take only the values 1, 2 and 3, the respective probabilities of these values being θ, θ, and $1-2\theta$, where $0 < \theta < 1/2$. Determine the mean and the variance of X in terms of θ.

    In a random sample of n values of X, let $\overline{X}$ denote the mean of the sample values, and let R denote the number of 3's among the sample values. Show that $p_1 = 1-(\overline{X}/3)$ and $p_2 = 1/2\{1-(R/n)\}$ are both unbiased estimators of θ. Determine which of p_1 and p_2 has the smaller standard error.

(WJEC)

Section 9.6

18. A random sample of 10 observations $x_1, x_2, \ldots, x_{10}$ is taken from a normal population with mean μ and variance σ^2. Specify the sampling distribution of

$$\frac{1}{10}\sum_{r=1}^{r=10} x_r$$

Find also the sampling distribution of the statistic, m_1, where

$$m_1 = \frac{x_4 + x_5 + x_7}{3}$$

and show that m_1 is an unbiased estimator of μ.

Let $$m_2 = x_{10} - x_1$$

and $$m_3 = \frac{x_1 + 2x_2 + 3x_3}{6}$$

Determine whether (i) m_2 and (ii) m_3 are unbiased estimators of μ, giving your reasons.

Determine which of the unbiased estimators named in this question is the more efficient. Give your reasons.

(JMB)

Section 9.7

19. Let X_i $(i = 1, \ldots, n)$ be independent random variables each with mean μ and variance σ^2, and let $\overline{X} = \Sigma X_i/n$. Show that

    (i) $E[X_i^2] = \mu^2 + \sigma^2$

    (ii) $E[\overline{X}^2] = \mu^2 + (\sigma^2/n)$

    (iii) $\Sigma(X_i - \overline{X})^2 = \sum X_i^2 - n\overline{X}^2$

    Hence, or otherwise, show that $\Sigma(X_i - \overline{X})^2/(n-1)$ is an unbiased estimator of σ^2.

(OLE) ►

Section 9.8

20. A seed merchant sells trays, each containing 10 flower seeds in a specifically prepared soil. The number of seeds germinating in each of 100 trays was noted and the following results were obtained:

No. of seeds germinating	5	6	7	8	9	10
No. of trays	1	12	27	33	23	4

(i) Find the median number of seeds germinating per tray in this sample.
(ii) Find an unbiased estimate of the proportion of these seeds that germinate and also find, correct to three decimal places, an approximate value of the standard error of this estimate.

(JMB)

Section 9.9

21. Distinguish between the expressions,

$$\sum \frac{(x_i - \bar{x})^2}{n} \quad \text{and} \quad \sum \frac{(x_i - \bar{x})^2}{n-1}$$

which are both used in connection with a set of n observations, $x_1, \ldots, x_n$.
A random sample of 50 observations is taken from a distribution and used to obtain unbiased estimates of the mean and variance of the distribution; these estimates are 33.2 and 15.7 respectively. A second independent random sample of 30 observations leads to estimates of 32.9 and 14.8 respectively. Give an unbiased estimate of the distribution mean based on the two samples combined. Find the sum of the squares of all the 80 observations, and hence derive an unbiased estimate of the variance based on the combined samples.

(OLE)

22. (a) Let $X_1, X_2, \ldots, X_n$ be independent observations from a distribution having mean μ and variance σ^2, and let $\bar{X} = \left(\sum_{i=1}^{n} X_i\right)/n$. Show that

$$E(\bar{X}) = \mu, \ \text{Var}(\bar{X}) = \frac{\sigma^2}{n}$$

(b) Let $\bar{X}_1$ be the mean of a random sample of n_1 observations from a distribution having mean μ and variance σ_1^2 and let $\bar{X}_2$ be the mean of an independent random sample of n_2 observations from a distribution having mean μ and variance σ_2^2, where the common mean μ is unknown but the variances σ_1^2 and σ_2^2 are known. Show that for any value of λ, $\lambda\bar{X}_1 + (1-\lambda)\bar{X}_2$ is an unbiased estimator of μ and obtain its variance. Determine the value of λ which you consider will give the best estimator of μ.

(IOS) ►

23. Let X and Y be independent random variables each with mean μ but with variances σ_x^2, σ_y^2 respectively. Show that the weighted mean $\omega X + (1-\omega) Y$, where ω is a constant, is an unbiased estimator of μ.

    Find the value of ω which minimizes the variance of the weighted mean; find the minimum variance.

    Two laboratories made determinations of the amount of an element present in separate collections of 1 g samples of a substance. The following results were presented:

Laboratory	*Mean amount (mg per g substance)*	*Standard error of mean*
A	6.19	0.12
B	6.24	0.24

    Use the method developed in the first part of the question to combine the results and give an overall estimate of the mean amount of the element present, together with a standard error.

(OLE)

Section 9.10

24. Explain what is meant by the sampling distribution of an estimator for a parameter p. State the mean of this distribution if the estimator is unbiased.

    A random variable X has a binomial distribution $B(n, p)$, where the parameter p is unknown. Show that the mean and variance of the sampling distribution of the estimator X/n are p and $p(1-p)/n$, respectively.

    State why X/n is a consistent estimator of p.

    A manufacturer produces a large batch of components. Each component in the batch has constant probability p of being defective, independently of any other component. The manufacturer wishes to estimate p. An inspector selects a random sample of size 10 from the batch, and the random variable X_1 denotes the number of defectives he finds. Another inspector selects independently a random sample of 20 components, and the random variable X_2 denotes the number of defectives he finds. Show that the two estimators $\hat{p}_1$ and $\hat{p}_2$, where

$$\hat{p}_1 = \frac{1}{2}\left(\frac{X_1}{10} + \frac{X_2}{20}\right) \quad \text{and} \quad \hat{p}_2 = \frac{X_1 + X_2}{30},$$

    are both unbiased.

    Determine which of these two estimators is the more efficient and find the relative efficiency of $\hat{p}_1$ with respect to $\hat{p}_2$.

(JMB)

10

Interval estimation

▼

10.1 INTRODUCTION

The methods described in Chapter 9 enable us to calculate *point* (single value) estimates of some population parameters. In this chapter we discuss *interval* estimation, which involves calculating *confidence intervals* for a number of population parameters. For example, we will see how to calculate what is called a 95% confidence interval for the mean of a normally distributed population. Such an interval will be specified by two values called 95% confidence limits.

Essentially a confidence interval is a statement which combines a point estimate with the precision of that estimate. For example, in estimating the mean of a normally distributed population using the sample mean, $\bar{X}$, a point estimate based on 100 values has a smaller variance, and hence a higher precision, than one based on only 10 values ($\sigma^2/100$ compared with $\sigma^2/10$). In confidence interval estimation this translates into the fact that a 95% confidence interval based on 100 values is not as wide as one based on only 10 values.

10.2 CONFIDENCE INTERVAL FOR THE MEAN OF A NORMAL DISTRIBUTION WITH KNOWN VARIANCE

Suppose that a random variable, X, has a normal distribution with unknown mean, μ, and known variance, σ^2. It follows from Section 9.5 that $\bar{X} \sim N(\mu, \sigma^2/n)$, where $\bar{X}$ denotes the mean of a random sample of size n. Since 95% of the area of any normal distribution lies within 1.96 standard deviations of the mean (see Section 6.6), it follows that 95 % of the area of the distribution of $\overline{X}$ lies within 1.96 standard deviations of the mean of the distribution of $\overline{X}$ (see Fig. 10.1).

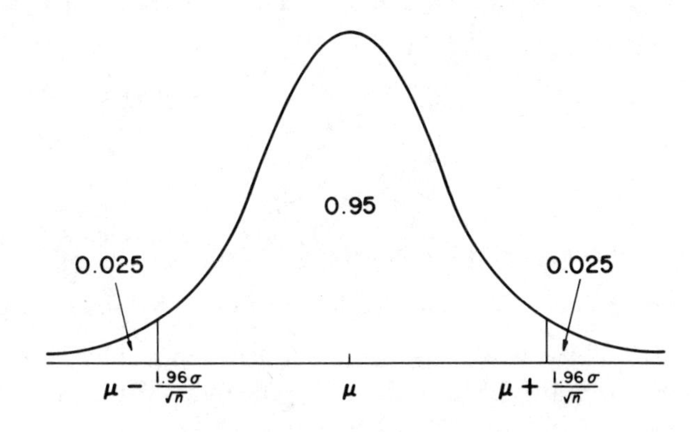

Fig. 10.1 The distribution of $\overline{X}$ if $X \sim N(\mu, \sigma^2)$

As a probability statement we may write:

$$P\left[\mu - \frac{1.96\sigma}{\sqrt{n}} \leqslant \overline{X} \leqslant \mu + \frac{1.96\sigma}{\sqrt{n}}\right] = 0.95$$

This statement may be rewritten with the unknown parameter μ in the middle:

$$P\left[\overline{X} - \frac{1.96\sigma}{\sqrt{n}} \leqslant \mu \leqslant \overline{X} + \frac{1.96\sigma}{\sqrt{n}}\right] = 0.95$$

Because σ is known, if we now take a random sample of size n and calculate the sample mean $\bar{x}$, we can then calculate what are called 95 % *confidence limits for* μ, using the formulae:

$$\bar{x} - \frac{1.96\sigma}{\sqrt{n}} \quad \text{and} \quad \bar{x} + \frac{1.96\sigma}{\sqrt{n}} \tag{10.1}$$

We say that the interval $\bar{x} - \dfrac{1.96\sigma}{\sqrt{n}}$ to $\bar{x} + \dfrac{1.96\sigma}{\sqrt{n}}$ is a 95 % confidence interval for μ.

Example 10.1

In an investigation to estimate the mean weight in kg of 15-year-old children in a particular geographical region, a random sample of 100 children is selected. Previous studies indicate that the variance of the weight of such children is 30 kg^2. Suppose the sample mean weight is $\bar{x} = 38.4$ kg.

Estimate, μ, the weight of all 15-year-old children in the region assuming that these weights are normally distributed.

Our point estimate of μ is 38.4 kg.

Since $\sigma^2 = 30$, $\sigma = 5.477$ kg, and a 95% confidence interval for μ is

$$38.4 - \frac{1.96 \times 5.477}{\sqrt{100}} \quad \text{to} \quad 38.4 + \frac{1.96 \times 5.477}{\sqrt{100}}$$

$$38.4 \pm 1.1 \qquad \text{(1 d.p.)}$$

$$37.3 \quad \text{to} \quad 39.5$$

We are 95% confident that μ lies between 37.3 and 39.5 kg.

Some important points arise from what has been discussed so far in this section:

1. Why do we use the word 'confidence' rather than 'probability'? Why do we not say in the above example, 'there is a 95% probability that μ lies between 37.3 and 39.5 kg'?

   The answer is that, although we can make probability statements about the random variable, $\bar{X}$, we cannot make useful probability statements about known numbers such as 38.4 (the observed value of $\bar{X}$). Also, μ, the weight of all 15-year-old children, has a fixed value which happens to be unknown to the investigator who has only taken a sample of 100 weights. So either μ does lie between 37.3 and 39.5 (with a probability of 1) or it doesn't (with a probability of 0). We must think of another way of expressing our uncertainty, which is why the word 'confidence' is used.
2. What does 95% confidence mean? It means that in 95% of cases (where 95% confidence intervals are quoted), the value of the parameter being estimated will actually lie inside the stated interval. However, in any particular investigation, the investigator will not know whether he/she has 'captured' the value inside the stated confidence interval.
3. Why 95% confidence; why not some other *confidence level*? The use of 95% is partly convention, but levels such as 90%, 98% and sometimes 99.9% are also used. We should not attempt to be certain (100% confident) about the value of a population parameter if we have taken only a sample of the population.

The following examples show what happens when we vary (i) the confidence level, and (ii) the sample size. □

Example 10.2

Using the data from the previous example, calculate a 98% confidence interval for μ.

In Fig. 10.1, if we wanted an area of 0.98 instead of 0.95, we would have to move further than 1.96 standard deviations away from the mean. The connection between the confidence level and the number of standard deviations is via the normal distribution (see Table C.3(b)).

z_α is the number of standard deviations for a confidence level of $100(1-2\alpha)\%$. For example,

(i) when $\alpha = 0.025$, $z_\alpha = 1.96$, confidence level $= 100(1-2\times 0.025) = 95\%$
(ii) when $\alpha = 0.01$, $z_\alpha = 2.33$, confidence level $= 100(1-2\times 0.01) = 98\%$

So a 98% confidence interval for μ is provided by the formula

$$\bar{x} - \frac{2.33\sigma}{\sqrt{n}} \quad \text{to} \quad \bar{x} + \frac{2.33\sigma}{\sqrt{n}} \qquad (10.2)$$

Using the data, $\bar{x} = 38.4$, $n = 100$, a 98% confidence interval for μ is:

$$38.4 \pm \frac{2.33 \times 5.477}{\sqrt{100}}, \quad \text{or } 37.1 \text{ to } 39.7 \text{ (1 d.p)}$$

We are 98% confident that μ lies between 37.1 and 39.7 kg.

We see that the 98% confidence interval is wider than the 95% confidence interval ($39.7-37.1 = 2.6$ compared with $39.5-37.3 = 2.2$ kg). If we wish to be more confident that we have 'captured' μ we must specify a wider interval, given the same sample data. It is left as an exercise for the reader to show that a 100% confidence interval will be infinitely wide, leading to the useless statement that 'we are 100% confident that the mean weight of the children lies between $-\infty$ and $+\infty$!' □

Example 10.3

Suppose we had obtained a sample mean of 38.4 kg from a sample of only 10 values (instead of 100). The 95% confidence interval for μ is then

$$38.4 \pm \frac{1.96 \times 5.477}{\sqrt{10}} \quad \text{or } 35.0 \text{ to } 41.8,$$

much wider than the 95% confidence interval based on 100 values. As we would expect from common sense, reducing our sample size increases the width of the 95% confidence interval. □

In this section we have concentrated on estimating the mean of a normal distribution assuming that the variance, σ^2, is known. The intervals (10.1) and (10.2) apply only to this situation. If we do not know σ^2, however, we may still obtain confidence intervals for μ by making use of some theory involving what is called the t distribution. The next section is a short introduction to this important distribution (which we have not met so far in this book) and the concept of 'degrees of freedom'.

10.3 THE t DISTRIBUTION AND DEGREES OF FREEDOM

If we take samples of size n from a normal distribution, $N(\mu, \sigma^2)$, we know (from Section 9.5) that the random variable $\bar{X} \sim N(\mu, \sigma^2/n)$. It follows that the random variable

$$\frac{\bar{X}-\mu}{\sigma/\sqrt{n}}$$

has an $N(0, 1)$ distribution (the standardized normal). Replacing σ by S, where S^2 is the unbiased estimator of σ^2 (Section 9.7), the distribution of the random variable $\dfrac{\bar{X}-\mu}{S/\sqrt{n}}$ is called the t distribution with $\nu = n-1$ degrees of freedom.

The t distribution[†] is continuous, symmetrical and similar in shape to the normal distribution. Figure 10.2 shows a sketch of a standardized normal distribution and t distributions for $\nu = 4$ and $\nu = 9$ degrees of freedom.

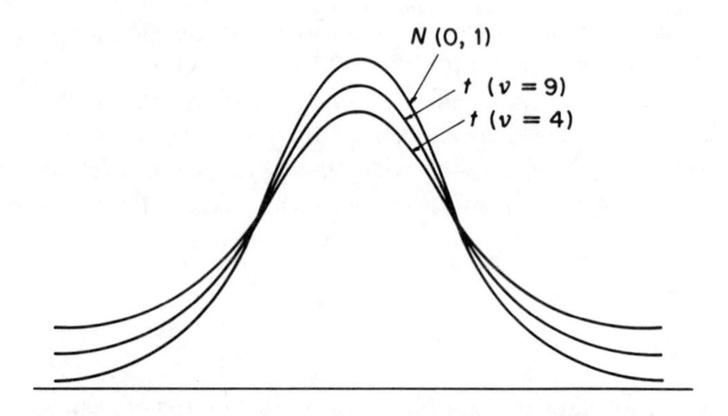

Fig. 10.2 The standardized normal and two t distributions

The t distribution for $\nu = 4$ degrees of freedom is flatter than the normal and has longer tails, while the t distribution for $\nu = 9$ degrees of freedom is more like the normal. In fact, as $n \to \infty$ and hence $\nu = n-1 \to \infty$, the t distribution $\to$ the standardized normal.

The term 'degrees of freedom' requires some explanation at this stage. The unbiased estimator of σ^2 is $S^2 = \dfrac{\Sigma(X_i-\bar{X})^2}{n-1}$ (Section 9.7). There are n differences

[†]This distribution was discovered by W. S. Gosset while working for the Guinness brewery. He wrote his results in a paper using the pseudonym 'Student', which is why the distribution is often referred to as 'Student's t distribution'.

$X_1 - \overline{X}, X_2 - \overline{X}, \ldots X_n - \overline{X}$ which have to be squared and summed. However, if we know $(n-1)$ of these differences we can obtain the nth since $\overline{X} = \frac{\Sigma X_i}{n}$. So these are only $(n-1)$ independent differences, and we say we have only $(n-1)$ 'degrees of freedom' for the calculation of S^2.

10.4 CONFIDENCE INTERVAL FOR THE MEAN OF A NORMAL DISTRIBUTION WITH UNKNOWN VARIANCE

It was stated in the last section that, for samples taken from a normal distribution, $\frac{\overline{X}-\mu}{S/\sqrt{n}}$ has a t distribution with $\nu = n-1$ degrees of freedom. It follows that

$$P\left(-t_{0.025,\, n-1} \leqslant \frac{\overline{X}-\mu}{S/\sqrt{n}} \leqslant t_{0.025,\, n-1}\right) = 0.95 \qquad \text{(see Fig. 10.3).}$$

In general, $t_{\alpha,\,\nu}$ means the 'upper percentage point' of a t distribution with ν degrees of freedom corresponding to a right-hand tail area of α; see Table C.4.

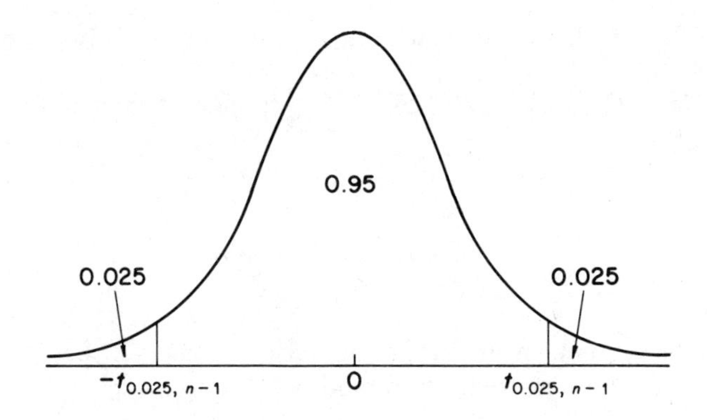

Fig. 10.3 A t distribution with $\nu = n-1$ degrees of freedom

The statement above may be rewritten with the unknown parameter μ in the middle:

$$P\left(\overline{X} - t_{0.025,\, n-1}\frac{S}{\sqrt{n}} \leqslant \mu \leqslant \overline{X} + t_{0.025,\, n-1}\frac{S}{\sqrt{n}}\right) = 0.95$$

If we now take a random sample of size n, calculate $\bar{x}$ and s, and look up the appropriate value of $t_{0.025,\, n-1}$ in Table C.4, we can calculate 95% confidence limits for μ using the formulae:

$$\bar{x} - t_{0.025,\, n-1}\frac{s}{\sqrt{n}} \quad \text{and} \quad \bar{x} + t_{0.025,\, n-1}\frac{s}{\sqrt{n}} \tag{10.3}$$

We say that $\bar{x} - t_{0.025,\, n-1}\dfrac{s}{\sqrt{n}}$ to $\bar{x} + t_{0.025,\, n-1}\dfrac{s}{\sqrt{n}}$ is a 95% confidence interval for μ.

Note
It can be shown that (10.3) applies even if the variable in question is only approximately normally distributed.

Example 10.4

In an investigation to estimate the mean weight in kg of 15-year-old children in a particular geographical region, a random sample of 10 children is selected. Suppose the sample statistics are $\bar{x} = 38.4$, $s = 5.5$ kg.

Estimate μ, the mean weight of all 15-year-old children in the region, assuming that these weights are normally distributed, with (i) 95% confidence, (ii) 98% confidence.

(i) For 95% confidence, we need $t_{0.025,\, 9}$, since $n-1 = 10-1 = 9$ in this example.
From Table C.4, $t_{0.025,\, 9} = 2.26$. A 95% confidence interval for μ is:

$$38.4 - \frac{2.26 \times 5.5}{\sqrt{10}} \quad \text{to} \quad 38.4 + \frac{2.26 \times 5.5}{\sqrt{10}}$$

$$38.4 \pm 3.9 \qquad (1\ \text{d.p.})$$

$$34.5 \quad \text{to} \quad 42.3$$

We are 95% confident that μ lies between 34.5 and 42.3 kg.

(ii) For 98% confidence, we need to modify (10.3) to read:

$$\bar{x} - t_{0.01,\, n-1}\frac{s}{\sqrt{n}} \quad \text{to} \quad \bar{x} + t_{0.01,\, n-1}\frac{s}{\sqrt{n}} \tag{10.4}$$

From Table C.4, $t_{0.01,\, 9} = 2.82$, so a 98% confidence interval for μ is

$$38.4 - \frac{2.82 \times 5.5}{\sqrt{10}} \quad \text{to} \quad 38.4 + \frac{2.82 \times 5.5}{\sqrt{10}}$$

$$38.4 \pm 4.9$$

$$33.5 \quad \text{to} \quad 43.3$$

We are 98% confident that μ lies between 33.5 and 43.3 kg. Notice again that the higher confidence is associated with the wider interval. □

10.5 THE SAMPLE SIZE REQUIRED TO ESTIMATE THE MEAN OF A NORMAL DISTRIBUTION

Consider the statement made in Example 10.4(i), 'We are 95% confident that μ lies between 34.5 and 42.3 kg'. We may decide that this interval is too wide, in other words that our estimate of μ is not precise enough. What should we do? We could take a larger sample, but how large should it be? The answer is that it depends on how wide we wish to make the 95% confidence interval. A formula for this width is:

$$\left(\bar{x} + t_{0.025,\, n-1}\frac{s}{\sqrt{n}}\right) - \left(\bar{x} - t_{0.025,\, n-1}\frac{s}{\sqrt{n}}\right) = 2t_{0.025,\, n-1}\frac{s}{\sqrt{n}}$$

For our example, suppose we decide to reduce the width from 7.8 (= 42.3 − 34.5) to 3 kg. We must solve the equation $2t_{0.025,\, n-1}\dfrac{s}{\sqrt{n}} = 3$, to find the required sample size, n.

We can assume that s will remain unchanged at 5.5 kg. Further, if we assume that the resulting value of n will be large, we can take $t_{0.025,\, n-1}$ to be 2 since we notice that $t_{0.025,\, n-1}$ is approximately 2 for large n.

So

$$2 \times 2 \times \frac{5.5}{\sqrt{n}} = 3$$

$$n = \left(\frac{2 \times 2 \times 5.5}{3}\right)^2$$

$$n = 53.8$$

We require a sample size of 54. Notice that, in order to decide how large a sample to take, we needed to 'input' two pieces of information:

(i) an estimate of the variance, based on a small 'pilot' survey;
(ii) the required width of the 95% confidence interval for μ.

10.6 CONFIDENCE INTERVAL FOR THE DIFFERENCE BETWEEN THE MEANS OF TWO NORMAL DISTRIBUTIONS (UNPAIRED SAMPLES DATA)

In many practical investigations the aim is not to estimate the mean for one population, but to compare the means of two populations. In this section we discuss the case in which we wish to estimate the difference between the means of two normally distributed populations. In some investigations individuals in one population are 'matched' or 'paired' with those in the other population. We refer to these data as 'paired samples data'. In other investigations pairing is impracticable or inappropriate, and the data are referred to as 'unpaired samples data'. We discuss the latter in this section and the former in the next section.

Example 10.5

Thirty children are randomly selected from all 10-year-old children in a town to compare two methods, 1 and 2, of teaching children how to spell a list of 100 'difficult words'. 15 children were randomly selected to be taught by method 1 and the other 15 by method 2. Following the learning period each child is given a spelling test. Let X_1, X_2 denote the spelling test scores for methods 1 and 2 respectively. Suppose the sample statistics are:

$\bar{x}_1 = 65.3$, $s_1 = 4.2$, $n_1 = 15$, for the children taught by method 1,
$\bar{x}_2 = 62.4$, $s_2 = 3.9$, $n_2 = 15$, for the children taught by method 2.

Estimate $(\mu_1 - \mu_2)$, where μ_1 and μ_2 are the mean scores for methods 1 and 2 for all 10-year-old children in the town, assuming that X_1 and X_2 are normally distributed with the same, but unknown, variance σ^2.

A 95% confidence interval for $(\mu_1 - \mu_2)$ may be obtained using the formula

$$\bar{x}_1 - \bar{x}_2 \pm t_{0.025, n_1+n_2-2}\, s\sqrt{\frac{1}{n_1}+\frac{1}{n_2}} \qquad (10.5)$$

where $s^2 = \dfrac{(n_1-1)s_1^2 + (n_2-1)s_2^2}{n_1+n_2-2}$ is an unbiased estimator of the common variance, σ^2 (see Section 9.9).

The formula (10.5) is similar to (10.3) and stems from the fact that

$$\frac{\bar{X}_1 - \bar{X}_2}{s\sqrt{\dfrac{1}{n_1}+\dfrac{1}{n_2}}}$$

has a t distribution with $n_1 + n_2 - 2$ degrees of freedom.

Note

It can be shown that (10.5) applies even if X_1 and X_2 are only approximately normally distributed.

For the sample data above, $n_1 + n_2 - 2 = 28$, $t_{0.025, 28} = 2.04$ (from Table C.4), and

$$s^2 = \frac{(15-1)\times 4.2^2 + (15-1)\times 3.9^2}{15+15-2} = 16.425$$

$$s = 4.05.$$

So a 95% confidence interval for $(\mu_1 - \mu_2)$ is

$$(65.3 - 62.4) \pm 2.04 \times 4.05\sqrt{\frac{1}{15}+\frac{1}{15}}$$

$$2.9 \pm 3.0$$

$$-0.1 \quad \text{to} \quad 5.9$$

We are 95% confident that $(\mu_1 - \mu_2)$ lies between -0.1 and 5.9.

Note

Because $\mu_1 - \mu_2 = -0.1$ implies $\mu_1 < \mu_2$, while $\mu_1 - \mu_2 = 5.9$ implies $\mu_1 > \mu_2$, the interval indicates that the mean scores for methods 1 and 2 are 'not significantly different'. We will discuss the important concept of statistical significance fully in Chapter 11. □

The confidence interval described above is for the difference between two means if we know that the distributions have the same variance. What if we know or suspect that they are not equal? There are two methods we can use which are now briefly described:

(a) If we know the values of the variances σ_1^2 and σ_2^2, and we also know that the distributions are normal, then a $100(1-2\alpha)\%$ confidence interval for $(\mu_1 - \mu_2)$ may be obtained using the formula:

$$(\bar{x} - \bar{x}_2) + z_\alpha \sqrt{\frac{\sigma_1^2}{n_1} + \frac{\sigma_2^2}{n_2}}$$

For example, a 95% confidence interval may be obtained using

$$\alpha = 0.025 \quad \text{and hence} \quad z_\alpha = 1.96$$

(b) If we estimate σ_1^2 and σ_2^2 from data obtained from large samples ($n_1, n_2 > 30$, say) and we suspect that σ_1^2 and σ_2^2 are not equal (based on the values we calculate for their estimates s_1^2 and s_2^2), then in this case the assumption of normality is relatively unimportant and a $100(1-2\alpha)\%$ confidence interval for $(\mu_1 - \mu_2)$ may be obtained using the formula:

$$(\bar{x}_1 - \bar{x}_2) \pm z_\alpha \sqrt{\frac{s_1^2}{n_1} + \frac{s_2^2}{n_2}}$$

For example, a 95% confidence interval may be obtained using

$\alpha = 0.025$ and hence $z_\alpha = 1.96$

Example 10.6

A random sample of 100 boys of a certain age have a mean height of 152.3 cm and a standard deviation of 6 cm, while a random sample of 120 girls of the same age have a mean height of 148.6 cm and a standard deviation of 5 cm. Estimate $(\mu_1 - \mu_2)$, where μ_1 and μ_2 are the mean heights of boys and girls, respectively, for the age in question.

Using the formula in (b) above, a 95% confidence interval for $(\mu_1 - \mu_2)$ is given by:

$$152.3 - 148.6 \pm 1.96\sqrt{\frac{6^2}{100} + \frac{5^2}{120}}$$

$$3.7 \pm 1.5$$

$$2.2 \quad \text{to} \quad 5.2 \text{ cm}$$

We are 95% confident that $(\mu_1 - \mu_2)$ lies between 2.2 and 5.2 cm. □

10.7 CONFIDENCE INTERVAL FOR THE MEAN OF A NORMAL DISTRIBUTION OF DIFFERENCES (PAIRED SAMPLES DATA)

In some investigations the data occur in pairs. For example, we may measure the blood pressure of a number of hospital patients before and after some treatment aimed at reducing blood pressure. Each patient will provide a pair of sample values. Another example is when a group of individuals are matched in pairs, one of each pair being randomly allocated to one of two treatments while the other is allocated to the other treatment. In any analysis of such data we must take account of the fact that the data constitute a number of pairs of values.

Example 10.7

Fifteen children are randomly selected from all 10-year-old children in a town. Each child is then matched with another (previously unselected) child for age, sex, maturation and spelling ability. One child in each of the 15 pairs of children is randomly allocated to be taught how to spell a list of 100 'difficult' words using method 1. The other child in each pair is taught by method 2. Following the learning period each child is given a spelling test.

Estimate the mean difference in the scores for the two methods, assuming that these differences are normally distributed. Suppose the (sample) scores are as shown in Table 10.1.

Table 10.1 Spelling test scores for fifteen pairs of children

Pair No	1	2	3	4	5	6	7	8	9	10	11	12	13	14	15
Method 1 score	65	65	70	69	62	66	64	62	63	61	66	76	60	69	63
Method 2 score	61	66	66	63	59	66	57	57	60	59	65	69	58	65	65
d = difference in scores	4	−1	4	6	3	0	7	5	3	2	1	7	2	4	−2

The 15 differences, d, constitute a random sample of differences. We can imagine a population of such differences if we consider all possible matched pairs of 10-year-old children in the town. Let this population have a mean μ_d, say. Then a 95% confidence interval for μ_d is given by

$$\bar{d} \pm t_{0.025,\, n-1} \frac{s_d}{\sqrt{n}} \qquad (10.6)$$

where $\bar{d}$ and s_d are the mean and standard deviation obtained from the sample of n differences. (10.6) is essentially the same as (10.3), except for the use of d instead of x (to indicate that we are working with differences).

Note

It can be shown that (10.6) applies even if the differences are only approximately normally distributed.

For the sample data, $\sum d_i = 45, \quad \sum d_i^2 = 239,$

So

$$\bar{d} = \frac{\sum d_i}{n} = \frac{45}{15} = 3.0,$$

$$s_d = \sqrt{\frac{\sum d_i^2 - n\bar{d}^2}{n-1}} \qquad \text{using (9.6)}$$

$$= \sqrt{\frac{239 - 15 \times 3^2}{14}}$$

$$= 2.73.$$

Also $t_{0.025,\, 14} = 2.14$. So a 95% confidence interval for μ_d is

$$3.0 \pm \frac{2.14 \times 2.73}{\sqrt{15}}$$

$$3.0 \pm 1.5$$

$$1.5 \text{ to } 4.5$$

We are 95% confident that μ_d lies between 1.5 and 4.5.

Note

Because both values of μ_d, namely 1.5 and 4.5, imply that method 1 scores are on average higher than method 2 scores, we say the mean scores for the two methods are significantly different. We will discuss the concept of 'statistical significance' fully in Chapter 11. □

10.8 THE χ^2 DISTRIBUTION

Having introduced the t distribution in Section 10.3 because it was needed in Sections 10.4, 10.6 and 10.7, we now introduce the χ^2 distribution which has an application in Section 10.9, where we wish to find a confidence interval for the variance of a normal distribution.

Suppose we take samples of size n from a normal distribution with variance σ^2. Then the distribution of the random variable $(n-1)S^2/\sigma^2$ is called the χ^2 distribution with $\nu = n-1$ degrees of freedom. (χ^2 is pronounced 'chi-squared').

The χ^2 distribution is continuous. Its shape depends on the value of ν. For small ν it is positively skew, for large ν it approaches the shape of the normal distribution (see Fig. 10.4).

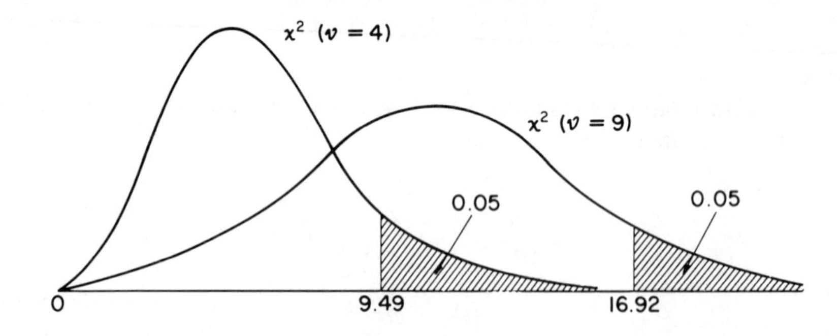

Fig. 10.4 Two χ^2 distributions with 4 and 9 degrees of freedom

10.9 CONFIDENCE INTERVAL FOR THE VARIANCE OF A NORMAL DISTRIBUTION

Since $(n-1)S^2/\sigma^2$ has a χ^2 distribution with $(n-1)$ degrees of freedom, it follows that

$$P\left(\chi^2_{0.975,\,n-1} \leqslant \frac{(n-1)S^2}{\sigma^2} \leqslant \chi^2_{0.025,\,n-1}\right) = 0.95 \qquad \text{(see Fig. 10.5)}$$

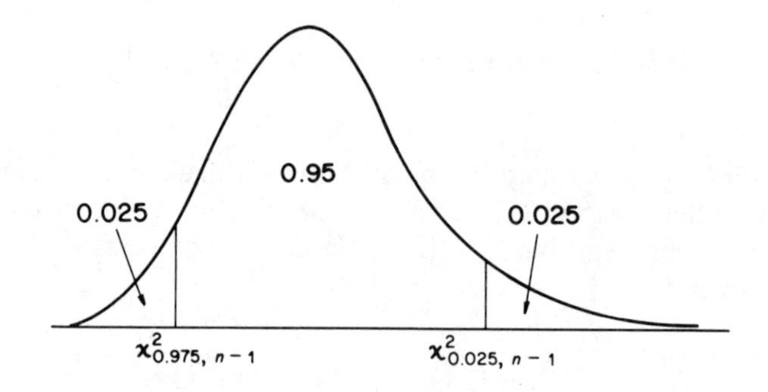

Fig. 10.5 A χ^2 distribution with $\nu = n-1$ degrees of freedom

This statement may be rewritten with the unknown parameter σ^2 in the middle:

$$P\left(\frac{(n-1)S^2}{\chi^2_{0.025,\,n-1}} \leqslant \sigma^2 \leqslant \frac{(n-1)S^2}{\chi^2_{0.975,\,n-1}}\right) = 0.95$$

If we now take a random sample of size n, calculate s^2, and look up the appropriate values of $\chi^2_{0.025,\,n-1}$ and $\chi^2_{0.975,\,n-1}$ in Table C.5, we can calculate 95% confidence limits for σ^2 using the formulae:

$$\frac{(n-1)s^2}{\chi^2_{0.025,\,n-1}} \quad \text{to} \quad \frac{(n-1)s^2}{\chi^2_{0.975,\,n-1}} \tag{10.7}$$

We say that

$$\frac{(n-1)s^2}{\chi^2_{0.025,\,n-1}} \quad \text{to} \quad \frac{(n-1)s^2}{\chi^2_{0.975,\,n-1}}$$

is a 95% confidence interval for σ^2.

Example 10.8

A random sample of size 10 is taken from a normal distribution with unknown variance σ^2. The sample statistics are $\bar{x} = 38.4$ and $s = 5.5$. Calculate a 95% confidence interval for σ^2.

From Table C.5, $\chi^2_{0.025,9} = 19.02$ and $\chi^2_{0.975,9} = 2.70$.
So a 95% confidence interval for σ^2 is

$$\frac{(10-1)\times 5.5^2}{19.02} \quad \text{to} \quad \frac{(10-1)\times 5.5^2}{2.70}$$

$$14.3 \quad \text{to} \quad 100.8$$

We are 95% confident that σ^2 lies between 14.3 and 100.8.
Notice how wide the confidence interval is. We could reduce the width by taking a larger sample, as the next example shows. □

Example 10.9

Using the data from the last example, what sample size would be required to reduce the width of the 95% confidence interval to 20 (instead of 100.8 − 14.3 = 86.5)?

We require

$$\frac{(n-1)\times 5.5^2}{\chi^2_{0.975,\,n-1}} - \frac{(n-1)\times 5.5^2}{\chi^2_{0.025,\,n-1}} \quad \text{to be 20.}$$

By trial and error, we find the values in Table 10.2.
By interpolating between the last two rows in Table 10.2 we see that a sample size of approximately 80 is required.

Table 10.2 How the width of a 95% confidence interval for σ^2 depends on the sample size

$v = n - 1$	*Width of 95% confidence interval*
20	45.4
40	29.1
60	23.0
100	17.4

10.10 CONFIDENCE INTERVAL FOR A BINOMIAL PARAMETER, p

If X is the number of successes in a binomial experiment having n trials, and if $np > 5$ and $n(1-p) > 5$, then we know from Section 7.3 that X is approximately normally distributed with mean np and standard deviation $\sqrt{np(1-p)}$. Similarly, X/n is approximately normally distributed with mean p and standard deviation $\sqrt{p(1-p)/n}$, where X/n is the proportion of successes. Using the property that 95% of the area of a normal distribution lies within 1.96 standard deviations of the mean, it follows that

$$P\left(p - 1.96\sqrt{\frac{p(1-p)}{n}} \leqslant \frac{X}{n} \leqslant p + 1.96\sqrt{\frac{p(1-p)}{n}}\right) = 0.95 \text{ (approx.)}$$

This statement may be rewritten as follows:

$$P\left(\frac{X}{n} - 1.96\sqrt{\frac{p(1-p)}{n}} \leqslant p \leqslant \frac{X}{n} + 1.96\sqrt{\frac{p(1-p)}{n}}\right) = 0.95 \text{ (approx.)}$$

If the parameter p is unknown, we can replace the p under the square root signs by X/n to give

$$\frac{X}{n} - 1.96\sqrt{\frac{\frac{X}{n}\left(1 - \frac{X}{n}\right)}{n}} \quad \text{to} \quad \frac{X}{n} + 1.96\sqrt{\frac{\frac{X}{n}\left(1 - \frac{X}{n}\right)}{n}}$$

as an approximate 95% confidence interval for p, under certain conditions. These conditions are $np > 5$ and $n(1-p) > 5$, but since p is unknown we must replace it by the point estimator, X/n (Section 9.8).

The above implies that if we carry out a binomial experiment consisting of n trials and the parameter p is unknown, and if we observe x successes, and if both $x > 5$ and $n - x > 5$, then an approximate 95% confidence interval for p is given by:

$$\frac{x}{n} \pm 1.96\sqrt{\frac{\frac{x}{n}\left(1-\frac{x}{n}\right)}{n}} \tag{10.8}$$

Example 10.10

A drawing-pin is tossed 100 times. It lands 'point-up' 64 times. Estimate p, the probability of 'point-up' in a single toss.

We can confirm that we have a binomial experiment by checking the four conditions (Section 6.2).

For our sample data $x = 64$, $n = 100$, so the conditions for using (10.8) apply since $64 > 5$ and $100 - 64 > 5$. So we may state that an approximate 95% confidence interval for p is:

$$\frac{64}{100} \pm 1.96\sqrt{\frac{0.64 \times 0.36}{100}}$$

$$0.64 \pm 0.09$$

$$0.55 \quad \text{to} \quad 0.73$$

We are 95% confident that p lies between 0.55 and 0.73. □

10.11 THE SAMPLE SIZE REQUIRED TO ESTIMATE A BINOMIAL PARAMETER, p

Consider the statement made in the last example, 'we are 95% confident that p lies between 0.55 and 0.73'. We may decide that this interval is too wide, in other words that our estimate of p is not precise enough. We can reduce the width by taking a larger sample, but how large should it be? The answer is that it depends on how wide we wish to make the 95% confidence interval. By (10.8) a formula for the width of this interval is:

$$2 \times 1.96\sqrt{\frac{\frac{x}{n}\left(1-\frac{x}{n}\right)}{n}}$$

For our example, suppose we decide to reduce the width from 0.18 (= 0.73 − 0.55) to 0.10. We must solve the equation

$$2 \times 1.96\sqrt{\frac{\frac{x}{n}\left(1-\frac{x}{n}\right)}{n}} = 0.10$$

to find the required sample size n. We can assume that the proportion of successes, x/n, will remain unchanged at 0.64.

$$2 \times 1.96 \sqrt{\frac{0.64 \times 0.36}{n}} = 0.10.$$

$$n = \left(\frac{2 \times 1.96}{0.10}\right)^2 \times 0.64 \times 0.36 = 354$$

We require a sample size of about 350. Notice that in order to decide how large a sample to take we needed to 'input' two pieces of information:

(i) an estimate of p, based on relatively small 'pilot' experiment;
(ii) the required width of the 95% confidence interval for p.

We could, however, obtain a conservative estimate of n without carrying out a 'pilot' experiment since $(x/n)(1-x/n)$ has a maximum value of 0.25 (an exercise left for the reader to solve). It follows that the maximum value of n to give a width of 0.10 is:

$$\left(\frac{2 \times 1.96}{0.10}\right)^2 \times 0.25 = 384$$

which is about 10% bigger than the 350 we calculated using the result of a pilot experiment.

10.12 CONFIDENCE INTERVAL FOR THE DIFFERENCE BETWEEN TWO BINOMIAL PARAMETERS

Suppose we carry out two binomial experiments, one having unknown parameter p_1, the other having unknown parameter p_2. If from n_1 trials in the first experiment we observe x_1 successes, and from n_2 trials in the second experiment we observe x_2 successes, what is our estimate of $(p_1 - p_2)$, the difference between the two binomial parameters?

A point estimate for $(p_1 - p_2)$ is $x_1/n_1 - x_2/n_2$, while a 95% confidence interval for $(p_1 - p_2)$ is given by:

$$\left(\frac{x_1}{n_1} - \frac{x_2}{n_2}\right) \pm 1.96 \sqrt{\frac{\frac{x_1}{n_1}\left(1 - \frac{x_1}{n_1}\right)}{n_1} + \frac{\frac{x_2}{n_2}\left(1 - \frac{x_2}{n_2}\right)}{n_2}} \qquad (10.9)$$

providing $x_1 > 5$, $n_1 - x_1 > 5$, $x_2 > 5$ and $n_2 - x_2 > 5$.

Example 10.11

Of a random sample of 25 boys and 20 girls, 10 boys and 6 girls said they intended to take Physics at A-level.

Estimate the difference in the proportions of boys and girls who intended to take Physics at A-level.

Let these proportions be p_1 and p_2, respectively. From the sample data, $x_1 = 10$, $n_1 = 25$, $x_2 = 6$, $n_2 = 20$. The conditions for using (10.9) apply to these data, so a 95% confidence interval for $(p_1 - p_2)$ is:

$$\left(\frac{10}{25} - \frac{6}{20}\right) \pm 1.96\sqrt{\frac{0.4 \times 0.6}{25} + \frac{0.3 \times 0.7}{20}}$$

$$0.10 \pm 0.28$$

$$-0.18 \quad \text{to} \quad +0.38$$

We are 95% confident that $(p_1 - p_2)$ lies between -0.18 and 0.38. As with previous examples, we could reduce the width of the 95% confidence interval by taking larger samples.

Note

Because $p_1 - p_2 = -0.18$ implies $p_1 < p_2$, while $p_1 - p_2 = 0.38$ implies the opposite, the interval indicates that the proportions of males and females who intended to take A-level Physics are not 'significantly' different. We will discuss the concept of 'statistical significance' fully in Chapter 11. □

10.13 CONFIDENCE INTERVALS BASED ON THE CENTRAL LIMIT THEOREM

The Central Limit Theorem (Section 9.5) states that, if $\overline{X}$ denotes the mean of samples of size n taken from a parent population with mean μ and variance σ^2, then $\overline{X}$ is approximately normally distributed, if n is large ($\geqslant 30$, say).
We also showed in Section 9.5 that $E(\overline{X}) = \mu$ and $\text{Var}(\overline{X}) = \sigma^2/n$.

In other words: $\overline{X} \sim N(\mu, \sigma^2/n)$, for large n.

Since we have made no assumptions about the distribution of the sampled variable X, this result is useful in obtaining confidence intervals for the parameters of various distributions (other than the normal and binomial which have already been covered in this chapter).

Example 10.12

The number of telephone calls received at a switchboard per minute is thought to have a Poisson distribution, but its parameter, a, is unknown.

For each minute in a 60-minute period the number of calls received is noted. Suppose, that the mean numbers of calls in this period is 2.53 per minute. Obtain a 95% confidence interval for a.

Since a is the mean and also the variance of the Poisson, we can say that:

$$\bar{X} \sim N(a, a/n), \text{ for large } n,$$

where $\bar{X}$ denotes the mean of samples taken from a Poisson distribution.

$$P\left(a - 1.96\sqrt{\frac{a}{n}} \leqslant \bar{X} \leqslant a + 1.96\sqrt{\frac{a}{n}}\right) = 0.95, \text{ approx.}$$

$$\therefore P\left(\bar{X} - 1.96\sqrt{\frac{a}{n}} \leqslant a \leqslant \bar{X} + 1.96\sqrt{\frac{a}{n}}\right) = 0.95, \text{ approx.}$$

We can replace the a under the square root sign by $\bar{X}$, since $\bar{X}$ is an unbiased point estimator of a, to give

$$\bar{x} - 1.96\sqrt{\frac{\bar{x}}{n}} \quad \text{to} \quad \bar{x} + 1.96\sqrt{\frac{\bar{x}}{n}}$$

as an approximate 95% confidence interval for a, if n is large. Here $\bar{x}$ is the observed value of $\bar{X}$.

For our sample data, $\bar{x} = 2.53$, $n = 60$, and a 95% confidence interval for a is

$$2.53 - 1.96\sqrt{\frac{2.53}{60}} \quad \text{to} \quad 2.53 + 1.96\sqrt{\frac{2.53}{60}}$$

$$2.13 \quad \text{to} \quad 2.93$$

We are 95% confident that a lies between 2.13 and 2.93. □

Example 10.13

At a building-society office, the time between successive arrivals of customers joining a queue for service was thought to have an exponential distribution with unknown parameter λ. Fifty such times were noted and the mean was 2.15 minutes. Obtain a 95% confidence interval for λ.

Since the mean and variance of the exponential distribution are $1/\lambda$ and $1/\lambda^2$, respectively (Section 6.7), we can say that:

$$\bar{X} \sim N\left(\frac{1}{\lambda}, \frac{1}{\lambda^2 n}\right), \qquad \text{for large } n$$

where $\bar{X}$ denotes the mean of samples taken from an exponential distribution.

$$P\left(\frac{1}{\lambda} - \frac{1.96}{\lambda\sqrt{n}} \leqslant \bar{X} \leqslant \frac{1}{\lambda} + \frac{1.96}{\lambda\sqrt{n}}\right) = 0.95, \quad \text{approx.}$$

i.e.

$$P\left(\frac{1 - 1.96/\sqrt{n}}{\bar{X}} \leqslant \lambda \leqslant \frac{1 + 1.96/\sqrt{n}}{\bar{X}}\right) = 0.95, \quad \text{approx.}$$

So if we take a sample of size n ($\geqslant 30$) and obtain the sample mean, $\bar{x}$, a 95% confidence interval for λ is given by:

$$\frac{1-1.96/\sqrt{n}}{\bar{x}} \quad \text{to} \quad \frac{1+1.96/\sqrt{n}}{\bar{x}}$$

For our sample data, $\bar{x} = 2.15$, $n = 50$, and a 95% confidence interval for is:

$$\frac{1-1.96/\sqrt{50}}{2.15} \quad \text{to} \quad \frac{1+1.96/\sqrt{50}}{2.15}$$

$$0.336 \quad \text{to} \quad 0.594$$

We are 95% confident that λ lies between 0.336 to 0.594. □

10.14 SUMMARY

Confidence intervals provide a way of estimating unknown parameters in the form of two values (called confidence limits) between which we can state, with a certain level of confidence, that the true value of the parameter lies.

Table 10.3 summarizes the main applications covered in this chapter.

It is also possible to calculate the required sample size if we can specify the required width of a 95% confidence interval and have the results of a small pilot experiment or survey.

Table 10.3 95% Confidence intervals for various parameters

Parameter	*Formula for a 95% confidence interval*	*Assumption or condition*
μ	$\bar{x} \pm \frac{1.96\sigma}{\sqrt{n}}$	X is 'normal', σ^2 is known
μ	$\bar{x} \pm t_{0.025, n-1}\frac{s}{\sqrt{n}}$	X is approx. 'normal'
$\mu_1 - \mu_2$	$(\bar{x}_1 - \bar{x}_2) \pm t_{0.25, n_1+n_2-2}\, s\sqrt{\frac{1}{n_1} + \frac{1}{n_2}}$, where $s^2 = \frac{(n_1 - 1)s_1^2 + (n_2 - 1)s_2^2}{n_1 + n_2 - 2}$	X_1, X_2 have same variance,† and are approx. 'normal' (unpaired samples data)
μ_d	$\bar{d} \pm t_{0.025, n-1}\frac{s_d}{\sqrt{n}}$	Differences are approx. 'normal' (paired samples data)
σ^2	$\frac{(n-1)s^2}{\chi^2_{0.025, n-1}}$ to $\frac{(n-1)s^2}{\chi^2_{0.975, n-1}}$	X is 'normal'
p	$\frac{x}{n} \pm 1.96\sqrt{\frac{\frac{x}{n}\left(1 - \frac{x}{n}\right)}{n}}$	X is 'binomial', $0.1 \leqslant x/n \leqslant 0.9$, $x > 5$
$p_1 - p_2$	$\left(\frac{x_1}{n_1} - \frac{x_2}{n_2}\right) \pm 1.96\sqrt{\frac{\frac{x_1}{n_1}\left(1 - \frac{x_1}{n_1}\right)}{n_1} + \frac{\frac{x_2}{n_2}\left(1 - \frac{x_2}{n_2}\right)}{n_2}}$	X_1, X_2 are 'binomial', $x_1 > 5, n_1 - x > 5$, $x_2 > 5, n_2 - x_2 > 5$.

† See other formulae at the end of Section 10.6 for the case of unequal variances

WORKSHEET 10

Section 10.2

1. How is the width of a confidence interval for the mean of a normal distribution with known variance affected if:

   (i) the confidence level is increased;
   (ii) the sample size is decreased;
   (iii) the confidence level and the sample size are both increased?

2. Random samples of size n are taken from a normal population with mean μ and variance σ^2. State the distribution of the means of these samples, and write down its mean and variance.

   Explain briefly, referring to your projects if possible, what is meant by 'a 98 % confidence interval for a population mean'.

   The error made when a certain instrument is used to measure the body length of a butterfly of a particular species is known to be normally distributed with mean 0 and standard deviation 1 mm. Calculate, to 3 decimal places, the probability that the error made when the instrument is used once is numerically less than 0.4 mm.

   Given that the body length of a butterfly is measured nine times with the instrument, calculate, to 3 decimal places, the probability that the mean of the nine readings will be within 0.5 mm of the true length.

   Given that the mean of the nine readings was 22.53 mm, determine a 98 % confidence interval for the true body length of the butterfly.

   (London)

Section 10.4

3. A random sample of eleven bags were selected from a machine packaging sugar in bags marked 1 kg. The actual weights of sugar in kg were:

   1.017 1.051 1.078 1.033 0.996 1.059 1.082 1.014 1.040 1.072 0.998

   Assuming that the weights are normally distributed, calculate a 95 % confidence interval for the mean weight of sugar in bags marked 1 kg.

Section 10.5

4. The diameter of a certain type of pine tree is normally distributed with a standard deviation of 8 cm. How many trees should be sampled if it is required to estimate the mean diameter to within ± 1.5 cm with

   (i) 95 % confidence (ii) 99 % confidence?

5. It is proposed to carry out a survey to estimate, with 95 % confidence, the mean weekly rent paid by college students in a certain town to within ± 50 p. A pilot survey indicates that the standard deviation of weekly rent is £2. How large a sample should be taken in the survey? What assumption have you made? ▶

Section 10.6

6. Two petrol additives are used in an experiment to determine the petrol consumption of cars using one or other of the additives. Of a random sample of 18 nominally identical cars, 9 were tested using Additive 1 and the other 9 were tested using Additive 2. The results in m.p.g. were as follows:

Additive 1	32.3	37.3	35.3	33.7	31.9	30.0	34.6	32.8	36.4
Additive 2	38.4	34.8	36.6	32.0	33.9	35.7	37.5	39.3	34.3

Calculate 95% confidence intervals for

(i) the mean petrol consumption for cars using Additive 1;
(ii) the mean petrol consumption for cars using Additive 2;
(iii) the difference between the mean petrol consumptions for the two additives.

In each case state the assumptions you have made.

7. Two independent samples of size 7 are taken, one from population 1 and the other from population 2. Assuming that the two populations are normally distributed and have the same variance, use the following summary statistics to calculate a 99% confidence interval for the difference between the means of the two populations.

Sample 1 Sum of the 7 observations, $\sum_{i=1}^{7} x_{1i} = 35$

Sum of squares of the 7 observations, $\sum_{i=1}^{7} x_{1i}^2 = 287$

Sample 2 Sum of the 7 observations, $\sum_{i=1}^{7} x_{2i} = 49$

Sum of squares of the 7 observations, $\sum_{i=1}^{7} x_{2i}^2 = 399$

8. Two variables X_1 and X_2 are normally distributed with unknown means μ_1 and μ_2, and known variances σ_1^2 and σ_2^2. If $\bar{x}_1$ and $\bar{x}_2$ denote the observed sample means of sizes n_1 and n_2, respectively, taken from the distributions of X_1 and X_2, show that a 95% confidence interval may be obtained using:

$$\bar{x}_1 - \bar{x}_2 \pm 1.96\sqrt{\frac{\sigma_1^2}{n_1} + \frac{\sigma_2^2}{n_2}}$$

9. (a) In an experiment to test the weight-reducing effect of a new diet, six women were weighed before and after following the diet. Their weight losses, in kg, were

1.6, 0.8, 0.1, 1.4, 0.5, 0.7

Stating clearly any assumptions that you make, calculate a 95% confidence interval for the mean loss in weight for women following the diet. State whether or not your interval supports the claim that the average loss in weight for women following this diet is 1 kg.

(b) In a large-scale experiment for comparing two diets A and B, 100 women followed diet A and 80 women followed diet B. From the weight losses of the 100 women who followed diet A it was concluded that the unbiased estimate of the mean weight loss was 2.97 kg and the unbiased estimate of the variance of the weight ▶

losses was 1.62 kg^2. From the weight losses of the 80 women who followed diet B the corresponding unbiased estimates were 2.31 kg and 1.55 kg^2, respectively. Stating clearly any assumptions that you make, calculate an approximate 95% confidence interval for the difference between the mean losses in weight for the two diets. What can you say about the relative weight-reducing effects of the two diets? (WJEC)

10. (a) The drained weights, in grams, of a random sample of eight cans of fruit of a particular brand were found to be:

    342, 344, 340, 339, 341, 338, 341, 343

    Given that the drained weights are normally distributed, calculate a 90% confidence interval for the mean drained weight of fruit per can of this particular brand. State, with your reason, whether or not your result is consistent with the claim on the can that the average weight of fruit is 340 g.

    (b) When an object is weighed on a certain weighing scale its recorded weight, in grams, is a random value from a normal distribution whose mean is the true weight of the object and whose standard deviation is 0.2 g.

    (i) Given that the mean of nine independently recorded weights of a particular object was 7.1 g, calculate a 99% confidence interval for the true weight of the object.

    A second object was weighed sixteen times on the same scale and the mean of the recorded weights was found to be 8.4 g.

    (ii) Determine a 95% confidence interval for the difference between the true weights of the two objects.

    (WJEC)

Section 10.7

11. A sleeping draught and an inactive control were tested in turn, in random order, on a random sample of 10 patients in a hospital, giving the following results:

Patient	*1*	*2*	*3*	*4*	*5*	*6*	*7*	*8*	*9*	*10*
Hours of sleep with drug	10.6	7.5	9.0	5.4	6.1	10.2	7.1	9.7	8.5	7.9
Hours of sleep with control	8.6	7.4	9.4	5.1	5.4	9.0	6.5	7.9	8.7	6.9

Estimate, with 99% confidence, the mean increase in hours of sleep due to the drug, stating any assumptions made.

12. In an experiment to test the effect of new rust inhibitor, seven pieces of metal were divided into halves. One half of each piece was treated with the new inhibitor while the other half was left untreated. After exposure to the atmosphere for one month, the degree of corrosion for each half was assessed with the following results:

Piece	*1*	*2*	*3*	*4*	*5*	*6*	*7*
Untreated	82	59	63	53	41	56	74
Treated	73	53	64	47	38	43	70

▶

Estimate the mean difference between the degree of corrosion for treated and untreated metal, with 95% confidence, stating any assumptions made.

Does your confidence interval support the claim that the inhibitor prevents corrosion?

Section 10.9

13. A researcher measured the wing-length to the nearest millimetre of a random sample of 50 house-flies and formed the following table:

Wing-length (mm)	*Number of house-flies*
3.5–3.9	5
4.0–4.4	14
4.5–4.9	23
5.0–5.4	7
5.5–5.9	1

(a) Draw a histogram and decide whether it would be reasonable to assume that the wing-length of house-flies is approximately normally distributed.
(b) Assuming that wing-length is normally distributed obtain a 95% confidence interval for:

(i) the mean wing-length of house-flies;
(ii) the variance of the wing-length of house-flies.

(c) How large a sample should be taken to reduce the confidence interval found in b(i) by 50%?

14. Let $X_1, X_2, \ldots, X_n$ denote a random sample of size n from a normal distribution with unknown mean μ and variance σ^2. For a random sample of 25 it was found that $\Sigma x_i = 3700$ and $\Sigma x_i^2 = 573{,}000$.

(a) Obtain unbiased point estimates of μ and σ^2.
(b) Obtain 95% confidence intervals for μ and σ^2.
(c) Calculate a new confidence interval for μ, assuming that σ^2 was known to be 800.

15. A random sample of fifteen workers from a vacuum flask assembly line was selected from a large number of such workers. Ivor Stopwatch, a work-study engineer, asked each of these workers to assemble a one-litre vacuum flask at their normal working speed. The times taken, in seconds, to complete these tasks are given below.

109.2, 146.2, 127.9, 92.0, 108.5, 91.1, 109.8, 114.9,

115.3, 99.0, 112.8, 130.7, 141.7, 122.6, 119.9.

Assuming that this sample came from an underlying normal population, calculate a 90% confidence interval for the variance of the population.

From his records of similar studies Ivor believes that the variance should equal 250. The assembly line manager disputes this and claims that 100 is the correct figure. Which of these figures is supported by the data?

Taking the variance as known and equal to the figure that you have chosen above, and assuming an underlying normal population again, calculate a 95% confidence interval for the mean of the population.

(AEB (1983))

16. (a) When an object is weighed on a chemical balance the readings obtained are subject to random errors which are known to be independent and normally distributed with mean zero and standard deviation 1 mg. A certain object is to be weighed nine times on such a balance and the mean of the nine readings is to be calculated. Find the probability that the mean of the readings will be within 0.5 mg of the true weight of the object.
(b) Another weighing device is undergoing tests to determine its accuracy. A certain object of known true weight 50 mg was weighed 10 times on this device and the readings in mg were:

49, 51, 49, 52, 49, 50, 52, 51, 49, 48.

(i) Calculate an unbiased estimate of the variance of the errors in readings using this device.
(ii) Calculate 95% confidence limits for the mean error in readings using this device.

(WJEC)

Section 10.10

17. Define random sampling numbers.

A firm has a fleet of 92 delivery vehicles. It is decided to test a random sample of 12 vehicles from the fleet. Explain, briefly, how random sampling numbers could be used to select the sample.

A set of random sampling numbers are divided into blocks of three digits. Find the probability that a block contains:

(a) three different digits;
(b) all three digits alike;
(c) a pair of like digits, the other one being different.

A vehicle tester claimed that the use of tables of random sampling numbers was unnecessary as he could write random digits down from his head. Challenged to do so he wrote down 300 digits. When these were split into 100 consecutive blocks of 3 digits, 88 of the blocks were found to contain 3 different digits. Find an approximate 95% confidence interval for the proportion of blocks containing 3 different digits which the vehicle tester would write down.

Comment briefly on your result.

(AEB (1984))

18. Assuming that the mean and variance of a random variable having a binomial distribution with parameters n and p are np and $np(1-p)$ respectively, prove that the mean and variance of a sample proportion, $\hat{p}$, based on a sample of size n, are p and $p(1-p)/n$ where p is the population proportion. What can you say about the probability distribution of $\hat{p}$ for large values of n? ►

A random sample of 100 fish was collected from a pond containing a large population of fish; each was marked and returned to the pond. A day later a random sample of 400 fish was taken and 42 were found to be marked. Obtain approximate 90% confidence limits for the proportion of marked fish in the pond. In what way are they approximate? Give similar limits for the size of the population of fish.

(SUJB)

Section 10.12

19. Of a random sample of 2000 people from those available for work in a large city, 1200 were male and 800 were female. Of the males, 150 were unemployed, while 80 of the females were unemployed. Calculate 95% confidence intervals for:

(a) the proportion of unemployed males;
(b) the proportion of unemployed females;
(c) the difference between the proportions of unemployed males and females.

Does your answer to (c) support the claim that the proportion of unemployed males is greater than the proportion of unemployed females?

20. The following table is from a readership survey carried out in England:

Region	*Number sampled*	*Percentage who read Daily Sunshine*
London and South-East	4128	15%
North-West	1551	24%

Assuming the samples were drawn randomly, estimate with 99% confidence the difference between the proportions who read the *Daily Sunshine* for the two regions. Do these data support the claim that the newspaper is read by a larger proportion in one region than the other?

21. A group of n seedlings is planted and the number, r, surviving after a week is counted. Assuming that the seedlings are independent and all have the same probability, p, of survival, state what distribution the number of survivors should follow. If n is large and p is close to $\frac{1}{2}$, what distribution can be used to approximate the proportion of survivors?

Two such independent groups each containing 100 seedlings are planted. In group 1, $r = 57$, and in group 2, $r = 48$. Give the following distributions with their parameters, or estimates of their parameters, and state clearly which of your values are estimates:

(a) the distribution of the number of survivors in group 1;
(b) an approximate distribution for the proportion of survivors in group 1;
(c) an approximate distribution for the difference between the proportions of survivors in the two groups.

Give an approximate 95% confidence interval for the difference between the probabilities of survival in the two groups. Use your interval to test for difference between the probabilities.

(OLE) ►

Section 10.13

22. Let $\overline{X}$ denote the mean of a random sample of n observations from a normal distribution whose mean is λ and whose standard deviation is also λ, where λ is an unknown positive constant.

    (i) Find the smallest value of n for which there is a probability of at least 0.95 that the numerical difference between $\overline{X}$ and λ will be less than $0.1\,\lambda$.

    (ii) For the case when $n = 100$, find a and b, in terms of λ, such that

    $$P(\overline{X} < a) = P(\overline{X} > b) = 0.025.$$

    Given that in this case $\overline{X} = 55$, deduce a 95% confidence interval for λ, giving each limit of the interval correct to the nearest integer.

    (JMB)

23. A continuous variable X is known to have a uniform distribution in the interval $(10, a)$, where $a > 10$, but is otherwise unknown. Write down an expression for $f(x)$, where f is the probability density function of X. Write down, in terms of a, the mean of X. Show from first principles that the variance is $(a-10)^2/12$.

    In one observation taken at random from this distribution a value of 21 for X was obtained. Write down an unbiased estimate of a.

    In 50 random observations of X the following results were obtained:

    $$\sum x_i = 1100, \qquad \sum x_i^2 = 35{,}000$$

    Use the value for Σx_i to obtain an unbiased estimate of a and give an approximate 95% symmetric confidence interval for a.

    Given that one of the 50 observations of X had value 60, state why a is certain to be greater than or equal to 60. Since 60 is outside the calculated confidence interval explain the apparent contradiction between this certainty and the 95% confidence interval.

    (JMB)

24. (a) Explain what is meant by the term *random sample*.

    For quality-control purposes, every twentieth refrigerator to come off the assembly line in a factory is tested. What criticism can be made of this method of sampling? Outline briefly how a random sample might be obtained in this case.

    (b) In a large city the distribution of incomes per family has a standard deviation of £5200. For a random sample of 400 families, what is the probability that the sample mean income per family is within £500 of the actual mean income per family? Given that the sample mean income was, in fact, £8300, calculate a 95% confidence interval for the actual mean income per family.

    (MEI)

11

Hypothesis tests for the mean and variance of normal distributions

▼

11.1 INTRODUCTION

Suppose we collect sample data from a normally distributed population with *unknown* mean, μ, but with *known* variance, σ^2. Instead of estimating μ by obtaining a confidence interval (using the methods of Section 10.2), we can set up a *hypothesis* specifying a particular numerical value of μ. For example, if we are interested in the mean IQ of a particular population, we may set up the hypothesis that $\mu = 100$.

The *Oxford English Dictionary* gives the following definition of *hypothesis*: 'Supposition made as basis for reasoning, or as starting point for further investigation from known facts'. So, for example, the hypothesis that $\mu = 100$ is a supposition.

In hypothesis testing we use the sample data (for example, the IQs of a random sample from the particular population) and our knowledge of statistical theory to decide whether the sample data support what is called the *null hypothesis*. If we so decide, we conclude that the null hypothesis is a reasonable supposition. On the other hand, if the sample data do not support the null hypothesis, we reject it as being an unreasonable supposition.

In this chapter we will consider various hypothesis tests concerning the mean, μ, and variance, σ^2, of normal distributions. Most of these tests require certain assumptions to be valid, otherwise the conclusions drawn from the tests may well be invalid. It is particularly important for the reader to be aware of these assumptions and not simply apply the various tests indiscriminately.

11.2 THE NULL AND ALTERNATIVE HYPOTHESES

In carrying out any hypothesis test, we actually need to consider *two* hypotheses. There are the *null hypothesis*, which we denote by H_0, and the *alternative hypothesis*, which we denote by H_1. The null hypothesis usually expresses the idea of 'no difference', and can often be expressed in terms of an equality.

H_0: $\mu = 100$ is an example of a null hypothesis.

The alternative hypothesis usually expresses the idea of a difference. An alternative hypothesis may be either one-sided or two-sided.

H_1: $\mu > 100$ is an example of a *one-sided* alternative hypothesis,

since > 100 implies only one side of the value 100.

H_1: $\mu < 100$ is an example of a *one-sided* alternative hypothesis,

since < 100 implies only one side of the value 100.

H_1: $\mu \neq 100$ is an example of a *two-sided* alternative hypothesis, since $\neq 100$ implies both sides of the value 100.

The decision as to which of these three alternative hypotheses we should choose depends on the particular application we are considering. The following examples are intended to clarify this point.

Example 11.1

A random sample of Members of Parliament were each given an IQ test to decide whether MPs have a mean IQ of 100 (the mean IQ of all adults in the UK) or whether their mean IQ is above 100.

We would use H_0: $\mu = 100$

and H_1: $\mu > 100$ (a one-side: H_1)

where μ is the mean IQ of all MPs. □

Example 11.2

Suppose that, according to national standards, five-year-old children of height 87.5–92.5 cm should have a mean weight of 13 kg. The weights of a random sample of five-year-old refugee children of height 87.5–92.5 cm are collected. Are these children below average weight for their height?

We would use H_0: $\mu = 13$

and H_1: $\mu < 13$ (a one-sided H_1)

where μ is the mean weight for the sampled population of refugees of height 87.5–92.5 cm. □

Example 11.3

A machine produces components of mean length 10 cm if it is correctly set up. A random sample of components is taken from those produced by the machine, and the length of each component is measured. Is the machine correctly set up?

We would use H_0: $\mu = 10$

H_1: $\mu \neq 10$ (a two-sided H_1)

where μ is the population mean length of components produced by the machine.

Note how the null hypothesis, H_0, expresses the idea of 'no difference' in each example. □

11.3 HYPOTHESIS TEST FOR THE MEAN OF A NORMAL DISTRIBUTION WITH KNOWN VARIANCE

In this section we will develop a method for testing a null hypothesis about the mean, μ, of a normal distribution of a random variable X when the variance, σ^2, is known. In doing so the concepts of a significance level and a critical region will be introduced.

Suppose $X \sim N(\mu, \sigma^2)$. Then we know that $\overline{X} \sim N(\mu, \sigma^2/n)$ from Section 9.5, where $\overline{X}$ denotes the mean of a sample of size n.

Suppose we wish to test:

$$H_0\colon \mu = \mu_0$$

against

$$H_1\colon \mu > \mu_0$$

where μ_0 is a particular value of μ.

If H_0 is true, it follows that $\overline{X} \sim N(\mu_0, \sigma^2/n)$, which by standardizing we can rewrite as

$$\frac{\overline{X} - \mu_0}{\sigma/\sqrt{n}} \sim N(0, 1).$$

This result is illustrated in Fig. 11.1.

Note the use of Z to represent a variable with a standardized normal distribution (first used in Section 6.6).

If we collect a sample of n values, calculate the sample mean, $\bar{x}$, and then calculate the value of the *test statistic*

$$z = \frac{\bar{x} - \mu_0}{\sigma/\sqrt{n}},$$

we can decide whether to reject H_0. Large positive values of z will result from large values of $\bar{x}$ and will make us lean towards the conclusion that the mean is really greater than μ_0. (For example, if we were testing $\mu = 100$ for the IQ example, and a random sample of 100 MPs had a mean of 150 we would tend to

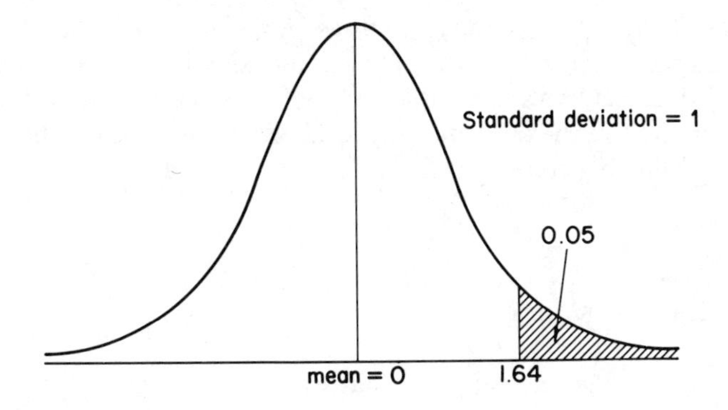

Fig. 11.1 The distribution of $Z = \dfrac{\bar{X} - \mu_0}{\sigma/\sqrt{n}}$, if H_0 is true

think that μ was really larger than 100). In other words, large values of $\bar{x}$ will tend to lead to the rejection of H_0 and the acceptance of H_1.

We have to decide on a *critical value* of z above which we will decide to reject H_0, and below which we will decide not to reject H_0. However, since we are trying to draw conclusions about the mean of a population on the basis of only a sample from this population, we must always run a small risk that we will reject H_0 when H_0 is a true hypothesis.

Since the probability that $z > 1.64$ is 0.05 (from Table C.3(b) or Fig. 11.1) then the critical value of z is 1.64 if we decide to run a risk of 0.05 (or 5%) of rejecting H_0 when H_0 is true.

Definition 11.1

> The risk of rejecting H_0 when H_0 is true is called the *significance level* of a test and is usually denoted by α.

Definition 11.2

> The values of a test statistic which lead to the rejection of H_0 are said to form a *critical region.*

So, if we decide that our critical region is $z > 1.64$ in this case we are carrying out this hypothesis test at the 5% level of significance. The critical value of the test statistic z, is 1.64.

Why do we choose a significance level of 5%? The answer is similar to one we asked in Chapter 10, namely 'Why choose a 95% confidence interval?' Just as 95% implies that that there is a 5% chance that the parameter we are trying to estimate lies outside the interval we calculate, so a significance level of 5% implies a 5% chance of wrongly rejecting H_0. Just as 95% (for a confidence interval) is a conventionally chosen level, so is 5% for a significance level. The implications of choosing significance levels other than 5% will be discussed in Section 11.10.

We may summarize the one-tailed† hypothesis test described in this section in terms of the following steps:

$H_0: \mu = \mu_0$

$H_1: \mu > \mu_0$

5% level of significance

$z = \dfrac{\bar{x} - \mu_0}{\sigma/\sqrt{n}}$ is the calculated test statistic

If $z > 1.64$, reject H_0, and conclude that

'The mean is significantly greater than μ_0 (5% level)'

If $z \leqslant 1.64$, do not reject H_0, and conclude that

'The mean is not significantly greater than μ_0 (5% level)'

Example 11.4

Numerical example of a one-tailed hypothesis test for the mean, μ, of a normal distribution, when the variance, σ^2, is known.

Suppose the distribution of IQ for Members of Parliament is known to be normal with variance $\sigma^2 = 225$. If the mean IQ for a random sample of 20 MPs is 109, can we conclude that the mean IQ for all MPs is significantly greater than 100? Use a 5% level of significance.

H_0: $\mu = 100$, where μ is the mean IQ of all MPs

H_1: $\mu > 100$, a one-sided H_1 as implied by the above question.

5% level of significance.

$z = \dfrac{\bar{x} - \mu_0}{\sigma/\sqrt{n}}$ is the calculated test statistic.

$$\bar{x} = 109, \quad \mu_0 = 100, \quad \sigma = \sqrt{225} = 15, \quad n = 20$$

therefore,
$$z = \frac{109 - 100}{15/\sqrt{20}} = 2.68$$

Since $2.68 > 1.64$, we reject H_0 (see Fig. 11.2) and conclude that:

'the mean IQ is significantly greater than 100 (5% level)'

†We say a hypothesis test is one-tailed if H_1 is one-sided, and two-tailed if H_1 is two-sided.

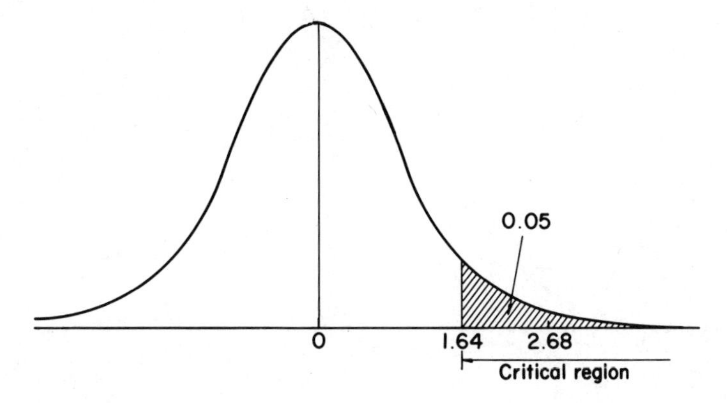

Fig. 11.2 Example of a critical region for a one-tailed test □

Example 11.5

Numerical example of a two-tailed hypothesis test for the mean, μ, of a normal distribution, when the variance, σ^2, is known.

Suppose that the lengths of components produced by a machine are normally distributed with standard deviation, $\sigma = 0.13$ cm. The machine is required to produce components with a mean length of 10 cm. A random sample of 15 components is taken and the lengths of the components in cm are as follows:

10.0 9.9 10.1 9.8 9.7 10.0 9.8 10.1 9.7 9.8
9.8 9.9 10.0 9.9 10.0

Is the machine producing components to the required mean length? Use a 5% level of significance.

$H_0: \mu = 10$

$H_1: \mu \neq 10$, a two-sided H_1, since the question implies that differences from 10 in either direction are equally important to detect.

5% level of significance

$$z = \frac{\bar{x} - \mu_0}{\sigma/\sqrt{n}} \quad \text{is the calculated test statistic.}$$

Since H_1 is two-sided, this is a two-tailed test.

The significance level, 5%, is divided equally between the two tails of the $N(0, 1)$ distribution. Since $P(z < -1.96) = 0.025$ and $P(z > +1.96) = 0.025$, the critical values are -1.96 and $+1.96$ (Fig. 11.3). So our decision rule is:

Reject H_0 if $|z| > 1.96$, where $z = \dfrac{x - \mu_0}{\sigma/\sqrt{n}}$.

For the data given, $\bar{x} = 9.90$, $\mu_0 = 10$, $\sigma = 0.13$, $n = 15$

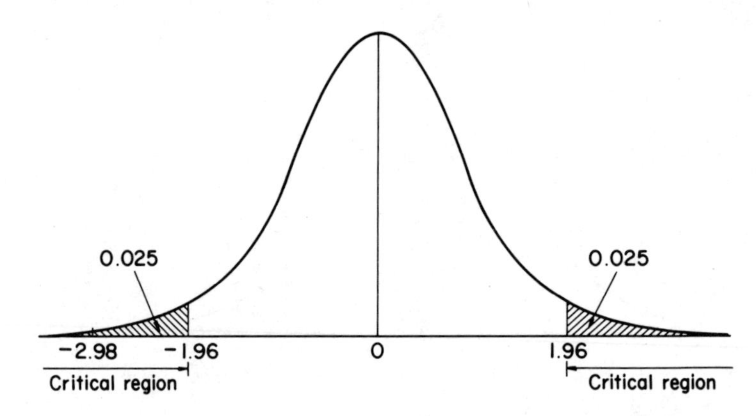

Fig. 11.3 Example of a critical region for a two-tailed test

therefore $$z = \frac{9.90 - 10}{0.13/\sqrt{15}} = -2.98$$

Since $|-2.98| > 1.96$, we reject H_0, and conclude that:

'the mean length is significantly different from 10 cm (5 % level)'

Note

For a one-tailed test for which $H_1: \mu < \mu_0$, we reject H_0 if $z < -1.64$, where $z = \dfrac{\bar{x} - \mu_0}{\sigma/\sqrt{n}}$, when the significance level is 5 %; see Fig. 11.4.

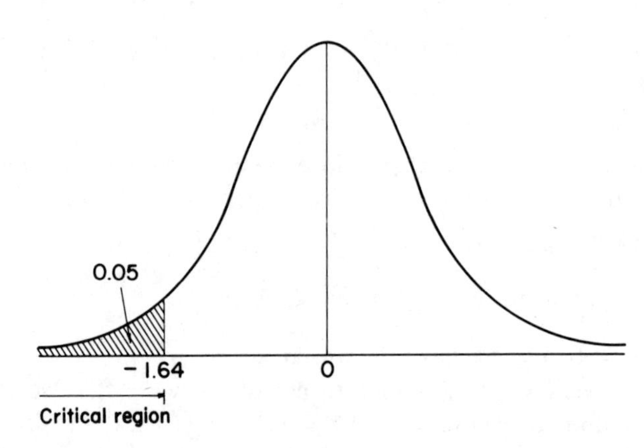

Fig. 11.4 Example of a critical region for a one-tailed test □

Table 11.1 summarizes the critical regions for the 'z tests' described in this section. A significance level of 5 % has been used in each case.

Table 11.1 Critical regions for three '*z* tests'

H_0	H_1	*Critical region*	*Assumption*
$\mu = \mu_0$	$\mu > \mu_0$	$\dfrac{\bar{x} - \mu_0}{\sigma/\sqrt{n}} > 1.64$	Variable is normally distributed, σ is known
$\mu = \mu_0$	$\mu < \mu_0$	$\dfrac{\bar{x} - \mu_0}{\sigma/\sqrt{n}} < -1.64$	as above
$\mu = \mu_0$	$\mu \neq \mu_0$	$\dfrac{\lvert\bar{x} - \mu_0\rvert}{\sigma/\sqrt{n}} > 1.96$	as above

11.4 HYPOTHESIS TEST FOR THE MEAN OF A NORMAL DISTRIBUTION WITH UNKNOWN VARIANCE

If we wish to test a hypothesis about the mean, μ, of a normal distribution when the variance, σ^2, is *unknown*, the test statistic we calculate is:

$$t = \frac{\bar{x} - \mu_0}{s/\sqrt{n}}$$

This follows from the theory of the *t* distribution introduced in Section 10.3. Table 11.2 summarizes the critical regions for three '*t* tests' (with the same null and alternative hypotheses as in Table 11.1). A significance level of 5 % has been used in each case.

Table 11.2 Critical regions for three '*t* tests'.

H_0	H_1	*Critical region*	*Assumption*
$\mu = \mu_0$	$\mu > \mu_0$	$\dfrac{\bar{x} - \mu_0}{s/\sqrt{n}} > t_{0.05,n-1}$	Variable is normally distributed
$\mu = \mu_0$	$\mu < \mu_0$	$\dfrac{\bar{x} - \mu_0}{s/\sqrt{n}} < -t_{0.05,n-1}$	as above
$\mu = \mu_0$	$\mu \neq \mu_0$	$\dfrac{\lvert\bar{x} - \mu_0\rvert}{s/\sqrt{n}} > t_{0.025,n-1}$	as above

Notes

(i) $t_{0.05,\,n-1}$ is the value of *t* obtained from Table C.4 for a single tail area of 0.05 and $\nu = n - 1$ degrees of freedom.

(ii) See Section 11.11 for a discussion of the importance of the assumption of normality for this test.

Example 11.6

Numerical example of a one-tailed t-test

The weights of five-year-old children of height 87.5–92.5 cm are normally distributed and should have a mean of 13 kg according to a national standard. Suppose a random sample of 10 five-year-old refugee children of height 87.5–92.5 cm had weights in kg as follows:

12.6 9.4 9.6 11.3 8.6 12.4 11.5 9.9 10.3 13.5

Are these children on average below the standard weight for their height? Use a significance level of 1 %.

H_0: $\mu = 13$

H_1: $\mu < 13$, a one-sided H_1

1 % level of significance

$t = \dfrac{\bar{x} - \mu_0}{s/\sqrt{n}}$ is the calculated test statistic

For the data given, $\bar{x} = 10.91$, $\mu_0 = 13$, $s = 1.60$, $n = 10$,

$$t = \frac{\bar{x} - \mu_0}{s/\sqrt{n}} = \frac{10.91 - 13}{1.60/\sqrt{10}} = -4.13$$

Noting H_1 and the 1 % significance level, the critical value, is $t_{0.01,9} = -2.82$ (Table C.4).
Since $-4.13 < -2.82$, we reject H_0 (see Fig. 11.5), and conclude that:

'the mean weight for their height of the refugee children is significantly below the national standard (1 % level)'

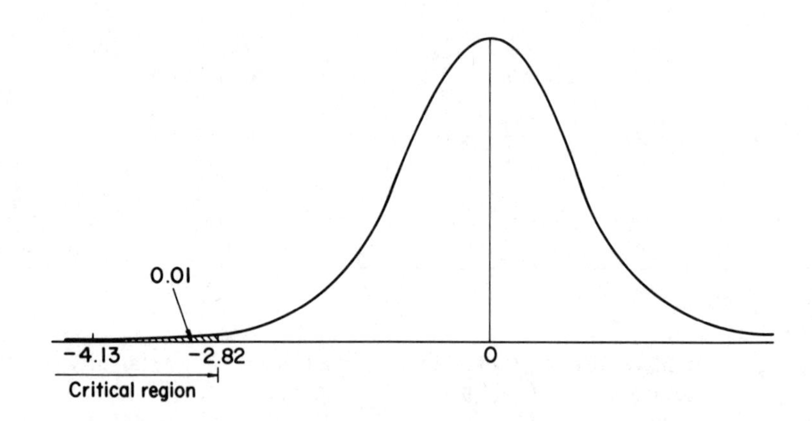

Fig. 11.5 Example of a critical region for a t test with 9 d.f. □

11.5 HYPOTHESIS TEST FOR THE DIFFERENCE BETWEEN THE MEANS OF TWO NORMAL DISTRIBUTIONS (UNPAIRED SAMPLES DATA)†

Suppose that we are interested in comparing two distinct populations in terms of some variable. For example, we might wish to compare the mean scores in a spelling test for children of a particular age taught by one of two different methods. We could collect data from random samples of children taught by each method. If μ_1 and μ_2 are the means of the distributions of scores for each method for children of the age in question; then our null hypothesis might be H_0: $\mu_1 = \mu_2$ with an alternative hypothesis H_1: $\mu_1 \neq \mu_2$.

More generally, in comparing two populations we might wish to set up a null hypothesis H_0: $\mu_1 - \mu_2 = \mu_0$ (where μ_0 is a particular numerical value of interest in a given application) with a one- or two-sided alternative hypothesis. If we know that the variable of interest is normally distributed in each population, and we also know that the two distributions have the same variance, then a hypothesis test may be carried out using a test statistic based on the t-distribution. Table 11.3 summarizes the critical regions for three 't tests', each with a 5% level of significance.

Notes

(i) μ_0 is the 'null hypothesis value of μ', and will be equal to zero in the usual case when we simply wish to test whether $\mu_1 = \mu_2$, i.e. whether $\mu_1 - \mu_2 = 0$.

(ii) $\bar{x}_1$, s_1, n_1 refer to the sample taken from the population with mean μ_1. Similarly, $\bar{x}_2$, s_2, n_2 refer to the sample taken from the population with mean μ_2.

(iii) The value of s is calculated using $s^2 = \dfrac{(n_1 - 1)s_1^2 + (n_2 - 1)s_2^2}{n_1 + n_2 - 2}$, where s^2 is the unbiased estimator of the common variance, σ^2, say (see also Section 10.6).

(iv) See Section 11.11 for a discussion of the two assumptions of the above t tests, and Section 11.8 for a test of the assumption of equality of variance.

Example 11.7

Thirty ten-year-old children were randomly selected for an investigation to compare two methods of teaching children how to spell. Fifteen of the children were randomly allocated to each method. Following the learning period, each child was scored for spelling ability. The scores of the 30 children are summarized in Table 11.4.

Test, at the 5% level of significance, whether the mean scores for the two methods are equal. What assumptions have you made?

† See Section 10.6 for a discussion of the difference between 'unpaired' and 'paired' samples data.

Table 11.3 Critical regions for three t tests for the difference between two means

H_0	H_1	*Critical region*	*Assumption*
$\mu_1 - \mu_2 = \mu_0$	$\mu_1 - \mu_2 > \mu_0$	$\dfrac{(\bar{x}_1 - \bar{x}_2) - \mu_0}{s\sqrt{\dfrac{1}{n_1} + \dfrac{1}{n_2}}} > t_{0.05, n_1 + n_2 - 2}$	The variable is normally distributed for each population, and the distributions have the same variance
$\mu_1 - \mu_2 = \mu_0$	$\mu_1 - \mu_2 < \mu_0$	$\dfrac{(\bar{x}_1 - \bar{x}_2) - \mu_0}{s\sqrt{\dfrac{1}{n_1} + \dfrac{1}{n_2}}} < -t_{0.5, n_1 + n_2 - 2}$	as above
$\mu_1 - \mu_2 = \mu_0$	$\mu_1 - \mu_2 \neq \mu_0$	$\dfrac{\lvert(\bar{x}_1 - \bar{x}_2) - \mu_0\rvert}{s\sqrt{\dfrac{1}{n_1} + \dfrac{1}{n_2}}} > t_{0.025, n_1 + n_2 - 2}$	as above

Table 11.4 Scores from two methods of teaching spelling

Method 1	*Method* 2
$\bar{x}_1 = 65.3$	$\bar{x}_2 = 62.4$
$s_1 = 4.2$	$s_2 = 3.9$
$n_1 = 15$	$n_2 = 15$

Let μ_1 and μ_2 be the means of the distributions of scores for methods 1 and 2 respectively.

H_0: $\mu_1 - \mu_2 = 0$ (here $\mu_0 = 0$)

H_1: $\mu_1 - \mu_2 \neq 0$

5 % level of significance

$$t = \frac{\bar{x}_1 - \bar{x}_2}{s\sqrt{\dfrac{1}{n_1} + \dfrac{1}{n_2}}}$$ is the calculated test statistic (since $\mu_0 = 0$)

For the data given

$$s^2 = \frac{(n_1 - 1)s_1^2 + (n_2 - 1)s_2^2}{n_1 + n_2 - 2}$$

$$= \frac{14 \times 4.2^2 + 14 \times 3.9^2}{28}$$

$$= 16.43$$

$$s = 4.05$$

$$t = \frac{65.3 - 62.4}{4.05\sqrt{\dfrac{1}{15} + \dfrac{1}{15}}} = 1.96$$

The critical value is $t_{0.025, 28} = 2.05$, from Table C.4.
Since $|1.96| < 2.05$, we do not reject H_0 (see Fig. 11.6), and we conclude that:

'the mean scores of the two methods are not significantly different (5 % level)'

We have assumed that the distributions of Method 1 and Method 2 scores are both normal and have the same variance. □

The tests described above are for the difference between two means if we know that the distributions have the same variance. What if we know or suspect that they are not equal? There are tests we can perform, which are briefly described below.

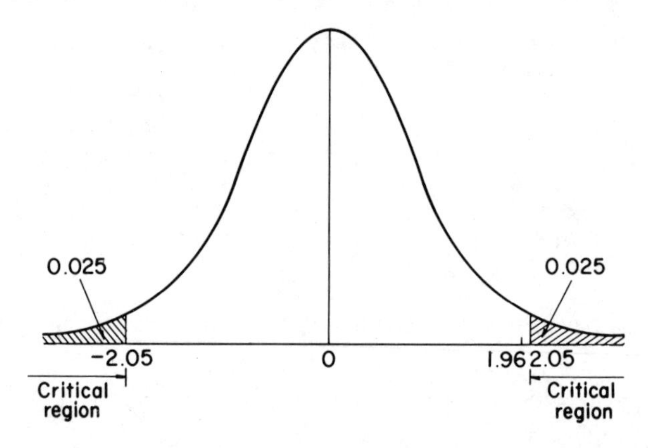

Fig. 11.6 Example of a critical region for a two-tailed t test with 28 d.f.

(a) If we know the values of the two variances, σ_1^2 and σ_2^2, and we also know the distributions are normal, then the calculated test statistic is:

$$z = \frac{(\bar{x}_1 - \bar{x}_2) - \mu_0}{\sqrt{\dfrac{\sigma_1^2}{n_1} + \dfrac{\sigma_2^2}{n_2}}}$$

The critical value of z corresponding to the three cases in Table 11.3 are 1.64, −1.64 and 1.96 respectively.

(b) If we estimate σ_1^2 and σ_2^2 from the data obtained from large samples ($n_1, n_2 > 30$, say) and we suspect that σ_1^2 and σ_2^2 are not equal (based on the values we calculate for their estimators s_1^2 and s_2^2), then in this case the assumption of normality is relatively unimportant.† The test statistic is

$$z = \frac{(\bar{x}_1 - \bar{x}_2) - \mu_0}{\sqrt{\dfrac{s_1^2}{n_1} + \dfrac{s_2^2}{n_2}}}$$

The critical values of z corresponding to the three cases in Table 11.3 are also 1.64, −1.64 and 1.96 respectively. □

Example 11.8

A random sample of 100 boys of a certain age have a mean height of 152.3 cm and a standard deviation of 6 cm, while a random sample of 120 girls of the same age have a mean height of 148.6 cm and a standard deviation of 5 cm. Test the hypothesis, at the 5 % level of significance, that boys are on average 2 cm taller than girls for the age in question.

†However, if we *know* that the variable is normally distributed we can test the hypothesis that $\sigma_1^2 = \sigma_2^2$; see Section 11.8.

H_0: $\mu_1 - \mu_2 = 2$ where μ_1 is the mean height of boys and μ_2 is the mean height of girls, for the age in question.

H_1: $\mu_1 - \mu_2 \neq 2$

5 % level of significance

$$z = \frac{(\bar{x}_1 - \bar{x}_2) - \mu_0}{\sqrt{\frac{s_1^2}{n_1} + \frac{s_2^2}{n_2}}}$$ is the calculated test statistic.

For the data given,

$$z = \frac{(152.3 - 148.6) - 2}{\sqrt{\frac{6^2}{100} + \frac{5^2}{120}}} = 2.26$$

The critical value of z is 1.96.
Since $|2.26| > 1.96$, reject H_0, and conclude that:

'the difference in the mean heights of boys and girls is significantly different from 2 cm (5 % level)' □

11.6 HYPOTHESIS TEST FOR THE MEAN OF A NORMAL DISTRIBUTION OF DIFFERENCES (PAIRED SAMPLES DATA)

Sometimes we have to deal with data which consist of pairs of values. An example was given in Section 10.7 where matched pairs of children were selected for an investigation. Another type of example of paired data arises when two measurements are made on each of a number of individuals, one measurement is made before, and the other measurement after, some 'treatment'. (We can consider this example as one in which each individual is matched with himself or herself.)

The differences, d, are calculated for each individual. From these we calculate the mean $\bar{d}$ and the standard deviation s_d using:

$$\bar{d} = \frac{\sum d_i}{n} \quad \text{and} \quad s_d = \sqrt{\frac{\sum (d_i - \bar{d})^2}{n-1}} = \sqrt{\frac{\sum d_i^2 - n\bar{d}^2}{n-1}}$$

where n is the number of differences (= the number of individuals). In practice we may well use the mean and standard deviation facilities on our calculators to obtain $\bar{d}$ and s_d.

Suppose the null hypothesis we wish to test is that the treatment has had 'no effect'. Then if μ_d is the mean of the distribution of differences, our null hypothesis is H_0: $\mu_d = 0$. The alternative hypothesis is that there has been some effect, the direction of which will depend on the particular application.

The critical regions are similar to those of Section 11.4, since we have *one* distribution (of differences). The notation differs to indicate the fact that we are working with differences (so we have used $\bar{d}$ instead of $\bar{x}$ and s_d instead of s).

Table 11.5 summarizes the critical regions for three t tests, each for a 5% level of significance.

Table 11.5 Critical regions for t tests for mean of differences

H_0	H_1	*Critical region*	*Assumption*
$\mu_d = 0$	$\mu_d > 0$	$\dfrac{\bar{d}}{s_d/\sqrt{n}} > t_{0.05,\, n-1}$	Differences are normally distributed
$\mu_d = 0$	$\mu_d < 0$	$\dfrac{\bar{d}}{s_d/\sqrt{n}} < -t_{0.05,\, n-1}$	as above
$\mu_d = 0$	$\mu_d \neq 0$	$\dfrac{\lvert\bar{d}\rvert}{s_d/\sqrt{n}} > t_{0.025,\, n-1}$	as above

Note

See Section 11.11 for a discussion of the assumption of normality for this test.

Example 11.9

A random sample of ten students was tested before and after being coached in a subject. Their marks were as shown in Table 11.6.

Table 11.6 Marks before and after coaching for 10 students

Student number	1	2	3	4	5	6	7	8	9	10
Before coaching	53	59	61	48	39	56	75	45	81	60
After coaching	60	57	67	52	61	71	70	46	93	75
d = '*after* − *before*'	7	−2	6	4	22	15	−5	1	12	15

Is the coaching effective? Use a 0.1% level of significance.

Note that in Table 11.6 we calculated the differences, d, using

$$d = (\text{marks after coaching}) - (\text{marks before coaching})$$

Then $\mu_d = 0$ indicates that the coaching has no effect on average one way or the other, while $\mu_d > 0$ indicate that the coaching has increased the marks on average. These are the H_0 and H_1 we will use:

H_0: $\mu_d = 0$

H_1: $\mu_d > 0$

0.1% level of significance

$$t = \frac{\bar{d}}{s_d/\sqrt{n}} \quad \text{is the calculated test statistic}$$

For the data given, $\bar{d} = 7.50$, $s_d = 8.48$, $n = 10$

$$t = \frac{7.50}{8.48/\sqrt{10}} = 2.80$$

The critical value is $t_{0.001,\,9} = 4.30$

Since $2.80 < 4.30$ we do not reject H_0 (see Fig. 11.7), and conclude that:

'the mean marks after coaching are not significantly higher than before coaching (0.1% level)'

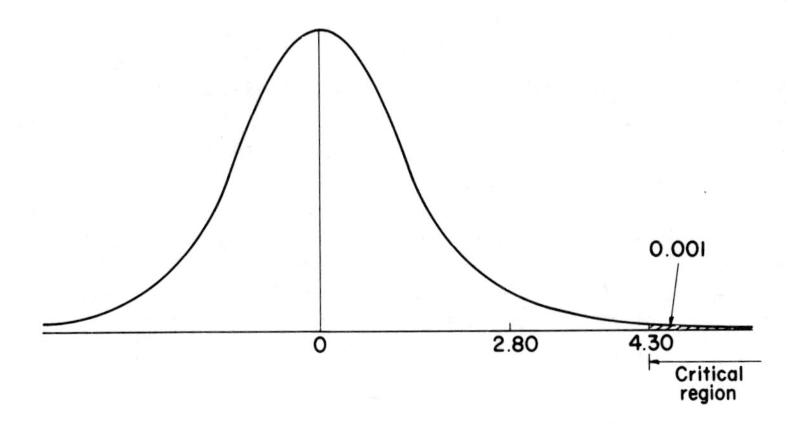

Fig. 11.7 Example of a critical region for a t test with 9 d.f.

Notes

(i) We have assumed that the differences are normally distributed.

(ii) Choosing such a low level of significance, 0.1%, has made it less likely that we will reject H_0 when it is true. The implications of choosing low significance levels are discussed in Section 11.10. □

Note on the difference between paired and unpaired sample data

In both the paired and unpaired samples cases there are two populations of measurements and a sample is taken from each. In the 'paired' example just considered, the two populations are the marks before coaching and the marks after coaching. In the 'unpaired' example of Section 11.5 the two populations are the scores of children taught by method 1 and the scores of children taught by method 2.

In a paired samples cases there must be a reason to associate a particular measurement in one sample with a measurement in the other sample (for example, both were collected from the same 'individual' or from the same matched pair). If there is no reason to pair measurements in this way, the data are treated as unpaired.

Unequal sample sizes are usually indicative of unpaired samples data. Equal sample sizes, however, are not indicative of either paired or unpaired samples data.

11.7 HYPOTHESIS TEST FOR THE VARIANCE OF A NORMAL DISTRIBUTION

We noted in Section 10.9 that

$$\frac{(n-1)S^2}{\sigma^2} \sim \chi^2_{n-1}$$

where σ^2 is the variance of a normal distribution, S^2 is the unbiased estimator of σ^2, and n is the sample size. We can use this result to test hypotheses about the value of σ^2. Table 11.7 summarizes the critical regions for three such tests. A significance level of 5% has been used in each case.

Table 11.7 Critical region for three χ^2 tests for a variance

H_0	H_1	*Critical region*	*Assumption*
$\sigma^2 = \sigma_0^2$	$\sigma^2 > \sigma_0^2$	$\frac{(n-1)s^2}{\sigma_0^2} > \chi^2_{0.05,\,n-1}$	Variable is normally distributed
$\sigma^2 = \sigma_0^2$	$\sigma^2 < \sigma_0^2$	$\frac{(n-1)s^2}{\sigma_0^2} < \chi^2_{0.95,\,n-1}$	as above
$\sigma^2 = \sigma_0^2$	$\sigma^2 \neq \sigma_0^2$	$\frac{(n-1)s^2}{\sigma_0^2} > \chi^2_{0.025,\,n-1}$ or $< \chi^2_{0.975,\,n-1}$	as above

Note

(i) See Section 11.11 for a discussion of the importance of the assumption of normality for this test.

Example 11.10

Consider the weights of a random sample of five-year-old children of height 87.5–92.5 cm (as used in the example of Section 11.4).

Are these data consistent with the hypothesis that the variance in the weights of such children equals 3 kg^2? Use a 5% level of significance.

The sample data were summarized as follows:

$\bar{x} = 10.91$, $s = 1.60$, $n = 10$

H_0: $\sigma^2 = 3$

H_1: $\sigma^2 \neq 3$ a two-sided H_1, since the question is not directional

5% level of significance.

$$\chi^2 = \frac{(n-1)s^2}{\sigma^2}$$ is the calculated test statistic.

From the data given, $\chi^2 = \dfrac{(10-1)1.60^2}{3} = 7.68$.

The critical values are:

$$\chi^2_{0.025,\,9} = 19.02 \quad \text{and} \quad \chi^2_{0.975,\,9} = 2.70 \qquad \text{(Table C.5)}$$

Since 7.68 is not in the critical region, we do not reject H_0 (see Fig. 11.8), and we conclude that:

'the variance in the weights is not significantly different from 3 (5% level)'. We have assumed that the distribution of weight is normal.

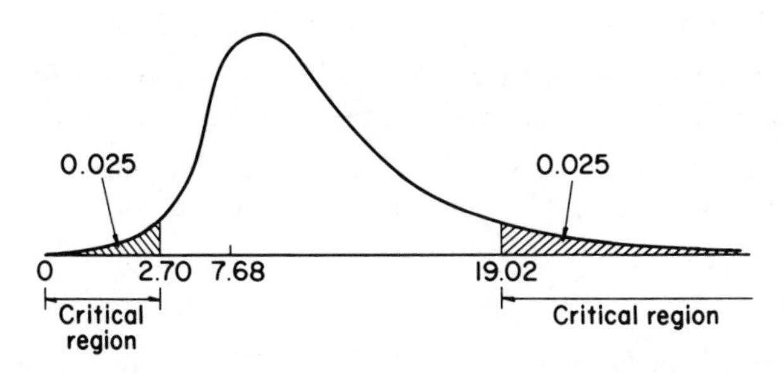

Fig. 11.8 Example of a critical region for a χ^2 test with 9 d.f.

11.8 HYPOTHESIS TEST FOR THE EQUALITY OF THE VARIANCES OF TWO NORMAL DISTRIBUTIONS

In Section 11.5 we tested a hypothesis that the means of two normally distributed populations were equal, assuming that the distributions had the same variance, σ^2. The F test provides a method of testing the 'equality of variance' assumption. (The F test is important in the statistical analysis of designed experiments; see Chapter 17). If σ_1^2 and σ_2^2 are the variances of the two distributions, and s_1^2 and s_2^2 are our unbiased estimates of σ_1^2 and σ_2^2, respectively, such that $s_1^2 > s_2^2$ (see note (i) below), Table 11.8 summarizes the critical region for three such tests. A significance level of 5% has been used in each case.

Notes

(i) If $s_1^2 < s_2^2$, simply interchange the suffixes in Table 11.8, since it is arbitrary which we call distribution 1 and which we call distribution 2.

(ii) If $s_1^2 > s_2^2$ we will *never* wish to reject $H_0: \sigma_1^2 = \sigma_2^2$ in favour of $H_1: \sigma_1^2 < \sigma_2^2$.

(iii) See Section 11.11 for a discussion of the importance of the assumption of normality for this test.

Table 11.8 Critical region for three F tests for the equality of two variances

H_0	H_1	*Critical region*	*Assumption*
$\sigma_1^2 = \sigma_2^2$	$\sigma_1^2 > \sigma_2^2$	$\frac{s_1^2}{s_2^2} > F_{0.05, n_1-1, n_2-1}$	The variable is normally distributed for each population
$\sigma_1^2 = \sigma_2^2$	$\sigma_1^2 < \sigma_2^2$	see note (ii)	—
$\sigma_1^2 = \sigma_2^2$	$\sigma_1^2 \neq \sigma_2^2$	$\frac{s_1^2}{s_2^2} > F_{0.025, n_1-1, n_2-1}$ or $< F_{0.975, n_1-1, n_2-1}$	as above

□

Example 11.11

Suppose that two samples of sizes 10 and 9 respectively are randomly selected from two normally distributed populations with variances σ_1^2 and σ_2^2. Suppose we calculate $s_1 = 4.2, s_2 = 3.9$. Test the hypothesis that $\sigma_1^2 = \sigma_2^2$ against a two-sided alternative at the 5% level of significance.

$H_0: \sigma_1^2 = \sigma_2^2$

$H_1: \sigma_1^2 \neq \sigma_2^2$

5% level of significance

$F = \frac{s_1^2}{s_2^2}$ is the calculated test statistic, since $s_1^2 > s_2^2$ for the above data

$$F = \frac{s_1^2}{s_2^2} = \frac{4.2^2}{3.9^2} = 1.16$$

The critical values are:

$F_{0.025, 9, 8} = 4.37$ (interpolating in Table C.6(b) to 2 d.p.)

and $F_{0.975, 9, 8} = \frac{1}{F_{0.025, 8, 9}} = \frac{1}{4.10} = 0.24$

Since $0.24 < 1.16 < 4.37$, we do not reject H_0 (see Fig. 11.9), and conclude that:

'the variances are not significantly different (5% level)' □

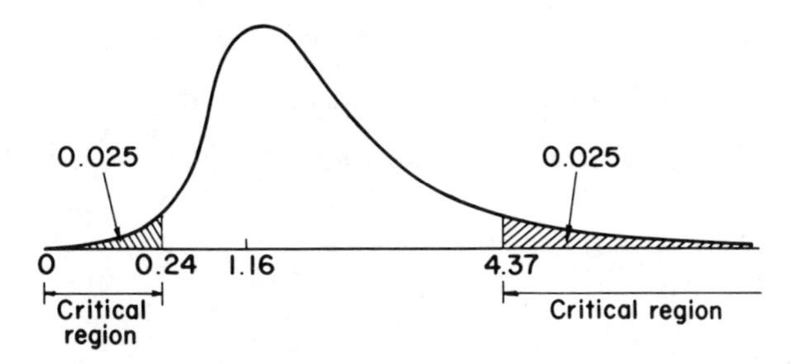

Fig. 11.9 Example of a critical region for an F test with 9 and 8 d.f.

11.9 HOW A CONFIDENCE INTERVAL CAN BE USED TO TEST HYPOTHESES

The reader who has studied Chapters 10 and 11 in sequence will have noticed that, whenever we wished to test a hypothesis in this chapter about a particular parameter, we employed exactly the same theory as that used in Chapter 10 to obtain a confidence interval for the particular parameter. Also we used exactly the same sample statistics.

What, then, is the relationship between the two types of statistical inference, namely *confidence interval estimation* and *hypothesis testing*?

We know that confidence intervals provide a range of values within which we are, say, 95 % confident that the true value of a parameter lies. We also know that hypothesis tests help us to decide whether a specified value of a parameter should be rejected in favour of some alternative values, with a chance of, say, 5 %, that we will wrongly reject the specified value. Could a confidence interval be used to test hypotheses? The answer is yes.

Suppose, for example, that a 95 % confidence interval for the mean, μ, of a normal distribution is 65 to 75. If a null hypothesis H_0: $\mu_0 = 72$ is suggested, we would not wish to reject it in favour of H_1: $\mu \neq 72$ at the 5 % level of significance. On the other hand, we would reject H_0: $\mu = 79$ in favour of H_1: $\mu \neq 79$, again at the 5 % level. The reasons for these two decisions should be obvious:

72 lies inside the confidence interval, but 79 does not.

Thus, a 95 % confidence interval can be used to test a null hypothesis against a two-sided alternative hypothesis, at the 5 % level of significance. Likewise a 99 % confidence interval can be used to test hypotheses at the 1 % level of significance.

11.10 TYPE I AND II ERRORS, AND THE POWER OF A TEST

The significance level, α, is also known as the Type I error. We recall from Section 11.3 that $\alpha = P$ (rejecting H_0 when H_0 is true). There is another type of error, the

Type II error, which is given the symbol β, and is defined as follows:

$$\beta = P(\text{rejecting } H_1 \text{ when } H_1 \text{ is true})$$

The *power* of a test is defined as $P(\text{accepting } H_1 \text{ when } H_1 \text{ is true})$

and so

$$\text{power} = 1 - \beta = P(\text{accepting } H_1 \text{ when } H_1 \text{ is true}).$$

In carrying out a hypothesis test we choose a small value for α, conventionally 5%. It seems natural to choose a small value for the Type II error, β, and hence make $(1-\beta)$ relatively large. However, this is easier said than done, because each alternative hypothesis, H_1, discussed in this chapter has specified a range of values for the parameters tested (whereas only a single value is specified in the null hypothesis). We can only choose a particular value for β when H_1 specifies only one value for the parameter, for example, H_1: $\mu = 110$. In other words:

β is a function of the value of the parameter

Likewise, power $= (1-\beta)$ is a function of the value of the parameter, called the *power function.*

Example 11.12

Suppose we wish to carry out the following hypothesis test for the mean, μ, of a normal distribution, with known variance, $\sigma^2 = 225$. And suppose we decide to take a sample of $n = 20$ and we decide that the significance level, α, is 5%.

If H_0: $\mu = 100$
and H_1: $\mu > 100$,

(i) what is the critical region?
(ii) what is the power of the test at $\mu = 110$, say?

(i) The critical region is

$$\frac{\bar{x} - \mu_0}{\sigma/\sqrt{n}} > 1.64 \qquad \text{(see Table 11.1)}$$

Since $\mu_0 = 100$, $\sigma = \sqrt{225} = 15$, $n = 20$ in our example, the critical region is $z = \dfrac{\bar{x} - 100}{15/\sqrt{20}} > 1.64$. In other words, we will reject H_0 if:

$$\bar{x} > 100 + 1.64 \times \frac{15}{\sqrt{20}}$$

$$\text{i.e. if} \quad \bar{x} > 105.5$$

But we know that the random variable $\bar{X} \sim N(\mu_0, \sigma^2/n)$ if H_0 is true (Section 11.3), which for this example becomes $\bar{X} \sim N(100, 15^2/20)$, if $\mu = 100$.

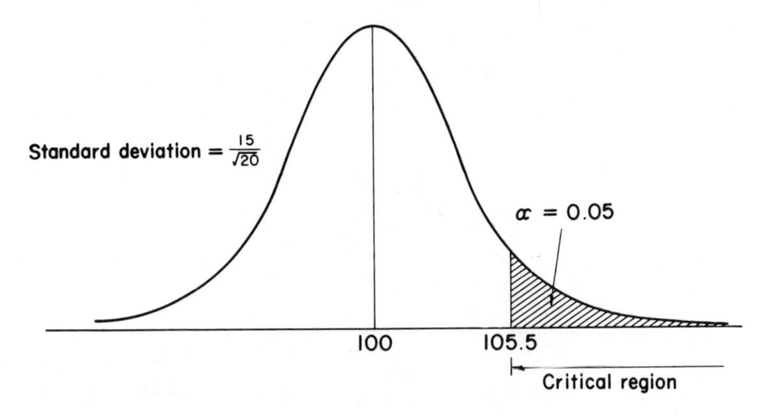

Fig. 11.10 The critical region in terms of the distribution of $\overline{X}$

Figure 11.10 shows the distribution of $\overline{X}$ when $\mu = 100$, and how the critical region relates to it.

Figure 11.10 is another way of representing the critical region, namely in terms of the distribution of $\overline{X}$ instead of in terms of the standardized normal distribution.

(ii) When $\mu = 110$, it follows that $\overline{X} \sim N(110, 15^2/20)$. The distribution in Fig. 11.10 must be shifted 10 units to the right (note that it is still a normal distribution and that its variance is unchanged).

Figure 11.11 incorporates both distributions.

Since power $= (1-\beta) = P$(accepting H_1 when H_1 is true),

then the power of the test when $\mu = 100$

$= P$(accepting H_1: $\mu = 110$ when $\mu = 110$)

$= P$(rejecting H_0: $\mu = 100$ when $\mu = 110$) since accepting H_1 implies rejecting H_0

$= P(\overline{X} > 105.5$ when $\mu = 110)$

= area to the right of 105.5 for the right-hand distribution of Fig. 11.11

$$= P\left(Z > \frac{105.5 - 110}{15/\sqrt{20}}\right)$$ using normal distribution theory (see Section 6.6).

$= P(Z > -1.34)$

$= 0.91$ (2 d.p.) from Table C.3(a)

The power of the test at $\mu = 110$ is 0.91.

Since power $= 1 - \beta$, it follows that $\beta = 0.09$ when $\mu = 110$. □

Notice that β is the area to the left of the critical value for the right-hand distribution and that α is the area to the right of the critical value for the left-hand distribution. This point is well worth remembering!

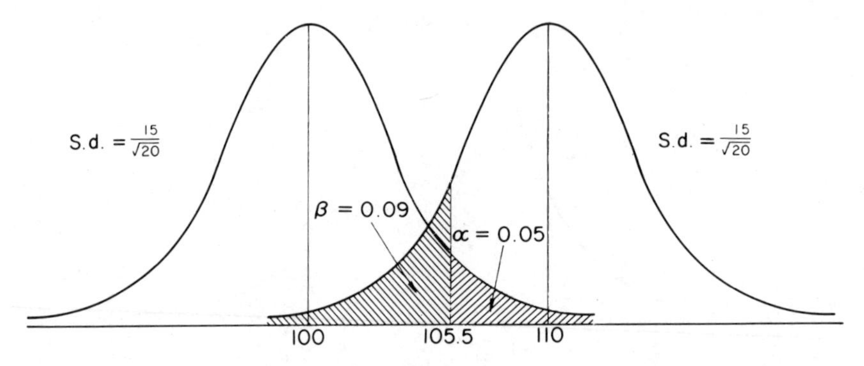

Fig. 11.11 The distributions of $\overline{X}$ when $\mu = 100$ and when $\mu = 110$

We can calculate the power of this test for values of μ other than 110 using the method above. Remember, however, that H_1 specifies $\mu > 100$, so we are only interested in values greater than 100. Table 11.9 summarizes the results of such calculations, which you should check for yourself.

Table 11.9 An example of how the power of a test varies for different values of μ

μ	100	102	104	106	108	110	112
power $= (1-\beta)$	0.05	0.15	0.33	0.56	0.77	0.91	0.97
β	0.95	0.85	0.67	0.44	0.23	0.09	0.03

Figure 11.2 shows the power function curve for this test, namely a plot of power versus μ.

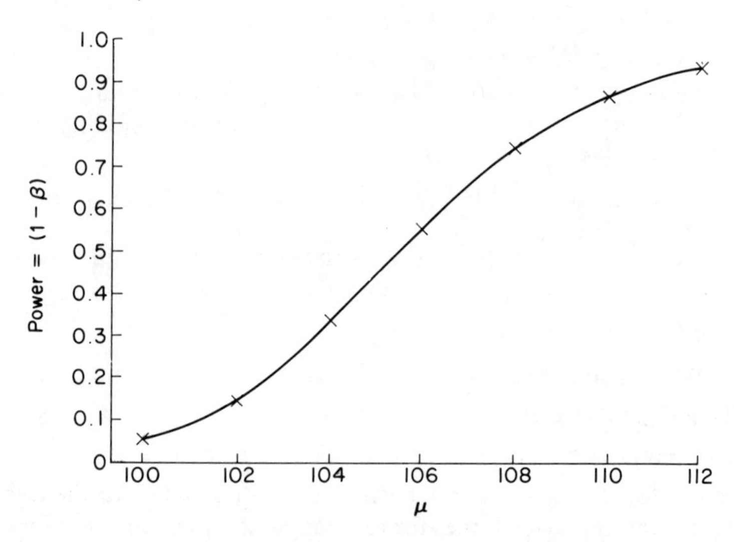

Fig. 11.12 Example of a power function curve

We note from Fig. 11.12, remembering that H_0 specifies $\mu = 100$, that the power of the test increases as the value of μ in the alternative hypothesis increases. In other words, we are more likely to accept the alternative hypothesis the further the true value of μ is from the value specified in the null hypothesis.

The effect of increasing power as μ increases can also be seen by reference to Fig. 11.11 in which we noted that power = area to the right of 105.5 for the right-hand distribution.

Figure 11.11 is also useful in three other ways:

1. Suppose we wish to carry out the same hypothesis test, except that we wish to reduce the significance level, α. If everything else stayed the same, the effect would be to increase the critical value of $\bar{X}$, and hence to increase β and decrease the power for a particular value of μ. So a gain in reducing the probability of a Type I error is offset by an increased probability of a Type II error and consequent loss of power.
2. Suppose we wish to carry out the same hypothesis test, except that we wish to reduce both α and β. From what was stated in 1 this would seem to be impossible. However, it can be achieved, at some expense, by increasing the sample size, n. The effect of increasing n is to reduce the variability of both distributions in Fig. 11.11 (since each has a standard deviation of $\sigma/\sqrt{n}$). If we keep the same critical value of 105.5, both α and β will be reduced.
3. A third and important use of Fig. 11.11 is that it will enable us to decide how large a sample to take if we can specify:

   (a) the value of α we would like;
   (b) the value of β we would like for a particular value of μ.

Clearly, deciding how much data we should collect is a very important part of any investigation. □

Example 11.13

What sample size is required to test

H_0: $\mu = 100$

where μ is the mean of a normal distribution with known variance, $\sigma^2 = 225$, and we wish the following Type I and II errors:

$$\alpha = 0.05$$

and $$\beta = 0.01 \text{ when } \mu = 110$$

(i.e. giving a power of 0.99 when $\mu = 110$)?

We draw Fig. 11.13 (like Fig. 11.11), and realize that there are two unknowns:

(i) the sample size, n;
(ii) the critical value of $\bar{X}$, let's call it $\bar{x}^*$.

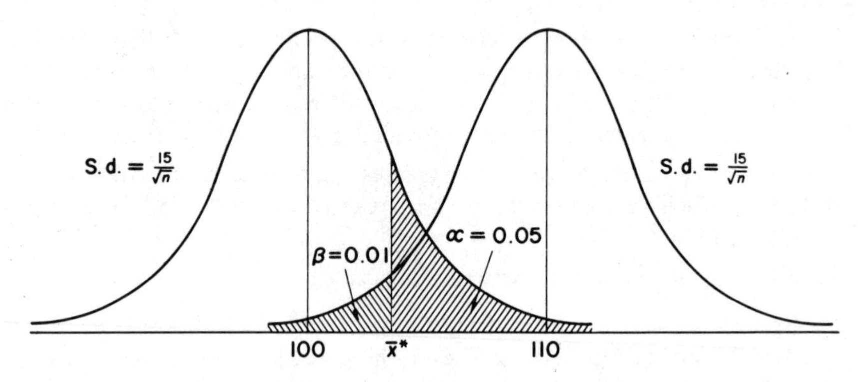

Fig. 11.13 The distributions of $\bar{X}$ when $\mu = 100$ and when $\mu = 110$

Considering the left-hand distribution, we know that:

$$\frac{\bar{x}^* - 100}{15/\sqrt{n}} = 1.64, \quad \text{since } \alpha = 0.05$$

Considering the right-hand distribution, we know that:

$$\frac{\bar{x}^* - 110}{15/\sqrt{n}} = -2.33, \text{ since } \beta = 0.01$$

From these equations, $\bar{x}^* = 100 + 1.64 \times \dfrac{15}{\sqrt{n}} = 110 - 2.33 \times \dfrac{15}{\sqrt{n}}$

Hence
$$n = \frac{(1.64 + 2.33)^2 \times 15^2}{(110 - 100)^2} = 35.5$$

So we would need a sample size of at least 36. □

From this example we can deduce a general formula for n for testing H_0: $\mu = \mu_0$ against H_1: $\mu = \mu_1$ with specified values of α and β, and assuming σ^2 is known.

The formula is
$$n = \frac{(z_\alpha + z_\beta)^2 \sigma^2}{(\mu_1 - \mu_0)^2}$$

where z_α and z_β are obtained from Table C.3(b) for tail areas of α and β, and $\mu_1 > \mu_0$ respectively.

11.11 NOTE ON ASSUMPTIONS MADE IN HYPOTHESIS TESTS

Several hypothesis tests have been described in this chapter. Most of these tests require certain assumptions to be valid in order for the test to give valid conclusions. Some of the assumptions are more important than others, because some tests are said to be *robust* against departures from their assumptions. In particular the t test (see Sections 11.4, 11.5 and 11.6) is robust against departures from the assumption of normality, especially when sample sizes are large. The unpaired samples t test (Section 11.5) requires not only the normality assumption but also the assumption of equality of variance. This test is also robust against departures from the assumption of equality of variance particularly if the sample sizes, n_1 and n_2 are roughly equal.

However, neither the χ^2 test (Section 11.7) nor the F test (Section 11.8) are robust against departures from normality. The assumption of normality is therefore particularly important for these tests.

Some methods of testing for normality are described in Section 13.9.

If we decide that the conclusions we reach may be invalid because of violation of the assumptions, we may, in some instances, use other tests which require fewer assumptions to be made. Three of these 'distribution-free' or 'non-parametric' tests are described in Chapter 14. However, 'non-parametric' tests are generally less *powerful* than the corresponding 'parametric' tests (such as those described in this chapter).

For the moment it is important for you, the reader, to realize that performing a hypothesis test is not simply a question of putting numbers into a formula and drawing a conclusion. The assumptions underlying each test must be carefully considered and the most appropriate test chosen in the light of the available information.

11.12 SUMMARY

To test hypotheses about the parameters of a distribution we must specify a null hypothesis, an alternative hypothesis, and a significance level. Using statistical theory, it may then be possible to specify a critical region, which provides a method for deciding whether to reject H_0, given appropriate sample data.

Tests were considered for:

(a) the mean of a normal distribution with known variance;
(b) the mean of a normal distribution with unknown variance;
(c) the difference between the means of:

    (i) two normal distributions with the same but unknown variance;
    (ii) two normal distributions with different but known variances;
    (iii) two distributions with different but unknown variances (large samples test only);

(d) the mean of a distribution of differences;
(e) the variance of a normal distribution;
(f) the equality of the variances of two normal distributions.

The critical regions for these tests are given in various tables in Sections 11.3–11.8.

We also saw some examples of how confidence intervals could be used to test hypotheses.

Type I and II errors, power and power function were discussed by reference to an example of (a) above. The effect of reducing the Type I error, the effect of increasing sample size, and the method for deciding the sample size of an investigation, were all discussed by reference to the same example.

Finally, there was a discussion about the importance of the assumptions underlying the various hypothesis tests described in the chapter.

Section 11.4

1. Identify the null and alternative hypotheses in each of the following:
   (a) A tobacco company prints the following claim on its cigarette packets: 'Average tar content is 5 mg per cigarette'. A consumer testing group suspects that the true average tar content may be higher than 5 mg and the group intends to analyse a random sample of cigarettes and measure the tar content of each.
   (b) The research department of a tobacco company believes that the tar content of a new brand of cigarettes will be less than 5 mg per cigarette, and tests a random sample of cigarettes to decide whether their belief is justified.
   (c) Records for the past several years show that the mean mark obtained by first-year undergraduate students who take a course in statistics is 55 out of 100. The lecturer has just finished marking the examination papers for a class of 20 students who have recently completed the course, and now wishes to test whether the mean mark for this class is typical of those of previous years.

2. Carry out the appropriate tests for Exercise 1 (a), (b) and (c) above given the following data and levels of significance:
   (a) The tar contents (mg) of a random sample of 20 cigarettes is:

4.3	5.2	4.1	6.3	5.1	3.7	4.0	5.1	3.7	4.3
2.5	6.3	4.8	3.7	2.9	6.3	5.1	4.2	3.9	2.8

   Use a 5% level of significance, and assume tar content is normally distributed.
   (b) The tar contents (mg) of a random sample of 100 cigarettes were summarised as follows:

   $\bar{x} = 4.8, \quad s = 1.2$ mg.

   Use a 1% level of significance.
   (c) The marks for the 20 students were arranged in a grouped frequency table as follows:

Mark range	*Number of students*
20–29	1
30–39	3
40–49	4
50–59	7
60–69	2
70–79	3

   Use both 5% and 1% levels of significance, and comment on the conclusions drawn. Assume that marks are normally distributed.

►

3. An oil company wishes to assess the effect on fuel consumption of using a new additive intended to improve the performance of cars. A random sample of ten cars of a particular model were selected from the week's production of cars made by a car company. The results in miles per gallon (mpg) for a standard test run were as follows:

   32.6 27.4 25.7 28.1 31.7 29.6 24.8 28.5 30.6 29.4

   Can the oil company claim a significant increase in miles per gallon over the previous long-term average of 28 mpg? Use a 5% level of significance and state any assumptions made.

4. A random sample of eleven bags were selected from a machine packaging sugar in bags marked 1 kg. The weights of sugar, which may be assumed to be normally distributed, in grams were:

   1017 1051 1078 1033 996 1059 1082 1014 1040 1072 998

   Test the hypothesis that the mean weight of sugar in bags marked 1 kg is 1 kg. Use a 10% level of significance.

5. The mean of a random sample of n observations $x_1, x_2, \ldots, x_n$ from a normal distribution is $\bar{x}$. It is proposed to test the null hypothesis that the population mean is μ_0, against some alternative hypothesis concerning the value of the mean. For such a test, state the circumstances under which you would use (i) a normal distribution, (ii) a t distribution.

   A production line is designed to produce 100 finished articles per hour. Observations over nine separate hourly periods give the following numbers of finished articles:

   96 102 86 78 101 94 93 92 103

   The manager wishes to know whether the output from the production line is lower than could be expected from random variation. Carry out a suitable test, at the 5% level of significance, in order to answer the manager's question, showing all necessary details of your working. State clearly your null and alternative hypotheses concerning the value of the mean. State any assumptions which you need to make in order to carry out the test.

   (JMB)

6. The diameters of rivets produced by a machine before an overhaul are normally distributed with a mean of 2 cm and a standard deviation of 0.05 cm.

   After an overhaul, which may change the mean of this distribution but not its standard deviation, five rivets are selected at random from the machine's production, and the diameters (in cm) are recorded as follows:

   2.02 2.08 1.99 2.11 2.05

   Test whether there is significant evidence at the 5% level that the mean has changed.

   Provide a suitable 99% confidence interval for the mean after overhaul, and explain carefully what this confidence interval means.

   Describe what modifications to the above test would have to be made if it could not be assumed that the standard deviation 0.05 cm remained unchanged.

   (JMB) ►

7. When a standard lubricant is used on the bearings of a wheel, the length of time for which the wheel will spin in an experiment has mean 200 s and standard deviation 10 s. A 'new improved' lubricant is tried out and five spins take times (in seconds):

   209 224 196 219 212

   Test, assuming that the standard deviation is unchanged and stating clearly all the other assumptions you need to make, whether the new lubricant has significantly increased the spinning times.

   (SMP)

8. The weights of steaks sold by a supermarket are distributed normally with mean μ and standard deviation 0.02 lb. A quality control inspector tests the hypothesis that $\mu = 1$ lb at the 5% level of significance. He takes a random sample of five steaks whose weights (in lb) are:

   0.977 1.014 0.989 0.972 0.968

   His null hypothesis is that $\mu = 1$ lb, and he performs a two-tailed test. State his alternative hypothesis and perform the test.

   Another inspector is employed to check that customers are not (on average) sold underweight steaks. If he had conducted a one-tailed test using the same random sample, the same level of significance and the same null hypothesis, what would have been his alternative hypothesis, and his conclusion?

   (MEI)

9. Describe briefly what is meant by the sampling distribution of an estimator, illustrating your remarks by reference to the sampling distribution of the mean, $\overline{X}$, of a random samples of n observations from a normal distribution with mean μ and variance σ^2.

   Explain also what is meant by saying that the statistic $\overline{X}$ gives an *unbiased* and *consistent* estimate of the mean.

   Weights of a certain type of biscuit have a normal distribution with mean 8 g and standard deviation 0.8 g. Packets, of nominal weight 200 g, are made up of 25 randomly selected biscuits. Calculate the probability that the mean weight of the contents of three such packets is less than 195 g.

   The moulds used for making these biscuits are replaced. The new moulds produce biscuits whose weights are also normally distributed with standard deviation 0.8 g and are also sold in packets containing 25 biscuits. For a random sample of 30 packets of biscuits from the new moulds, the contents had a mean weight per packet of 198.0 g. Examine whether there is significant evidence that the mean weight of the contents of the packets of biscuits produced in the new moulds differs from 200.0 g.

   (JMB)

Section 11.6

10. Eight randomly selected rats were matched with eight other rats. One rat in each pair was randomly allocated to receive a treatment, D, and the other rat in each pair ►

received a neutral control, C. The responses of each rat to the treatments were as follows:

Pair No.	*1*	*2*	*3*	*4*	*5*	*6*	*7*	*8*
Treatment, D	87.2	83.3	65.7	97.9	109.1	97.6	87.8	107.0
Control, C	89.6	72.1	61.5	94.4	94.1	83.8	79.5	108.0

Is there a significant difference at the 5% level between the mean responses for the two treatments? State any assumption you have made.

11. The following results are the weights of eggs (in grams) produced by eight hens fed on ordinary corn and eight fed on vitamin-enriched corn pellets. The allocation of hen to type of corn was done randomly.

Ordinary corn	31	32	34	33	32	32	31	33
Enriched corn	30	34	34	33	35	33	35	34

Use a 1% level of significance to decide whether the mean weight of eggs for hens receiving enriched corn is greater than for hens receiving ordinary corn. Assume that the weight of eggs is normally distributed for both types of corn. What other assumption have you made?

12. The following data refer to the incomes of a random sample of 120 families from region A and 280 families from region B of the USA in 1971.

Incomes ($000)	*0–5*	*5–10*	*10–15*	*15–20*	*Total*
Region A	28	42	30	20	120
Region B	42	78	80	80	280

Is there a significant difference between the mean incomes for the two regions? Use a 5% level of significance, and assume that the variance in incomes is the same for the two regions.

13. On entering a graduate school, a random sample of eight psychologists were given a statistics test. After a course of instruction in statistics they were re-tested and the scores were as follows:

Psychologist No.	*1*	*2*	*3*	*4*	*5*	*6*	*7*	*8*
Score before instruction	24	26	41	13	18	24	27	32
Score after instruction	31	20	44	19	29	35	28	33

Do these data support the hypothesis that the scores following instruction are significantly higher on average than the scores before instruction? Use both a 5% and 1% significance and compare the conclusions drawn. What assumptions have you made?

14. A sleeping draught and an inactive control were tested in turn in random order on a random sample of ten hospital patients. Their hours of sleep were as follows:

Patient	*1*	*2*	*3*	*4*	*5*	*6*	*7*	*8*	*9*	*10*
Hours of sleep with drug	10.6	7.5	9.0	5.4	6.1	10.2	7.1	9.7	8.5	7.9
Hours of sleep with control	8.6	7.4	9.4	5.1	5.4	9.0	6.5	7.9	8.7	6.9

Test the hypothesis, at the 5% level of significance, that the mean hours of sleep with the drug are the same as with the control, using a one-tailed test. Assume that the difference in the hours of sleep for each patient are normally distributed.

15. Tests were made to determine the performance of cars in miles per gallon of petrol using two different additives. Eighteen nominally identical cars took part, nine using Additive 1 and the other nine using Additive 2. The miles per gallon covered by each car in a standard test were as follows:

Additive 1	32.3	37.3	35.5	33.7	31.9	30.0	34.6	32.8	36.4
Additive 2	38.4	34.8	36.6	32.0	33.9	35.7	37.5	39.3	34.3

Do these data support the hypothesis that cars using Additive 2 give on average more miles to the gallon than cars using Additive 1? Use a 5% level of significance. What assumptions have you made?

16. A variable X is normally distributed in one population with unknown mean μ_1 and variance σ^2. The same variable is also normally distributed in another population with unknown mean μ_2 and the same variance as population 1. A random sample of seven values is taken from each population.

    The sum, and sum of squares of the seven values from the first sample are 35 and 387 respectively. The corresponding values for the second sample are 49 and 399. Test the hypothesis that $\mu_1 = \mu_2$ with a 1% level of significance using a two-tailed test.

17. Random samples of 15 men and 15 women are selected from the large number of people living in a certain region, and the height of each individual is measured. Let x_i, $i = 1, 2, \ldots, 15$ denote the heights (cm) of the men and y_i, $i = 1, 2, \ldots, 15$ the heights (cm) of the women. The following results are obtained:

$$\bar{x} = \sum x_i/15 = 175.7, \qquad \sum (x_i - \bar{x})^2 = 378.56,$$
$$\bar{y} = \sum y_i/15 = 168.2, \qquad \sum (y_i - \bar{y})^2 = 609.84,$$

It is known that for the people who lived in the region at a certain time in the past, the mean height of men was 12 cm greater than the mean height of women. Is there evidence that the difference in height has changed between then and now? ►

State any assumptions that you make in your analysis.

(IOS)

18. (a) Compare the uses of the two-sample t-test and the paired t-test showing clearly why each method should be used in preference to the other in particular situations.
(b) Eighteen steel bars of similar hardness were strength-tested after undergoing one of two treatments, quenching and tempering, and their strengths (kN/mm^2) were as given:

Quenched (x)	1.17	1.21	1.15	1.23	1.18	1.13	1.24	1.24	1.16	1.19
Tempered (y)	1.20	1.15	1.13	1.24	1.17	1.15	1.18	1.16		

(*Note*: $\sum x = 11.9$, $\sum y = 9.38$, $\sum x^2 = 14.1746$, $\sum y^2 = 11.0064$)

Analyse the data to conclude whether quenched or tempered steel is generally the stronger, comment on the limitations of your conclusions and describe the conditions that would have to apply for this to be considered a properly designed, controlled experiment.

(IOS)

19. The only visible difference between wild and farmed trout is the greater average length of the latter. I have kept a record of the lengths of six wild trout caught recently as follows:

Length (cm) 26.8 26.0 25.7 25.6 24.9 24.3.

I am offered four trout which are claimed to be farm-reared and their lengths (cm) are:

27.1 27.5 26.4 26.1.

Apply a suitable t-test, stating clearly your null and alternative hypotheses, to determine whether I can reasonably accept the statement that the four are farm-reared.

(Cambridge)

20. Two methods of determining the starch content of potatoes are to be compared. Twelve potatoes are each cut in half. From each potato, a half is chosen at random and allocated to method 1, the other half being allocated to method 2. The starch percentages obtained are as follows:

Potato	*1*	*2*	*3*	*4*	*5*	*6*	*7*	*8*	*9*	*10*	*11*	*12*
Method 1	23.2	20.3	18.7	20.1	20.8	21.4	19.5	18.9	22.0	19.8	21.1	22.5
Method 2	23.1	20.0	18.7	19.9	20.6	21.2	19.6	18.8	22.1	19.6	21.3	22.4

You may assume that starch content is normally distributed.

(i) Determine whether the data provide evidence, at the 5% significance level, that the two methods give significantly different results.
(ii) Given that it is known that the difference, if any, between the methods will result in method 1 giving higher readings, in general, than method 2, determine whether the data provide evidence, at the 5% significance level, for such a difference.

(Cambridge) ►

21. The random sample of six observations given below was obtained from a normally distributed population with unknown mean μ and unknown variance σ^2:

$$5.5 \quad 6.3 \quad 7.4 \quad 5.5 \quad 5.1 \quad 6.6.$$

(i) Provide unbiased estimates of μ and σ^2.
(ii) Test the hypothesis that $\mu = 7$ using a 5% significance level.

Before this sample was obtained, μ had been estimated as 7.4 with a standard error of 0.5 based on a random sample of three observations. Test the hypothesis that the two samples come from the same population against the alternative hypothesis that they come from two populations with different means but with the same variance.

(O&C)

22. In order to compare two methods for finding the percentage of iron in a compound, each of ten different compounds was analysed by both methods, and the results are given below.

Compound number	*1*	*2*	*3*	*4*	*5*	*6*	*7*	*8*	*9*	*10*
Method A: % iron	13.3	17.6	4.1	17.2	10.1	3.7	5.1	7.9	8.7	11.6
Method B: % iron	13.4	17.9	4.1	17.0	10.3	4.0	5.1	8.0	8.8	12.0

Use Student's t distribution to test, at the 5% level of significance, the hypothesis that the two methods of analysis differ.

State the assumptions that are necessary to justify the use of this distribution.

(MEI)

23. To test a new chicken food additive, eight hens were given the normal food for three weeks and then were given the normal food together with the special additive for the next three weeks.

The number of eggs laid by each hen was as follows:

Hen	*1*	*2*	*3*	*4*	*5*	*6*	*7*	*8*
Fed with normal food, number of eggs	14	15	16	15	16	15	17	18
Fed with food + additive, number of eggs	15	16	16	16	17	17	18	18

Perform a paired-sample t-test at the 5% level to investigate whether or not the additive results in a greater mean number of eggs.

(SUJB)

24. State what you understand by the terms (i) null hypothesis, (ii) alternative hypothesis, (iii) significance level.

To investigate the effectiveness of a new petrol additive on the performance of cars the following test was carried out. Twenty cars of varying ages were selected randomly and divided into two groups of ten at random. One group (A) was tested with the additive and the other group (B) without the additive. The twenty tests were carried out under similar conditions. The number of miles per gallon was calculated for each car and the results summarized in the table: ►

Group	*A*	*B*
Sample size	10	10
Mean m.p.g.	43.2	40.9
Standard deviation	2.32	2.47

Stating any necessary assumptions and your null and alternative hypotheses, test (using a 5% significance level) whether or not the additive results in improved petrol consumption.

How might the test have been better designed using the same twenty cars?

(SUJB)

Section 11.9

25. In a comparative study of the effect of oxygen on the peripheral nerve in cats and rabbits, the survival time of a nerve under anoxic conditions was measured. The times in minutes were as follows:

Cat no.	1	2	3	4
Time	45	43	33	25

Rabbit no.	1	2	3	4	5	6	7	8	9	10	11	12	13	14
Time	35	35	30	30	28	28	23	22	22	20	17	16	16	15

(a) Test the hypothesis that the variance of the survival time of the nerve is the same for cats and rabbits. Use a 5% level of significance.

(b) Also test whether the mean survival time of the nerve is the same for cats and rabbits. Use a 1% level of significance.

You may assume that the survival times for cats and for rabbits are both normally distributed.

26. Each of the n independent random variables X_i $(i = 1, 2, \ldots, n)$ has mean μ and variance σ^2. Find expressions for:

$$E(X_i^2),\ E\left(\sum_{i=1}^{n} X_i\right) \text{ and } E\left[\left(\sum_{i=1}^{n} X_i\right)^2\right] \text{ in terms of } n,\ \mu \text{ and } \sigma^2.$$

Hence, or otherwise, show that

$$E\left[\sum_{i=1}^{n} X_i^2 - \frac{1}{n}\left(\sum_{i=1}^{n} X_i\right)^2\right] = (n-1)\sigma^2.$$

The thermal conductivity of a material can be measured by two methods. Using method A, seven independent measurements gave results x_i for which:

$$\sum_{i=1}^{7} (x_i - \bar{x})^2 = 25 \times 10^{-4}.$$

▶

Assuming that these are a random sample from a normal distribution, find 90% confidence limits for its standard deviation.

Using method B, four independent measurements were made and the standard deviation of the normal population from which they came was estimated to be 0.011. Show that these data are not sufficient to prove that method B is more accurate (i.e. has smaller variance) than method A, using a 5% level of significance.

(O & C)

27. (a) The weights of all pupils in a school are taken and are distributed normally with standard deviation 6 kg. The separate boys' and girls' weights are also distributed normally, the boys with standard deviation 5 kg and the girls 8 kg.

The weights of a random sample of ten children are taken (in kg):

Boys	46,	42,	52,	44,	56,	36
Girls	45,	53,	38,	36.		

Determine 95% confidence limits for the mean weight of (i) the boys; (ii) the girls; (iii) the whole school.

(b) The mean number of letters delivered to the School Office and to the Principal's Office were recorded over varying periods of time:

	No. of days recorded	*Mean no. of letters*	*Standard deviation*
School Office	63	48.1	6.24
Principal's Office	52	44.6	8.56

Is there a significant difference between the means? (Use a 5% level of significance.)

(SUJB)

28. A weak source of radioactivity give the following numbers of counts per minute over a period of 100 minutes.

Counts/minute	*Number of minutes*
200–209	9
210–219	15
220–229	28
230–239	21
240–249	18
250–259	9

(a) Calculate a 95% confidence interval for the mean number of counts per minute and use it to test the hypothesis that the mean number of counts per minute is 200. Assume that the number of counts per minute is normally distributed.

(b) Calculate a 95% confidence interval for the variance of the number of counts per minute and use it to test the hypothesis that the variance of the number of counts per minutes is 200.

►

29. A physical model suggests that the mean temperature increase in water used as a coolant in a certain process should be 5 °C. Temperature increases in the coolant measured on eight individual runs of the process were as follows (°C):

    6.4   4.3   5.7   4.9   6.5   5.9   6.4   5.1

    Calculate a 95% confidence interval for the mean increase in the temperature of the coolant, stating any assumptions made. Use your result to test the hypothesis that mean temperature increase is 5 °C, using a two-tailed test and a 5% level of significance.

    Is the mean increase in temperature significantly greater than 5 °C for a 5% level of significance?

30. A research worker measured the wing-lengths of a random sample of 50 wings of house-flies and formed the following table:

Wing-length (mm)	*Number of wings*
3.5–3.9	5
4.0–4.4	14
4.5–4.9	23
5.0–5.4	7
5.5–5.9	1

    (a) Test the hypothesis that the variance in the wing-length for house-flies is 0.2 mm^2. Use a 5% level of significance, and assume that wing-length is normally distributed.
    (b) Also test whether the mean wing-length of house-flies is 5 mm against the alternative hypothesis that it is less than 5 mm. Again use a 5% level of significance.

31. State under what circumstances you would use a pooled (or combined) estimate of variance when testing for the difference between means of two large samples.

    A firm has produced an additive for petrol which, it claims, will increase the number of miles per gallon obtained by a car. To test this claim an independent agency tests 120 cars without the additive and 150 cars with the additive. Letting x_i be the number of miles per gallon obtained by each of the first 120 cars and y_j the number of miles per gallon obtained by each of the second 150 cars, the following results were obtained:

$$\sum x_i = 3984, \quad \sum x_i^2 = 133{,}373, \quad \sum y_j = 5115, \quad \sum y_j^2 = 175{,}727.$$

    Estimate the increase in miles per gallon given by the additive and test whether or not this increase is significant. State clearly your null and alternative hypotheses and the significance level at which you are testing.

    Assuming that the additive gives a particular motorist an increase from 33.2 to 34.1 miles per gallon and that petrol costs £1.75 per gallon, calculate the maximum cost of the additive per gallon of petrol if it is to be economically worthwhile for this motorist to use it.

    (JMB)

32. A random variable X has unknown mean μ and unknown variance σ^2. A random sample of 50 observations of X has sum 2625 and sum of squares 137,950. Calculate an unbiased estimate of μ and an unbiased estimate of σ^2. ▶

Determine an approximate 95 per cent symmetrical confidence interval for μ. Test the hypothesis that $\mu = 52$ against the alternative hypothesis that $\mu \neq 52$:

(i) at the 5 per cent level of significance;
(ii) at the 2 per cent level of significance.

A random sample of 100 observations on another random variable Y has sum 5310 and sum of squares 285,000. Test, at the 5 per cent level of significance, whether or not the distribution means of X and Y are equal.

(JMB)

33. A random sample of size ten was chosen from London schoolchildren of the same age. The heights (in cm) of the children chosen were:

							Total
Boys	142	125	127	122			516
Girls	130	122	117	125	123	121	738

On the assumption that the variance of boys' and girls' heights is the same, estimate the variance. Hence, assuming that the heights are normally distributed, find a 95% confidence interval for the difference in mean height of boys and girls of the age sampled.

What would be the result of a significance test of the null hypothesis that the mean height of boys and girls, of the chosen age, is equal?

(OLE)

34. Two independent normal populations have the same known variance σ^2, but unknown, possibly different, means μ_1 and μ_2. Random samples, each of size n, are drawn from each population, and the sample means are $\overline{X}_1$ and $\overline{X}_2$. Obtain symmetric 95% confidence intervals for μ_1 and μ_2. Also state the distribution of $\overline{X}_1 - \overline{X}_2$.

The following test of the hypothesis that $\mu_1 = \mu_2$ is proposed. The hypothesis will be rejected only if the 95% confidence intervals for μ_1 and μ_2 have no point in common.

(i) What conclusion is reached when $\sigma = 2.5, n = 100$ and the values taken by $\overline{X}_1$ and $\overline{X}_2$ are 10 and 11 respectively?
(ii) Suppose that, in fact, $\mu_1 = \mu_2$. Show that the probability that the test gives a wrong conclusion is:

$$P\left(|\overline{X}_1 - \overline{X}_2| > \frac{3.92\sigma}{n^{1/2}}\right)$$

Evaluate this probability, showing that it is independent of σ and n.

(MEI)

Section 11.10

35. A bus company decides to investigate complaints by the public that its buses are running late. A random sample of 20 bus journeys are selected in a given week and the hypothesis that buses do not run on average either early or late is to be tested against ►

the alternative that they run late on average. Suppose previous surveys of the company's buses have shown that the number of minutes late is normally distributed with a standard deviation of three minutes.

(a) What critical region should the company use if it wishes to carry out a test at the 5% level of significance?
(b) If the average number of minutes late is really two minutes:
    (i) what is the probability that the bus company will wrongly decide that its buses do not run on average either early or late;
    (ii) what is the power of the test?

36. Let $X_1, X_2, \ldots, X_{100}$ denote a random sample of size 100 taken from an $N(\mu, 9)$ distribution.

    (a) What critical value of $\overline{X}$ should be used to test $H_0: \mu = 20$ against $H_1: \mu > 20$, for a 5% level of significance?
    (b) Calculate the power of the test for values of μ between 20 and 21 (in steps of 0.2), and draw the power function curve.
    (c) Suppose the sample size is doubled to 200, what is the new critical value of $\overline{X}$? Will the new test be more powerful? Verify your answer by calculating the new value of power at $\mu = 20.6$, say.
    (d) What sample size would be required to give a power of 0.99 at $\mu = 20.6$?

37. The operating efficiency of a process is known to be normally distributed with a mean of 86 units and a standard deviation known to be 3 units. An experiment is planned to test a modification to the process which is expected to produce an improved mean efficiency. The risk of concluding that an improvement has occurred when none has occurred is set at 1% and the risk of not detecting an improvement to 90 units is set at 10%.

    (a) Draw a sketch showing the distribution of the sample mean efficiency, $\overline{X}$, under the two hypotheses:

        (i) that the mean efficiency is 86 units;
        (ii) that the mean efficiency has improved to 90 units.

        Assume that the standard deviation remains unchanged at 3 units. Find the number of experiments which should be performed, and the corresponding critical value of $\overline{X}$.

    (b) Suppose the operating efficiencies measured in twelve tests following the modification were as follows:

        88.4 88.2 91.5 91.7 90.5 88.0 91.0 88.3 91.0 88.3 91.6 90.2

        Taking the number of results as indicated by your answer to (a), decide whether an improvement in average operating efficiency to 90 units has occurred.

38. The average time required for the manufacture of an item in a production process should be $\mu = 100$ minutes, with a standard deviation of 5 minutes. However, it is suspected that the process has slowed down so that is now greater than 100 minutes. Given a sample of nine items and a risk of 0.01 of wrongly deciding the process has slowed, what is the critical value of the sample mean? Calculate the Type II error and the power of the test when $\mu = 107$ minutes. You may assume that the time to manufacture an item is normally distributed.

39. Experimental data concerning a variable X, which measures the reliability of a certain electronic component, are summarized as follows:

    $\Sigma x = 1164.2$, $\Sigma x^2 = 13{,}911.60$, $n = 100$, where n is the sample size.

    Calculate unbiased estimates of the population mean and variance. Explain whether, on the evidence of this sample, you would reject the hypothesis that the mean of X is 12.

    (You may treat the sample as a large sample).

    Figures collected over a long period have established that the mean and standard deviation of X are 12 and 2 respectively. After a change in the manufacturing process it is expected that the mean will have been increased but it may be assumed that the standard deviation remains equal to 2. A large sample of n values of X is taken, with sample mean m. If m is greater than some critical value then it will be accepted that the mean has in fact increased; if m is less than the critical value then the increase is not established. State carefully appropriate null and alternative hypotheses for this situation and find, in terms of n, the critical value for a 1% significance level.

    (Cambridge)

40. The diameters of some rods made by a machine have a normal distribution with a mean μ and standard deviation $\sigma = 0.05$ mm. When the machine is working correctly, $\mu = 2.50$ mm. At regular intervals an inspector takes a random sample of n rods and finds their mean diameter $\bar{x}$. If $\bar{x}$ exceeds 2.50 mm by more than h, the machine is stopped for adjustment. Otherwise it continues making rods. Find the values n and h must have to ensure that the probability of stopping the machine when it is working correctly is less than 0.015, the probability of stopping it when $\mu = 2.54$ mm is greater than 0.90 and n is as small as possible, consistent with these requirements.

    Taking h to be 0.025 and using the value of n found above:

    (a) find the probability of stopping the machine when $\mu = 2.53$ mm and when $\mu = 2.55$ mm;

    (b) sketch a graph showing the probability of stopping the machine for values of μ from 2.50 mm to 2.55 mm.

    If μ remains equal to 2.50 mm but now σ increases to σ', find the value of σ' which will cause the probability of stopping the machine to be 0.10.

    Suggest very briefly how increases in σ might be distinguished from changes in μ.

    (O&C)

41. A company packs sugar into bags. The masses of the packed bags are normally distributed with a standard deviation of 0.05 kg. The mean mass varies from day to day, but always lies between 0.94 kg and 1.02 kg. Each day a random sample of n bags is taken from the output, and the day's output is rejected if the sample mean mass is less than m kg, where n and m are to be chosen to satisfy two criteria. These criteria are:

    (a) When the mean mass is actually 1.02 kg, there is a probability of at least 95% that the sample mean mass is greater than m kg.

    (b) When the mean mass is actually 0.94 kg, there is a probability of at most 5% that the sample mean mass is greater than m kg.

    Determine suitable values for n and m.

▶

The company advertises that its bags of sugar have a mass of 1 kg. Suggest one reason each for the adoption by the company of each of the above criteria.

(MEI)

42. A manufacturer claims that the average lifetime of his candles is 12 hours.

    The lifetimes in hours of six of his candles were 11.4, 11.6, 11.1, 11.2, 12.1, 11.7. Perform a two-tailed t test at the 5% level of significance to examine the correctness of the manufacturer's claim. What assumptions need to be made about (i) the distribution of candle lifetimes, and (ii) the candles providing the data used in the test?

    Subsequent investigation shows that the distribution of lifetimes can be supposed normal with a standard deviation of 0.40 hours, but with a mean lifetime that varies from day to day. A random sample of six candles from one day's output is taken, and the candle lifetimes are measured. The results are used in a two-tailed test at the 5% level of significance. What is the probability that the manufacturer's claim, stated above, will be accepted if the mean lifetime is, in fact, 11.5 hours?

(MEI)

12

Hypothesis tests for the binomial parameter, p

▼

12.1 INTRODUCTION

In this chapter we will look at hypothesis tests for the parameter, p, of a binomial distribution. All the concepts used in Chapter 11 will be taken as read in this chapter, so it will be assumed that the reader is familiar with null and alternative hypotheses, significance level, critical region, one- and two-tailed tests, Type I and II errors, and power.

Three applications will be discussed:

1. An *exact* test for p, when the number of trials, n, is small.
2. An *approximate* test for p, when the conditions for using the normal approximation to the binomial apply.
3. An *approximate* test for $(p_1 - p_2)$, the difference between two binomial parameters, when the conditions for using the normal approximation to the binomial apply.

12.2 AN EXACT TEST FOR A BINOMIAL PARAMETER

Suppose we wish to test the null hypothesis, H_0: $p = p_0$ (where $0 < p_0 < 1$) against an alternative hypothesis, $H_1: p > p_0$, for a level of significance of α. If X is the number of successes resulting from a binomial experiment consisting of n trials, then it would seem reasonable to reject H_0 in favour of H_1 if

$$X > \text{critical value},$$

since we note that H_1 specifies larger values of p than that specified in H_0.

What critical value should we choose? The answer is the value which will result in the required significance level of α. The argument is as follows. Since the significance level, α, is defined by the equation:

$$\alpha = P(\text{rejecting } H_0 \text{ when } H_0 \text{ is true}) \quad \text{(see Section 11.3)}$$

then, for the above test,

$$\alpha = P(X > \text{critical value when } p = p_0)$$

The critical value therefore depends on the significance level, the number of trials, and the value of p_0. Given numerical values for these three quantities we can evaluate the critical value, using our knowledge of binomial probabilities (Section 6.2).

Example 12.1

In a multiple-choice test consisting of 10 questions, each question has five alternative answers, only one of which is correct. What is the critical number of questions for deciding whether a candidate is guessing the answers or whether the candidate is performing better than a 'guesser'? Use a 5% level of significance.

Here H_0: $p = 0.2$ (a candidate who guesses has a 1 in 5 chance of choosing the correct answer)

H_1: $p > 0.2$ (a candidate is doing better than a 'guesser')

5% level of significance (implying a risk of wrongly concluding that a candidate is doing better than a guesser).

$$\therefore 0.05 = P(X > \text{critical value when } p = 0.2)$$

Since X is the number of successes in $n = 10$ trials, we can find the critical value using binomial probabilities. In this case, since $n = 10$ and $p = 0.2$, we can use tables of cumulative binomial probabilities (Table C.1).

For example, the above statement is equivalent to

$0.95 = P(X \leqslant \text{critical value when } p = 0.2)$

In Table C.1 we look for the probability closest to 0.95 in the column of probabilities corresponding to $n = 10$, $p = 0.2$ and find:

$0.967 = P(X \leqslant 4 \text{ when } p = 0.2)$, and this is equivalent to
$0.033 = P(X > 4 \text{ when } p = 0.2)$

The critical value is 4, so we will reject the null hypothesis that the candidate is guessing if the candidate gets more than 4 (i.e. 5 or more) answers correct out of 10. The significance level for this test is 3.3%. (We cannot choose a significance level of *exactly* 5% because the binomial is a discrete distribution). □

The power of tests like the one in the example above depends on the actual value of

p. We can calculate the power for various values of p and hence determine a power function curve (in a similar way to the method used in Section 11.10).

Example 12.2

Find the power of the test at $p = 0.3$ for the previous example in this section. Hence determine the power function curve for that example.

$$\begin{aligned}\text{Power} &= P(\text{accepting } H_1 \text{ when } H_1 \text{ is true}) \quad \text{(Section 11.10)}\\ &= P(\text{rejecting } H_0 \text{ when } H_1 \text{ is true}) \quad \text{(logically equivalent)}\\ &= P(X > 4 \text{ when } p = 0.3) \quad \text{for this example}\\ &= 1 - P(X \leqslant 4 \text{ when } p = 0.3)\\ &= 1 - 0.850, \text{ from Table C.1 } (n = 10)\\ &= 0.150\end{aligned}$$

Since power $= 1 - \beta$, the Type II error at $p = 0.3$ is

$$\beta = 1 - 0.150 = 0.850$$

For other values of p we can similarly calculate the power and β. Table 12.1 (which you should check for yourself) shows the results of such calculations for $p = 0.2$ up to 1.0 in steps of 0.1.†

Table 12.1 An example of how the power of a test varies for different values of p

p	0.2	0.3	0.4	0.5	0.6	0.7	0.8	0.9	1.0
power $= (1 - \beta)$	0.03	0.15	0.37	0.62	0.83	0.95	0.99	1.00	1.00
β	0.97	0.85	0.63	0.38	0.17	0.05	0.01	0.00	0.00

Figure 12.1 is a plot of power versus p, using Table 12.1. □

12.3 AN APPROXIMATE TEST FOR A BINOMIAL PARAMETER

We saw in Section 7.3 that there are cases when the calculation of binomial probabilities can become very tedious, or when tables of cumulative binomial probabilities cannot be used. Instead we use the 'normal approximation to the binomial' to obtain approximate probabilities when:

$$np > 5 \quad \text{and} \quad n(1-p) > 5$$

†For values of $p > 0.5$, Table C.1 may still be used, for example at $p = 0.6$,

$$\begin{aligned}P(X > 4 \text{ when } p = 0.6) &= P(Y \leqslant 5) \text{ where } Y \sim B(10, 0.4)\\ &= 0.834\end{aligned}$$

Note that X is the number of correct answers and Y is the number of incorrect answers out of 10.

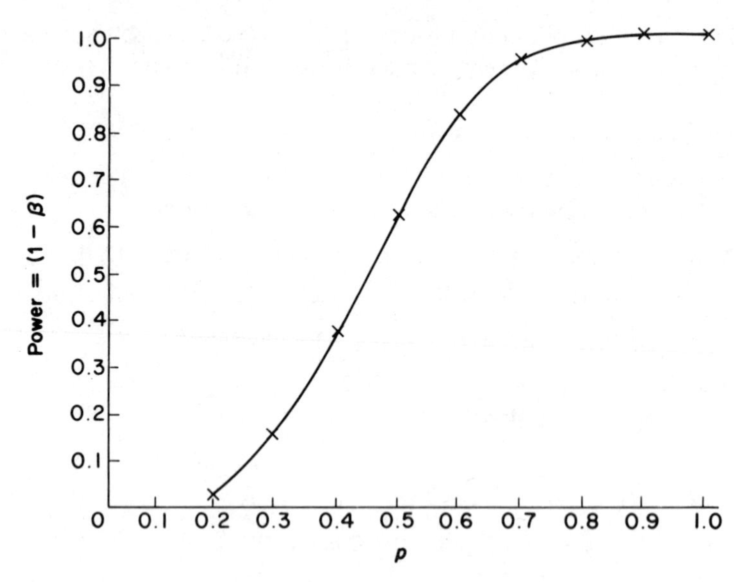

Fig. 12.1 Example of a power function curve

The same applies when we wish to test a binomial parameter, and the above inequalities apply. Instead of a test like the one described in the previous section, we carry out an approximate z test (see Table 11.1). The justification is as follows:

If X denotes the number of successes in n trials, and if $np > 5$ and $n(1-p) > 5$,

then $\quad X \sim N\,[np, np(1-p)] \quad$ approximately

Similarly, $\dfrac{X}{n} \sim N\left[p, \dfrac{p(1-p)}{n}\right]$ approximately,

where X/n denotes the proportion of successes.

So, for example, we will reject H_0: $p = p_0$ in favour of H_1: $p > p_0$, at the 5% level of significance, if

$$\frac{\dfrac{x}{n} - p_0}{\sqrt{\dfrac{p_0(1-p_0)}{n}}} > 1.64, \qquad \text{provided } np_0 > 5 \text{ and } n(1-p_0) > 5$$

Here x denotes the observed number of successes in n trials. This approximate test and two others (involving different alternative hypotheses) are summarized in Table 12.2, where a 5% level of significance has been used in each case.

Table 12.2 Critical regions for three approximate tests for p

H_0	H_1	*Critical region*	*Condition*
$p = p_0$	$p > p_0$	$\dfrac{\dfrac{x}{n} - p_0}{\sqrt{\dfrac{p_0(1-p_0)}{n}}} > 1.64$	$np_0 > 5,\ n(1-p_0) > 5$
$p = p_0$	$p < p_0$	$\dfrac{\dfrac{x}{n} - p_0}{\sqrt{\dfrac{p_0(1-p_0)}{n}}} < -1.64$	as above
$p = p_0$	$p \neq p_0$	$\dfrac{\left\lvert\dfrac{x}{n} - p_0\right\rvert}{\sqrt{\dfrac{p_0(1-p_0)}{n}}} > 1.96$	as above

Example 12.3

In a multiple-choice test consisting of 25 questions, each question has five alternative answers, only one of which is correct. A candidate gets nine correct answers. Test the hypothesis that the candidate is guessing, using:

(i) a 5% level of significance;
(ii) a 1% level of significance;

against the alternative that the candidate is doing better than a 'guesser'.

H_0: $p = 0.2$
H_1: $p > 0.2$

$$\frac{\dfrac{x}{n} - p_0}{\sqrt{\dfrac{p_0(1-p_0)}{n}}} = \frac{\dfrac{9}{25} - 0.2}{\sqrt{\dfrac{0.2 \times 0.8}{25}}} = 2.0$$

(i) Since $2.0 > 1.64$ we reject H_0, at the 5% level of significance.
(ii) Since $2.0 < 2.33$ we do not reject H_0, at the 1% level of significance.

These conclusions of this example are *not* contradictory. In the second we decide to run a smaller risk of a Type I error, i.e. of wrongly rejecting H_0, and so we are less likely to reject H_0. The consequence of a smaller Type I error, however, is a larger Type II error, β, and hence lower power. □

12.4 AN APPROXIMATE TEST FOR THE DIFFERENCE BETWEEN TWO BINOMIAL PARAMETERS

Suppose we carry out two binomial experiments, one having unknown parameter p_1, the other having unknown parameter p_2. If from n_1 trials in the first experiment we observe x_1 successes, and from n_2 trials in the second experiment we observe x_2 successes, we can carry out approximate tests for the difference, $(p_1 - p_2)$, using the 'normal approximation to the binomial', provided $n_1 p_1 > 5$, $n_1(1-p_1) > 5$, $n_2 p_2 > 5$, $n_2(1-p_2) > 5$.

The justification is as follows:

If X_1 and X_2 denote the number of successes in each experiment, then

$$\frac{X_1}{n_1} \sim N\left[p_1, \frac{p_1(1-p_1)}{n}\right] \text{ and } \frac{X_2}{n_2} \sim N\left[p_2, \frac{p_2(1-p_2)}{n}\right] \text{ approximately.}$$

$$\text{Hence } \frac{X_1}{n_1} - \frac{X_2}{n_2} \sim N\left[(p_1 - p_2), \frac{p_1(1-p_1)}{n_1} + \frac{p_2(1-p_2)}{n_2}\right] \text{ approximately.}$$

The null hypothesis we usually test is $H_0: p_1 = p_2$, i.e. that the two parameters are equal, to some value p, say. Under the null hypothesis,

$$\frac{X_1}{n_1} - \frac{X_2}{n_2} \sim N\left[0, p(1-p)\left(\frac{1}{n_1} + \frac{1}{n_2}\right)\right]$$

We can estimate p by using data from both binomial experiments, this estimate being

$$\hat{p} = \frac{x_1 + x_2}{n_1 + n_2} \qquad \text{(see Section 9.10)}$$

and we reject $H_0: p_1 = p_2$ in favour of $H_1: p_1 > p_2$, at the 5% level of significance, if:

$$\frac{\dfrac{x_1}{n_1} - \dfrac{x_2}{n_2}}{\sqrt{\hat{p}(1-\hat{p})\left(\dfrac{1}{n_1} + \dfrac{1}{n_2}\right)}} > 1.64$$

this being a valid approximate test if $(n_1 + n_2)\hat{p} > 5$ and $(n_1 + n_2)(1 - \hat{p}) > 5$.

This test and two others (involving different alternative hypotheses) are summarized in Table 12.3, where a 5% level of significance has been used in each case.

Example 12.4

Of a random sample of 25 boys and 20 girls, 10 boys and 6 girls said they intended to take Physics at A-level. Test, at the 5% level, whether these proportions are significantly different.

Table 12.3 Critical regions for three approximate tests for $(p_1 - p_2)$

H_0	H_1	*Critical region*	*Condition*
$p_1 = p_2$	$p_1 > p_2$	$\dfrac{\dfrac{x_1}{n_1} - \dfrac{x_2}{n_2}}{\sqrt{\hat{p}(1-\hat{p})\left(\dfrac{1}{n_1} + \dfrac{1}{n_2}\right)}} > 1.64$	$(n_1 + n_2)\hat{p} > 5$, $(n_1 + n_2)(1 - \hat{p}) > 5$, where $\hat{p} = \dfrac{x_1 + x_2}{n_1 + n_2}$.
$p_1 = p_2$	$p_1 < p_2$	$\dfrac{\dfrac{x_1}{n_1} - \dfrac{x_2}{n_2}}{\sqrt{\hat{p}(1-\hat{p})\left(\dfrac{1}{n_1} + \dfrac{1}{n_2}\right)}} < -1.64$	as above
$p_1 = p_2$	$p_1 \neq p_2$	$\dfrac{\left\lvert\dfrac{x_1}{n_1} - \dfrac{x_2}{n_2}\right\rvert}{\sqrt{\hat{p}(1-\hat{p})\left(\dfrac{1}{n_1} + \dfrac{1}{n_2}\right)}} > 1.96$	as above

If p_1, p_2 are the population proportions of boys and girls, respectively, who intend to take Physics at A-level,

H_0: $p_1 = p_2$
H_1: $p_1 \neq p_2$ since the above implies a difference rather than a one-directional effect

$$\hat{p} = \frac{10 + 6}{25 + 20} = 0.3556, \text{ so } 0.1 < \hat{p} < 0.9.$$

Both conditions hold, so we may use the third result in Table 12.3.

The value of the test statistic is

$$\frac{\dfrac{10}{25} - \dfrac{6}{20}}{\sqrt{0.3556\,(0.6444)\left(\dfrac{1}{25} + \dfrac{1}{20}\right)}}$$

$$= \frac{0.1}{0.1436}$$

$$= 0.696$$

Since $0.696 < 1.96$, we do not reject H_0, at the 5% level of significance.

The proportions of boys and girls who intend to take A-level Physics are not significantly different (5% level). □

12.5 SUMMARY

An exact hypothesis test for a binomial parameter p was described. When the binomial calculations become tedious or tables are not available, an approximate test may be carried out (see Table 12.2).

An approximate test for the difference between two binomial parameters may also be carried out (see Table 12.3).

Section 12.2

1. A subject claims that he can distinguish between two brands of cigarette when blindfolded. The null hypothesis that he guesses, and hence has a probability of 0.5 of correctly identifying the brand of each cigarette, is to be tested against the alternative hypothesis that he has a probability of 0.8 of being correct. The null hypothesis is to be rejected if the subject correctly identifies the brands of 7 or more out of 10 cigarettes presented to him.
   (a) Find values for α and β, the Type I and II errors.
   (b) Show how α can be decreased to approximately 0.05 by changing the critical value of the number of correctly identified cigarettes, but with a consequent increase in β. Find the new value of β.
   (c) Show that if the number of cigarettes presented to the subject is increased to 20, α can be maintained at approximately 0.05 and β can be decreased from the value found in (b) to approximately 0.09.

2. (a) Explain clearly the following terms used in significance tests:
      (i) null hypothesis and alternative hypothesis;
      (ii) significance level and critical region;
      (iii) the power of a test;
      (iv) one-sided and two-sided tests.

   (b) A trial has two possible outcomes, success and failure. To test the null hypothesis that the probability, p, of a success is 0.48, an experiment consisting of 10 independent trials is performed. If the trials result in 8 or more successes it will be concluded that p is greater than 0.48. Calculate the significance level of the test.

   Find the power of the test when $p = 0.5$.
   $[(0.48)^8 = 0.002818]$

   (JMB)

3. Explain the terms *critical region* and *significance level* as used in tests of hypotheses.

   It is claimed that a new material for contact lenses reduces irritation. In a controlled trial, each of 20 contact lens wearers tries out the standard material and the new material. Of these wearers, the number X who prefer the new material because it reduces irritation is recorded. Assume that the distribution of X is of binomial form $B(20, p)$.

   It is proposed to test the null hypothesis that $p = \frac{1}{2}$ against the alternative hypothesis that $p > \frac{1}{2}$. The null hypothesis is to be rejected if $X \geqslant 14$. Derive the significance level of the test. Find the power of the test when p is 0.7.

   Determine a critical region for a test of the above hypotheses when the significance level is at most 0.01.

   Explain why this claim is better tested by a one-sided test rather than a two-sided test.

   (JMB)

4. Explain what is meant by a 'Type I error' and a 'Type II error' in hypothesis testing. ▶

A manufacturer of the cat food, Mush, claims that cats prefer his product to another well-known brand. This claim is to be investigated using the following procedure: each of 20 cats is to be presented independently with a free choice between the two brands and if at least 15 of the cats choose Mush then the manufacturer's claim will be accepted. Writing down appropriate null and alternative hypotheses, calculate the probability of making a Type I error.

If, in fact, the probability is 0.7 that a cat selected at random will choose Mush in preference to the other brand, calculate the probability that the manufacturer's claim will not be accepted.

(JMB)

5. For the 1981/82 football season in England the number of points awarded to teams winning a match in the Football League was changed from two to three. One purpose of this change was to encourage teams to be more attacking and so score more goals. During that season the total number of goals scored by the 92 teams was 5277. The same number of teams played the same number of games as in previous seasons.

   Over the previous few seasons the number of goals scored per team had a distribution with mean 55.26 and variance 143.5. Taking these as the values of the mean and variance of the underlying distribution of goals per team per season and, assuming this distribution to be approximately normal, test whether the mean number of goals scored per team in 1981/82 was significantly more than in previous years. State clearly your null hypothesis, the significance level of your test and whether you are using a one- or two-tail test.

   A football match may result in a win for either team or in a draw. A draw may be a *score-draw* (when both sides score in the same number of goals) or a *no-score draw* (when neither side scores a goal). It was suggested that one effect of the new points system would be to decrease the proportion of no-score draws. State briefly and clearly what data you would need to collect and how you would use these data to test the truth of this suggestion.

   (JMB)

6. A standard sleep-inducing drug is such that 50% of experimental animals injected with the drug sleep for at least 3.2 hours and the remaining 50% waken within this period. A new drug is obtained by chemical modification of the standard one. Explain how you would decide whether the new drug was better than the standard if you could inject it into 11 randomly selected experimental animals.

   If subsequent extensive trials were to show that the new drug caused 60% of experimental animals to sleep for at least 3.2 hours, what is the probability that your decision procedure would have indicated that the new drug was better?

   (OLE)

7. A machine I produces glass marbles with median diameter 1 cm. Write down the probability that a marble chosen at random from those produced by this machine will have diameter less than 1 cm.

   Anne, Brian and Charles are each given ten glass marbles and are asked to decide whether or not they are produced by this machine by considering the value of X, the number of marbles with diameter less than 1 cm. They take their null hypothesis to be that the marbles are produced by machine I. Anne chooses as her critical region the ▶

values $X = 0, 1, 2, 8, 9$ and 10. Brian chooses as his critical region the values $X = 7, 8, 9$ and 10. Charles, who has not studied statistics, chooses as his critical region the value $X = 6$. Write down expressions for the exact probabilities that X falls in each of these critical regions if the marbles are from machine I, and evaluate them. Hence write down the significance levels of each of these three test procedures.

The marbles may alternatively come from machine II for which marbles of diameter 1 cm are at the 65th percentile. Write down the probability that a marble taken at random from this machine will have a diameter less than 1 cm. Find the power of each of the tests of Anne, Brian and Charles when the alternative hypothesis is that the marble comes from machine II. Discuss briefly which is the most sensible test to use, giving reasons for your answer.

(JMB)

Section 12.3

8. A new method of treating a skin complaint is to be tested against a standard treatment. The success rate of the standard treatment is known to be 50%. Of a random sample of 100 subjects given the new treatment, 60 respond successfully. Is the new treatment significantly better than the old treatment? Use 5% level of significance.

9. Of 500 throws of a die, 300 resulted in an even number. Do these data support the hypothesis that the die is fair? Use a 1% level of significance.

10. The random variable X has binomial distribution $B(n, p)$ and so has mean np and variance $np(1-p)$. Given that the random variable Y is such that $Y = X/n$, derive the mean and variance of Y.

    The Post Office claims that 94% of all first-class letters posted for delivery in England and Wales are delivered the next day (counting Monday as the day following Saturday). In answer to an advertisement for a teaching post in South West England, 295 applicants from various parts of the country sent first-class letters. A total of 253 of these letters arrived on the day after posting. Show that the proportion of those arriving on the day after posting is significantly different from 0.94 at the 0.1% significance level, stating clearly your null and alternative hypotheses.

    Give **two** reasons why this result does not necessarily invalidate the Post Office's claim and describe briefly an experiment which could be carried out to test this claim properly.

    (JMB)

11. In each of n independent trials, the probability of success is p. Show that the probability of k successes is given by the binomial probability function

$$\binom{n}{k} p^k (1-p)^{n-k}, \qquad k = 0, 1, \ldots, n$$

    When A and B play table tennis, the probability that A wins any game is 0.6. Find, to three decimal places, the probability that A wins at least three games out of six.

    C and D play ten games, of which C wins eight and he then asserts that this confirms that he is the better player. With the help of the table of cumulative binomial probabilities, test, at the five per cent significance level, the null hypothesis that the probability of C winning any game is 1/2, against the alternative hypothesis that it is greater than 1/2. ►

E and F play 25 games, of which E wins 19. Use the normal approximation to the binomial distribution to test, at the five per cent level, whether this provides significant evidence of E's superiority over F at table tennis.

(JMB)

12. Illustrate the role of the null hypothesis with reference, if possible, to one of your projects, making sure that you mention the alternative hypothesis used. Explain how you decided whether a one-tail or a two-tail test was appropriate.

    The sex ratio in rabbits bred in captivity is known to be 11 males to 10 females. A random sample of six rabbits bred in captivity is taken. Find, to 3 decimal places, the probability that:

    (a) exactly one will be male;
    (b) at least two will be male.

    To investigate whether the sex ratios are the same for rabbits bred in captivity and for wild rabbits, a random sample of 100 wild rabbits was obtained and it was found that this contained 63 males. Set up an alternative hypothesis and carry out a 5% significance test, using a normal approximation, to determine whether or not the sex ratios are equal for the two types of rabbit. If you conclude that there is a difference, state the direction of the difference.

    (London)

13. A market researcher has found that only 192 out of the 800 people he has interviewed are in favour of a new hypermarket being built near their town, which has a population of 15,000. How many inhabitants should he expect from this information to want the development? The researcher claims that it is unlikely that as many as 4000 of the town's residents actually want the new development. Test his claim (stating carefully any assumptions you need to make).

    (SMP)

Section 12.4

14. A drug for the prevention of colds was tested on a random sample of 100 individuals and a placebo was given to a control group of another 100 individuals. Of these receiving the drug 70% caught one or more colds over the period of treatment, while 80% of those receiving the placebo caught one or more colds over the treatment period. What conclusions can be drawn about the effectiveness of the drug?

15. A geneticist is interested in the proportion of males to females in the population who have a certain minor blood disorder. In random samples of 100 males and 100 females, 31 males and 24 females were found to have the disorder. Test at the 1% level of significance whether the proportions of males and females having the disorder are equal.

16. A factory has two production lines A and B for the manufacture of a certain article. A random sample of 250 articles from line A is found to contain 14 defectives while a random sample of 350 articles from line B is found to contain 28 defectives. Is there evidence of a real difference between the proportions of defective articles produced by the two lines?

▶

Assuming that the proportion of defective articles produced is the same for both production lines, give an approximate 95% confidence interval for this proportion.

(IOS)

17. In a new course last year a university department received applications from 450 home students and 50 overseas students. This year, after an increase in the fees charged to overseas students, the corresponding numbers were 550 and 40. Test whether there is a significant difference in the proportion of overseas students applying for admission between the two years.

    Recalculate your answer if the figures of 450 and 50 are average figures over a long period.

    (SMP)

13

Hypothesis tests for independence and goodness-of-fit

▼

13.1 INTRODUCTION

The hypothesis tests discussed in this chapter mainly involve the χ^2 test (see Section 10.8) in which observed frequencies (O) and expected frequencies (E) are compared using the test statistic

$$\sum \frac{(O-E)^2}{E}$$

The O values are the frequencies we actually count in an experiment or survey, and the E values are the frequencies we would expect if the null hypothesis we are testing is true.

Two types of null hypothesis to be discussed in this chapter are:

1. the hypothesis of independence between two categorical (non-numerical) variables (Sections 13.2 and 13.3);
2. the hypothesis that a set of observed frequencies 'fit' (are consistent with) a theoretical probability distribution or model (Sections 13.4–13.7).

The corresponding hypothesis tests are consequently often referred to as the χ^2 test for independence and the χ^2 goodness-of-fit test, respectively.

Graphical methods for testing for particular distributions, namely the Poisson and the normal distribution, will also be discussed.

13.2 χ^2 TEST FOR INDEPENDENCE, CONTINGENCY TABLE DATA

Suppose we record data concerning two categorical variables (Section 1.2) for a sample of individuals chosen randomly from a population. It is convenient to set out the data in the form of a two-way or *contingency table*. The values in the table are the observed frequencies for the various cross-categories of the variables.

Example 13.1

A random sample of 400 people is selected from all the 16-year-olds in a town. The variables recorded were:

(i) temper (vile or mild);
(ii) hair colour (red, brown or black).

The observed frequencies are displayed in Table 13.1.

Table 13.1 A 3 × 2 contingency table

	Temper			
Colour of hair	*Vile*	*Mild*	*Total*	
Red	40	20	60	
Brown	80	100	180	
Black	60	100	160	
Total	180	220	400	*Grand total*

What frequencies would we expect if the null hypothesis that temper and hair colour are independent is true?

The formula we use for the expected frequencies (E) is:

$$E = \frac{\text{row total} \times \text{column total}}{\text{grand total}} \tag{13.1}$$

and we apply this formula to each of the six cells of Table 13.1. (The rationale for this formula is as follows: the proportion of vile-tempered individuals is 180/400. If temper and colour of hair are independent, this proportion would apply to the 60 red-headed individuals. We would expect $60 \times 180/400 = 27$ individuals to have red hair and a vile temper. Notice that applying (13.1) to the 'red and vile' cell in Table 13.1 gives the same answer.)

It is convenient to write the E values in brackets next to the corresponding O values, as in Table 13.2.

Notice that the expected frequencies in each row and column sum to the corresponding observed totals, which is a useful arithmetical check.

Table 13.2 The observed and expected frequencies

	Temper		
Colour of hair	*Vile*	*Mild*	
Red	40 (27)	20 (33)	60
Brown	80 (81)	100 (99)	180
Black	60 (72)	100 (88)	160
	180	220	400

□

It can be shown that, for a contingency table with r rows and c columns, $\sum \frac{(O-E)^2}{E}$ has an approximate χ^2 distribution with $(r-1)(c-1)$ degrees of freedom.† If we carry out a test using the 5% level of significance we reject H_0 if:

$$\sum \frac{(O-E)^2}{E} > \chi^2_{0.05,(r-1)(c-1)} \tag{13.2}$$

where $\chi^2_{0.05,(r-1)(c-1)}$ is obtained from Table C.5.

Example 13.2

Using the data from the previous example, test the hypothesis that temper and hair colour are independent at the 5% level of significance.

H_0: Temper and hair colour are independent
H_1: Temper and hair colour are not independent
5% level of significance

$$\chi^2 = \sum \frac{(O-E)^2}{E}$$

$$= \frac{(40-27)^2}{27} + \frac{(20-33)^2}{33} + \frac{(80-81)^2}{81} + \frac{(100-99)^2}{99} + \frac{(60-72)^2}{72} + \frac{(100-88)^2}{88}$$

$$= 15.04$$

Since $r = 3$, $c = 2$, $(r-1)(c-1) = 2$ and the critical value is

$$\chi^2_{0.05,2} = 5.99, \qquad \text{see Table C.5 and Fig. 13.1}$$

Since $15.04 > 5.99$, we reject H_0, and conclude that temper and hair colour are not independent (5% level). In order to decide the direction of dependence, we

† Except when $r = 2$, $c = 2$, when a slightly different formula for χ^2 applies; see (13.3) in Section 13.3.

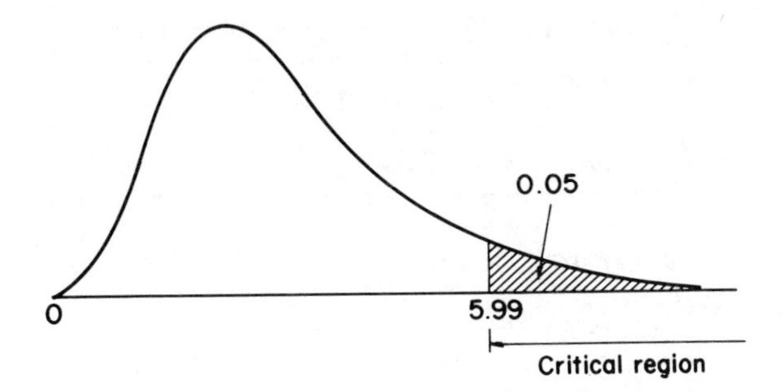

Fig. 13.1 χ^2 distribution with 2 degrees of freedom

need to go back to Table 13.2. For example, we note that more red-headed individuals have vile tempers than expected under the null hypothesis of independence. We can conclude that red hair and vile temper are 'positively associated'. □

In carrying out a χ^2 test for independence, the following important points should be noted:

(a) The observed and expected values must be frequencies and not percentages or proportions. (Percentages and proportions may of course easily be converted to frequencies if the total sample size is known).
(b) The conclusion that two categorical variables are dependent does not imply cause and effect. There may be other factors at work.
(c) The calculation of $\sum \frac{(O-E)^2}{E}$ should not be performed if any E value is < 5.

    In contingency tables with more than two rows or more than two columns, it may be possible and sensible to combine rows or columns to avoid low E values. Example 13.3 illustrates this method.
(d) The observations which are summarized in a contingency table must themselves be independent. This may be achieved by random sampling but Example 13.4 illustrates what would be classified as dependent observations.

Example 13.3

The attendance and examination results of a random sample of 60 pupils are given in Table 13.3 along with (in brackets) the expected frequencies to 1 d.p.

We note that one of the E values is less than 5. By combining rows 2 and 3, we can form a 2×2 contingency table (see Table 13.4).

Now all E values are greater than 5 and we may carry out a test for independence, see Section 13.3.

Table 13.3 A 3 × 2 contingency table

Attendance	*Exam result* *Pass*	*Fail*	
Excellent	25 (23.3)	10 (11.7)	35
Satisfactory	10 (10)	5 (5)	15
Poor	5 (6.7)	5 (3.3)	10
	40	20	60

Table 13.4 A 2 × 2 contingency table†

Attendance	*Exam result* *Pass*	*Fail*
Excellent	25 (23.3)	10 (11.7)
Satisfactory or poor	15 (16.7)	10 (8.3)

□

Example 13.4

Twenty matched pairs of patients took part in a clinical trial. One patient in each pair was allocated to be given a drug, while the other was treated by a neutral control (placebo). It was noted whether each patient improved with treatment; see Table 13.5.

Table 13.5 A 2 × 2 contingency table

	Drug	*Control*
Improvement	12	4
No improvement	8	16

The 40 observations are *not* independent, since the 40 patients were matched in pairs. The χ^2 test is invalid here. Similar conclusions would hold if 20 patients were treated twice, once with the drug and once with the neutral control (so each patient is matched with him/herself). □

13.3 2 × 2 CONTINGENCY TABLE, χ^2 TEST

The smallest contingency table is one with $r = 2$ rows and $c = 2$ columns, and the

† In this example the variable 'Attendance' has ordered categories. The χ^2 test does not use this fact. For methods which do, see Everitt, *The Analysis of Contingency Tables*, Chapman and Hall, 1977.

number of degrees of freedom for the χ^2 test in this case is $(2-1)(2-1) = 1$. It can be shown that for the '1 degree of freedom' case, we should use the test statistic:

$$\chi^2 = \sum \frac{(|O - E| - \frac{1}{2})^2}{E} \tag{13.3}$$

where $|O - E|$ means the magnitude of $(O - E)$. This formula incorporates what is called Yates's continuity correction.

Example 13.5

Use the data in Table 13.4 to test, at the 5% level of significance, whether examination result is independent of attendance.

H_0: Examination result and attendance are independent
H_1: Examination result and attendance are not independent
5% level of significance

$$\chi^2 = \sum \frac{(|O - E| - \frac{1}{2})^2}{E}, \qquad \text{since Table 13.4 is a } 2 \times 2 \text{ table.}$$

$$= \frac{(|25 - 23.3| - \frac{1}{2})^2}{23.3} + \frac{(|10 - 11.7| - \frac{1}{2})^2}{11.7} + \frac{(|15 - 16.7| - \frac{1}{2})^2}{16.7} + \frac{(|10 - 8.3| - \frac{1}{2})^2}{8.3}$$

$$= \frac{1.2^2}{23.3} + \frac{1.2^2}{11.7} + \frac{1.2^2}{16.7} + \frac{1.2^2}{8.3} \qquad \text{(note equal numerators)}$$

$$= 0.44$$

The critical value is $\chi^2_{0.05, 1} = 3.84$ (see Table C.5)

Since $0.44 < 3.84$, we do not reject H_0, and we conclude that examination result and attendance are independent. (5% level). □

The points (a) to (d) made at the end of Section 13.2 of course apply to the 2×2 contingency table. However, we cannot combine rows or columns (point (c)) if any E value is < 5 in a 2×2 table. Luckily we can perform another test in this situation, namely the Fisher exact test, which is beyond the scope of this book (see Bailey, *Statistical Methods in Biology*, Hodder and Stoughton, 1981).

13.4 χ^2 GOODNESS-OF-FIT TEST FOR A SIMPLE PROPORTION DISTRIBUTION

In this section we begin a discussion of goodness-of-fit tests, in which we test whether an observed set of frequencies 'fit' (are consistent with) a theoretical probability distribution or model.

The first type of model we consider is the 'simple proportion distribution' which leads to a null hypothesis that the frequencies should occur in certain numerical proportions.

Example 13.6

A genetic theory indicates that, for a certain species of flower, white, red and blue flowers should occur in the ratio 5:3:1. Suppose that, of a random sample of 180 flowers, 90 are white, 65 are red and 25 are blue. What frequencies would we expect if the theory is correct?

We would expect the proportions of white, red and blue to be 5/9, 3/9 and 1/9 respectively (since $5+3+1=9$). Hence the expected frequencies are $5/9 \times 180 = 100$, $3/9 \times 180 = 60$ and $1/9 \times 180 = 20$. □

It can be shown that, for a simple proportion distribution with k categories, $\sum \frac{(O-E)^2}{E}$ has a χ^2 distribution with $(k-1)$ degrees of freedom.† If we carry out a test at the α% level of significance we reject H_0 if

$$\sum \frac{(O-E)^2}{E} > \chi^2_{\alpha,\, k-1} \tag{13.4}$$

Example 13.7

Use the data from the previous example to test, at the 1% level of significance, the genetic theory stated in that example.

H_0: The genetic theory is correct
H_1: The genetic theory is incorrect
1% level of significance

From Table 13.6, $\chi^2 = 2.67$

Table 13.6 Calculation of χ^2 for a simple proportion distribution

Category	O	E	$\frac{(O-E)^2}{E}$
White	90	100	1.00
Red	65	60	0.42
Blue	25	20	1.25
			$\chi^2 = 2.67$

† unless $k-1=1$, in which case we should incorporate Yates's continuity correction, and reject H_0 if $\sum \frac{(|O-E|-\frac{1}{2})^2}{E} > \chi^2_{\alpha,\, 1}$

Since $k = 3$, and the significance level is 1 %, the critical value is

$$\chi^2_{0.01,\,2} = 9.21 \qquad \text{(see Table C.5)}$$

Since $2.67 < 9.21$, we do not reject H_0, and we conclude that the observed data support the genetic theory (1 % level of significance). □

As with the χ^2 test for independence (Section 13.2) the observed values must be frequencies (not percentages or proportions), and all E values must be at least 5. If the latter condition is not satisfied it may be sensible to combine categories to increase E values.

13.5 χ^2 GOODNESS-OF-FIT TEST FOR A BINOMIAL DISTRIBUTION

Suppose we carry out an experiment consisting of n trials, and that there are only two possible outcomes ('success' and 'failure') for each trial. Suppose we wish to decide whether a binomial distribution would be a suitable model for the discrete random variable, namely 'the number of successes in n trials'. One way is to repeat the experiment a large number of times and carry out a χ^2 goodness-of-fit test. Two cases will be considered:

Case 1: the value of the binomial parameter p is specified without reference to the experimental data.

Case 2: the value of p is not so specified, but must be estimated from the experimental data.

Example 13.8

Case 1

In 100 families each containing three children the numbers of girls are shown in Table 13.7.

Table 13.7 The number of girls in 100 families

Number of girls	0	1	2	3
Number of families (O)	8	27	45	20

What frequencies would we expect if the number of girls in families with three children has a $B(3, 0.5)$ distribution? Are these data consistent with this distribution (use a 5 % level of significance)?

The expected frequencies (E) for the categories 0, 1, 2, 3 girls are obtained by multiplying the corresponding probabilities, assuming that a $B(3, 0.5)$ applies (see note (i) below), by the total of the observed frequencies; see Table 13.8.

Table 13.8 also shows the calculation of the test statistic, $\chi^2 = \sum \dfrac{(O-E)^2}{E}$

Table 13.8 χ^2 for a $B(3, 0.5)$ goodness-of-fit test

Number of girls (x)	$p(x)$	$E = p(x)\sum O$	O	$\frac{(O-E)^2}{E}$
0	0.125	12.5	8	1.62
1	0.375	37.5	27	2.94
2	0.375	37.5	45	1.50
3	0.125	12.5	20	4.50
			$\sum O = 100$	$\chi^2 = 10.56$

Notes

(i) For $B(3, 0.5)$, $p(x) = \binom{3}{x}(0.5)^x(1-0.5)^{3-x}$, for $x = 0, 1, 2, 3$.

(ii) Once again all E values must be $\geqslant 5$, otherwise it would be necessary to combine adjacent categories (see example in Section 13.8).

We now formally set out the steps of the hypothesis test:

H_0: $B(3, 0.5)$ is an appropriate model for the data in Table 13.7
H_1: $B(3, 0.5)$ is not an appropriate model.
5% level of significance
$\chi^2 = 10.56$, from Table 13.8

Since there are four categories (0, 1, 2, 3 girls), and p was not estimated from the experimental results, the number of degrees of freedom for the critical value of χ^2 is $4 - 1 = 3$ (but see the discussion of the degrees of freedom in the following example). Hence the critical value is $\chi^2_{0.05, 3} = 7.81$.

Since $10.56 > 7.81$, we reject H_0 and conclude that $B(3, 0.5)$ is not an appropriate model (5% level). □

Example 13.9

Case 2

Use the data from Table 13.7 to test their consistency with a $B(3, p)$ distribution where p = probability of a girl at each birth. Use a 5% level of significance.

Since p is unspecified, we estimate it from the data in Table 13.7. Since the proportion of girls is given by

$$\frac{\text{Number of girls}}{\text{Number of children}} = \frac{27 \times 1 + 45 \times 2 + 20 \times 3}{100 \times 3} = 0.59,$$

we will use 0.59 as our estimate of p in the calculation of E values. Table 13.9 is formed in the same way as Table 13.8, this time using:

$$p(x) = \binom{3}{x}(0.59)^x(1-0.59)^{3-x}, \quad \text{for } x = 0, 1, 2, 3$$

Table 13.9 χ^2 for a $B(3, p)$ goodness-of-fit test

Number of girls	$p(x)$	$E = p(x)\sum O$	O	$\frac{(O-E)^2}{E}$
0	0.0689	6.9	8	0.18
1	0.2975	29.8	27	0.26
2	0.4282	42.8	45	0.11
3	0.2054	20.5	20	0.01
			$\sum O = 100$	$\chi^2 = 0.56$

We note that all E values are > 5, so it is unnecessary to combine categories.

H_0: $B(3, p)$ is an appropriate model for the data in Table 13.7
H_1: $B(3, p)$ is not an appropriate model
5% level of significance
$\chi^2 = 0.56$, from Table 13.9

The number of degrees of freedom is $4 - 1 - 1 = 2$, one more degree of freedom being lost than in the previous example, because one parameter p has been estimated from the experimental data. The critical value is $\chi^2_{0.05, 2} = 5.99$.

Since $0.56 < 5.99$, we do not reject H_0, and conclude that $B(3, p)$ is an appropriate model for the data in Table 13.7 (5% level). □

13.6 χ^2 GOODNESS-OF-FIT TEST FOR A POISSON DISTRIBUTION

The general method for testing whether a Poisson distribution is an appropriate model for a set of experimental data is very similar to that used in the previous section.

Suppose we collect data concerning the occurrence of events in time (or space), for example the times when vehicles pass a particular spot on a lonely country road, and we wish to test whether the events are occurring randomly. This is equivalent to testing whether the number of events per unit time (or space) are consistent with a Poisson distribution.

We can consider two cases (as we did with the binomial):

Case 1: the value of the Poisson parameter a is specified without reference to the experimental data.

Case 2: the value of a is not so specified, but must be estimated from the experimental data.

We will describe an example of Case 2 and leave Case 1 as an exercise for the reader.

Example 13.10

Case 2

On a lonely country road the number of vehicles passing a particular spot is noted for 60 consecutive minutes, as recorded in Table 13.10.

Table 13.10 The number of vehicles passing in 60 minutes

Number of vehicles	0	1	2	3	4	5
Number of minutes (O)	25	15	10	5	3	2

What frequencies would we expect if the number of vehicles per minute has a Poisson distribution? Are these data consistent with this distribution? Use a 5% level of significance.

Since the value of the parameter, a, is unspecified, we estimate it from the data in Table 13.10. Since a is also the mean of the Poisson, our estimate of a is the mean number of vehicles per minute, i.e.

$$\frac{\text{Number of vehicles}}{\text{Number of minutes}} = \frac{15 \times 1 + 10 \times 2 + 5 \times 3 + 3 \times 4 + 2 \times 5}{25 + 15 + 10 + 5 + 3 + 2}$$

$$= 1.2.$$

The expected frequencies (E) for the categories 0, 1, 2, 3, 4 and 5 *or more* (see note (ii) below) are obtained by multiplying the corresponding probabilities, assuming that a Poisson distribution applies (see note (i) below), by the total of the observed frequencies, see Table 13.11. This table also shows the calculation of the test statistic, $\chi^2 = \sum \frac{(O-E)^2}{E}$

Table 13.11 χ^2 for a Poisson goodness-of-fit test

Number of vehicles (x)	$p(x)$	$E = p(x) \sum O$		O		$\frac{(O-E)^2}{E}$
0	0.3012	18.1		25		2.63
1	0.3614	21.7		15		2.07
2	0.2169	13.0		10		0.69
3	0.0867	5.2		5		
4	0.0260	1.6	7.3	3	10	1.00
5 or more	0.0078	0.5		2		
				$\sum O = 60$		$\chi^2 = 6.39$

Notes

(i) Since the estimated value of a is 1.2,

$$p(x) = \frac{e^{-1.2} 1.2^x}{x!} \qquad \text{for } x = 0, 1, 2, \ldots.$$

(ii) Calculating the probability of 5 or more (rather than 5) ensures that the total of the E and O values are equal, apart from rounding errors. Use p (5 or more) $= 1 - p(0) - \ldots - p(4)$.

(iii) The bottom three categories have been combined to ensure that all E values are $\geqslant 5$.

The steps in the hypothesis test are:

H_0: A Poisson distribution is an appropriate model for the data in Table 13.10.
H_1: A Poisson distribution is not an appropriate model.
5% level of significance
$\chi^2 = 6.39$, from Table 13.11

After combining the bottom three categories, there are only four categories. Also a has been estimated from the experimental data. The number of degrees of freedom is $4 - 1 - 1 = 2$, and so the critical value is $\chi^2_{0.05,\,2} = 5.99$.

Since $6.39 > 5.99$, we reject H_0, and conclude that a Poisson distribution is not an appropriate model. (5% level).

As an exercise for the reader, test the hypothesis at the 5% level of significance, that the data in Table 13.9 are consistent with a Poisson distribution with (parameter) $a = 1$. □

13.7 GRAPHICAL METHOD OF TESTING FOR A POISSON DISTRIBUTION

Poisson probability paper (see Fig. 6.2 in Section 6.3) can be used for testing whether a Poisson distribution is an appropriate model for a set of experimental data. Estimates of cumulative probabilities are calculated from the data and represented as points on the graph paper. If these points appear to be on a vertical line, this indicates that a Poisson distribution is a suitable model, and the approximate value of the parameter a can also be read from the graph.

Example 13.11

Use the data from Table 13.10 to test for a Poisson distribution by a graphical method.

Table 13.12 is formed from Table 13.10.

The final column probabilities (apart from the first) are plotted on the corresponding 'c curves' (see Fig. 13.2). The impression given by the five points is not one of a vertical line, and indicates that a Poisson distribution is not a suitable model (this agrees with the conclusion reached using a χ^2 test in Section 13.5 for the same data).

Table 13.12 Estimating cumulative probabilities

Number of vehicles c	*Number of minutes (frequency)*	*Probability,* $P(X = c)$ *using rel. frequency*	*Cumulative probability* $P(X \geq c)$
0	25	0.4167 ($= \frac{25}{60}$)	1
1	15	0.2500	0.5833
2	10	0.1667	0.3333
3	5	0.0833	0.1667
4	3	0.0500	0.0833
5	2	0.0333	0.0333

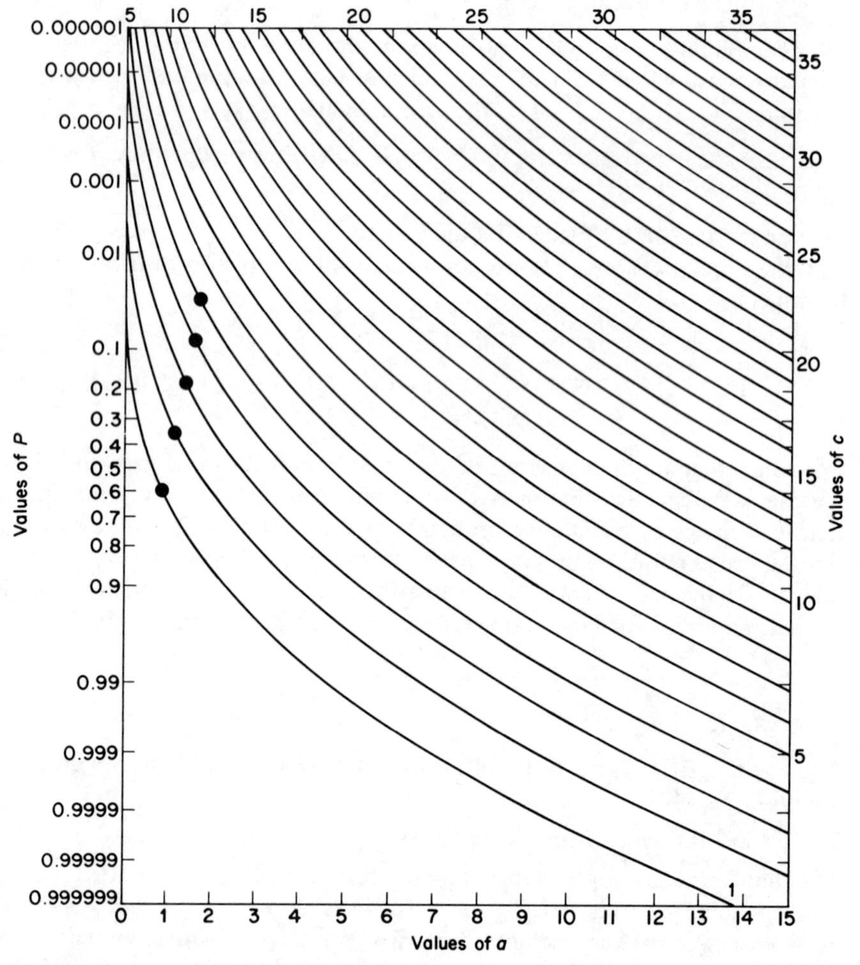

Fig. 13.2 Example of a goodness-of-fit test for a Poisson distribution, using the data from Table 13.12

The graphical method is clearly more subjective than the χ^2 method, but has the advantage that it is quicker. You are asked to use both methods in some of the Worksheet 13 exercises, so that you can decide for yourself which method you prefer. □

13.8 χ^2 GOODNESS-OF-FIT TEST FOR A NORMAL DISTRIBUTION

Many of the methods used in Chapters 11 and 12 require that the variable in question is at least approximately normally distributed. Instead of blindly asserting that this is the case, it would be an advantage to carry out a test for normality. If we have a reasonable amount of data (at least 50 values, say) then a χ^2 goodness-of-fit test may be carried out. (A graphical method of testing for normality is also possible, and is discussed in Section 13.9.)

In order to carry out a χ^2 test it is necessary to form a frequency distribution (as in Section 1.3), and essentially follow the methods used in Sections 13.5 and 13.6 for the (discrete) binomial and Poisson distributions.

Example 13.12

The weights of coffee in 70 jars were as shown in Table 13.13.

Table 13.13 Frequency distribution for the weights of coffee

Weight (g)	*Number of jars*
200 upto but not including 201	13
201 " " " " 202	27
202 " " " " 203	18
203 " " " " 204	10
204 " " " " 205	1
205 " " " " 206	1

What frequencies would we expect if the weight of coffee was normally distributed? Are these data consistent with a normal distribution? Use a 10% level of significance.

Since parameters, μ and σ^2, of the normal distribution are unspecified they must be estimated from the data. From Section 9.7 we know that unbiased estimates of μ and σ^2 are given by (2.2) and (9.7), i.e.

$$\bar{x} = \frac{1}{n}\sum f_i x_i \quad \text{and} \quad s^2 = \frac{\sum f_i x_i^2 - n\bar{x}^2}{n-1}$$

The same data were used in Table 3.1, so we calculate that

$$\bar{x} = \frac{14{,}137}{70} = 202.0 \qquad \text{and} \qquad s = \sqrt{\frac{2{,}855{,}149.5 - 70 \times (\frac{14{,}137}{70})^2}{69}} = 1.1 \text{ g.}$$

We now calculate the probabilities of getting values within each group for the $N(202, 1.1^2)$ distribution (see Fig. 13.3), using Table C.3(a).

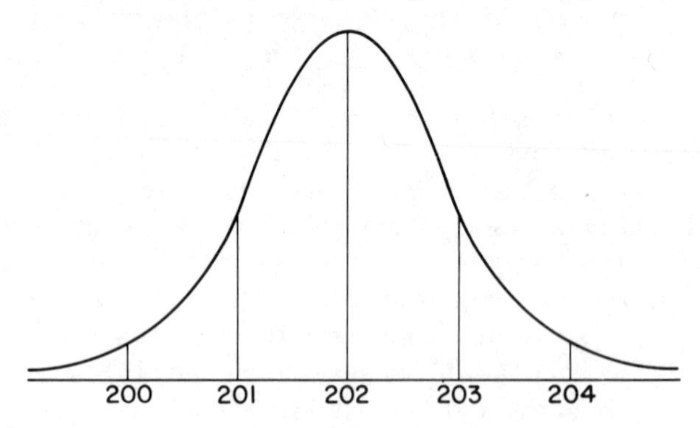

Fig. 13.3 $N(202, 1.1^2)$ distribution

The expected frequencies (E) are obtained by multiplying these probabilities by the total of the observed frequencies; see Table 13.14. This table also shows the calculation of the test statistic $\chi^2 = \sum \frac{(O-E)^2}{E}$.

Table 13.14 χ^2 for a normal distribution goodness-of-fit test

Weight	*Prob.*	$E = Prob. \times \sum O$		O		$\frac{(O-E)^2}{E}$
< 201	1.0000 − 0.8186 = 0.1814	12.7		13		0.01
201–202	0.5000 − 0.1814 = 0.3186	22.3		27		0.99
202–203	0.8186 − 0.5000 = 0.3186	22.3		18		0.83
203–204	0.9656 − 0.8186 = 0.1470	10.3		10		
204–205	0.9968 − 0.9656 = 0.0312	2.2	12.7	1	12	0.04
> 205	1.0000 − 0.9968 = 0.0032	0.2		1		
	1.0000			$\sum O = 70$		$\chi^2 = 1.87$

Notes

(i) for example, the probability for 201–202 is:

$$\Phi\left(\frac{202-202}{1.1}\right) - \Phi\left(\frac{201-202}{1.1}\right) = \Phi(0) - \Phi(-0.91)$$
$$= 0.5 - 0.1814$$
$$= 0.3186.$$

(ii) the two end-groups are treated as open-ended to ensure that the total of the E and O values are equal, apart from rounding errors.
(iii) the bottom three categories have been combined to ensure that all E values are $\geqslant 5$.

The steps in the hypothesis test are:

H_0: A normal distribution is an appropriate model for the data in Table 13.13.
H_1: A normal distribution is not appropriate.
10% level of significance
$\chi^2 = 1.87$, from Table 13.14.

There are four groups after combinations. Also two parameters (μ and σ) were estimated from the data. Hence the number of degrees of freedom $= 4 - 1 - 2 = 1$. The critical value is $\chi^2_{0.1,\,1} = 2.71$.

Since $1.87 < 2.71$, we do not reject H_0, and conclude that a normal distribution is an appropriate model. (10% level). □

13.9 GRAPHICAL METHODS OF TESTING FOR A NORMAL DISTRIBUTION

If we wish to test for normality and we have less than 50 values it may be impossible to carry out a χ^2 goodness-of-fit test. The reason for this is that we may have to combine so many categories (to ensure that all E values are $\geqslant 5$) that we are left with no degrees of freedom! (In any case the χ^2 test is not particularly sensitive to modest departures from normality in the tails).

Graphical methods exist for testing for normality. These use normal probability paper. Two examples will be given, one with ungrouped data and the other for grouped data, i.e. when there are sufficient data to form a frequency distribution.

For *grouped* data the method is to form a cumulative relative frequency distribution and plot this on normal probability paper. If the points lie on a straight line this indicates that the data are consistent with a normal distribution. The mean and standard deviation of this distribution may also be obtained from the graph.

Example 13.13

Use a graphical method and the data in Table 13.12 to decide whether the weight of coffee in jars marked 200 g is normally distributed. From Table 13.13 we can form Table 13.15.

Note
Cu % stands for 'Cumulative relative frequency expressed a percentage'. Weight is plotted against Cu % (see Fig. 13.4). The plot looks roughly linear (but see note

Table 13.15 Cumulative frequency distribution for weights of coffee

Weight (g)	*Cu* %
less than 201	18.6 $(=\frac{13}{70}\times 100)$
" " 202	57.1 $(=\frac{13+27}{70}\times 100)$
" " 203	82.9
" " 204	97.1
" " 205	98.6
" " 206	100.0

below), and a line has been drawn by eye through the points. Also since, for a normal distribution,

(i) 50 % of the values are less than the mean, μ, our estimate of μ is 201.9 g (at Cu % = 50);
(ii) 84 % of the values are less than $\mu+\sigma$, our estimate of $\mu+\sigma$ is 203.0 g (at Cu % = 84), and our estimate of (the standard deviation) σ is $203.0-201.9=1.1$ g.

These results agree well with those obtained in the previous section using other methods.

Note
It has to be said that judging whether the points plotted on normal probability paper lie on a straight line is subjective. In this author's view a χ^2 test is preferable. □

For *ungrouped* data the sample values are first ranked in increasing order of magnitude. A cumulative probability is then calculated, and each value is plotted on normal probability paper against the corresponding probability. If the plotted points lie approximately on a straight line this indicates that a normal distribution is a suitable model. The approximate values for the mean, μ, and the standard deviation, σ, can also be obtained from the graph.

Example 13.14

The heights (cm) of a random sample of 13 male students are: 176, 173, 180, 178, 175, 169, 181, 183, 181, 182, 171, 184, 179.
Are these data consistent with a normal distribution?

In rank order the 13 values are:

169, 171, 173, 175, 176, 178, 179, 180, 181, 181, 182, 183, 184.

Referring to these values as $y_1, y_2, \ldots, y_{13}$, we now calculate

$$P(y_i)=\frac{\text{Number of values}\leqslant y_i}{n+1},\ \text{for } i=1,2,\ldots,13,$$

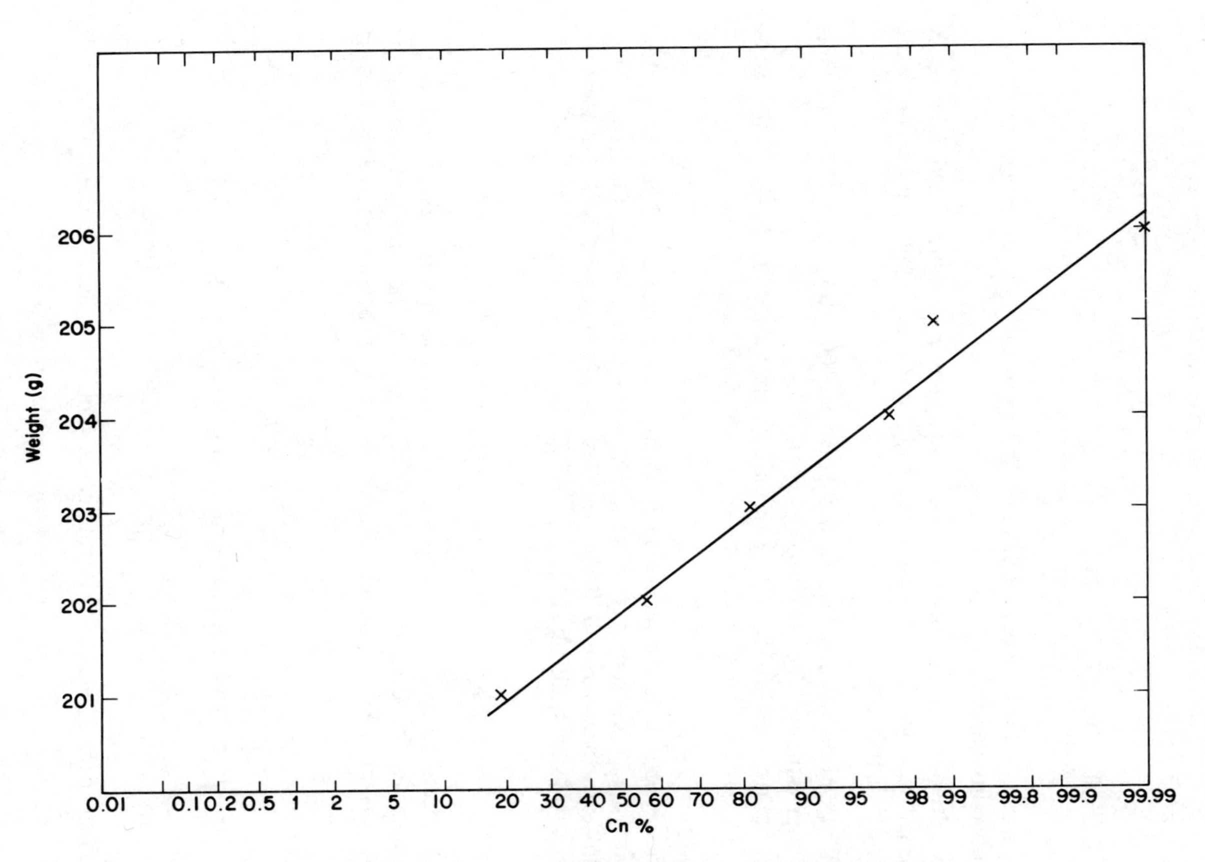

Fig. 13.4 Example of a graphical test for normality (grouped data)

where n is the sample size. Since $n = 13$ in this example, we obtain the values in Table 13.16.

Table 13.16 Probabilities for a graphical test of normality

i	1	2	3	4	5	6	7	8	9	10	11	12	13
y_i	169	171	173	175	176	178	179	180	181	181	182	183	184
$P(y_i) \times 100$	7	14	21	29	36	43	50	57	71	71	79	86	93

Figure 13.5 shows a plot of y_i against $P(y_i)$, expressed as a percentage, on normal probability paper.

If we decide that the points are approximately linear (see note below), indicating a normal distribution, we can estimate μ and σ, using the following properties of the normal distribution:

(i) 50% of the values are less than μ. At $P(y_i) \times 100 = 50\%$, $y_i = 177.8$ and this is our estimate of μ.
(ii) Approximately 84% of the values are less than $\mu + \sigma$. At $P(y_i) \times 100 = 84\%$, $y_i = 183.3$. Our estimate of σ is $183.3 - 177.8 = 5.5$.

Note

Once again a subjective judgement is required concerning how 'linear' the plotted points are. A less subjective method, but one which is outside the scope of this book, is the Shapiro–Wilk W test (see Wetherill, *Intermediate Statistical Methods*, Chapman and Hall, 1981). □

13.10 SUMMARY

Categorical variables may be tested for independence by first recording the frequencies for each cross-category in a contingency table. Provided the observations are independent and all expected values are all $\geqslant 5$, the test statistic is:

$$\chi^2 = \sum \frac{(O - E)^2}{E},$$

with $(r-1)(c-1)$ degrees of freedom, provided $(r-1)(c-1) > 1$,

or
$$\chi^2 = \sum \frac{(|O - E| - \frac{1}{2})^2}{E}, \quad \text{if } (r-1)(c-1) = 1.$$

The expected values for a contingency table are calculated using

$$E = \frac{\text{row total} \times \text{column total}}{\text{grand total}}$$

A χ^2 test may also be used to test whether a set of observed frequencies are consistent with a particular probability model. In this chapter, the simple

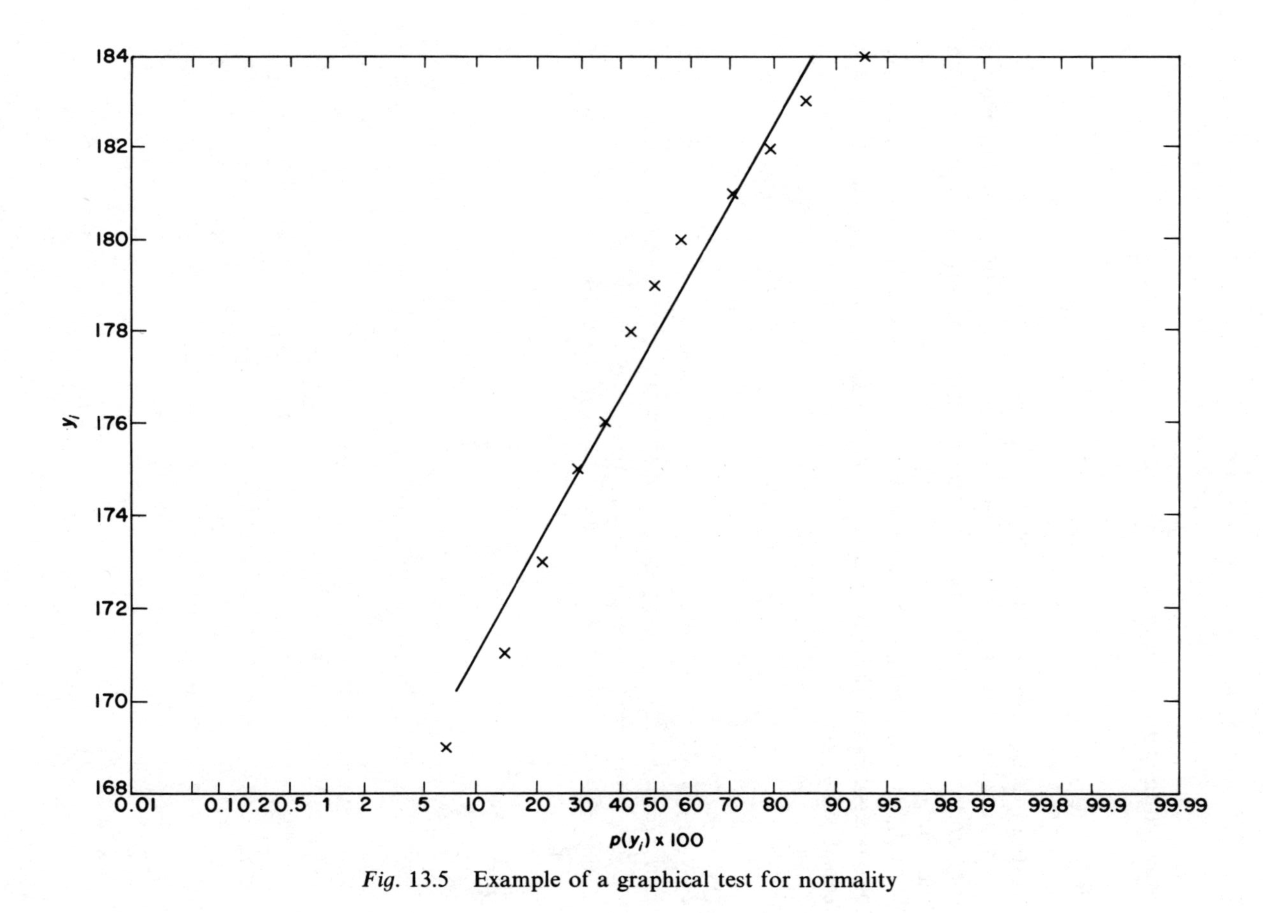

Fig. 13.5 Example of a graphical test for normality

proportion, binomial, Poisson and normal distributions were given as applications of this goodness-of-fit test. The test statistic is:

$$\chi^2 = \sum \frac{(O - E)^2}{E},$$

provided the number of degrees of freedom is greater than 1,

or

$$\chi^2 = \sum \frac{(|O - E| - \frac{1}{2})^2}{E},$$

for the 1 degree of freedom case.

The number of degrees of freedom in a goodness-of-fit test is (number of categories after combinations) $-1-$ (the number of parameters estimated from the data).

In the case of the Poisson and normal distributions, graphical methods may also be used to test for 'goodness-of-fit', but the conclusions reached are, to some extent, subjective.

Section 13.2

1. The following are the numbers of pupils from two schools who gained honours, passed or failed an examination in a certain subject.

	Honours	*Pass*	*Fail*
School A	12	18	20
School B	20	16	14

   (a) Test, at the 5% level of significance, whether the examination results are independent of school.
   (b) Suppose it is subsequently discovered that the figures above are in fact percentages out of the total number of pupils from both schools. Repeat (a) assuming that 150 pupils from each school took the examination.
   (c) Comment on your answers to (a) and (b).

2. A local council carried out a small survey to help them decide whether to build a fountain in a town-centre complex. Each randomly selected person who took part in the survey was told of the idea and asked to say whether they were in favour, against, or indifferent. The replies for male and female adults were as follows:

	For	*Against*	*Indifferent*
Men	12	13	15
Women	8	17	35

   Test, at the 5% level, whether the views of adults are independent of sex. In addition to the adult survey, the views of 100 schoolchildren were sought. Of these, 40 were for, 20 were against, and 40 were indifferent to the idea. Test, at the 1% level of significance, whether the views of children and adults are the same.

3. A random sample of the employees in an industrial organization were asked to indicate a preference for one of three pension plans. Of the 42 supervisors in the sample, 9 preferred Plan 1, 13 preferred Plan 2, and 20 preferred Plan 3. The corresponding preferences among the 108 clerical workers in the sample were 39, 50, and 19 respectively. Among the 150 shop-floor workers in the sample the preferences were 52, 57, and 41 respectively.

   Are the preferences of the employees independent of their type of job?

4. The following data show place of wedding ceremony against length of courtship, for a random sample of 250 couples.

▶

Place of ceremony	Length of courtship				
	< 6 *months*	6–12	13–24	25–36	> 36 *months*
Registry Office	39	20	14	10	2
Church or Chapel	16	13	35	62	39

Can it be said that place of marriage is independent of length of courtship?

(IOS)

5. Two drugs, denoted by A and B, were tested for their effectiveness in treating a certain common mild illness. Of 1000 patients suffering from the illness, 700 were chosen at random and given drug A, and the remaining 300 were given drug B. After one week, 100 of the patients were worse, 400 showed no change in their condition, and 500 were better. On the assumption that the two drugs are identical in their effect, complete a table similar in the form to that below to show, for each drug, the expected number of patients getting worse, showing no change, and becoming better.

   The given table shows the observed number of patients in each category. Carry out a χ^2-test, at the 1% level, to determine whether the six observed frequencies are consistent with the assumption of identical effects. State your conclusion briefly.

   (Cambridge)

	Number of patients		
	Becoming worse	*Showing no change*	*Becoming better*
Drug A	64	255	381
Drug B	36	145	119

6. The observed frequencies in a contingency table are as follows.

$a-b$	$a-c$	$2a+b+c$
$a+b$	$a+c$	$2a-b-c$

Show that the value of the chi-squared test statistic is

$$\frac{3(b+c)^2-4bc}{a}$$

A plant species may be broken into three genetic classes, A, B, C. The table shows how two samples, each of size 100, are distributed among these classes.

	A	*B*	*C*
1st sample	20	29	51
2nd sample	30	21	49

►

Find values a, b, c to make this table conform to the previous one, and hence test whether the proportions of A, B, C vary significantly between the two samples.

(OLE)

7. A survey was made of the arrival times of rush hour trains on a particular day at Queen's Cross and Bakerloo stations. At Queen's Cross 72 of the trains arrived on or before time, 49 were late by less than five minutes, and 33 were later than this. The corresponding figures for Bakerloo were 63, 45, and 58. Investigate whether the time-keeping records for the two stations are (on this evidence) comparable.

(SMP)

Section 13.3

8. A market research company that wished to determine whether it was more usual in the North for working people to come home for a mid-day meal than in the South interviewed random samples of 500 working people in each region. In the sample from the North 330 people returned home for a mid-day meal, compared with 280 in the sample from the South.

   Is coming home for a mid-day meal independent of region? Use a 5% level of significance.

9. Of a random sample of 40 rented television sets, the tubes of 9 sets burnt out within the guarantee period of 3 years. Of a random sample of 60 bought sets, the tubes of 5 sets burnt out within 3 years. Test the hypothesis that the probability of burning out within 3 years is independent of the type of set.

10. Records from a maternity hospital for 91 births in a given week indicated the following:

	Type of birth	
Mother's age (years)	*No complications*	*With complications*
Under 25	42	3
25–29	14	1
30–34	8	4
35 and over	12	7

Test, at the 5% level of significance, the hypothesis that the incidence of complications is independent of the age of the mother.

11. Among 19 snails, all of the same species, found on a garden rubbish heap, 12 had a dark band on their shells. Among 16 snails of the same species found in some nearby grass, only 5 had dark bands. Use χ^2 and a 5% level of significance to test the hypothesis that these were independent random samples from the same population.

(O & C)

12. The observed frequencies in a 2×2 contingency table are represented by the notation: ▶

a	b
c	d

Show that the χ^2 statistic $\sum \frac{(O-E)^2}{E}$ can be written as:

$$\frac{n(ad-bc)^2}{(a+b)(c+d)(a+c)(b+d)}, \quad \text{where } n = a+b+c+d.$$

Section 13.4

13. In a survey a random sample of 1000 adult males who lived in a particular town, the social class and season of birth were noted.

Social class	*Season of birth*			
	Spring	*Summer*	*Autumn*	*Winter*
A	130	90	80	90
B	170	140	150	150

(a) Do these data support the hypothesis that social class and season of birth are independent? Use a 5% level of significance.
(b) Assuming that the answer to (a) is 'Yes', test the hypothesis, at the 1% level of significance, that the proportion of male births is the same for all seasons of the year.

14. The number of fatal road accidents in one year in a city were tabulated according to their time of occurrence.

Time	0.00–04.00	04.00–08.00	08.00–12.00	12.00–16.00	16.00–20.00	20.00–24.00
Number	25	10	20	12	20	33

(a) Show that the hypothesis that the number of fatal road accidents is uniformly distributed in time should be rejected, at the 1% level of significance.
(b) Suppose that it is known that the numbers of road miles covered by all vehicles for the six time periods above are in the ratio 2:1:4:3:5:3. Test the hypothesis that the number of fatal road accidents is distributed in these ratios, again using a 1% level of significance.

15. To test the hypothesis that family car ownership of 0, 1, 2 or more cars is in the ratio 1:2:1, a random sample of 100 families was asked how many cars they owned. The results were:

Number of cars	0	1	2 or more
Number of families	35	45	20

Do these data support the hypothesis? Use a 5% level of significance.

16. The following table shows the number of deaths in a hospital in four quarterly periods:

Quarter	Jan–Mar	Apr–Jun	Jul–Sep	Oct–Dec
Number of deaths	142	52	34	72

Do these data support the hypothesis that deaths in the hospitals are in the ratio 3:1:1:2 for the four quarters? Use a 0.1% level of significance.

17. A die is thrown 200 times with the following results:

Score	1	2	3	4	5	6
Frequency	32	35	28	32	37	36

Is the die fair? Use a 5% level of significance.

18. In Mendel's experiments with peas he observed:

Type of pea	*Number of peas*
Round and yellow	315
Round and green	108
Wrinkled and yellow	101
Wrinkled and green	32

Do these data support his genetic theory that these types should occur in the ratio 9:3:3:1?

19. The positioning of nests in shrubs was noted for a species of bird:

Nest position	N	NE	E	SE	S	SW	W	NW
Number of nests	65	73	67	51	47	45	45	48

Test the hypothesis that the birds have no directional preference in positioning their nests.

20. Genetic theory predicts that, for each blue-winged fly among the offspring of a certain type of cross, there will be three red-winged flies. If 76 red-winged flies and 22 blue-winged flies result from such a cross, do these data support the genetic theory? ►

21. A firm uses three similar machines to produce a large number of components. On a particular day a random sample of 99 from the defective components produced on the early shift were traced back to the machines which produced them. The same was done with a random sample of 65 from the defectives produced on the late shift. The table shows the number of defectives found to be from each machine on each shift.

	Machine		
Shift	*A*	*B*	*C*
Early	37	29	33
Late	13	16	36

(a) Use the χ^2 distribution to test, at the 5% significance level, the hypothesis that the probability of a defective coming from a particular machine is independent of the shift on which it was produced.
(b) Using the 'Early' shift figures *only*, test, at the 5% significance level, the hypothesis that a defective is equally likely to have come from any machine.
(c) Comment briefly on your results given the following additional information:

All machines produce components at a similar rate. During the Early shift machines B and C were overhauled and so were in production for a much shorter time than A. All machines were in production for the same time on the Late shift.

(AEB (1984))

22. The table below shows the number of employees and the number of accidents occurring in a year in four factories:

Factory	*A*	*B*	*C*	*D*
Employees (thousands)	4	3	1	2
Accidents	22	13	11	17

Use χ^2 and a 5% level of significance to test the hypothesis that the expected number of accidents is proportional to the number of employees in each factory.

(O&C)

23. In order to determine whether or not the number of books borrowed from a school library depends on the day of the week, the librarian noted the number of books borrowed daily during a particular week, with the following results.

	Mon.	Tu.	Wed.	Thu.	Fri.
Number borrowed	68	74	78	85	92

Not being a statistician, she concluded that, since all the figures were different, the numbers borrowed do depend on the day of the week. Why is the argument incorrect? Perform a significance test at the 5% level to resolve the problem. ►

During another week the figures were:

Mon.	Tu.	Wed.	Thu.	Fri.
70	80	83	92	98

Determine if the extra data alter your conclusion.

(SUJB)

24. The numbers of individuals possessing each of four blood types should be in the proportions:

$$q^2 : p^2 + 2pq : r^2 + 2qr : 2pr, \qquad \text{where} \quad p + q + r = 1$$

(a) Check that these proportions add up to 1.
(b) Given the observed frequencies are 180, 360, 132, and 98 respectively, test whether these data support the hypothesis that
$p = 0.4$, $q = 0.5$, $r = 0.1$.
Use a 0.1% level of significance.

Section 13.5

25. Eight hundred seeds were tested in 80 batches of 10 seeds. The numbers germinating in each batch were recorded and the results were as follows:

Number of germinating seeds	0	1	2	3	4	5	6 or more
Number of batches	6	20	28	12	8	6	0

(a) Test the hypothesis that the number of germinating seeds per batch has a $B(10, 0.2)$ distribution. Use a 5% level of significance.
(b) Also test the hypothesis that the number of germinating seeds per batch has a $B(10, p)$ distribution, by estimating p from the data in the above table. Again use a 5% level of significance.
(c) Draw on overall conclusion from the two tests.

26. Observations of 80 litters of rabbits, each containing three rabbits, reveal the following numbers of male rabbits:

Number of males	0	1	2	3
Number of litters	19	32	23	6

(a) Test the hypothesis, at the 1% level of significance, that the number of male rabbits per litter has a binomial distribution.
(b) Calculate the number of male and female rabbits in the 80 litters. Test the hypothesis, at the 5% level of significance, that male and female rabbits occur in equal proportions in three-rabbit litters.

▶

27. Two hundred rats were each run five times in a T-maze and the number of times each rat turned right was recorded. Do the results below appear to conform to the $B(5, 0.5)$ distribution and, if so, what does this imply?

Number of right turns	0	1	2	3	4	5
Number of rats	10	25	60	65	35	5

28. In an ESP experiment four cards were used, with either A, B, C or D written on each card. The experimenter shuffled the cards and selected one at random. The subject could not see the cards and had to guess which card was selected. This procedure was repeated five times for each of 50 subjects. The number of correct guesses was recorded for each subject, and the results are displayed in the following table.

Number of correct guesses	0	1	2	3	4	5
Number of subjects	15	18	8	5	3	1

(a) If a subject was guessing, what is the probability that he would correctly guess the letter written on a particular card selected?
(b) Are the data above consistent with a binomial distribution with $n = 5$ and $p =$ the value you obtained in (a)?

29. A marksman fires five shots at a target and records the number r of bull's-eyes hit. After a series of 100 such trials he analvses his scores, giving the following distribution:

r	0	1	2	3	4	5
Frequency	6	31	36	15	8	4

Assuming that the probability p of hitting a bull's-eye remains constant throughout the trials, estimate p. Use a χ^2 test to determine whether these results are consistent with the appropriate binomial distribution.

Using the normal approximation to the binomial distribution, provide a two-sided 95% confidence interval for p.

(O&C)

30. A survey was carried out of 320 five-puppy litters* to test the hypothesis that the probability of a puppy being born a female is 0.5. The following frequency distribution was obtained.

Number of female puppies in litter	0	1	2	3	4	5
Frequency of occurrence	18	57	110	88	40	7

* A litter is the set of puppies born to a female dog on one occasion.

What probability model might be suitable for predicting frequencies to compare with these data? What factor might limit this model's applicability in the above situation? Carry out a comparison of observed and predicted frequencies and state whether or not the data collected support the initial hypothesis.

If, in your model, the probability of p of a puppy being born a female, instead of being assumed to be 0.5, is estimated from the above data, what factor might limit the resulting model's applicability? Evaluate the predicted frequencies that correspond to this estimated value of p, and test how well these agree with the collected data.

(SMP)

31. A biased coin is tossed five times and the number of heads obtained is recorded. This is repeated 200 times. The table below summarizes the results.

Number of heads	x	0	1	2	3	4	5
Frequency	f	5	39	70	52	25	9

Find the frequencies of the number of heads given by a binomial distribution having the same mean and total as the observed distribution.

Instead of calculating the usual goodness-of-fit statistic

$$X^2 = \sum \frac{(O-E)^2}{E}$$

calculate the values of the following:

(a) $X_1^2 = \sum \frac{(O-E)}{E}$;

(b) $X_2^2 = \sum \frac{|O-E|}{E}$.

How useful do you think these new statistics may be in measuring the goodness-of-fit in the above situation?

(AEB (1983))

32. An experiment consists of three independent tosses of a coin, and the number of heads observed is i. The experiment is conducted independently $8n$ times, and on exactly r_i of these occasions the outcome of the experiment is i ($i = 0, 1, 2, 3$). If a chi-squared test is conducted to investigate whether the coin is fair, write down expected values for each r_i, and hence show that the test statistic may be written:

$$\frac{3r_0^2 + r_1^2 + r_2^2 + 3r_3^2}{3n} - 8n$$

The experiment is conducted 160 times, and the values of r_0, r_1, r_2, r_3 are respectively 15, 70, 50, 25. Do you consider the coin is a fair one?

(OLE) ►

Section 13.6

33. A subject was observed for 250 periods of 5 seconds and the number of eye-blinks made by the subject was recorded, as follows:

Number of blinks	0	1	2	3	4	5
Number of periods	33	53	84	49	23	8

Are these data consistent with a Poisson distribution?

34. One hundred small samples of the same volume were taken from a large cylinder filled with a gas. For each sample the number of dust particles was counted, and the results were summarized as follows:

Number of dust particles	0	1	2	3	4	5
Number of samples	24	32	26	10	4	4

(a) Use a χ^2 test, at the 1% level of significance, to test whether the above data are consistent with a Poisson distribution.

(b) Another test for a Poisson distribution is to see if the mean and the variance of the number of dust particles per sample are approximately equal (since the mean and variance of a Poisson distribution are equal). Carry out this test and compare your conclusion with that obtained in (a).

35. A sample of 500 values of a discrete variable Z has the following frequency distribution:

Z	0	1	2	3	4	5	6	7	8 or more
f	50	98	120	113	61	29	19	10	0

It is postulated that Z is distributed as a Poisson variable.

Show that the mean of the sample is 2.5. Some of the expected frequencies for the Poisson distribution with this mean are given in the following table:

Z	0	1	2	3	4	5	6	7	8 or more
f	—	—	—	107	67	33	14	5	2

Compare this table and test at the 5% level whether the sample supports the belief that Z is a Poisson variable.

(Cambridge) ▶

36. The random variable X has Poisson distribution with:

$$P(X = r) = \frac{e^{-\lambda}\lambda^r}{r!}, \qquad r = 0, 1, 2, \ldots$$

Show that the mean of X is λ.

State the conditions under which the Poisson distribution gives a good approximation to the binomial distribution.

During the Second World War records were kept of the number of flying bombs falling in South London and their exact points of impact. The bombs were launched from the Continent and did not have sophisticated guidance systems. The particular part of the city studied was divided into 576 regions each having an area of 0.25 km^2. The table gives the number of regions experiencing x hits.

Number of hits (x)	0	1	2	3	4	5	*Total*
Frequency	229	211	93	35	7	1	576

Give reasons why these data might be expected to fit a Poisson distribution. Calculate the mean of these data.

Test the above data for goodness of fit to a Poisson distribution with mean 0.95, listing the expected frequencies.
(Take $e^{-0.95}$ as 0.3867.)

An important factory covered two neighbouring regions. It was estimated that two direct hits would cripple the factory. Find the probability that it was crippled.

(JMB)

37. The numbers of customers entering a shop in forty consecutive periods of one minute are given below.

3	0	0	1	0	2	1	0	1	1
0	3	4	1	2	0	2	0	3	1
1	0	1	2	0	2	1	0	1	2
3	1	0	0	2	1	0	3	1	2

Draw up a frequency table, and illustrate it by means of a bar chart. Calculate values for the mean and variance of the number of customers entering the shop in a one-minute period. Fit a Poisson distribution to the data, and comment briefly on the degree of agreement between the calculated and observed frequencies.

Estimate the probability that no customers enter the shop in a given two-minute period.

(MEI)

38. A random variable X has a Poisson distribution of the form:

$$\Pr(X = x) = \frac{e^{-\mu}\mu^x}{x!} \qquad \text{for} \quad x = 0, 1, 2, \ldots$$

Prove that the mean of X is μ.

▶

One hundred electrical components are tested to see how many defects each has. The results are:

No. of defects	0	1	2	3	4	5	6	7 or more
No. of components	11	22	26	24	9	5	3	0

(i) Calculate the mean of the distribution.
(ii) Calculate the frequencies (to 1 decimal place) of the associated Poisson distribution having the same mean.
(iii) Perform a χ^2 goodness-of-fit test to determine whether or not the above results are likely to have come from a Poisson distribution, using a 5% significance level.

(SUJB)

39. In an investigation into the behaviour of rats the following experiment was carried out.

A rat was placed in a box with four doors, one leading to food, the others leading to no food. The experiment was stopped as soon as the rat found the food. The number of doors tried, up to and including the one leading to the food, was recorded for each of the 50 rats. The results were as follows:

Number of doors tried	1	2	3	4	5	6	7	8	9 or more
Number of rats	19	11	8	6	3	1	1	1	0

The investigator's theory is that the number of doors tried has a geometric distribution with probability function:

$$p(x) = (1 - p)^{x-1}p, \qquad \text{for} \quad x = 1, 2, \ldots$$

where p is the constant probability that the door leading to the food will be tried.

(i) Calculate the mean number of doors tried, and use this to estimate p (*Hint*: the mean for the above geometric distribution is $1/p$).
(ii) Test, at the 5% level of significance, whether the above data support the investigator's theory.

Section 13.7

40. Use Poisson probability paper to decide whether the data in Exercise 33 above are consistent with a Poisson distribution, and estimate the mean of the distribution. Compare your conclusions with those from Exercise 33.

41. Use Poisson probability paper to decide whether the data in Exercise 34 above are consistent with a Poisson distribution, and estimate the mean of the distribution. Compare your conclusions with those from Exercise 34.

Section 13.8

42. The marks obtained in a large public examination by 500 candidates were summarized as follows:

Mark range	10–19	20–29	30–39	40–49	50–59	60–69	70–79
Number of candidates	19	50	47	125	132	100	27

(i) Show that the mean and standard deviation of the number of marks per candidate are 48.7 and 14.9 respectively (to 1 d.p.).
(ii) Test, at the 5% level of significance, whether these data are consistent with a normal distribution.

43. (a) Two independent readings X_1 and X_2 are taken from a population with mean μ and variance σ^2. Show that

$$(X_1 - \overline{X})^2 + (X_2 - \overline{X})^2 = \tfrac{1}{2}(X_1 - X_2)^2,$$

where $\overline{X} = \frac{1}{2}(X_1 + X_2)$. By considering the mean and variance of $(X_1 - X_2)$, or otherwise, determine the expectation of $(X_1 - \overline{X})^2 + (X_2 - \overline{X})^2$.
(b) The marks in a large public examination may be taken to be distributed normally with mean μ and standard deviation σ. Each candidate is given one of five grades A, B, C, D, E, depending on his marks X as shown below:

Mark	*Grade*
$X \geqslant \mu + \frac{3}{2}\sigma$	A
$\mu + \frac{3}{2}\sigma > X \geqslant \mu + \frac{1}{2}\sigma$	B
$\mu + \frac{1}{2}\sigma > X \geqslant \mu - \frac{1}{2}\sigma$	C
$\mu - \frac{1}{2}\sigma > X \geqslant \mu - \frac{3}{2}\sigma$	D
$\mu - \frac{3}{2}\sigma > X$	E

One school, with 283 candidates, has 25 A's, 77 B's, 114 C's, 48 D's and 19 E's. Perform the χ^2 test of goodness-of-fit, at the 5% level of significance, to determine whether the results from this school are typical of those of the whole entry.

(MEI)

Section 13.9

44. In Exercise 2 of Worksheet 11 the assumption was made that tar content was normally distributed. Use normal probability paper to decide whether this assumption is reasonable given the tar contents (mg) of a random sample of 20 cigarettes from the exercise referred to above, as follows:

4.3	5.2	5.1	6.3	5.1	3.7	4.0	5.1	3.7	4.3
2.5	6.3	4.8	3.7	2.9	6.3	5.1	4.2	3.9	2.8

Do *not* form a frequency distribution. If the data appear to be consistent with a normal distribution, estimate its mean and standard deviation.

▶

45. A manufacturer of fishing lines uses a machine to produce lengths of line whose breaking strain is normally distributed with standard deviation 1.50 newtons. Before buying a large batch of fishing lines, a retailer tests a random sample of 10 lengths of line from the batch, and accepts the batch provided the mean breaking strain of the sample is greater than 39.5 newtons. What is the probability of accepting a batch with a mean breaking strain of 40.00 newtons?

    The manufacturer installed a new machine to produce lines. The breaking strain of a random sample of 180 lengths of line produced by the new machine is given in the table below:

Breaking strain (newtons)	38.1–38.5	38.6–39.0	39.1–39.5	39.6–40.0	40.1–40.5
Frequency	6	11	26	46	43

Breaking strain (newtons)	40.6–41.0	41.1–41.5	41.6–42.0
Frequency	31	13	4

    Use arithmetic probability paper to verify that it is reasonable to assume that the sample comes from a normal population. Estimate from your graph the mean and standard deviation of the distribution.

    Subsequent batches produced by the new machine are all found to have the same standard deviation, but not necessarily the same mean breaking strain.

    The retailer continues to test a sample of 10 lengths of line from each batch. Use your estimate of the standard deviation to calculate the minimum acceptable size of the sample mean to give a probability of at least 0.85 of accepting a batch with mean breaking strain of 40.00 newtons.

    (AEB (1984))

46. Use normal probability paper to decide whether the marks in Exercise 42 above are consistent with a normal distribution.

14

Non-parametric hypothesis tests

▼

14.1 INTRODUCTION

In Chapter 11 we saw that a number of hypothesis tests require assumptions in order for the tests to give valid conclusions. In particular:

1. the 'paired-samples t test' (Section 11.6) requires the assumption that the differences between the pairs are approximately normally distributed;
2. the 'unpaired-samples t test' (Section 11.5) requires the assumptions that (i) both samples are from approximately normal distributions, and (ii) these distributions have the same variance.

Tests which do not require such stringent distributional assumptions are called 'distribution-free' tests, but they are more commonly referred to as *non-parametric* tests.

In this chapter three non-parametric tests will be discussed:

1. Sign test, for paired samples data and for a test for the median of a continuous distribution.
2. Wilcoxon signed rank test, also for paired sample data.
3. Mann–Whitney U test, for unpaired samples data.

It is important to realize that non-parametric tests are generally less powerful† than the equivalent parametric tests, if the assumptions required by the parametric tests can be justified.

† Power = P(accepting H_1 when H_1 is true); see Section 11.10.

14.2 SIGN TEST

Consider an example of paired-samples data in which the differences between the pairs are not quantifiable, but we can decide whether the differences are positive, negative or zero. Such data can arise in preference testing.

Example 14.1

Two brands of cat food, A and B, were fed in random order on two consecutive days to a random sample of 12 cats. The results are set out in Table 14.1, where + indicates a preference for A, − indicates a preference for B, and 0 indicates no preference.

Table 14.1 Preference test results for 12 cats

Cat number	1	2	3	4	5	6	7	8	9	10	11	12
Preference	+	+	+	+	−	0	+	+	0	0	+	+

How can we decide whether these data support the hypothesis that cats (in the general population) show equal preference for brands A and B, against the alternative that cats prefer one of the brands to the other?

First we ignore the data for the cats who showed no preference. This leaves the data for 9 cats, namely 8 + signs and 1 − sign. Under the null hypothesis of equal preference, the probabilities of a + sign and a − sign will be equal to 1/2. The various steps in the hypothesis test are as follows:

$H_0: p(+) = p(-) = \frac{1}{2}$ equal preference

$H_1: p(+) \neq p(-)$ unequal preference, two-sided H_1

5% level of significance

Calculate P (8 *or more* + in 9 trials, where $p(+) = \frac{1}{2}$). This is a binomial probability (Section 6.2) and refers to the $B(9, \frac{1}{2})$ distribution, so

$$p(x) = \binom{9}{x}\left(\frac{1}{2}\right)^x\left(1-\frac{1}{2}\right)^{9-x}, \qquad x = 0, 1, \ldots, 9$$

The required probability is:

$$p(8) + p(9) = 9\left(\frac{1}{2}\right)^9 + \left(\frac{1}{2}\right)^9 = 0.0195$$

We reject H_0 if the calculated probability is less than

$$\frac{\text{significance level}}{2} \quad \text{or} \quad \frac{\text{significance level}}{1},$$

depending on whether H_1 is two-sided or one-sided.

For our example, $0.0195 < \frac{0.05}{2}$, so we reject H_0.

We conclude that there is a significant difference between the preferences of cats for brands A and B. (5% level).

Notes

(i) We calculate P (8 or more +) rather than P (8 +) because 'more than 8 +' is a more extreme result than '8 +'.

(ii) Instead of P (8 or more +) we could have calculated P (1 or fewer −). The answer would have been the same (left as an exercise for the reader). □

A second example of a sign test will now be given to illustrate a one-sided alternative hypothesis, and hence a one-tailed test. The example also illustrates how the sign test can be used to test hypotheses about the median of a continuous distribution.

Example 14.2

At a pick-your-own farm a customer filled 10 punnets with blackcurrents. These were weighed on a balance which indicated only if the weight of the blackcurrents in each punnet was greater or less than 500 g. Seven of the punnets were overweight, three were underweight. Can we conclude that the median weight is greater than 500 g?

Under the null hypothesis that the median weight equals 500 g, and an alternative that the median weight is greater than 500 g, the following test may be made:

H_0: $p(+) = p(-) = \frac{1}{2}$, where + indicates overweight and − indicates underweight

H_1: $p(+) > p(-)$, a one-sided H_1

5% level

$$P(7 \textit{ or more } + \text{ in 10 trials, where } p(+) = \tfrac{1}{2})$$
$$= p(7) + p(8) + p(9) + p(10)$$
$$= (120 + 45 + 10 + 1)\left(\tfrac{1}{2}\right)^{10}$$
$$= 0.1719$$

(or using Table C.1, $P(\geqslant 7) = 1 - P(\leqslant 6) = 1 - 0.828 = 0.172$)

Since $0.1719 > \frac{0.05}{1}$, we do not reject H_0.

Data are consistent with a median weight of 500 g. (5% level).

Note

We calculate P (7 or more) because 8, 9, 10 + signs are more extreme in the direction indicated by H_1. □

In cases in which $n > 10$, we can continue to calculate probabilities using the binomial distribution, but it may be more convenient to calculate approximate probabilities using a 'normal approximation to the binomial' method (see Section 7.3) by putting:

$$\mu = \frac{n}{2} \quad \text{and} \quad \sigma = \frac{\sqrt{n}}{2}.$$

14.3 WILCOXON SIGNED RANK TEST

In the paired-samples case where the magnitudes of the differences between pairs are available, but we cannot realistically assume that these differences are normally distributed, the Wilcoxon signed rank test may be used. This test requires only the assumption that the differences are from a continuous and symmetrical distribution. The Wilcoxon test is more powerful than the sign test.

For the Wilcoxon test the null hypothesis is that the median of the differences is zero.

To calculate the test statistic, T, the method is as follows:

> Ignoring any zero differences, the remaining differences are ranked without regard to sign. Then T_+ and T_-, the sum of the ranks of the positive and negative differences, respectively, are calculated. T is the smaller of T_+ and T_-.

The null hypothesis is rejected if $T \leqslant$ critical value of T given in Table C.7 (note that H_0 is rejected even when the equality holds).

Example 14.3

A psychologist tested a random sample of eight subjects using aptitude tests A and B. The scores were as shown in Table 14.2.

Table 14.2 Results of two aptitude tests on eight subjects

Subject	*1*	*2*	*3*	*4*	*5*	*6*	*7*	*8*
A	135	103	129	96	122	140	110	91
B	125	102	120	94	120	130	110	92
Difference A − B	10	1	9	2	2	10	0	−1

Do these scores suggest that the median score for test A is higher than for test B? Use a 5% level of significance. Since we have paired samples data here we might be tempted to perform a t test. However, a simple 'cross-diagram' of the differences (see Fig. 14.1) indicates no bunching in the middle as one would expect

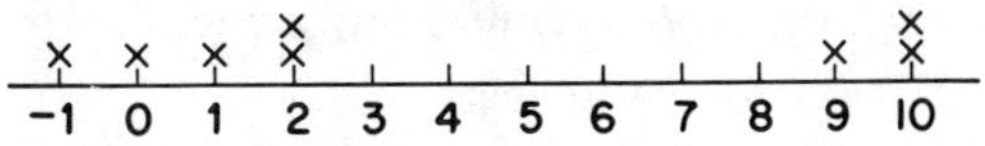

Fig. 14.1 A cross-diagram for the differences in test scores

if the data were normally distributed. We conclude that we would be wrong to assume that the differences are normally distributed. (A plot could be made on normal probability paper (Section 13.9), but it is unlikely that our conclusion concerning normality would be contradicted.)

We should carry out a Wilcoxon signed rank test instead of a t test:

H_0: Median of differences $= 0$, indicating median scores for tests A and B are equal.

H_1: Median of differences > 0, indicating median score for test A is higher than for test B.

5% level of significance

First we ignore the difference of zero for Subject 8. Ranking the remaining 7 differences without regard to sign we obtain:

Values	1	−1	2	2	9	10	10
Ranks	$1\frac{1}{2}$	$1\frac{1}{2}$	$3\frac{1}{2}$	$3\frac{1}{2}$	5	$6\frac{1}{2}$	$6\frac{1}{2}$

Note

Since we ignore signs when we rank, 1 and -1 have the same rank, this being an example of 'tied ranks'. There are two other pairs of tied ranks. Tied ranks are averaged.

$T_- = 1\frac{1}{2}$, since T_- is the sum of the ranks of the negative differences.

$T_+ = 1\frac{1}{2} + 3\frac{1}{2} + 3\frac{1}{2} + 5 + 6\frac{1}{2} + 6\frac{1}{2} = 26\frac{1}{2}$

(As a numerical check we should always find that

$$T_+ + T_- = \frac{n(n+1)}{2},$$

where n is the number of differences remaining after zero differences are ignored.

For example,

$$T_+ + T_- = 28, \quad \frac{n(n+1)}{2} = \frac{7 \times 8}{2} = 28, \text{ which checks.)}$$

$T = 1\frac{1}{2}$, the smaller of T_+ and T_-.

The critical value of T from Table C.7 is 3, for $n = 7$, a 5% level of significance and a one-sided H_1. Since $1\frac{1}{2} < 3$, we reject H_0.

The median score for test A is significantly higher than for test B (5% level).

□

For large samples, $n > 25$ say, Table C.7 cannot be used. However, we may use a normal approximation method in this event, since it can be shown that, under the null hypothesis, T is approximately normally distributed with mean, μ_T, and standard deviation, σ_T, where:

$$\mu_T = \frac{n(n+1)}{4} \quad \text{and} \quad \sigma_T = \sqrt{\frac{n(n+1)(2n+1)}{24}}$$

14.4 MANN–WHITNEY U TEST

In order to carry out an unpaired-samples t test we must be able to justify the assumptions of normality and equality of variance (Section 11.5). If we cannot justify these assumptions, the Mann–Whitney U test may be used. This test requires only the assumption that the variable of interest is continuous.

For the Mann–Whitney U test the null hypothesis is usually that the two populations from which random samples have been drawn are identically distributed. The alternative hypothesis is that the two medians are either different or one is greater than the other (depending on the particular application), but otherwise the distributions are identical.

To calculate the test statistic, U, the method is as follows:

If the sample sizes are n_1 and n_2, the $(n_1 + n_2)$ values are ranked as one group. Then R_1 and R_2, the sums of the ranks for the samples of size n_1 and n_2, respectively, are calculated. U_1 and U_2 are then calculated using:

$$U_1 = n_1 n_2 + \tfrac{1}{2} n_1 (n_1 + 1) - R_1$$
$$\text{and } U_2 = n_1 n_2 + \tfrac{1}{2} n_2 (n_2 + 1) - R_2$$

U is the smaller of U_1 and U_2.

The null hypothesis is rejected if $U \leqslant$ critical value of U given in Table C.8 (note that H_0 is rejected even when the equality holds).

Example 14.4

A random sample of 30 students at a university was asked the distance to their home town. The distances in miles for 13 male and 17 female students were as shown in Table 14.3.

Table 14.3 The distances from home town of 30 students

Male	80	485	176	224	141	259	120	80	278	240	192	35	45				
Female	3	90	272	80	8	10	72	294	22	144	160	50	64	480	56	96	104

Is it reasonable to conclude that the distributions of the distances from home are the same for male and female students (thus indicating the same median distances)? Use a 1% level of significance.

Since we have unpaired samples data we might be tempted to perform a t test. However, a 'cross-diagram' for the female student data indicates positive skewness, and a Mann–Whitney U test will be performed instead:

H_0: The distributions of the distances from home for male and female students are identical, thus indicating that the medians are the same.

H_1: The distributions have different medians but are otherwise identical.

1% level of significance.

Ranking all $(n_1 + n_2) = 13 + 17 = 30$ values as one group is made easier by the diagrams in Fig. 14.2 (not drawn to scale).

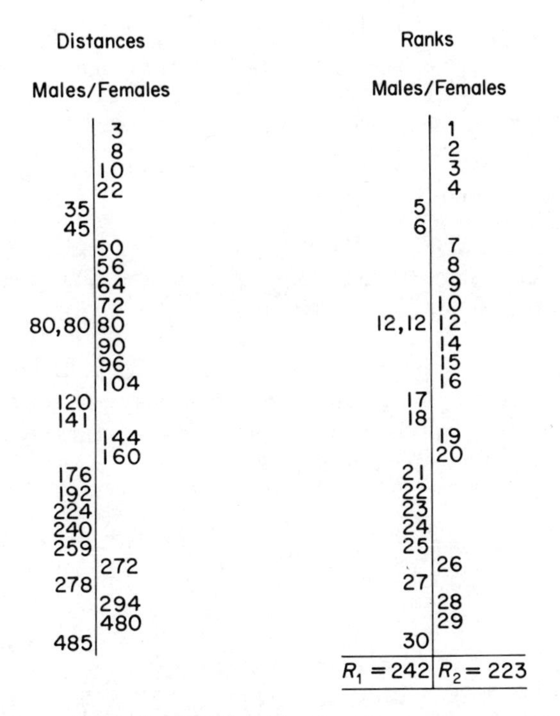

Fig. 14.2 Ranking 30 values in two stages

In the left-hand diagram the 30 values are listed and the equivalent ranks may then be listed on the right-hand diagram. There are three values of 80 which therefore receive the same average rank of 12 $\left(= \dfrac{11 + 12 + 13}{3}\right)$.

From Fig. 14.2, we read that $R_1 = 242$, $R_2 = 223$

Hence
$$U_1 = n_1 n_2 + \frac{1}{2} n_1 (n_1 + 1) - R_1$$

$$= 13 \times 17 + \frac{1}{2} \times 13 \times 14 - 242$$

$$\therefore U_1 = 70$$

and
$$U_2 = n_1 n_2 + \frac{1}{2} n_2 (n_2 + 1) - R_2$$

$$= 13 \times 17 + \frac{1}{2} \times 17 \times 18 - 223$$

$$\therefore U_2 = 151$$

(As a numerical check we should always find that $U_1 + U_2 = n_1 n_2$)

For this example, $U_1 + U_2 = 70 + 151 = 221$,

$n_1 n_2 = 13 \times 17 = 221$, which checks

$\therefore U = 70$, the smaller of U_1 and U_2

The critical value of U from Table C.8 is 49, for $n_1 = 13$, $n_2 = 17$, a 1% level of significance and a two-sided H_1.

Since $70 > 49$, we do not reject H_0.

The median distances from home of male and female students are not significantly different (1% level). □

For large samples, $n_1 > 20$ and $n_2 > 20$, say, Table C.8 cannot be used. However, we may use a normal approximation method in this event, since it can be shown that, under the null hypothesis, U is approximately normally distributed with mean μ_U and standard deviation σ_U, where:

$$\mu_U = \frac{n_1 n_2}{2} \quad \text{and} \quad \sigma_U = \sqrt{\frac{n_1 n_2 (n_1 + n_2 + 1)}{12}}$$

14.5 SUMMARY

Three non-parametric tests were described for paired and unpaired samples. The sign test may be used as a test for the median of a continuous distribution. For paired samples, the sign test may be used when only the signs of the differences between pairs is known. The test statistic is the probability of the result obtained or a more extreme result (in the direction indicated by the alternative hypothesis). The Wilcoxon signed rank test is preferred to the sign test when the magnitudes (as well as the signs) of the differences between pairs are known. The test statistic T is the smaller of the sums of the ranks of the positive and negative differences.

For unpaired samples the Mann–Whitney U test may be used. The test statistic U is the smaller of U_1 and U_2, where:

$$U_1 = n_1 n_2 + \tfrac{1}{2} n_1 (n_1 + 1) - R_1$$

and

$$U_2 = n_1 n_2 + \tfrac{1}{2} n_2 (n_2 + 1) - R_2,$$

and R_1, R_2 are the sums of the ranks of the two samples, when all $(n_1 + n_2)$ values are ranked as one group.

The Wilcoxon and Mann–Whitney tests require fewer assumptions than the corresponding t tests, but are less powerful when the assumptions required by these t tests are valid.

Section 14.2

1. Show that if the sign test is applied to $n = 5$ pairs the null hypothesis will never be rejected in favour of a two-sided alternative hypothesis at the 5% level of significance, no matter how extreme the results are. Why does this imply that there is no point in carrying out this test when $n = 5$ (or less)?

   What is the corresponding value of n in the case of a one-sided alternative?

2. A convenience food, known as 'Quicknosh', was introduced into the British market in January 1982. After a poor year for sales the manufacturers initiated an intensive advertising campaign during January 1983. The table below records the sales, in thousands of pounds, for a one-month period before and a one-month period after the advertising campaign, for each of eleven regions.

Region	*A*	*B*	*C*	*D*	*E*	*F*	*G*	*H*	*I*	*J*	*K*
Sales before campaign	2.4	2.6	3.9	2.0	3.2	2.2	3.3	2.1	3.1	2.2	2.8
Sales after campaign	3.0	2.5	4.0	4.1	4.8	2.0	3.4	4.0	3.3	4.2	3.9

   Determine, at the 5% level of significance, whether an increase in sales has occurred, by using:

   (a) the sign test;
   (b) the t test for paired values.

   Plot the differences of the paired values on arithmetical probability paper and hence comment on the appropriateness of the above tests.

   (AEB)

Section 14.3

3. In the Wilcoxon signed rank test, why should $T_+ + T_- = \frac{1}{2}n(n+1)$?

4. Ten students were tested before and after being coached in a subject. Their marks were as follows:

Student number	*1*	*2*	*3*	*4*	*5*	*6*	*7*	*8*	*9*	*10*
Before coaching	53	59	61	48	39	56	75	45	81	60
After coaching	60	57	67	52	61	71	70	46	93	75

   Test, at the 5% level of significance, the hypothesis that coaching has no effect on average marks against a suitable alternative hypothesis using:

   (i) a sign test;
   (ii) a Wilcoxon signed rank test.

Draw one overall conclusion from these tests.

What 'parametric' test might have been used instead, and what assumption would be necessary for this to be a valid test? Use a graphical method to decide whether this assumption is likely to be valid.

5. Ten pairs of identical twins have the following birthweights in kg.

Pair	*1*	*2*	*3*	*4*	*5*	*6*	*7*	*8*	*9*	*10*
First-born	3.95	3.41	3.73	4.13	3.48	4.28	3.98	4.18	4.04	3.73
Second-born	3.93	3.35	3.72	4.18	3.44	4.15	3.89	4.20	4.00	3.72

Use a non-parametric test, at the 1 % level of significance, to test the hypothesis that the birthweights of first and second-born identical twins are equal.

6. Two rats were randomly selected from each of 12 litters of rats. One rat in each pair was fed on diet A while the other was fed on diet B. After six months the weights (g.) of all the rats were recorded as follows:

Litter	*1*	*2*	*3*	*4*	*5*	*6*	*7*	*8*	*9*	*10*	*11*	*12*
A	70	65	66	70	68	68	70	71	66	70	69	69
B	68	65	66	69	67	66	70	71	65	70	68	69

Do these results suggest that the median weight of the rats fed on diet A are significantly different, at the 5 % level, from the median weight of the rats fed on diet B? Use a suitable non-parametric test.

7. The weights of fish in two populations were compared by analysing the differences in the weights of a sample of pairs of fish (one from each population) matched by length. The weight differences (g) for 10 pairs of fish were as follows:

   12 −13 −125 −210 −73 2 3 −147 −12 −4

   Why is a parametric test unsuitable? Analyse these data using a test which:

   (i) ignores the magnitudes of the differences;
   (ii) uses both the signs and magnitudes of the differences.

   Draw an overall conclusion from (i) and (ii).

8. A salesman for visual aids claims that visual memory is more effective than aural memory. To test this claim, nine members of the public are examined for visual and aural perception and the results are listed in the table below.

Visual and aural memory test results (maximum 80)

Visual	69	66	73	67	54	48	55	76	77
Aural	58	65	64	65	42	44	67	73	56

►

Discuss briefly the relative merits of using a Wilcoxon signed rank test and a sign test for investigating the salesman's claim. State, explicitly, suitable null and alternative hypotheses and carry out a test of significance at the 5% level using the Wilcoxon method. State your conclusion concerning the salesman's claim.

(Cambridge)

9. Twenty students were divided into two groups of ten. Each member of one group was told to memorize a large list of words using Method A, and each member of the other group was given a similar task using Method B. The efficacy of each method was measured by counting the number of words each student could remember. The next day the students used the opposite method to the one they had used the day before. Each student was asked to report the method he had found to be more successful. Sixteen said A and four said B. Make a sign test of whether the methods differed in effectiveness.

   Let the number of words remembered by student i ($i = 1, \ldots, 20$) be a_i by Method A, and b_i by Method B. Outline another method of testing the difference in effectiveness of the two methods of memorizing, using these data. Compare the relative merits of this second test and the sign test.

(OLE)

Section 14.4

10. In the Mann–Whitney U test, why should:

    (i) $R_1 + R_2 = \frac{1}{2}(n_1 + n_2)(n_1 + n_2 + 1)$;
    (ii) $U_1 + U_2 = n_1 n_2$?

    You may quote the formulae for U_1 and U_2 in terms of R_1 and R_2.

11. The following data are the results of fly-spray tests using two preparations, A and B. The numbers are the percentages of flies killed in eight tests using A and eight tests using B on 16 groups of flies. Use a non-parametric test to test, at the 5% level, whether the median percentage mortality is the same for the two preparations.

A	68	68	59	72	64	67	70	74
B	60	67	61	62	67	63	56	58

    What parametric test might have been used instead? What assumptions would be required by this test?

12. Two brands of car tyres were tested for length of life (in thousands of kilometres) on 16 cars, chosen at random from all the cars of a particular model. Do the results below show that the median lengths of life for the two brands differ significantly at the 1% level of significance? Use a non-parametric test.

Brand 1	38	45	37	44	39	41	39	32
Brand 2	41	37	43	33	40	38	31	39

13. Two tanks containing a well-mixed liquid were sampled and density measurements were recorded (in coded units) as follows:

Tank 1	8	17	18	13	16	15	4
Tank 2	4	8	9	12	4	5	

Use a non-parametric test, at the 5% level, to decide whether these data support the hypothesis that the median densities of liquid in the two tanks are the same.

14. A bank manager claims that the number of errors made by bank clerks depends on whether they are given a tea-break in the afternoon. Twelve clerks were randomly divided into two groups. Group 1 had an afternoon tea break, Group 2 did not. The number of errors made in the working period following the period of the tea-break were recorded:

Group 1	4	3	4	5	1	3
Group 2	7	8	4	5	7	4

Use a non-parametric test to decide whether the median number of errors is significantly less for the group having the afternoon tea-break. Also criticize the design of this experiment and suggest a better experiment.

15. Use a non-parametric test to re-analyse the data in Exercise 6 above, this time assuming that the 24 rats were not paired by litter, but instead were randomly selected from a group of nominally identical rats. Assume also that each rat was randomly allocated to one of the two diets.

16. Sixteen subjects were randomly selected to take part in a reaction-time task. Eight subjects were randomly allocated to perform the task after they had each drunk three double whiskies, the other eight subjects were completely sober when they performed the task. The following scores were obtained:

Whisky group	320	360	320	540	360	300	680	1120
Sober group	340	290	270	370	330	310	320	330

Is there a significant difference between the median scores of the two groups?

17. Ten boys and ten girls were chosen at random from a group of children of comparable age in a school. The weekly pocket money of each child was recorded. The investigator ►

found the median (£1.95) of the complete set of data and tabulated the data as follows.

		Weekly pocket money (£)	*Number in group*
Boys	Less than median	0.4, 0.9, 1.1, 1.4, 1.5, 1.6, 1.7	7
	Greater than median	2.3, 3.5, 8.6	3
Girls	Les than median	1.0, 1.0, 1.8	3
	Greater than median	2.1, 2.4, 3.5, 5.5, 6.0, 6.3, 7.3	7

He analysed the data as a 2×2 contingency table and calculated $\chi^2 = 3.2$.

State two other tests, one parametric and one non-parametric, which might be used to test whether the boys' pocket money differed from the girls'. Discuss the relative merits of the three tests for the given set of data.

Analyse the data by the non-parametric test you have named, and give your conclusions.

(OLE)

15

Correlation

▼

15.1 INTRODUCTION

We frequently hear reports in the media that researchers have found a correlation between two variables. For example, we hear that there is a correlation between cigarette smoking and lung cancer, implying that the person who smokes cigarettes has a higher risk than a non-smoker of contracting lung cancer. The amount of animal fats we eat is reported to be associated with our chances of suffering from heart disease. In economics, too, we find politicians arguing that if only this variable was decreased then this other variable would also decrease, presumably because the two variables are highly correlated (think of the rate of inflation and the number of people unemployed).

It is not the purpose of this book to discuss politics, but to give some idea of what is meant by correlation, how to calculate numerical measures of correlation which are called *correlation coefficients*, and how to interpret these coefficients.

We will mainly discuss the correlation between two variables X and Y, say, for two cases:

1. When X and Y are normally distributed variables, in which case the calculation of Pearson's product moment correlation coefficient (Pearson's r) is appropriate.
2. When X and Y are not normally distributed but can take values which may be ranked, in which case the calculation of Spearman's coefficient of rank correlation (Spearman's r'_s) or Kendall's coefficient of rank correlation (Kendall's τ) is appropriate.

Having calculated the appropriate correlation coefficient, we may test it for statistical significance. However, the wider interpretation of significant correlation coefficients is fraught with problems, and these will also be discussed.

15.2 THE CORRELATION COEFFICIENT BETWEEN TWO VARIABLES

$\mathrm{Cov}(X, Y)$, the covariance between two random variables X and Y, was defined in Chapter 8 (see (8.3)) as follows:

$$\mathrm{Cov}(X, Y) = E(XY) - E(X)E(Y)$$

The *Pearson correlation coefficient* between X and Y which is denoted by $\mathrm{Corr}(X, Y)$, but more usually by ρ, is defined as follows:

$$\rho = \mathrm{Corr}(X, Y) = \frac{\mathrm{Cov}(X, Y)}{\sqrt{\mathrm{Var}(X)\mathrm{Var}(Y)}} \tag{15.1}$$

This result may be written in terms of expectation (using (8.3) and (5.3)) as follows:

$$\rho = \frac{E(XY) - E(X)E(Y)}{\sqrt{[E(X^2) - \mu_X^2][E(Y^2) - \mu_Y^2]}} \tag{15.2}$$

where $\mu_X = E(X)$ and $\mu_Y = E(Y)$ are the means of X and Y respectively.

We can think of ρ as a measure of the correlation between all the possible pairs of values of X and Y for a bivariate population. For example, X and Y could denote the height and weight of adult males in the UK. If we knew all these heights and weights we could calculate ρ, the correlation coefficient for the whole population, using (15.2). The more practical situation is that we can only afford to take a sample from the population. If we denote the sample estimate of ρ by r, how should we calculate r given the sample heights and weights? The next section provides the answer to this question.

15.3 CALCULATION AND INTERPRETATION OF PEARSON'S CORRELATION COEFFICIENT, r

The following is a useful formula for calculating r, assuming our sample consists of n pairs of values of X and Y.

$$r = \frac{\sum x_i y_i - \dfrac{\sum x_i \sum y_i}{n}}{\sqrt{\left[\sum x_i^2 - \dfrac{(\sum x_i^2)}{n}\right]\left[\sum y_i^2 - \dfrac{(\sum y_i)^2}{n}\right]}} \tag{15.3}$$

(We can see the similarity between (15.3) and (15.2) by dividing the numerator and denominator of (15.2) by n.)

An alternative formula for r is:

$$r = \frac{\sum (x_i - \bar{x})(y_i - \bar{y})}{\sqrt{\sum (x_i - \bar{x})^2 \sum (y_i - \bar{y})^2}} \tag{15.4}$$

but (15.4) is not recommended for use with a hand calculator because of possible rounding errors, and because the calculation takes longer! A proof that (15.3) and (15.4) are equivalent is given in the Appendix to this chapter.

Example 15.1

The heights (cm) and weights (kg) of a random sample of eight adult males are shown in Table 15.1. Calculate r and draw a scatter diagram.

Table 15.1 The heights and weights of eight adult males

Height (x)	177	163	168	174	184	182	171	173
Weight (y)	71	67	62	73	80	85	69	77

Using these data, we calculate:

$$\sum x_i = 177 + 163 + \ldots \qquad + 173 = 1{,}392$$

$$\sum x_i^2 = 177^2 + 163^2 + \ldots \qquad + 173^2 = 242{,}548$$

$$\sum y_i = 71 + 67 + \ldots \qquad + 77 = 584$$

$$\sum y_i^2 = 71^2 + 67^2 + \ldots \qquad + 77^2 = 43{,}018$$

$$\sum x_i y_i = 177 \times 71 + 163 \times 67 + \ldots + 173 \times 77 = 101{,}916$$

$$n = 8$$

$$r = \frac{101{,}916 - \dfrac{1{,}392 \times 584}{8}}{\sqrt{\left[242{,}548 - \dfrac{1{,}392^2}{8}\right]\left[43{,}018 - \dfrac{584^2}{8}\right]}}$$

$$= \frac{300}{\sqrt{340 \times 386}}$$

$$r = 0.8281$$

Note
In calculating r it is arbitrary which variable we denote by X and which by Y, since the formula for r, (15.3), is unchanged if x_i and y_i are interchanged.

The impression given by Fig. 15.1 is that of a trend in which increasing height is associated with increasing weight. However, the trend is not 'perfect', since at any given height Fig. 15.1 indicates a variability in weight and vice versa. In the same

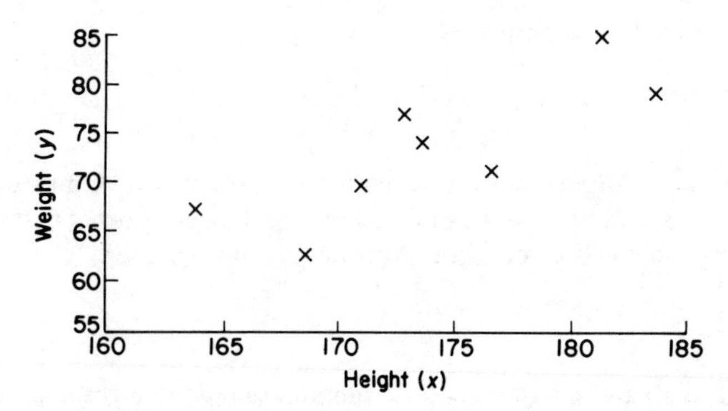

Fig. 15.1 Scatter diagram for height and weight of eight adults

way, the value of 0.8281 for r is not 'perfect', a point which will now be enlarged on.

It can be shown that if the points in a scatter diagram all lie on a straight line with a positive gradient, then $r = +1$. When the gradient of the straight line is negative, $r = -1$. When the points indicate no trend, then r will be near the middle of the range from -1 to $+1$, i.e. close to 0.

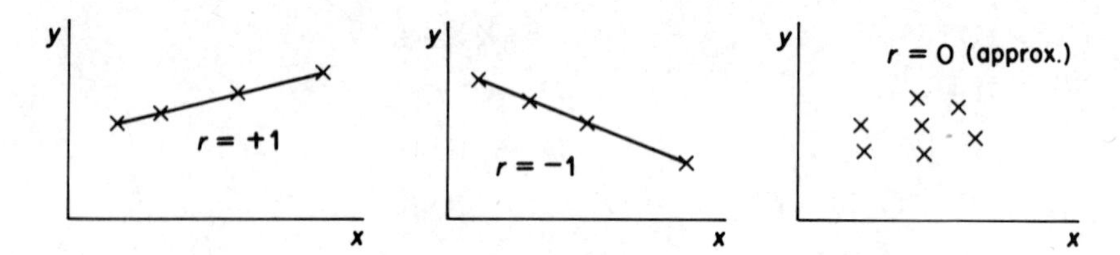

Fig. 15.2 Perfect positive, perfect negative and zero correlation

What conclusions can we draw from a correlation coefficient we have calculated? For example, what conclusion can we draw from the value of r obtained in Example 15.1? We can say that $r = 0.8281$ is high positive correlation, but this doesn't take account of the fact that $n = 8$ is a small sample size. A more useful idea is to test the hypothesis that ρ, the correlation coefficient between height and weight in the population, is zero. The method involves the use of a t test (Section 11.4). The assumption required for this test is that both variables are normally distributed. □

Example 15.2

Test, at the 5% level of significance, the hypothesis that $\rho = 0$ against the alternative that $\rho > 0$ using the data in Table 15.1. (Common sense indicates that if height and weight are significantly correlated, then this correlation will be positive). Here ρ is the population correlation coefficient for the variables height and weight.

H_0: $\rho = 0$

H_1: $\rho > 0$

5 % level of significance

The appropriate test statistic is

$$t = r\sqrt{\frac{n-2}{1-r^2}} \tag{15.5}$$

For the data in Table 15.1, we already know that

$r = 0.8281$, $n = 8$, and so

$$t = 0.8281\sqrt{\frac{8-2}{1-0.8281^2}} = 3.62 \quad (2\text{ d.p.})$$

The critical value of t is $t_{\alpha,\nu}$, where

$$\alpha = \frac{\text{sig. level}}{1} \quad \text{or} \quad \frac{\text{sig. level}}{2},$$

depending on whether H_1 is one- or two-sided, and $\nu = n-2$ degrees of freedom.

For this example, $\alpha = 0.05$, $\nu = 8-2 = 6$, so the critical value of t is 1.94 (Table C.4).

Since $3.62 > 1.94$, we reject H_0 (see Fig. 15.3), and conclude that there is significant positive correlation between height and weight (5 % level).

Note

We have assumed that both height and weight are normally distributed.

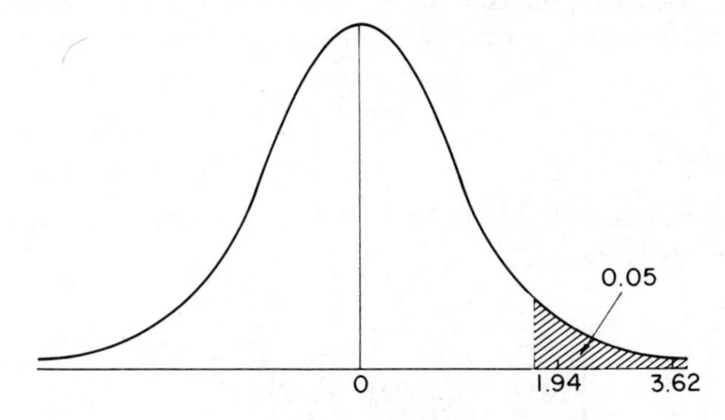

Fig. 15.3 A t distribution with 6 d.f.

We now make three important points which are relevant when we try to interpret the values of correlation coefficients in a wider context: □

1. A statistically significant correlation coefficient does not necessarily imply cause and effect. For the example just discussed it is meaningless to talk about increasing height 'causing' increasing weight. We realize that our height and weight are determined by a number of other factors, such as heredity, age, sex, diet and exercise. In other instances the argument may not be so clear-cut. For example, a politician may point out that it is possible to show a positive correlation between the rate of inflation and the number of people unemployed, based on an analysis of historical data. The argument that reducing the rate of inflation will *cause* a decrease in the number of people unemployed is attractive. Whether such an argument is justified depends on whether all the other factors affecting unemployment act in the same way in the future as they did in the past.

   There are instances when a cause-and-effect conclusion *is* justified, for example, if we keep a number of identical blocks of ice in the same conditions except that we vary the temperature from block to block, then we would probably find a significant negative correlation between the weight of ice (after a given period of time) and temperature. We could safely conclude that increasing the temperature had caused more ice to melt.

   'Nonsense correlations', in which we conclude that it is nonsense to draw a conclusion of cause and effect, often arise when two variables are measured over a period of time. Think of recording, for 10 years say, the average salary of an MP in the UK and the amount of wine drunk per capita in the UK. A scatter diagram of 10 points (one for each year) would probably indicate positive correlation, both variables increasing over the period, but it would clearly be nonsense to imply cause and effect.
2. Pearson's r measures *linear* correlation. We saw in Chapter 8 that, if two variables are independent then their covariance is zero. It follows from (15.1) that, if two variables are independent then their correlation coefficient, ρ, is also zero. However, the converse is not true. We quote this result without proof, but the following example is intended to make the result seem at least plausible.

Example 15.3

Show that, for the sample data in Table 15.2, $r = 0$, but that the variables do not appear to be independent.

$$r = \frac{385 - \dfrac{35 \times 77}{7}}{\sqrt{\left[203 - \dfrac{35^2}{7}\right]\left[931 - \dfrac{77^2}{7}\right]}} = \frac{0}{\sqrt{28 \times 84}} = 0$$

We see that, although the estimates of the covariance and correlation coefficients are zero, Fig. 15.4 indicates non-linear correlation and hence that X and Y are not independent.

Table 15.2 A sample of seven pairs of values of X and Y

x	2	3	4	5	6	7	8
y	16	11	8	7	8	11	16

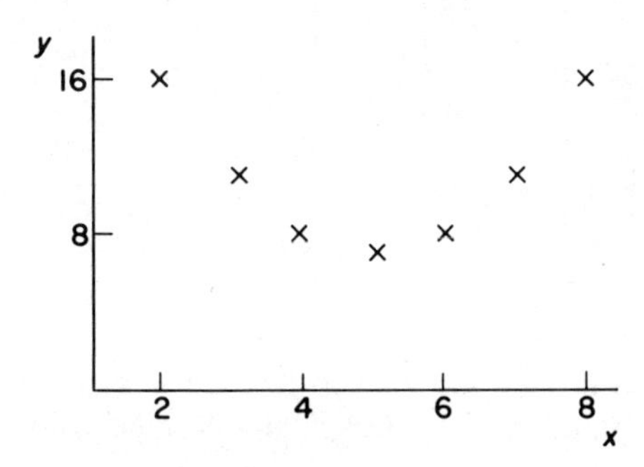

Fig. 15.4 Non-linear correlation

3. The value of r may be restricted because the ranges of X and Y are themselves restricted. This is another way of saying that we must be very careful to define the population from which we take our sample. For example, the fact that little correlation has been observed between the A-level count and the academic achievement of university students may be because universities only take those students with above a certain A-level count. A higher correlation might be found if university education was available to all those who had studied a minimum of one subject at A-level.

One final note; the value of r is independent of the units in which we measure the variables X and Y. We say that r is 'dimensionless' i.e. it has no units. It is left as an exercise for the reader to change the units in Table 15.1 to metres for height and grams for weight, and show that r is still equal to 0.8281. □

15.4 THE CODING METHOD OF CALCULATING PEARSON'S r

If we have so much data that we decide to record them as frequencies in a two-way table, then we may use a special coding method for calculating r. An example of such data is given in Table 15.3, which gives the heights (cm) and weights (kg) of a random sample of 100 adult males.

Notice that the five groups for height are all equally wide, each having a width of 5 cm. The four weight groups are also equally wide, each having a width of 5 kg. These properties enable us to code the variables height and weight before we calculate the correlation coefficient r.

Table 15.3 The heights and weights of 100 adult males

	Height (x)					
Weight (y)	160–	165–	170–	175–	180–	*Total*
60–	3	4	2	1	0	10
65–	5	15	14	5	1	40
70–	2	0	11	14	3	30
75–	0	1	3	10	6	20
Total	10	20	30	30	10	100

Example 15.4

Use the coding method to calculate r for the data in Table 15.3. Test the hypothesis H_0: $\rho = 0$, at the 5% level of significance, using a two-sided H_1.

Let
$$u_x = \frac{x - 172.5}{5}, \qquad u_y = \frac{y - 67.5}{5},$$

where 172.5 and 67.5 are the 'assumed means' of X and Y, respectively. Notice that they are both group mid-points. We divide by 5 in each case because 5 is the group width for both height and weight. This helps to simplify the arithmetic, which is the main purpose of the coding method. Table 15.4 shows the coded variables u_x and u_y and nearly all the calculations which are required in order to calculate r, using the following formula:

$$r = \frac{\sum f u_x u_y - \dfrac{\sum f_x u_x \sum f_y u_y}{n}}{\sqrt{\left[\sum f_x u_x^2 - \dfrac{(\sum f_x u_x)^2}{n}\right]\left[\sum f_y u_y^2 - \dfrac{(\sum f_y u_y)^2}{n}\right]}} \tag{15.6}$$

All the summations required for (15.6), except $\sum f u_x u_y$, are performed in Table 15.4.

$$\sum f u_x u_y = 3 \times -2 \times -1 + 4 \times -1 \times -1 + \ldots + 6 \times 2 \times 2 = 67$$

$$r = \frac{67 - \dfrac{10 \times 60}{100}}{\sqrt{\left[130 - \dfrac{10^2}{100}\right]\left[120 - \dfrac{60^2}{100}\right]}} = \frac{61}{\sqrt{129 \times 84}} = 0.586.$$

Testing H_0: $\rho = 0$ against H_1: $\rho \neq 0$, at the 5% level of significance, gives:

$$t = 0.586\sqrt{\frac{100 - 2}{1 - 0.586^2}} = 7.16,$$

Table 15.4 An example of the coding method of calculating *r*

u_y \ u_x	−2	−1	0	1	2	f_y	$f_y u_y$	$f_y u_y^2$
−1	3	4	2	1	0	10	−10	10
0	5	15	14	5	1	40	0	0
1	2	0	11	14	3	30	30	30
2	0	1	3	10	6	20	40	80
f_x	10	20	30	30	10	$n = 100$	$\sum f_y u_y = 60$	$\sum f_y u_y^2 = 120$
$f_x u_x$	−20	−20	0	30	20	$\sum f_x u_x = 10$		
$f_x u_x^2$	40	20	0	30	40	$\sum f_x u_x^2 = 130$		

which leads to rejection of H_0, since the critical value of t is 1.98 and $7.16 > 1.98$. We conclude that there is significant correlation between X and Y, based on the sample data in Table 15.3 (5 % level). □

15.5 HYPOTHESIS TEST FOR NON-ZERO VALUES OF ρ (FISHER'S TRANSFORMATION)

The test for ρ described in Section 15.3 is only appropriate if the null hypothesis is that $\rho = 0$. If we wish to test H_0: $\rho = \rho_0$, where $\rho_0 \neq 0$, we cannot use a t test. However, we can use Fisher's transformation. This states that, if r is the sample estimate of the correlation coefficient between two normally distributed variables, X and Y, and if H_0: $\rho = \rho_0$ is true, then

$$\left.\begin{array}{l}\dfrac{1}{2}\log_e\left(\dfrac{1+r}{1-r}\right) \text{ is approximately normally distributed with mean} \\ \mu = \dfrac{1}{2}\log_e\left(\dfrac{1+\rho_0}{1-\rho_0}\right) \text{ and standard deviation } \sigma = \dfrac{1}{\sqrt{n-3}}\end{array}\right\} \quad (15.7)$$

Note

We could alternatively use $\tanh^{-1} r$ and use Table C.9 instead of $\frac{1}{2}\log_e(1+r)/(1-r)$, but logs are usually more convenient for calculation on basic hand calculators.

Example 15.5

A sample of 100 pairs of observations of two normally distributed variables, X and Y, gave $r = 0.586$. Test the hypothesis that $\rho = 0.5$ against the alternative that $\rho \neq 0.5$, at the 5 % level of significance.

Using Fisher's transformation, $\quad \dfrac{1}{2}\log_e\left(\dfrac{1+0.586}{1-0.586}\right) = 0.6716,$

$$\mu = \frac{1}{2}\log_e\left(\frac{1+0.5}{1-0.5}\right) = 0.5493, \qquad \sigma = \frac{1}{\sqrt{100-3}} = 0.1015$$

Standardizing, $$z = \frac{0.6716 - 0.5493}{0.1015} = 1.205$$

For a two-sided alternative at the 5 % level, the critical value is 1.96 (Table C.3(b)). Since $1.205 < 1.96$, do not reject H_0.

We conclude that ρ is not significantly different from 0.5 (5 % level). □

15.6 CONFIDENCE INTERVAL FOR ρ

Rather than performing a hypothesis test for a particular value of ρ, Fisher's transformation can be used to obtain a confidence interval for ρ. (We may of course use this confidence interval for hypothesis testing, as in Section 11.9.)

For example, a 95 % confidence interval for $\frac{1}{2}\log_e\left(\frac{1+\rho}{1-\rho}\right)$ is given by

$$\frac{1}{2}\log_e\left(\frac{1+r}{1-r}\right) \pm 1.96\frac{1}{\sqrt{n-3}} \qquad (15.8)$$

Here r is the sample estimate of the correlation coefficient between two normally distributed variables, based on a sample of size n.

Example 15.6

A sample of 100 pairs of observations of two normally distributed variables, X and Y gave $r = 0.586$.

Calculate a 95 % confidence interval for ρ.

Using these data:

$$\frac{1}{2}\log_e\left(\frac{1+r}{1-r}\right) \pm 1.96\frac{1}{\sqrt{n-3}} = 0.6716 \pm 1.96 \times 0.1015$$

$$= 0.4727 \text{ and } 0.8705$$

Setting $\frac{1}{2}\log_e(1+\rho)/(1-\rho)$ equal to each of these values in turn gives 0.4404 to 0.7016 as our 95 % confidence interval for ρ. (Alternatively, enter Table C.10 with $z = 0.473$ and $z = 0.871$ to obtain similar values after interpolation.) □

15.7 HYPOTHESIS TEST FOR THE DIFFERENCE BETWEEN TWO CORRELATION COEFFICIENTS, ρ_1 AND ρ_2

Suppose we take samples of two normally distributed random variables X and Y from two distinct populations. We may then use Fisher's transformation to test whether ρ_1 and ρ_2, the correlation coefficients in the two populations, are equal. The result we need is that, under the null hypothesis H_0: $\rho_1 = \rho_2$,

$$\left.\begin{array}{l} \frac{1}{2}\log_e\left(\frac{1+r_1}{1-r_1}\right)-\frac{1}{2}\log\left(\frac{1+r_2}{1-r_2}\right) \text{ is approximately} \\ \text{normally distributed with mean zero and standard deviation} \\ \sqrt{\frac{1}{n_1-3}+\frac{1}{n_2-3}} \end{array}\right\} \quad (15.9)$$

Here r_1 and r_2 are sample estimates of ρ_1 and ρ_2, based on samples of sizes n_1 and n_2, respectively.

Example 15.7

A sample of normally distributed variables X and Y of size 100 is taken from each of two populations. Test the hypothesis that the two population correlation coefficients are equal if the sample estimates are: $r_1 = 0.5$ and $r_2 = 0.6$.

Using (15.9) for these data,

$$\frac{1}{2}\log_e\left(\frac{1+0.5}{1-0.5}\right)-\frac{1}{2}\log_e\left(\frac{1+0.6}{1-0.6}\right) = -0.1438$$

$$\sqrt{\frac{1}{100-3}+\frac{1}{100-3}} = 0.1436$$

Standardizing, $$z = \frac{-0.1438-0}{0.1436} = -1.00 \quad (2 \text{ d.p.})$$

For a two-sided alternative at the 5% level, the critical value is 1.96 (Table C.3(b)).

Since $|-1.00| < 1.96$, we do not reject H_0.

We conclude that the correlation coefficients are not significantly different (5% level). □

15.8 SPEARMAN'S COEFFICIENT OF RANK CORRELATION, r_s

When we are interested in the correlation between two variables X and Y, but we cannot be sure that both variables are normally distributed, distribution-free correlation coefficients may be calculated and tested for significance. In this section we will discuss the best known such coefficient, namely Spearman's coefficient of rank correlation, r_s. (Another coefficient, Kendall's tau (τ), is discussed in Section 15.9.)

To calculate r_s we first rank the n sample values of X and separately rank the n sample values of Y. The differences d_i, $i = 1, 2, 3, \ldots, n$ between the n pairs of ranks are calculated and r_s is then given by:

$$r_s = 1 - \frac{6\sum d_i^2}{n(n^2 - 1)} \tag{15.10}$$

It can be shown that, like Pearson's r, r_s must lie in the range -1 to $+1$. It can also be shown that if the formula for Pearson's r (15.3) is applied to the ranks of the variables X and Y (rather than to the values of X and Y), then we get the same value as if we had applied (15.10) to the ranks (see Worksheet 15, Exercise 14).

Example 15.8

In an ecological survey nine small areas were investigated and the numbers of plants, X and Y, of two plant species were counted; see Table 15.5.

Table 15.5 Ranking X and Y and calculating $\sum d_i^2$

	Number of plants of species				
Area	X	Y	*Ranks for X*	*Ranks for Y*	d_i^2
1	99	10	9	2	40
2	95	4	8	1	49
3	83	22	7	5	4
4	82	13	6	3	9
5	68	35	5	7	4
6	64	26	4	6	4
7	62	21	3	4	1
8	49	36	2	8	36
9	46	37	1	9	64
					$\sum d_i^2 = 220$

Calculate r_s and test its significance at the 5% level, using an alternative hypothesis that large numbers of one species are associated with small numbers of the other species (negative correlation).

$$r_s = 1 - \frac{6 \times 220}{9(9^2 - 1)} = -0.833$$

The steps in this hypothesis test for r_s are:

H_0: The ranks of X and Y are uncorrelated

H_1: There is negative correlation, e.g. high ranks of X are associated with low ranks of Y. This is a one-sided alternative.

5% level of significance.

$r_s = -0.833$

From Table C.11 we read that the critical value of r_s is 0.600. Since the calculated value is negative and larger in magnitude then the critical value, and hence tends to support H_1 (i.e. since $-0.833 < -0.600$), we reject H_0.

We conclude that there is significant negative correlation between X and Y (5% level). (It seems that each species inhibits the growth of the other.)

In some cases we start with data which are already in the form of ranks, for example in preference testing and competitions, as in the following example.

□

Example 15.9

Two judges are asked to rank each of six choirs taking part in a singing competition. Test whether the rankings of the two judges are uncorrelated, against the alternative that they are positively correlated. Use a 1% level of significance.

The data are given in Table 15.6.

Table 15.6 Ranks in a singing competition

	Ranks of judge		
Choir	*1*	*2*	d_i^2
A	4	4.5	0.25
B	1	3	4.00
C	2	1	1.00
D	3	2	1.00
E	6	4.5	2.25
F	5	6	1.00
			$\sum d_i^2 = 9.5$

Note

Judge 2 gave A and E as 'equal fourth place'.

$$r_s = 1 - \frac{6 \times 9.5}{6(6^2 - 1)} = 0.7286$$

The steps in this hypothesis test for r_s are:

H_0: The rankings of the judges are uncorrelated

H_1: The rankings are positively correlated, e.g. high ranks for judge 1 are associated with high ranks for judge 2 (one-sided H_1)

1% level of significance

$r_s = 0.7286$

From Table C.11 we read that the critical value of r_s is 0.943 for a one-sided H_1 at the 1% level of significance. Since $0.7286 < 0.943$, do not reject H_0.

We conclude that the rankings of the two judges are not significantly positively correlated (1% level). □

15.9 KENDALL'S TAU (τ)

Kendall's tau (τ) is another correlation coefficient which may be used on non-normally distributed bivariate data and (like Spearman's r_s) involves the use of ranks. The value of τ must also lie in the range -1 to $+1$, but it measures correlation in a different way from Spearman's r_s and so the values of τ and r_s for a given set of data will generally differ.

To calculate τ, we define

$$\tau = \frac{S}{\frac{1}{2}n(n-1)} \tag{15.11}$$

where n is number of pairs of ranks, and S is found as follows: place the n ranks of one variable, X say, in a row from 1 to n. Beneath each of these place the corresponding ranks for the other variable, Y. For each of the Y ranks, calculate:

(the number of Y ranks to the right which exceed it) − (the number of Y ranks to the right which are less than it)

Then S is the total of these values summed over all the ranks of Y.

Example 15.10

Using the data from Table 15.6, calculate Kendall's τ and test its significance at the 1% level.

We place the ranks for judge 1 in increasing order; see Table 15.7.

Table 15.7 Ranks in a singing competition

Choir	*B*	*C*	*D*	*A*	*F*	*E*
Ranks, judge 1	1	2	3	4	5	6
Ranks, judge 2	3	1	2	4.5	6	4.5

Considering the ranks of judge 2 in turn, there are three ranks to the right of the value 3 greater than 3, and two ranks to the right of the value 3 less than 3, and so on.

$$S = (3-2) + (4-0) + (3-0) + (1-0) + (0-1)$$

$$S = 8$$

$$\tau = \frac{S}{\frac{1}{2}n(n-1)} = \frac{8}{\frac{1}{2} \times 6 \times 5} = 0.533$$

The steps in this hypothesis test for τ are:

H_0: The rankings of the judge are uncorrelated

H_1: High ranks of judge 1 are associated with high ranks of judge 2 (one-sided H_1)

1% level of significance

$\tau = 0.533$

The critical value of τ is 0.867 for a one-sided H_1 at the 1% level (see Table C.12). Since $0.533 < 0.867$, we do not reject H_0.

We conclude that the rankings of the two judges are not significantly positively correlated (1% level) □

15.10 CORRELATION COEFFICIENTS BETWEEN LINEAR FUNCTIONS OF TWO VARIABLES

We are sometimes interested in the value of Pearson's ρ, not between two random variables X and Y, but between linear functions of X and Y. The calculations involve the use of (15.1) and some expectation algebra (see Chapter 5). The methods are best illustrated with the following examples.

Example 15.11

Two random variables X and Y have means μ_X, μ_Y and variances σ_X^2, σ_Y^2 respectively.
The correlation coefficient between X and Y is ρ.
Calculate the correlation coefficient between:

(i) cX and Y, where c is a constant;
(ii) X and $Y+c$;
(iii) X and $X+Y$.

The solutions are as follows:

(i) Let $Z = cX$

$$\begin{aligned}
\text{Cov}(Z, Y) &= E(ZY) - E(Z)E(Y) && \text{by (8.3)} \\
&= E(cXY) - E(cX)E(Y) && \\
&= cE(XY) - cE(X)E(Y) && \text{by (A5.2) (Appendix to Chapter 5)} \\
&= c\,\text{Cov}(X, Y) && \text{by (8.3)}
\end{aligned}$$

$$\text{Var}(Z) = \text{Var}(cX) = c^2\text{Var}(X) \quad \text{by (5.22)}$$

$$\begin{aligned}
\therefore \text{Corr}(Z, Y) &= \frac{\text{Cov}(Z, Y)}{\sqrt{\text{Var}(Z)\,\text{Var}(Y)}} && \text{by (15.1)} \\
&= \frac{c\,\text{Cov}(X, Y)}{\sqrt{c^2\,\text{Var}(X)\,\text{Var}(Y)}} && \text{by above} \\
&= \rho && \text{by (15.1)}
\end{aligned}$$

(ii) Let $W = Y + c$

$$\begin{aligned}
\text{Cov}(X, W) &= E[X(Y+c)] - E(X)E(Y+c) \\
&= E(XY) + cE(X) - E(X)E(Y) - cE(X) && \text{by (8.1), (A5.1) and (A5.2)} \\
&= \text{Cov}(X, Y) && \text{by (8.3)}
\end{aligned}$$

$$\begin{aligned}
\text{Var}(W) &= \text{Var}(Y+c) \\
&= \text{Var}(Y) && \text{by (5.22)}
\end{aligned}$$

$$\begin{aligned}
\therefore \text{Corr}(X, W) &= \frac{\text{Cov}(X, W)}{\sqrt{\text{Var}(X)\text{Var}(W)}} && \text{by (15.1)} \\
&= \frac{\text{Cov}(X, Y)}{\sqrt{\text{Var}(X)\,\text{Var}(Y)}} && \text{by above} \\
&= \rho && \text{by (15.1)}
\end{aligned}$$

(iii) Let $U = X + Y$

$$\begin{aligned}
\text{Cov}(X, U) &= E(XU) - E(X)E(U) \\
&= E[X(X+Y)] - E(X)E(X+Y) \\
&= E(X^2) + E(XY) - E(X)E(X) - E(X)E(Y) \\
&= E(X^2) - \mu_X^2 + E(XY) - E(X)E(Y) \\
&= \sigma_X^2 + \text{Cov}(X, Y),
\end{aligned}$$

where μ_X and σ_X^2 are the mean and variance of X

Also $\mu_U = E(U) = E(X+Y) = \mu_X + \mu_Y$

$$\begin{aligned}
\therefore \text{Var}(U) &= \text{Var}(X+Y) \\
&= \text{Var}(X) + \text{Var}(Y) + 2\,\text{Cov}(X, Y) && \text{by (8.2)} \\
&= \sigma_X^2 + \sigma_Y^2 + 2\text{Cov}(X, Y)
\end{aligned}$$

$$\begin{aligned}
\therefore \text{Corr}(X, U) &= \frac{\text{Cov}(X, U)}{\sqrt{\text{Var}(X)\,\text{Var}(U)}} \\
&= \frac{\sigma_X^2 + \text{Cov}(X, Y)}{\sqrt{\sigma_X^2(\sigma_X^2 + \sigma_Y^2 + 2\text{Cov}(X, Y)}} && \text{by above}
\end{aligned}$$

This result may also be expressed in terms of ρ, since $\text{Cov}(X, Y) = \rho\sigma_X\sigma_Y$ from (15.1)

$$\begin{aligned}
\text{i.e. Corr}(X, U) &= \frac{\sigma_X^2 + \rho\sigma_X\sigma_Y}{\sqrt{\sigma_X^2(\sigma_X^2 + \sigma_Y^2 + 2\rho\sigma_X\sigma_Y)}} \\
&= \frac{\sigma_X + \rho\sigma_Y}{\sqrt{\sigma_X^2 + \sigma_Y^2 + 2\rho\sigma_X\sigma_Y}}
\end{aligned}$$

We will now repeat this example, but this time we will assume that X and Y are independent random variables. In this case: Cov $(X, Y) = 0$ (Section 8.2), and from (15.1) it follows that $\rho = 0$, also □

Example 15.12

Repeat the previous example, assuming that X and Y are independent.

(i) Let $Z = cX$. Since we showed Corr $(Z, Y) = \rho$ it follows that Corr $(Z, Y) = 0$ if X and Y are independent.
(ii) Let $W = Y + c$. Since we showed Corr $(X, W) = \rho$ it follows that Corr $(X, W) = 0$ if X and Y are independent.
(iii) Let $U = X + Y$. Since we know $\rho = 0$,

$$\text{Corr}\,(X, U) = \frac{\sigma_X}{\sqrt{\sigma_X^2 + \sigma_Y^2}}.$$ □

15.11 SUMMARY

Three correlation coefficients, Pearson's r, Spearman's r_s and Kendall's τ, were discussed, including how to test their significance. All three coefficients can only take values between -1 and 1.

Pearson's r is used when we are dealing with normally distributed variables. A t test is used to test the null hypothesis that ρ, the population correlation coefficient, is zero. Fisher's transformation is used for testing non-zero values of ρ, for calculating a confidence interval for ρ, and for testing whether two population correlation coefficients are equal.

For non-normal variables which are capable of being ranked, Spearman's r_s and Kendall's τ are appropriate.

Great care must be taken in the interpretation of correlation coefficients (the pitfalls were discussed in Section 15.3).

APPENDIX TO CHAPTER 15

Proof that (15.3) and (15.4) are equivalent

$$\begin{aligned}\sum (x_i - \bar{x})(y_i - \bar{y}) &= \sum (x_i y_i - \bar{x} y_i - x_i \bar{y} + \bar{x}\bar{y}) \\ &= \sum x_i y_i - \bar{x} \sum y_i - \bar{y} \sum x_i + n\bar{x}\bar{y},\end{aligned}$$

summations are from $i = 1$ to $i = n$, so neither $\bar{x}$ nor $\bar{y}$ change as i changes.

$$= \sum x_i y_i - \frac{\sum x_i}{n} \sum y_i - \frac{\sum y_i}{n} \sum x_i + n \frac{\sum x_i}{n} \frac{\sum y_i}{n} \qquad \text{by (2.1)}$$

$$= \sum x_i y_i - \frac{\sum x_i \sum y_i}{n}$$

Hence the numerators of (15.3) and (15.4) are equal.

Similarly,
$$\sum (x_i - \bar{x})^2 = \sum x_i^2 - \frac{(\sum x_i)^2}{n},$$

putting $y_i = x_i$ for all i in the previous result.

Similarly,
$$\sum (y_i - \bar{y})^2 = \sum y_i^2 - \frac{(\sum y_i)^2}{n}.$$

Hence the denominators of (15.3) and (15.4) are also equal.

Section 15.3

1. If the value of Pearson's r is $+1$ (or -1) if the points on the scatter diagram lie on a straight line with a positive (or negative) gradient, what is the value of r if the straight line has a gradient of zero? (*Hint*: plot five points on such a straight line at convenient coordinates, read off their coordinates and calculate r.)

2. A random sample of 10 subjects was timed in the task of writing the letters of the alphabet first with their right hand and then with their left hand. Is there a significant correlation between the two sets of times below? Use a 5% level of significance and assume that the time to complete the task for 'right hands' and for 'left hands' are both normally distributed.

	Subject	*1*	*2*	*3*	*4*	*5*	*6*	*7*	*8*	*9*	*10*
Time	Right hand	12	9	17	15	13	18	12	16	18	16
(secs)	Left hand	25	22	30	29	33	29	28	31	32	34

3. Ten pairs of identical twins have the following birthweights (kg):

		1	*2*	*3*	*4*	*5*	*6*	*7*	*8*	*9*	*10*
Weight	1st born	3.95	3.41	3.73	4.13	3.48	4.28	3.98	4.18	4.04	3.73
(kg)	2nd born	3.93	3.35	3.72	4.18	3.44	4.15	3.89	4.20	4.00	3.72

(a) Plot a scatter diagram.
(b) The value of Pearson's r for these data is one of the following:

-0.105 0.505 -0.985 0.985 0.264

Choose the correct one, giving a reason for your choice, and test its significance at the 5% level. You may assume that birthweight is normally distributed.

Section 15.4

4. The following data for 40 students have been extracted from Table 1.1. Use the coding method to calculate the correlation coefficient between height and A-level count. Are you surprised at the result?

A-level	*Height* (cm)			
count	*149.5–*	*159.5–*	*169.5–*	*179.5–*
0–4	0	6	3	4
5–9	7	9	3	4
10–14	1	1	0	0
15–19	0	1	1	0

▶

Section 15.5

5. Use the answer to Exercise 4 to test the hypothesis that the correlation coefficient between height and A-level count is 0.2 against a two-sided alternative. Choose a 5% level of significance, and assume that height and A-level count are normally distributed.

Section 15.6

6. Use the answer to Exercise 4 to calculate a 95% confidence interval for the correlation coefficient between height and A-level count. Does your interval contain the number 0.2? Is your answer consistent with your answer to Exercise 5?

7. Twelve representative Cheviot ewes and their pure-bred lambs were weighed at the time of the birth of the lambs; the weights of each (in kg) were recorded as shown.

Ewe (x)	44	41	43	40	41	37	38	36	44	43	35	40
Lamb (y)	3.5	2.8	3.2	2.7	2.9	2.5	2.8	2.6	3.6	2.6	2.4	2.9

(*Note:* $\sum x = 482$, $\sum x^2 = 19466$, $\sum y = 34.5$, $\sum y^2 = 100.77$, $\sum xy = 1396.0$)

(i) Represent the data graphically.
(ii) Calculate the product-moment correlation coefficient, and test whether the population correlation coefficient between the weights of Cheviot ewes and their pure-bred lambs is zero.
(iii) Find a 95% confidence interval for the population correlation coefficient between Cheviot ewe and lamb weights on the basis of the data.

(You may assume that $\tanh^{-1} r$, where r is the sample correlation coefficient, is distributed approximately normally with mean $\tanh^{-1} \rho$ and variance $1/(n-3)$, where ρ is the true correlation coefficient and n is the sample size).

(iv) One of the lambs in the sample was born prematurely, the others were born after full-term pregnancies, but the records are not available. Which of the observations would you suspect for the premature birth?

(IOS)

Section 15.7

8. Let X_1, X_2 denote the heights of male and female students, and let Y_1, Y_2 denote the A-level counts of male and female students. Using the data in Table 1.1, the following summary statistics may be obtained:

$\sum x_1 = 2319$ $\sum x_1^2 = 413{,}961$ $\sum y_1 = 65$ $\sum y_1^2 = 403$ $\sum x_1 y_1 = 11{,}555$
$n_1 = 13$

$\sum x_2 = 4411$ $\sum x_2^2 = 721{,}597$ $\Sigma y_2 = 174$ $\sum y_2^2 = 1350$ $\sum x_2 y_2 = 28{,}339$
$n_2 = 27$

►

Calculate r_1 and r_2, the sample estimates of ρ_1 and ρ_2, the population correlation coefficients between X_1 and Y_1 and X_2 and Y_2 respectively.

Test the hypothesis that $\rho_1 = \rho_2$, using a 5% level of significance.

Section 15.8

9. The marks awarded by two judges to 10 contestants at a music festival were as follows:

	Contestant									
	A	*B*	*C*	*D*	*E*	*F*	*G*	*H*	*I*	*J*
Judge 1	10	13	8	7	3	4	15	6	17	18
Judge 2	5	4	6	5	7	8	3	7	2	2

Calculate Spearman's r_s and test the hypothesis that the marks of the judges are uncorrelated against a suitable one-sided alternative. Use a 5% level of significance.

10. From a large number of families living in a rural area each having a youngest child aged between 5 and 7 years, a researcher selected a random sample of six families and measured the IQ of the youngest child. The researcher also recorded the total number of children in each family and drew up the following table:

Number of children in family	2	3	3	3	4	5
IQ of youngest child	110	100	100	80	80	70

The researcher calculated the value of Pearson's r for these data to be -0.875, and concluded that the cause of lower IQ in larger families is a direct result of the parents having to divide their attention between a larger number of children.

Verify the researcher's value for r, and suggest another correlation coefficient which might be more appropriate for these data. Calculate its value and test its significance at the 5% level.

Do you agree with the researcher's conclusion?

11. An experimenter has undertaken to measure the degree of association between intelligence and the ability to think laterally. The scores of eight subjects in both a standard intelligence test and in a test of lateral thinking are shown in the table below.

Subject	*1*	*2*	*3*	*4*	*5*	*6*	*7*	*8*
Intelligence test score	120	158	110	135	140	125	130	121
Lateral thinking score	6	5	7	2	1	8	6	4

Calculate a non-parametric correlation coefficient for these data, and test its significance at the 1% level, using a two-sided alternative.

►

12. Sixteen subjects were matched in pairs for their ability to perform a certain task. Then one subject in each pair was given three double whiskies and asked to perform the task again a few minutes later. The other subjects were completely sober when they performed the task again. The following scores were obtained:

Pair	*1*	*2*	*3*	*4*	*5*	*6*	*7*	*8*
Sober group	340	290	270	370	330	310	320	320
Whisky group	320	360	320	540	360	300	680	1180

Give a reason why Spearman's correlation coefficient would be more appropriate here, rather than Pearson's r. Calculate Spearman's r_s and test its significance at the 5% level.

13. Explain briefly how Spearman's rank correlation is related to product moment correlation.

Ten 16-year-old pupils in a class take a chemistry test and are ranked in order of merit as shown in the table. Also shown is their order of merit in a standard IQ test when they were 13 years old. Is there any evidence to reject the hypothesis that there is no correlation between the rankings in the two tests?

Order of merit \ *Pupil*	*A*	*B*	*C*	*D*	*E*	*F*	*G*	*H*	*I*	*J*
Chemistry test	5	2	8	3	1	7	10	4	9	6
IQ test	3	7	6	5	2	8	9	4	10	1

(The appropriate two-tailed 5% and 10% points of the distribution of Spearman's rank correlation coefficient are 0.648 and 0.564 respectively.)

(O&C)

14. Each of the sequences $x_1, x_2, \ldots, x_n$ and $y_1, y_2, \ldots, y_n$ is a permutation of $1, 2, \ldots, n$. Show that.

$$\sum_{i=1}^{n} x_i y_i = \frac{n(n+1)(2n+1)}{6} - \frac{1}{2}\sum_{i=1}^{n} (x_i - y_i)^2$$

Hence show that

$$\frac{\sum_{i=1}^{n} x_i y_i - (1/n)\sum_{i=1}^{n} x_i \sum_{i=1}^{n} y_i}{\sqrt{\left\{\left[\sum_{i=1}^{n} x_i^2 - (1/n)\left(\sum_{i=1}^{n} x_i\right)^2\right]\left[\sum_{i=1}^{n} y_i^2 - (1/n)\left(\sum_{i=1}^{n} y_i\right)^2\right]\right\}}}$$

$$= 1 - \frac{6}{n(n^2-1)}\sum_{i=1}^{n} (x_i - y_i)^2$$

►

At a sports meeting nine competitors (A, B, . . ., I) take part in the pentathlon. The table below shows the results of the first two events in the pentathlon.

Order of merit	*1*	*2*	*3*	*4*	*5*	*6*	*7*	*8*	*9*
Event 1	B	I	G	H	D	C	A	F	E
Event 2	I	C	E	F	D	H	B	A	G

Find and comment on the value of Spearman's coefficient of rank correlation between performance in event 1 and performance in event 2.

(O & C)

15. What criteria determine the choice of correlation coefficient (product-moment or rank) to be used to investigate the relation between two variables? What can be said about the relation if:

(i) the product-moment correlation coefficient has the value 1;
(ii) a rank correlation coefficient has the value 1?

The following are the grades (A to F) received by twelve candidates in Mathematics and French examinations, together with an overall grade based on all examinations taken.

Candidate	*1*	*2*	*3*	*4*	*5*	*6*	*7*	*8*	*9*	*10*	*11*	*12*
Overall	A	B	B	C	C	C	D	D	D	E	F	F
Mathematics	B	C	A	D	C	B	C	E	D	D	E	F
French	C	A	C	B	D	E	C	C	E	D	F	E

By calculating suitable correlation coefficients, decide which of Mathematics or French is the better predictor of the overall result.

(SUJB)

Section 15.9

16. Compare briefly the advantages and disadvantages of a rank correlation coefficient and a linear (product-moment) correlation coefficient for bivariate distributions.

In an investigation into baldness, ten adult males were ranked according to head size, and their ages (in years) and amount of hair (in arbitrary units) were recorded. The results are shown in the table.

Head size (rank)	*1*	*2*	*3*	*4*	*5*	*6*	*7*	*8*	*9*	*10*
Age	36	42	70	68	23	39	55	49	61	32
Amount of hair	38	27	15	19	34	37	7	12	9	28

▸

Rank the data from smallest to largest and calculate rank correlation coefficients between head size and amount of hair, and between age and amount of hair.

Determine whether your results are significant at the 10% level and comment upon your findings.

(Cambridge)

Section 15.10

17. (a) If X and Y are random variables and a, b, c, d are positive constants, show that $\text{Cov}(a+bX, c+dY) = bd \text{ Cov}(X, Y)$. Hence show that $\text{Corr}(a+bX, c+dY) = \text{Corr}(X, Y)$.
    (b) If X is measured in centimetres and Y is measured in kilograms, what values of a, b, c and d would be required to give equivalent values of $(a+bX)$ in metres and $(c+dY)$ in grams. What is the significance of the result proved in (a)?

18. (a) Let X_1 and X_2 be two random variables with the same mean and variance, and with correlation coefficient ρ. Show that $X_2 - X_1$ and X_1 have correlation $-(1-\rho)^{1/2}/\sqrt{2}$ and that $X_2 - X_1$ and $(X_1 + X_2)/2$ are uncorrelated.
    (b) In a clinical trial 10 patients had their systolic blood pressure measured before and after administration of a certain drug. The results were as follows (mm Hg):

Patient number	*1*	*2*	*3*	*4*	*5*	*6*	*7*	*8*	*9*	*10*
Blood pressure before (x_1)	142	143	136	117	140	132	120	123	128	103
Blood pressure after (x_2)	124	147	101	132	128	104	122	130	102	122

Calculate the sample correlation coefficient between $X_2 - X_1$ and X_1 and also between $(X_2 - X_1)$ and $(X_2 + X_1)/2$. Test these for significance using a t test, to determine if there is a relationship between change in blood pressure and level of blood pressure.

[*Note*: $\sum x_1 = 1284$, $\sum x_2 = 1212$, $\sum x_1^2 = 166{,}344$, $\sum x_2^2 = 148{,}882$, $\sum x_1 x_2 = 155{,}709$]

(IOS)

19. The random variables X_1, X_2 and X_3 are independent and each has mean μ and variance σ^2. Given that:

$$W = X_1 + X_2 - 2X_3,\ Y = X_1 + X_2, \text{ and } Z = X_1 + X_2 + X_3:$$

(i) find the mean and the variance of W;
(ii) show that the product-moment correlation coefficient of Y and Z is equal to $\sqrt{(2/3)}$

(JMB)

16

Regression

16.1 INTRODUCTION

In experimental work we may wish to find out how the values of one variable depend on another variable. For example, we may be interested in how the number of grams of a given salt which dissolve in 100 g of water depends on the temperature of the water. Generally we will let Y denote the dependent variable and X denote the independent variable, so that in the example 'the number of grams of salt' would be denoted by Y and the 'temperature of the water' by X. Moreover, we will assume in general that Y is a random variable but that X is a controlled variable which may be set at predetermined values.

Suppose we carry out a number of experiments at different values of X, observe the resulting values of Y, and plot the data on a scatter diagram, as in Fig. 16.1.

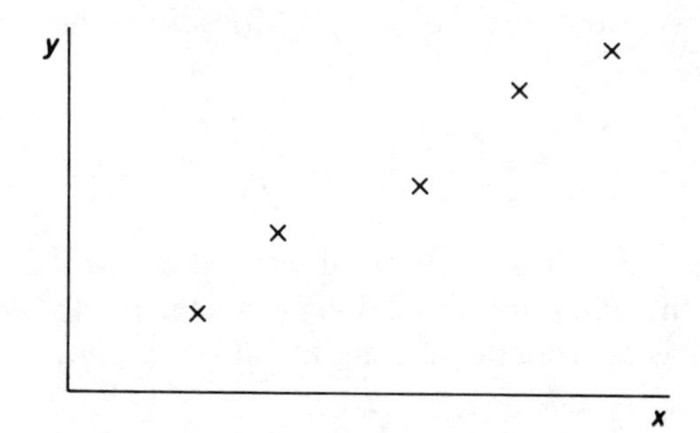

Fig. 16.1 The results of five experiments at different values of X

Then if the relationship between x and y appears to be reasonably linear we can postulate a linear equation for the line which 'best fits the sample data' of the form:

$$y = a + bx,$$

where a is the intercept and b is the slope of the line.

In Sections 16.2 and 16.9 we will see how to calculate a and b and make inferences about α and β, the intercept and slope of the underlying population line of 'best fit', and also inferences about predicted values of Y for given values of X.

Also in this chapter, in Section 16.10, we will deal with another problem involving two variables, X and Y, in which both variables are random. Here we may be interested in predicting Y given X, or X given Y.

16.2 METHOD OF LEAST SQUARES

The points on the scatter diagram Fig. 16.1 represent a sample of pairs of values of a controlled variable X and a random variable Y. If we consider all the experiments we might carry out, then the equation of a straight line which might best fit the population of points may be written as:

$$y = \alpha + \beta x$$

The practical problem is as follows: 'Given that we have only a sample of n values, say, of X and Y, how should we estimate α and β?'

The answer is that we use the criterion of *least squares*, namely we 'minimize the sum of squares of vertical† distances from the sample points to the so-called *regression line* $y = \alpha + \beta x$'.

This criterion can be justified theoretically if we can assume that the sample of distances are from a normally distributed population.

Suppose the n sample values of X and Y are (x_i, y_i), $i = 1, 2, \ldots, n$. Then the criterion of least squares implies that we wish to minimize $\sum_{i=1}^{n} e_i^2$ (see Fig. 16.2).

In Fig. 16.2 the general point (x_i, y_i) is shown. Notice that the point at which the vertical line through (x_i, y_i) meets the line $y = \alpha + \beta x$ also has an x coordinate of x_i and hence a y coordinate of $\alpha + \beta x_i$ (because this point is on the line $y = \alpha + \beta x$). It follows that

$$e_i = y_i - (\alpha + \beta x_i) \qquad i = 1, 2, \ldots, n$$

and hence that

$$y_i = \alpha + \beta x_i + e_i \qquad i = 1, 2, \ldots, n \qquad (16.1)$$

We now see that y_i consists of a fixed element $\alpha + \beta x_i$ (since x_i is fixed or controlled and both α and β are population parameters) and a random element e_i.

Equation (16.1) is an example of what is called a *linear statistical model.*

† 'vertical' here means 'parallel to the y axis'.

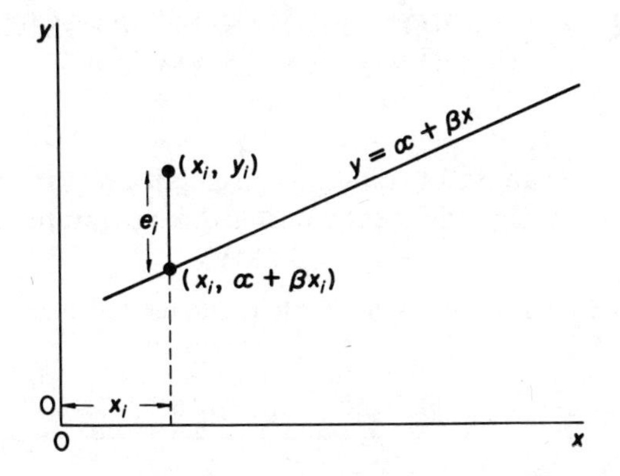

Fig. 16.2 The population regression line, $y = \alpha + \beta x$

Returning to the problem of estimating α and β we now see that minimizing $\sum_{i=1}^{n} e_i^2$ is the same as minimizing $\sum_{i=1}^{n} (y_i - \alpha - \beta x_i)^2$.

If we differentiate this summation partially with respect to the unknowns α and β in turn, set the derivatives equal to 0, and replace α by its estimator a and β by its estimator b, we obtain:

$$\sum_{i=1}^{n} -2(y_i - a - bx_i) = 0 \quad \text{and} \quad \sum_{i=1}^{n} -2x_i(y_i - a - bx_i) = 0$$

or $$\sum y_i - na - b\sum x_i = 0 \quad \text{and} \quad \sum x_i y_i - a\sum x_i - b\sum x_i^2 = 0$$

Eliminating a gives

$$b = \frac{\sum x_i y_i - \dfrac{\sum x_i \sum y_i}{n}}{\sum x_i^2 - \dfrac{(\sum x_i)^2}{n}} \tag{16.2}$$

Also we can divide $\Sigma y_i - na - b\Sigma x_i = 0$ by n to obtain a in terms of b:

$$a = \bar{y} - b\bar{x} \tag{16.3}$$

(An alternative formula for b (which is not recommended for calculation purposes) is

$$b = \frac{\sum (x_i - \bar{x})(y_i - \bar{y})}{\sum (x_i - \bar{x})^2}$$

That this formula is equivalent to (16.2) is easily established by reference to the Appendix to Chapter 15.)

The equation $y = a + bx$ is called the (sample) regression equation of Y on X and the line which has this equation is called the (sample) regression line of Y on X. From sample data we use (16.2) and (16.3) to calculate a and b. Then the regression equation may be used to predict values of Y for given values of X.

16.3 THE EQUATION OF THE REGRESSION LINE OF Y ON X: AN EXAMPLE

Example 16.1

In an experiment the number of grams of a given salt which dissolved in 100 g of water was observed at eight different controlled temperatures, as in Table 16.1.

Table 16.1 The weight of salt at different temperatures

Temp (°C)	0	10	20	30	40	50	60	70
Weight of salt (g)	51.5	61.5	67.2	72.6	73.5	82.2	83.5	88.0

Plot the data on a scatter diagram and comment. Estimate the coefficients of the regression equation which could be used to predict the weight of salt given the temperature. Plot the regression line on the scatter diagram. Predict the weight of salt which would dissolve at temperatures: (i) 25°C, (ii) $\bar{x} = 35$°C, (iii) 85°C.

The scatter diagram is shown in Fig. 16.3. Notice that temperature is denoted by X since it was the controlled variable, while weight of salt is denoted by Y. We comment that the relationship between X and Y looks reasonably linear.

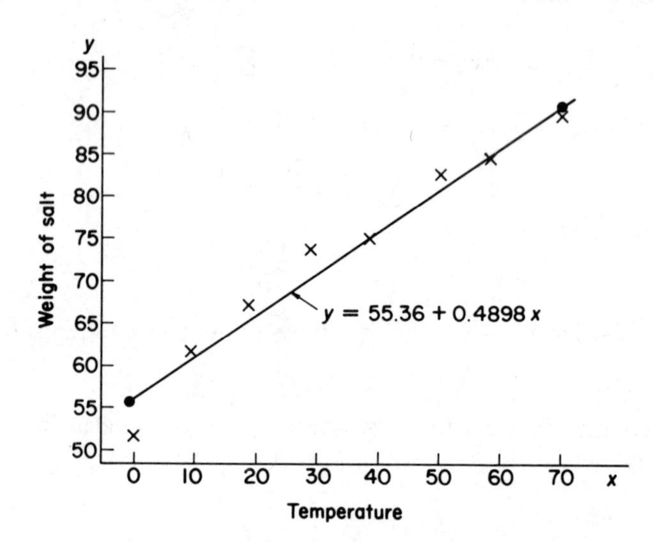

Fig. 16.3 The regression line of weight of salt on temperature

From Table 16.1 we calculate that:

$\sum x_i = 280 \quad \sum x_i^2 = 14{,}000 \quad \sum y_i = 580 \quad \sum y_i^2 = 43{,}096.44$

$\sum x_i y_i = 22{,}357 \quad n = 8$

Using (16.2), $$b = \frac{22{,}357 - \dfrac{280 \times 580}{8}}{14{,}000 - \dfrac{280^2}{8}} = \frac{2057}{4200} = 0.4898$$

Using (16.3), $$a = \frac{580}{8} - 0.4898 \times \frac{280}{8} = 72.5 - 17.14 = 55.36$$

NB Avoid premature rounding in the calculation of b.
The equation of the regression line of weight (Y) on temperature (X) is:

$$y = 55.36 + 0.4898\,x$$

We can plot the line on the scatter diagram by finding the values of Y for the minimum and maximum values of X. So, when $x = 0$, $y = 55.36 + 0.4898 \times 0 = 55.36$ and when $x = 70$, $y = 55.36 + 0.4898 \times 70 = 89.65$ (see Fig. 16.3).

The reason for drawing the line only between the minimum and maximum values of X is that the line may not be valid outside this range. (Using the line and the equation outside the valid range is called *extrapolation*). To predict the weight of salt for temperatures of 25°C, 35°C and 85°C we can use Fig. 16.3 or, more accurately the regression equation.

(i) For $x = 25$, $\hat{y} = 55.36 + 0.4898 \times 25 = 67.61$
(ii) For $x = \bar{x} = 35$, $\hat{y} = 55.36 + 0.4898 \times 35 = 72.5$
Notice that $\bar{y}$ also equals 72.5; see note (c) below.
(iii) For $x = 85$, $\hat{y} = 55.36 + 0.4898 \times 85 = 96.99$;
but see note (b) below.

Notes

(a) We use $\hat{y}$ to indicate a predicted value of Y. A predicted value is the mean value of Y for a given value of X. For example, the mean number of grams of salt expected to dissolve at 25°C is 67.61.
(b) We have no data beyond $x = 70$°C, so the prediction at $x = 85$°C may well be invalid.
(c) We notice that the line goes through the point $(\bar{x}, \bar{y})$. This will always be the case since we can rewrite (16.3) as $\bar{y} = a + b\bar{x}$. The regression equation is sometimes written as $(y - \bar{y}) = b(x - \bar{x})$, eliminating a altogether. □

16.4 A LINEAR STATISTICAL MODEL FOR REGRESSION

We return now to a discussion of the linear statistical model (16.1), namely:

$$y_i = \alpha + \beta x_i + e_i \qquad i = 1, 2, \ldots, n$$

If we are able to make certain assumptions about this model, then certain inferences may be made about α, β and predicted values of Y. These assumptions refer to the distribution of the random element e_i. We can think of the n sample values of e_i as coming from a larger population with its own distribution. A distribution which is often assumed for e_i is $N(0, \sigma^2)$ which is illustrated in Fig. 16.4 and implies that:

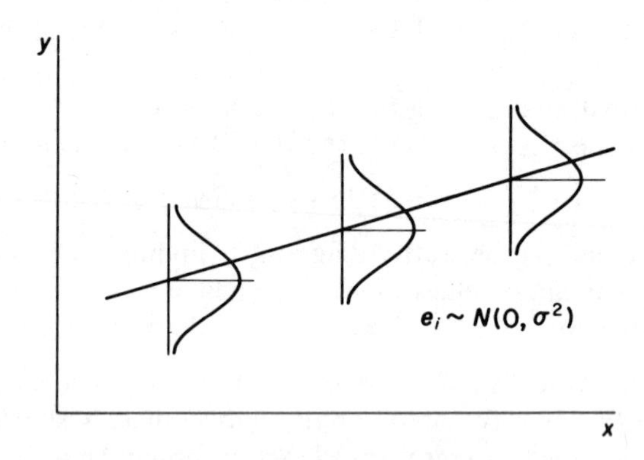

Fig. 16.4 The distribution of e_i

(i) The distances of the points on the scatter diagram are normally distributed about the regression line with zero mean, implying roughly as many points above the line as below the line.
(ii) The distribution is the same for all values of X, implying that variability (variance) about the line does not change with X.

16.5 INFERENCES ABOUT THE SLOPE, β, OF THE REGRESSION LINE, σ^2 KNOWN

We saw in Section 16.2 how to calculate the slope of the sample regression line, b, from sample data. Here b is a point (single-value) estimator of β, the slope of the population regression line. We may also find a confidence interval for β and test hypotheses about β, if we are able to assume that $e_i \sim N(0, \sigma^2)$.

In this section we will discuss the case in which σ^2 is known, while in the next section we deal with the case in which σ^2 is not known and must be estimated from the sample data.

We quote without proof that, if $e_i \sim N(0, \sigma^2)$ and σ^2 is known, then

$$b \sim N\left(\beta, \frac{\sigma^2}{\sum (x_i - \bar{x})^2}\right) \tag{16.4}$$

It follows (using arguments similar to those in Section 10.2) that a 95% confidence interval for β is provided by:

$$b \pm \frac{1.96\sigma}{\sqrt{\sum (x_i - \bar{x})^2}} \tag{16.5}$$

Example 16.2

For the data in Table 16.1, calculate a 95% confidence interval for β, assuming σ^2 is known to be 5.

We know from Section 16.3 that $b = 0.4898$ and that

$$\sum (x_i - \bar{x})^2 = \sum x_i^2 - \frac{(\sum x_i)^2}{n} \qquad \text{(Appendix to Chapter 15)}$$

$$= 4200 \text{ for the data in Table 16.1}$$

A 95% confidence interval for β is $0.4898 \pm \dfrac{1.96 \times \sqrt{5}}{\sqrt{4200}}$

i.e. 0.42 to 0.56 (2 d.p.) □

It also follows from (16.4) (using arguments similar to those in Section 11.3) that we can use the test statistic:

$$z = \frac{b - \beta_0}{\sigma/\sqrt{\sum (x_i - \bar{x})^2}} \tag{16.6}$$

to test the null hypothesis, H_0: $\beta = \beta_0$

Example 16.3

For the data in Table 16.1, test the hypothesis H_0: $\beta = 0.5$ against a two-sided alternative at the 5% level of significance. Assume that σ^2 is known to be 5.

H_0: $\beta = 0.5$
H_1: $\beta \neq 0.5$
5% level of significance

$$z = \frac{b - \beta_0}{\sigma/\sqrt{\sum (x_i - \bar{x})^2}} = \frac{0.4898 - 0.5}{\sqrt{5}/\sqrt{4200}} = -0.30$$

Since $|-0.30| < 1.96$, we do not reject H_0.
The slope is not significantly different from 0.5.

Notes

(i) One-sided alternative hypotheses may also be tested; see Table 11.1.
(ii) We could have reached the conclusion not to reject H_0 simply by noting that 0.5 is inside the 95% confidence interval for β; see Example 16.2. □

16.6 INFERENCES ABOUT β, σ^2 UNKNOWN

If σ^2 is unknown, we must estimate it from the sample data, and use the t statistic (instead of the z statistic) to obtain confidence intervals for β and test hypotheses about β.

Out estimator of σ^2 is s^2, where $s^2 = \dfrac{\sum (y_i - \bar{y})^2 - b^2 \sum (x_i - \bar{x})^2}{n-2}$, or the following which is more useful for calculation purposes:

$$s^2 = \frac{\left[\sum y_i^2 - \dfrac{(\sum y_i)^2}{n}\right] - b^2 \left[\sum x_i^2 - \dfrac{(\sum x_i)^2}{n}\right]}{n-2} \tag{16.7}$$

The numerator of this equation is in fact the sum of squares of distances of the points to the regression line of Y on X in the direction of the Y axis. It is also referred to as the *residual sum of squares* and s^2 is called the *residual variance.* Think of 'residuals' here as meaning the vertical distances (parallel to the Y axis) which are 'left over' after we have fitted the regression line through the points on the scatter diagram.

We now quote without proof that, if $e_i \sim N(0, \sigma^2)$ and σ^2 is unknown, then

$$\left.\begin{array}{l} \dfrac{b - \beta}{s/\sqrt{\sum (x_i - \bar{x})^2}} \quad \text{has a } t \text{ distribution} \\ \text{with } (n-2) \text{ degrees of freedom} \end{array}\right\} \tag{16.8}$$

It follows (using arguments to those in Section 10.4) that a 95% confidence interval for β is provided by:

$$b \pm \frac{t_{0.025,\, n-2}\, s}{\sqrt{\sum (x_i - \bar{x})^2}} \tag{16.9}$$

Example 16.4

For the data in Table 16.1, calculate a 95% confidence interval for β.

$$s^2 = \frac{\left[43{,}096.44 - \dfrac{580^2}{8}\right] - 0.4898^2 \left[14{,}000 - \dfrac{280^2}{8}\right]}{8-2}, \quad \text{by (16.7)}$$

$$= \frac{1046.44 - 0.4898^2 \times 4200}{6}$$

$$= 6.47$$

A 95% confidence interval for β is

$$0.4898 \pm \frac{2.45 \times \sqrt{6.47}}{\sqrt{4200}}, \quad \text{by (16.9) and since } t_{0.025,\,6} = 2.45$$

0.39 to 0.59 □

It also follows from (16.8) (using arguments similar to those in Section 11.4) that we can use the test statistic:

$$t = \frac{b - \beta_0}{s/\sqrt{\sum (x_i - \bar{x})^2}} \tag{16.10}$$

to test the null hypothesis $H_0\colon \beta = \beta_0$

Example 16.5

For the data in Table 16.1, test the hypothesis $H_0\colon \beta = 0.5$ against a two-sided alternative at the 5% level of significance.

$H_0\colon \beta = 0.5$
$H_1\colon \beta \neq 0.5$
5% level of significance

$$t = \frac{b - \beta_0}{s/\sqrt{\sum (x_i - \bar{x})^2}} = \frac{0.4898 - 0.5}{\sqrt{6.47}/\sqrt{4200}} = -0.26$$

The critical value of t is $t_{0.025,\,6} = 2.45$
Since $|-0.26| < 2.45$, we do not reject H_0.
The slope is not significantly different from 0.5. □

16.7 INFERENCES ABOUT α

Inferences about the intercept, α, can be made in a similar way to those about β. For example, if σ^2 is known,

$$a \sim N\left(\alpha, \frac{\sigma^2 \sum x_i^2}{n \sum (x_i - \bar{x})^2}\right) \tag{16.11}$$

and if σ^2 is unknown,

$$\left.\begin{array}{l} \dfrac{a - \alpha}{s\sqrt{\dfrac{\sum x_i^2}{n \sum (x_i - \bar{x})^2}}} \quad \text{has a } t \text{ distribution} \\ \text{with } (n-2) \text{ degrees of freedom} \end{array}\right\} \tag{16.12}$$

16.8 INFERENCES ABOUT PREDICTED MEAN VALUES OF Y

The main purpose of regression analysis is to make predictions about Y for given values of X. We have already seen how to make single-value predictions in Section 16.3, so that our prediction of Y when $x = x_0$, say, is $\hat{y} = a + bx_0$. If, now, we assume the linear statistical model of Section 16.4, namely that $e_i \sim N(0, \sigma^2)$, we can calculate confidence intervals for predicted mean values. For example, a 95% confidence interval for the predicted mean value of Y when $x = x_0$ is given by:

(i) $$(a+bx_0) \pm 1.96\sigma \sqrt{\frac{1}{n}+\frac{(x_0-\bar{x})^2}{\sum (x_i-\bar{x})^2}}, \quad \text{when } \sigma^2 \text{ is known} \qquad (16.13)$$

(ii) $$(a+bx_0) \pm t_{0.025,\, n-2}\, s \sqrt{\frac{1}{n}+\frac{(x_0-\bar{x})^2}{\sum (x_i-\bar{x})^2}}, \quad \text{when } \sigma^2 \text{ is unknown} \qquad (16.14)$$

Example 16.6

(*when σ^2 is unknown*). For the data in Table 16.1, calculate 95% confidence intervals for predicted values of the weight of salt for temperatures of 0, 35 and 70°C (the minimum, mean and maximum temperatures).

Using the following results already established in previous examples in this chapter, namely:

$a = 55.36$, $b = 0.4898$, $s^2 = 6.47$, $\bar{x} = 35$ and $\sum (x_i - \bar{x})^2 = 4200$, 95% confidence intervals for predicted mean values of Y are as follows:

$$\text{At } x_0 = 0, \ (55.36 + 0.4898 \times 0) \pm 2.45\sqrt{6.47}\sqrt{\frac{1}{8}+\frac{(0-35)^2}{4200}}$$

$$55.36 \pm 4.02$$

$$51.3 \text{ to } 59.4$$

At $x_0 = 35$, 70.3 to 74.7

At $x_0 = 70$, 85.6 to 93.7

The results of Example 16.6 are shown in Fig. 16.5, which also shows the approximate locus of all the confidence intervals for predicted values of weight of

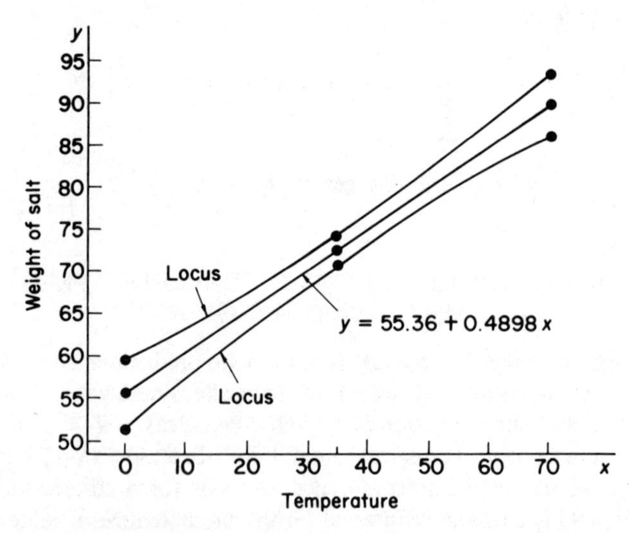

Fig. 16.5 95% confidence intervals for predicted values of weight of salt

salt between $x = 0$ and $x = 70$. It shows that the widest confidence intervals occur at the extremes of the data, as one would expect intuitively. It also shows that the narrowest interval occurs at $x = \bar{x} = 35\,^{\circ}\mathrm{C}$, as expected by inspection of the term under the square root sign in (16.14). □

16.9 INFERENCES ABOUT THE DIFFERENCE BETWEEN TWO PREDICTED MEAN VALUES OF Y

We can extend the ideas of the previous section to obtain a confidence interval for the difference between the predicted mean values of Y for two values of X, x_1 and x_2, say. For example, a 95% confidence interval for the difference between the predicted mean values of Y when $x = x_1$ and $x = x_2$ is given by:

$$b(x_1 - x_2) \pm 1.96\,\sigma \sqrt{\frac{(x_1 - x_2)^2}{\sum (x_i - \bar{x})^2}}, \qquad \text{if } \sigma^2 \text{ is known} \qquad (16.15).$$

Example 16.7

Calculate a 95% confidence interval for the difference between the predicted mean weights of salt for temperatures of 0 and 70 °C, using the data in Table 16.1, and assuming σ^2 is known to be 5.

The required confidence interval is:

$$0.4898\,(70 - 0) \pm 1.96\sqrt{5}\sqrt{\frac{(70 - 0)^2}{4200}}$$

$$34.29 \pm 4.73$$

$$29.6 \text{ to } 39.0$$

We are 95% confident that the predicted mean difference in the weight of salt dissolved at 0 and 70 °C lies between 29.6 and 39.0 g. □

16.10 REGRESSION WHEN BOTH VARIABLES ARE RANDOM

There are many situations in which we are interested in the relationship between two variables when both are random, so that we cannot control one variable as in the previous sections of this chapter. One example of such an 'observational study' (as opposed to a 'controlled experiment') is the possible relationship between the height and weight of adult males. In this example height and weight are both random variables. Suppose we arbitrarily denote height by X and weight by Y. If we can assume that both X and Y are normally distributed, and a scatter diagram based on a random sample of pairs of values of X and Y indicates a linear relationship between the variables, then:

1. If we are interested in predicting weight (Y) from height (X), we may use equations (16.2) and (16.3) to calculate the slope and intercept of the regression line of Y on X.
2. If we are interested in predicting height (X) from weight (Y), we may use modified versions of (16.2) and (16.3) by interchanging x and y in these equations to obtain the slope, b', and the intercept, a', of the regression line of X on Y, whose equation is:

$$x = a' + b' y$$

Thus the equations for a' and b' are:

$$b' = \frac{\sum x_i y_i - \frac{\sum x_i \sum y_i}{n}}{\sum y_i^2 - \frac{(\sum y_i)^2}{n}} \quad \text{and} \quad a' = \bar{x} - b' \bar{y} \tag{16.16}$$

Example 16.8

For the data in Table 15.1 calculate the coefficients of the regression lines of:

(i) Weight (Y) on height (X), and
(ii) Height (X) on weight (Y)

Show that both lines go through $(\bar{x}, \bar{y})$, but have no other points in common. We already know from Section 15.3 that:

$$\sum x_i = 1392 \qquad \sum x_i^2 = 242{,}548 \qquad \sum y_i = 584$$
$$\sum y_i^2 = 43{,}018 \qquad \sum x_i y_i = 101{,}916 \qquad n = 8$$

(i) Using (16.2) and (16.3),

$$b = \frac{101{,}916 - \frac{1392 \times 584}{8}}{242{,}548 - \frac{1392^2}{8}} = \frac{300}{340} = 0.8824$$

$$a = \frac{584}{8} - 0.8824 \times \frac{1392}{8} = -80.53$$

So the regression equation of weight (Y) on height (X) is:

$$y = -80.53 + 0.8824x$$

(ii) Using (16.16),

$$b' = \frac{101{,}916 - \dfrac{1392 \times 584}{8}}{43{,}018 - \dfrac{584^2}{8}} = \frac{300}{386} = 0.7772$$

$$a' = \frac{1392}{8} - 0.7772 \times \frac{584}{8} = 117.3$$

So the regression equation of height (X) on weight (Y) is:

$$x = 117.3 + 0.7772\,y$$

The regression line of X on Y also passes through $(\bar{x}, \bar{y})$ since $\bar{x} = a' + b'\bar{y}$ from (16.16). (We already know from Section 16.3 that the regression line of Y on X passes through $(\bar{x}, \bar{y})$). However, the slopes of the two regression lines, as calculated by dy/dx, are 0.8824 and 1/0.7772, which are clearly different. Hence the two lines have only one point in common.

Note that it is only when both X and Y are random variables that we can consider both regression lines. If only one variable is random (while the other is controlled) this must be the Y variable and only the regression of Y on X must be considered (as in Sections 16.2 to 16.9).

It is easy to show that, if the regression equations of Y on X and X on Y are $y = a + bx$ and $x = a' + b'y$, then

$$bb' = r^2, \quad \text{the square of Pearson's correlation coefficient.}$$

Proof From (16.2) and (16.16),

$$bb' = \left(\frac{\sum x_i y_i - \dfrac{\sum x_i \sum y_i}{n}}{\sum x_i^2 - \dfrac{(\sum x_i)^2}{n}}\right)\left(\frac{\sum x_i y_i - \dfrac{\sum x_i \sum y_i}{n}}{\sum y_i^2 - \dfrac{(\sum y_i)^2}{n}}\right)$$

$$= r^2, \quad \text{by (15.3)}$$

□

16.11 TRANSFORMATIONS TO PRODUCE LINEARITY

If the points plotted on the scatter diagram do not appear to show a linear trend, but instead show a non-linear trend, then it may be possible to 'transform' either or both variables to give a linear regression equation in the transformed variable(s).

Some examples are:

$y = a + b\log_e x$, which is linear in y and the transformed variable $u = \log_e x$;

$y = a + \dfrac{b}{x}$, which is linear in y and the transformed variable $v = \dfrac{1}{x}$;

$\log_e y = a + bx$, which is linear in x and the transformed variable $w = \log_e y$; $\log_e y = a + b \log_e x$, which is linear in the transformed variables $u = \log_e x$ and $w = \log_e y$.

We can test these possible transformations and others either by plotting the transformed variables on ordinary graph paper or, in some cases, by plotting the original variables on special graph paper (log-linear, log-log, and so on).

Example 16.9

Choose an appropriate linear regression equation for the five pairs of values of X and Y in Table 16.2.

Table 16.2 Five values of x and y, and some transformed variable

x	1	2	3	4	5
y	10	22	50	110	250
$\log_e x$	0.00	0.69	1.10	1.39	1.61
$\log_e y$	2.30	3.09	3.91	4.70	5.52
$\frac{1}{x}$	1.00	0.5	0.33	0.25	0.20

Various diagrams are shown in Fig. 16.6, and the corresponding values of Pearson's r (Section 15.3).

A comparison of the five plots and their corresponding values of r shows that the most suitable linear regression equation is:

$$\log_e y = a + bx$$

If we let $w = \log_e y$, this becomes $w = a + bx$, and a and b may then be calculated using:

$$b = \frac{\sum x_i w_i - \dfrac{\sum x_i \sum w_i}{n}}{\sum x_i^2 - \dfrac{(\sum x_i)^2}{n}}, \qquad a = \bar{w} - b\bar{x}$$

(The reader should check that $b = 0.805$, $a = 1.489$ for these data.) □

16.12 SUMMARY

In an experiment when one variable X is controlled at various values and another random variable Y is measured, and a plot of the resulting data on a scatter diagram exhibits a linear trend, we may postulate a linear statistical model:

$$y_i = \alpha + \beta x_i + e_i$$

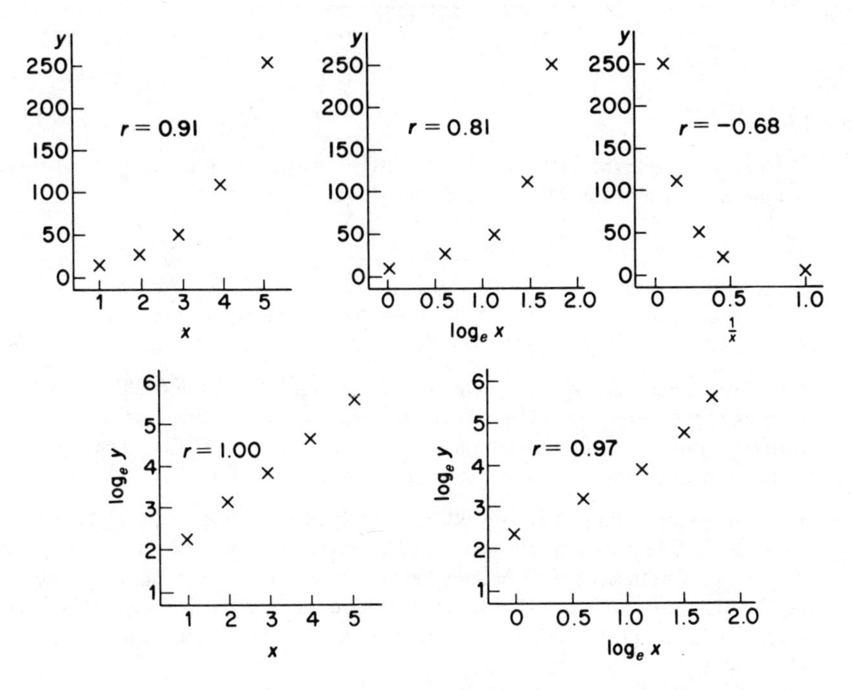

Fig. 16.6 Examples of plots of transformed variables

and estimate α and β using the method of least squares. If a and b are our estimators of α and β then $y = a + bx$ is called the (sample) regression equation for the regression of Y on X.

If certain assumptions hold regarding the distribution of the random element e_i we may make further inferences about α, β and predicted values of Y.

If X and Y are both random normally distributed variables we may use the same methods to obtain the regression equation of Y on X and modified methods for the regression of X on Y.

If the trend on the scatter diagram is non-linear it may be possible to transform either or both variables to produce a linear regression equation in the transformed variables.

Section 16.3

1. The following table shows the masses of a certain chemical substance which dissolved in a given mass of water at various temperatures.

Temperature (°C)	10	20	30	40	50	60	70	80	90
Mass (g)	4.5	4.6	5.0	5.6	5.9	6.3	6.4	6.7	7.4

Plot a scatter diagram and calculate, to 3 significant figures, the coefficients of the equation of the regression line of 'mass on temperature'. Plot the regression line on the scatter diagram and verify that it passes through the point given by the mean values of temperature and mass. Estimate mass for a temperature of 56 °C.

2. In an experiment to find the force required to pull a plough, the plough was towed by a tractor at a sequence of constant speeds (X kilometres per hour) and the force required (Y newtons) was measured. Originally the tractor was run at ten different speeds but the data on the force required at the highest speed of 11.2 km h^{-1} was mislaid. The results for the other nine speeds are shown in the table (and summarized below the table).

Speed (X km h^{-1})	1.5	2.0	3.2	4.3	5.5	6.6	8.4	8.8	9.7
Force (Y newtons)	96	95	109	112	122	133	138	156	154

$$\sum x = 50, \quad \sum y = 1115, \quad \sum x^2 = 350.88,$$
$$\sum y^2 = 142\,335, \qquad \sum xy = 6739.$$

Exhibit these results on a scatter diagram.

The original reading corresponding to the speed of 11.2 km h^{-1} was subsequently found, and was 162 newtons. Find, by calculation, the equation of the least squares regression line for the **ten** readings, showing all your working, and draw this line on your graph.

Estimate the force required with a tractor speed of 10 km h^{-1}.

Find, from your regression line, the value of y when $x = 0$ and interpret your result. (JMB)

Section 16.4

3. The model $y_i = a + bx_i + \varepsilon_i$ is to be fitted to a set of n points (x_i, y_i), $i = 1$ to n, where the x_i are the values of a controlled variable and the ε_i denote errors. Explain what is meant by 'the principle of least squares' for estimating the values of a and b.

In an experiment to test the effectiveness of a filtration system in removing waste solids from suspension in a carrying fluid, the flow was fixed at different rates (x_i gallons per ►

minute) and the resulting percentage (y_i) of waste solid removed was recorded. The results are shown in the table.

Rate of flow (x_i gallons per minute)	2	4	6	8	10	12	14
Percentage of waste removed (y_i)	14.3	19.7	17.8	14.0	12.3	7.2	5.5

Plot these data on a graph.

Calculate the appropriate least squares regression line, showing all your working, and plot this line on your graph.

Estimate the percentage of waste solid that would be removed at a flow of three gallons per minute.

Give **one** reason why the line cannot be a good fit for high values of flow.

(JMB)

Section 16.5

4. A study was made of the amount of converted sugar in a certain process at various temperatures. The coded data are as follows:

Temperature, x	1.0	1.1	1.2	1.3	1.4
Converted sugar, y	8.1	7.8	8.5	9.8	9.5

Find the least squares linear regression equation which could be used to predict the amount of converted sugar given the temperature. If the random element of the y values is known to be normally distributed with mean zero and variance 0.3, test the hypothesis, at the 5% level of significance, that the true slope of the regression line is 3. Also calculate a 95% confidence interval for the true slope.

Section 16.6

5. Rates of oxygen consumption were obtained for mammals running on a treadmill. Five mammals were randomly selected and assigned to one of five running speeds. Their rates of oxygen consumption were as follows:

Running speed, x (m/min)	2	4	6	8	10
Oxygen consumption, y (ml/g/h)	3.5	6.1	6.7	8.9	13.0

Assuming a linear regression model $y_i = \alpha + \beta x_i + e_i$, estimate α and β. Calculate a 99% confidence interval for β assuming $e_i \sim N(0, \sigma^2)$. Would you reject the hypothesis that $\beta = 0$ at the 1% level of significance? ►

6. The weight per cent, W, of nitrous oxide in a mixture of two chemicals at temperature, T, is given by the relation $W = \alpha + \beta/T$, where α and β are constants. In an experiment to estimate α and β the temperature was carefully controlled at four different values and three determinations of W were made at each temperature. The following calculations were made from the 12 observed values:

$$\sum x = 0.0432 \qquad \sum x^2 = 0.00016128$$
$$\sum w = 13.2 \qquad \sum xw = 0.04464$$

where $$X = \frac{1}{T}$$

(i) Determine the least squares estimate of the regression equation of W on X.
(ii) Estimate the value of W when the value of the T is 250.
(iii) Given that the determinations of W are subject to a normally distributed random error having mean zero and standard deviation 0.08, calculate 99% confidence limits for β.

(WJEC)

7. In an experiment to find the Young modulus for a brass wire the following 11 pairs of values of x (suspended mass in kg) and y (length of wire in mm − 7000 mm) are obtained. The equation connecting x and y is assumed to take the form $y = \alpha + \beta x$. Obtain the least squares estimates for α and β, showing your working clearly. Determine also a 95% symmetric confidence interval for β.

x	1	1.5	2	2.5	3	3.5	3	2.5	2	1.5	1
y	−1.1	−0.6	0	0.4	0.9	1.5	1.0	0.6	0.1	−0.5	−0.9

$$(\sum x^2 = 57.25, \sum y^2 = 7.22, \sum xy = 10.00)$$

(Cambridge)

Section 16.7

8. Calculate a 95% confidence interval for α using the data in Exercise 5. Test the hypothesis, at the 5% level of significance, that α is equal to zero.

Section 16.8

9. The mass y grams of a certain chemical substance which is dissolved in a certain mass of water at a temperature of x °C is given by the relation

$$y = \alpha + \beta x$$

where α and β are unknown constants. In an experiment to estimate α and β, the temperature was carefully controlled at 10 different values between 0 °C and 100 °C, and y was measured at each temperature value. The following calculations were made from the 10 observed pairs of values (x, y):

$$\sum x = 400, \quad \sum y = 688, \quad \sum x^2 = 19{,}200, \quad \sum xy = 38{,}720.$$

▶

(i) Determine the least squares estimate of the equation connecting x and y.
(ii) Estimate the mass of the chemical dissolved when the temperature is 60 °C.
(iii) Given that the measurements of y are subject to a normally distributed random error having mean zero and variance 0.4, calculate the 99 % symmetric confidence interval for the true mass of the chemical dissolved when the temperature is 60 °C.

(JMB)

Section 16.9

10. A chemist set up an experiment to determine how a variable y varied with an associated variable x. In the experiment, x was set at the five values 0, 1, 2, 3, 4, respectively, and the corresponding values of y were observed. The chemist noted that the values of y increased fairly steadily with the increasing values of x, and, on applying the method of least squares to the results, the chemist produced the equation:

$$y = 5.8 + 2.3x.$$

(i) Find the value of $\bar{y}$, the mean of the five observed values of y.

Suppose that the experimentally observed values of y are subject to independent random errors that are normally distributed with mean zero and standard deviation 1.1. Assuming that the true relationship connecting y and x is linear, calculate:

(ii) a 95 % confidence interval for the true value of y when $x = 4$;
(iii) a 90 % confidence interval for the difference between the true values of y corresponding to $x = 1$ and $x = 4$, respectively.

(WJEC)

11. An experiment was conducted to determine the mass y grams of a given amount of chemical that dissolved in glycerine at x °C. The results of the experiment are given in the following table.

Temperature (x °C)	0	10	20	30	40	50
Mass (y grams)	51.3	51.4	51.9	52.0	52.6	52.8

Assuming that the true value of y is linearly related to the value of x, obtain the least squares estimate of this relationship.

Assuming further that the temperatures used in the experiment were controlled accurately but that the measured values of y were subject to independent errors which are normally distributed with mean zero and standard deviation 0.2 g, calculate 95 % confidence intervals for:

(i) the mass of chemical that will dissolve in glycerine at 0 °C;
(ii) the additional mass of chemical that will dissolve in glycerine when the temperature is raised from 10 °C to 20 °C.

(WJEC)

12. An investigation of a possible linear relationship between two variables x and y was conducted with x having five prespecified values, the corresponding values of y being observed. The results obtained are given in the following table: ▶

x	5	10	15	20	25
y	55	52	50	48	45

You may assume that $\sum x^2 = 1375$ and that $\sum xy = 3630$.

Given that the true relationship between x and y is $y = \alpha + \beta x$, determine the least-squares estimates of α and β.

Suppose that the observed values of y are subject to independent random errors that are normally distributed with a mean of zero and a standard deviation of 0.5. Calculate 90% confidence limits for:

(i) tle true value of y when $x = 20$;

(ii) the difference between the true values of y when $x = 5$ and $x = 25$, respectively.

(WJEC)

Section 16.10

13. An aptitude test, designed to predict the potential productivity of new employees, was given to a random sample of nine employees. The table shows the test results together with a measure of the actual productivity of each employee:

Employee	*1*	*2*	*3*	*4*	*5*	*6*	*7*	*8*	*9*
Aptitude score, x	9	17	20	19	20	23	16	24	22
Productivity, y	20	25	29	33	43	32	24	33	47

Calculate the coefficients of a regression equation which could be used to predict productivity from aptitude score. Estimate productivity for aptitude scores of 10, 20 and 30. Which estimate is (a) the most reliable, (b) the least reliable? Give reasons.

14. A random sample of five families had the following annual incomes and annual savings in thousands of pounds.

Income	16	22	18	12	12
Savings	1.2	2.4	2.0	1.4	0.6

Draw a scatter diagram. Assuming both income and savings are measured with negligible measuring error, calculate the coefficients of the equations of the regression lines of (a) savings on income, (b) income on savings. Draw both lines on the scatter diagram.

Also calculate Pearson's r for these data and verify that the value of r^2 is equal to bb' (the product of the regression coefficients of the two regression lines). ►

15. The ages, x years (given to one place of decimals), and heights, y cm (to the nearest cm), of 10 boys were as follows:

x	6.6	6.8	6.9	7.5	7.8	8.2	10.1	11.4	12.8	13.5
y	119	112	116	123	122	123	135	151	141	141

Given that $\Sigma x^2 = 899.80$, $\Sigma y^2 = 166{,}091$ and $\Sigma xy = 12{,}023.3$, calculate the linear correlation coefficient between x and y and comment upon the result.

Calculate the equation of the regression line of y on x, and use it to estimate the height of a boy 9.0 years old.

State the value of y given by the regression line when $x = 30$ and comment upon your answer.

(Cambridge)

16. Two related variables, X and Y, were measured simultaneously n times and the values obtained are denoted by $(x_1, y_1), (x_2, y_2), \ldots, (x_n, y_n)$. The mean values of X and Y obtained from the data are $\bar{x}$ and $\bar{y}$, respectively. Given that the equation of the least squares regression line of Y on X is:

$$y - \bar{y} = m(x - \bar{x}),$$

where

$$m = \frac{\sum_{i=1}^{n} (x_i - \bar{x})(y_i - \bar{y})}{\sum_{i=1}^{n} (x_i - \bar{x})^2},$$

write down in a similar form the equation of the least squares regression line of X on Y. Define the product-moment correlation coefficient, r, and show that, if the two regression lines coincide, then $r = \pm 1$.

The diameters, X, and lengths, Y, of 10 tree trunks were measured. The data give, in certain units,

$$\sum_{i=1}^{10} x_i = 18.0, \quad \sum_{i=1}^{10} y_i = 366, \quad \sum_{i=1}^{10} x_i y_i = 853.2,$$

$$\sum_{i=1}^{10} x_i^2 = 41.50, \quad \sum_{i=1}^{10} y_i^2 = 17{,}568$$

Calculate the product-moment correlation coefficient and obtain the equation of the least squares regression line of Y on X. Use this regression line to estimate the length of a tree trunk whose diameter is 2.0.

State, giving a reason for your answer, whether it would here be appropriate to use the least squares regression line of Y on X to estimate the diameter of a tree trunk of given length.

(JMB)

17. The diagrams below show the two regression lines for three different bivariate distributions. The scales along the two axes $0V_1$ and $0V_2$ are the same in each diagram. ►

Explain, in each case, what the diagram tells us about the correlation between the variables V_1 and V_2. What does the point P represent?

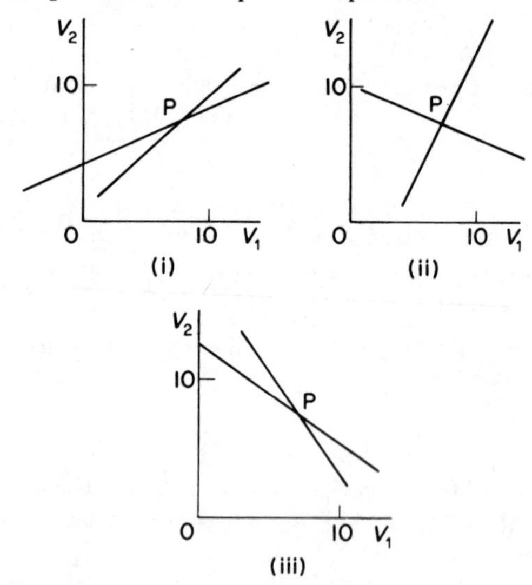

Ten students sat two physics tests, one practical and the other theoretical. Their marks out of 10 are recorded in the following table.

Practical test	8	6	10	8	5	6	8	10	7	7
Theoretical test	6	7	8	6	7	4	9	10	5	8

Draw a scatter diagram of the pairs of marks.

Show that the product moment correlation coefficient for the data is equal to $(\sqrt{15})/7$.

A student was absent from the theoretical test but obtained a mark of 6 in the practical test. Use the appropriate regression line to estimate a mark in the theoretical test for this student.

(London)

18. The table gives the heights, x inches, and the weights, y pounds, of a random sample of 10 men.

											Totals
x	63	71	72	68	75	66	68	76	71	70	700
y	145	158	156	148	163	155	153	158	150	154	1540
$x' = x - 70$	−7	1	2	−2	5	−4	−2	6	1	0	0
$y' = y - 154$	−9	4	2	−6	9	1	−1	4	−4	0	0
$(x')^2$	49	1	4	4	25	16	4	36	1	0	140
$(y')^2$	81	16	4	36	81	1	1	16	16	0	252
$x'y'$	63	4	4	12	45	−4	2	24	−4	0	146

▶

Plot the data on a scatter diagram, and comment on the sign, and degree, of the correlation between x and y.

The values of y have been measured to a high degree of accuracy, but the values of x are subject to error. Use the results of the method of least squares to calculate the appropriate line of regression for the data. Draw this line on the scatter diagram. Estimates are required of the weight of a man whose height is $68\frac{1}{2}$ inches, and of the height of a man whose weight is $152\frac{1}{2}$ pounds. State which of these estimates can justifiably be obtained, and find it.

Determine, correct to 2 decimal places, the value of the correlation coefficient.

(MEI)

19. The following table gives the marks (x) obtained by 10 students in a test in mathematics together with the marks (y) obtained in a test in science.

	A	*B*	*C*	*D*	*E*	*F*	*G*	*H*	*I*	*J*
x	3	24	16	25	5	10	12	20	15	8
y	1	25	8	20	10	5	7	16	17	14

(i) Another student was absent from the mathematics test but scored 15 in the science test. Which regression line, y on x or x on y, would be suitable for estimating her mark in mathematics?
(ii) Find the equation of the appropriate regression line. Work out a fair estimate for her mark in mathematics.
(iii) Determine the value of the product moment correlation coefficient.
(iv) Determine a coefficient of rank correlation between the two sets of marks.

(SUJB)

Section 16.11

20. Calculate the equation of the regression line of Y (height of a soya bean plant) on X (age of plant) given the following data for five plants.

x (weeks)	1	2	3	4	5
y (cm)	1	2	6	15	36

Suppose it is suggested that a more suitable regression line would be $\log_{10} Y$ on X. By drawing scatter diagrams for Y against X and for $\log_{10} Y$ against X, show that the latter is more suitable. Calculate the coefficients for this new regression equation and use it to predict, to 1 decimal place, the height of a soya bean plant after 2.5 weeks.

21. A research worker, Dr Lin Guistic, gave each of eight children a list of 100 words of varying difficulty, and asked them to define the meaning of each one. The table below gives for each child, their age in years and the number of correctly defined words. ►

Child		A	B	C	D	E	F	G	H
Age	x	2.5	3.1	4.3	5.0	5.9	7.1	8.1	9.4
Number of correct words	y	9	15	26	35	43	57	69	88

It is decided to fit a model of the form

$$\log_e y = a + b \log_e x.$$

(a) Verify that this is a reasonable course to take by plotting the above data on log–log paper.
(b) Transform the data in the table, recording your values correct to one decimal place, to a form appropriate for fitting the above model.
(c) Calculate the equation of the regression line of $\log_e y$ on $\log_e x$ as given above.
(AEB (1983))

22. The random variable Y has expectation $\alpha + \beta/x$ where α and β are unknown. The table below shows five observations y and Y for corresponding values of x:

y	1.7	2.8	3.2	3.0	3.7
x	1.0	1.5	2.0	2.5	4.0

(i) Plot these points on a graph of y against x.
(ii) Find the regression line of y on $1/x$ in the form $y = a + b/x$.
(iii) Find the ordinates of your regression curve for $x = 1$ and $x = 4$.
(iv) Sketch your regression curve on your graph.

(O&C)

23. The table shows the relationship between the height of a shrub and the number of weeks after it was planted.

Number of weeks, x	6	7	8	9	10	11
Height, y cm	105	113	124	136	142	151

Write down a coding formula to transform x to X in such a way that when $x = 6$, $X = -5$, and when $x = 11$, $X = 5$.

Use the method of least squares to fit a straight line relating y and X, taking X as the independent variable.

Use your coding formula to express y in terms of x.

(OLE)

24. The table gives the mass of a baby elephant from 6 to 14 weeks after birth. ▶

Time (weeks) t	6	8	10	12	14
Mass (kg) m	103	106	112	115	120

The data are to be coded as $T = \frac{1}{2}(t-10)$, $M = m - 100$. Plat these data on graph paper, and use the method of least squares to fit the following regression lines:

(a) regression line of M on T;
(b) regression line of m on t.

Use your regression lines to estimate the mass of the elephant after 16 weeks, and comment briefly on the reliability of this estimate.

(OLE)

25. When the model $y_i = a + bx_i + \varepsilon_i$ is to be fitted by the principle of least squares to a set of n points (x_i, y_i), $i = 1$ to n, where the x_i are the values of a controlled variable, state what assumptions are made about the errors ε_i.

The effect of a fertilizer on a particular plant is being tested under controlled conditions. The table shows the amount of fertilizer applied, x_i g, to nine different plots, each of area 1 m^2, and the eventual yield from these plots, y_i kg. It also gives the values of $\log_{10} x_i$ ($= z_i$).

Fertilizer applied (x_i g)	25	50	75	100	125	150	187.5	225	250
Yield (y_i kg)	2.03	2.42	2.69	2.95	3.04	3.18	3.28	3.41	3.40
z_i ($= \log_{10} x_i$)	1.40	1.70	1.88	2.00	2.10	2.18	2.27	2.35	2.40

(i) Plot the values for x and y on a graph.
(ii) State why it would not be sensible to fit a regression line for y on x from these data.
(iii) It was considered that for this range of values of x it might be reasonable to fit a regression line for y on $\log_{10} x$. Calculate the least squares regression line for y on z, showing all your working. Rewrite this as an equation connecting y and x.
(iv) Estimate the yield from a plot when 175 g of the fertilizer is applied ($\log_{10} 175 = 2.24$). Give reasons why it would not be sensible to use your equation to estimate the yield for very low values of x.

(JMB)

26. Two variables X and Y are thought to be related by the equation $y = \alpha\beta^x$ apart from random errors in values of Y. Given the following data, calculate estimates for $\log \alpha$ and $\log \beta$, and hence for α and β.

Assuming that the random element in $\log Y$ is normally distributed with mean zero and variance σ^2, derive a 95% confidence interval for $\log \beta$.

x	1	2	3	4	5
y	25	85	250	800	2500

17

Elements of experimental design and analysis

▼

17.1 INTRODUCTION

We will define a designed experiment as a series of trials in which a number of treatments are tested on a number of individuals or experimental units, and responses are measured which can be analysed to quantify and compare the effects of the treatments.

Two examples will now be described to show how this general definition can apply to experiments in widely different subject areas.

Example 17.1 *A medical experiment*

A doctor wishes to compare the effects of two medical drugs by carrying out (clinical) trials on a number of patients. The design of the experiment involves:

(a) Deciding how many patients should take part.
(b) Deciding whether
  (i) each patient should receive both drugs (one initially and the other after a suitable interval); or
  (ii) patients should be matched in pairs, one patient in each pair to be randomly allocated to one of the drugs, the other in the pair to receive the other drug; or

(iii) each patient should receive one drug only, the allocation being completely random.

(c) Deciding the drug dosage and the response (dependent variable). For example, if the drugs are designed to reduce blood pressure, the response might be (blood pressure before drug) – (blood pressure after drug). □

Example 17.2 *An agricultural experiment*

An agricultural scientist (agronomist) wishes to compare the effects of four types of fertilizer on the yield of potatoes. The design of the experiment involves:

(a) Deciding how many soil types should be included.
(b) Deciding how many plots of land to use and the size of each plot. (A plot is an example of an 'experimental unit').
(c) Deciding whether to allocate the type of fertilizer to each plot:
   (i) completely randomly; or
   (ii) in a more systematic way to take account of variations in soil type and drainage, for example.

(d) Deciding the amount of fertilizer to be applied per plot, and the response. For example, the response might be the yield of potatoes in kg per sq. metre of land.

Having carried out an experiment, the data must be analysed. The analysis should be considered *before* the experiment is begun, or the experiment may be a waste of time. The fact that a lot of data may have been collected is no guarantee that a lot of useful conclusions may be drawn. The purpose of an efficient experimental design is to produce the maximum amount of information for a given 'cost' in terms of effort, time and money.

It may be possible to analyse the data from the medical experiment described above using the methods of Chapters 10, 11 and 14, since only two treatments are involved. With more than two treatments, as in the agricultural experiment, a method called *analysis of variance* might be the appropriate technique. In this chapter we will revise some of the methods of these earlier chapters, and extend our knowledge of statistical methods by a discussion of one-way analysis of variance for the so-called 'completely randomized design', and two-way analysis of variance for the 'randomized block design'. □

17.2 A COMPLETELY RANDOMIZED DESIGN WITH TWO TREATMENTS: AN EXAMPLE

Example 17.3

Suppose a doctor decides to test drugs 1 and 2 on a number of patients. To avoid bias, he decides that each patient should be randomly allocated (by tossing a coin, so that 'heads equals drug 1', for example) to one of the drugs. This then becomes an example of what is called a *completely randomized design.*

Suppose, further, that he decides to include 20 patients in the experiment (basing this decision on considerations such as those described in Section 10.5 or Section 11.10). As a result of the random allocation 11 patients are to receive drug 1, and 9 patients are to receive drug 2. The response to be measured is the fall in blood pressure, i.e.

(blood pressure before drug) – (blood pressure after drug)

Notes

(i) The 'response' is sometimes called the 'dependent variable' since it depends on the treatment received.

(ii) The number of observations of a treatment is called the number of 'replications'. So in our example there are 11 replications for drug 1, and 9 replications for drug 2.

Suppose the responses (in mm of mercury) of the 20 patients to standard drug doses are as shown in Table 17.1

Table 17.1 The fall in blood pressure for 20 patients

Drug 1	11	20	5	10	3	12	5	15	4	7	8
Drug 2	14	11	22	12	18	10	24	12	20		

We may test the null hypothesis that the average response is the same for the two drugs, either:

(i) by carrying out an unpaired-samples t test (Section 11.5), if we can justify the assumptions required by this test; or

(ii) by carrying out a Mann–Whitney U test (Section 14.4), if these assumptions cannot be justified.

A cross-diagram (Section 1.3) for the responses to drug 1 and another for the responses to drug 2 seem to indicate that 'fall in blood pressure' is approximately normally distributed for those receiving drug 1 and also for those receiving drug 2. The second assumption, that the variance of the fall in blood pressure for those taking drug 1 is the same as the variance of the fall in blood pressure for those taking drug 2, seems to be justified since the sample estimates (s_1^2 and s_2^2) are almost equal; see Table 17.2. Hence a t test is appropriate.

H_0: $\mu_1 = \mu_2$, the two drugs are equally effective in reducing blood pressure, i.e. the mean response for drug 1 is the same as the mean response for drug 2.

H_1: $\mu_1 \neq \mu_2$, the two drugs are not equally effective.

5% level of significance.

Table 17.2 Summary of data in Table 17.1

Drug 1	*Drug* 2
$\bar{x}_1 = 9.09$	$\bar{x}_2 = 15.89$
$s_1 = 5.19$	$s_2 = 5.21$
$n_1 = 11$	$n_2 = 9$

Pooled estimate of common variance is:

$$s^2 = \frac{(n_1 - 1)s_1^2 + (n_2 - 1)s_2^2}{n_1 + n_2 - 2}$$

$$= \frac{10 \times 5.19^2 + 8 \times 5.21^2}{18}$$

$$= 27.03$$

$$= 5.20$$

$$t = \frac{|(\bar{x}_1 - \bar{x}_2) - \mu_0|}{s\sqrt{\frac{1}{n_1} + \frac{1}{n_2}}} = \frac{|(9.09 - 15.89) - 0|}{5.20\sqrt{\frac{1}{11} + \frac{1}{9}}} = 2.91$$

The critical value is $t_{0.025,18} = 2.10$

Since $2.91 > 2.10$, reject H_0. We conclude there is a significant difference between the effects of the two drugs. (5% level).

17.3 ANALYSIS OF VARIANCE FOR A COMPLETELY RANDOMIZED DESIGN WITH TWO TREATMENTS

In this section we will analyse the data from the experiment described in the previous section by a statistical method called *analysis of variance* (ANOVA for short). It is normally used when more than two treatments are being compared in a completely randomized design (and in more complex experiments), but the purpose of this section is to introduce the concepts of ANOVA with the aid of a simple example, and to show connections between this ANOVA and the unpaired-samples t test.

In the experiment to compare drugs 1 and 2 there are 20 observations of the response, namely fall in blood pressure. We can consider that the variation in the 20 values is attributable to two sources:

(i) difference *between* treatments (drugs) in their effect on the response;
(ii) the difference between patients taking the same drug, which we will refer to as the difference *within* treatments (drugs).

In ANOVA we can quantify how much of the total variation is attributable to each of the two possible sources of variation. Recall from earlier chapters (the earliest was Chapter 3) that variance is one measure of variation. Also recall (from Section 9.7) that the formula for s^2, the unbiased estimator of σ^2 (the variance of a normal distribution) is:

$$s^2 = \frac{\sum(x_i - \bar{x})^2}{n-1}$$

If we think of this as the ratio of the sum of squares, $\Sigma(x_i - \bar{x})^2$, to the number of degrees of freedom, $n-1$, we have the first idea of the ANOVA method:

> We partition (divide into parts) the total sum of squares among the various sources of variation. We similarly partition the total number of degrees of freedom. We then divide each sum of squares by the corresponding number of degrees of freedom to provide what are called 'mean squares' (just to be awkward the term 'mean square' is used in ANOVA instead of 'variance').
>
> We then compare these mean squares using an F test (Section 11.8) in order to decide whether the various effects are statistically significant.

This general method of ANOVA will now be illustrated by reference to a completely randomized design (CRD) with two treatments. We will then apply this theoretical example to the actual experiment described in the previous section.

Example 17.4

Suppose there are n_1 replications for treatment 1 and n_2 for treatment 2. Let x_{i1} be the ith response for treatment 1 and let x_{j2} be the jth response for treatment 2. The data may be set out as in Table 17.3.

Table 17.3 The results of a CRD with two treatments

	Treatment	
	1	*2*
	x_{11}	x_{12}
	x_{21}	x_{22}
	.	.
	.	.
	.	.
	$x_{n_1 1}$	$x_{n_2 2}$
Totals	T_1	T_2
Grand total, $G = T_1 + T_2$		

The formulae† for the various sum of squares (SS) and degrees of freedom (d.f.) are:

$$\text{Between-treatments SS} = \frac{T_1^2}{n_1} + \frac{T_2^2}{n_2} - \frac{G^2}{N}, \text{ where } N = n_1 + n_2$$

$$\text{Total SS} = \sum_{i=1}^{n_1} x_{i1}^2 + \sum_{j=1}^{n_2} x_{j2}^2 - \frac{G^2}{N}$$

$$\text{Within-treatments SS} = \text{Total SS} - \text{Between-treatment SS}$$

$$\text{Between-treatments d.f.} = (\text{Number of treatments}) - 1 = 2 - 1 = 1$$

$$\text{Total d.f.} = N - 1$$

$$\text{Within-treatments d.f.} = \text{Total d.f.} - \text{Between-treatment d.f.}$$
$$= N - 2$$

Notes

(i) Notice that each time a total is squared it is divided by the number of observed values which contribute to the total.

(ii) Notice that the usual '$n - 1$' formula for degrees of freedom applies to the between-treatment d.f. and also to the total d.f.

These results are set out in an ANOVA table (Table 17.4).

Table 17.4 ANOVA table for a CRD

Source of variation	*SS*	*d.f.*	*MS*	*F ratio*
Between treatments				
Within treatments				
Total				

The mean square (MS) column is formed by dividing the SS column by the d.f. column, and the F ratio is calculated as:

$$\frac{\text{Between-treatments MS}}{\text{Within-treatments MS}}$$

The critical value of F is found from Table C.6, and depends on the significance level of the test and also on ν_1 and ν_2, where

$$\nu_1 = \text{between-treatments d.f.}, \quad \nu_2 = \text{within-treatments d.f.}$$

The F test is a test of the null hypothesis that the mean response is the same for the two treatments against the alternative that the mean response is different for the two treatments.

†A justification for these formulae is given in the Appendix to this chapter.

Just as there are assumptions in a t test, there are similar assumptions in the F test just outlined. These are:

(i) σ^2, the variance within treatments, is the same for all treatments; and
(ii) the responses within each treatment are normally distributed.

In fact it can be shown that the within-treatment mean square is an estimate of σ^2; also that, if the null hypothesis is true, the between-treatment mean square is also an estimate of σ^2. Significantly large values of the F ratio consequently lead us to reject the null hypothesis.

We will now apply the above method to the results of the drug experiment of the previous section (see Table 17.1). □

Example 17.5

Analyse the data in Table 17.1 using ANOVA.

Table 17.5 The fall in blood pressure of 20 patients

Drug		
1	*2*	
11	14	
20	11	
5	22	
10	12	
3	18	
12	10	
5	24	
15	12	
4	20	
7		
8		
100	143	$G = 100 + 143 = 243$

From Table 17.5,

$$\text{Between-drugs SS} = \frac{100^2}{11} + \frac{143^2}{9} - \frac{243^2}{20} = 228.8$$

$$\begin{aligned}\text{Total SS} &= (11^2 + 20^2 + \ldots + 8^2) + (14^2 + 11^2 + \ldots + 20^2) - \frac{243^2}{20} \\ &= 1178 + 2489 - 2952.4 \\ &= 714.6\end{aligned}$$

Within-drugs SS $= 714.6 - 228.8 = 485.8$
Between-drugs d.f. $= 2 - 1 = 1$
Total d.f. $= 20 - 1 = 19$
Within-drugs d.f. $= 19 - 1 = 18$

Table 17.6 ANOVA table for Example 17.5

Source	*SS*	*d.f.*	*MS*	*F ratio*
Between drugs	228.8	1	228.8	8.47
Within drugs	485.8	18	27.0	
Total	714.6	19		

H_0: $\mu_1 = \mu_2$, the two drugs are equally effective in reducing blood pressure, i.e. the mean response for drug 1 is the same as the mean response for drug 2.

H_1: $\mu_1 \neq \mu_2$, the two drugs are not equally effective.

5% level of significance.

$F = 8.47$; see Table 17.6.

The critical value of F is 4.43, interpolating in Table C6(a) for $\nu_1 = 1, \nu_2 = 18$ (the d.f.s of the numerator and denominator of the F ratio).

Since $8.47 > 4.43$, we reject H_0, and conclude that there is a significant difference between the effects of the two drugs (5% level).

Of course, this is the same conclusion as we reached using the t test in the previous section. The reason for stating 'of course' is that it can be shown that $t^2_{0.025, \nu} = F_{0.05, 1, \nu}$. Checking this we see that:

(i) the calculated values of t and F are 2.91 and 8.47, and we note that $2.91^2 = 8.47$.

(ii) the critical values of t and F are 2.10 and 4.43, and we note that $2.10^2 = 4.43$, approximately.

Since we reject H_0 if $2.91 > 2.10$ in the t test, and we reject H_0 if $8.45 > 4.41$ in the F test, we see that the two tests are certain to lead to the same conclusion. Although this result applies only for the two-treatment case, it does show an interesting connection between two hypothesis tests.

When there are more than two treatments, only the F test should be used, rather than a number of t tests to compare all treatments in pairs. (We need ${}^kC_2$ t tests if k treatments are to be compared). There are two main reasons for this:

(i) The ${}^kC_2$ t tests are not independent, the data for each treatment being used in $(k - 1)$ tests.

(ii) The overall significance level is greater than the significance level of any one test.

In the next section we describe the analysis of variance for a completely randomized design with more than two treatments. □

17.4 ONE-WAY ANALYSIS OF VARIANCE FOR A COMPLETELY RANDOMIZED DESIGN WITH MORE THAN TWO TREATMENTS

When we perform an ANOVA which involves only one source of variation, apart from the variation due to replication at the same experimental conditions, we call this a *one-way* ANOVA. So for the medical experiment of the previous section, a one-way ANOVA was performed, drugs being the only source of variation apart from the variation due to replicated testing of the same drug on a number of patients.

Consider the following agricultural experiment designed to compare the effects of four fertilizers on the yield of potatoes.

Example 17.6

A field is divided into 20 equal plots, and one of four fertilizers A, B, C and D is randomly allocated to each plot, as in Fig. 17.1. The allocation may be done by choosing 20 one-digit random numbers from the set 0, 1, 2, . . . , 9 so that numbers 0 and 1 correspond to A, 2 and 3 correspond to B, and so on (numbers 8 and 9 are ignored if chosen).

D	C	B	A
A	C	D	B
B	A	C	B
A	B	C	D
D	B	C	C

Fig. 17.1 The allocation of four fertilizers to 20 plots

We notice that the numbers of plots assigned to each fertilizer turned out to be unequal (but this does not pose a problem in analysing the data). Having grown the crop, the yield of each plot (in kg) is recorded; see Table 17.7.

Before we carry out an ANOVA of these data we will state the assumptions of such an analysis. These are:

(i) That the variance within treatments is the same for all treatments. We will denote this variance by σ^2.
(ii) That the responses (yields) within each treatment (fertilizer) are normally distributed.

Table 17.7 The yields of potatoes (kg)

	Fertilizer			
	A	*B*	*C*	*D*
	25	30	23	27
	30	34	25	29
	21	37	26	23
	22	36	24	24
		39	25	
		35	23	
Totals	$T_A = 98$,	$T_B = 211$,	$T_C = 146$,	$T_D = 103$
	Grand total, $G = T_A + T_B + T_C + T_D = 558$			

It is beyond the scope of this book to describe how such assumptions could be tested. We will assume that we know from previous experience in similar experiments that these assumptions are reasonable. In fact the F test as used in ANOVA is robust against departures from normality. (If we do have grave doubts about the assumptions, however, there is a less powerful non-parametric test (Kruskal–Wallis) which may be used instead.)

The calculations follow very similar lines to those of the previous section (see the Appendix to this chapter for a justification of these formulae).

$$\text{Between-fertilizers SS} = \frac{T_A^2}{n_A} + \frac{T_B^2}{n_B} + \frac{T_C^2}{n_C} + \frac{T_D^2}{n_D} - \frac{G^2}{N}$$

(where $N = n_A + n_B + n_C + n_D$)

$$= \frac{98^2}{4} + \frac{211^2}{6} + \frac{146^2}{6} + \frac{103^2}{4} - \frac{558^2}{20}$$

$$= 16{,}026.1 - 15{,}568.2$$

$$= 457.9$$

$$\text{Total SS} = (\text{Sum of squares of all } N \text{ observations}) - \frac{G^2}{N}$$

$$= 16{,}152 - 15{,}568.2$$

$$= 583.8$$

$$\text{Within-fertilizers SS} = \text{Total SS} - \text{Between-fertilizers SS}$$

$$= 583.8 - 457.9$$

$$= 125.9$$

Between-fertilizers d.f. = Number of fertilizers − 1 = 3

Total d.f. $= n - 1 = 19$

Within-fertilizers d.f. = Total d.f. − Between-fertilizers d.f.

$= 19 - 3 = 16$

Table 17.8 ANOVA table for Example 17.6

Source of variation	*SS*	*d.f.*	*MS*	*F ratio*
Between fertilizers	457.9	3	152.6	19.4
Within fertilizers	125.9	16	7.87	
Total	583.8	19		

$H_0: \mu_A = \mu_B = \mu_C = \mu_D$, i.e. the mean yield is the same for all four fertilizers

$H_1: \mu_A, \mu_B, \mu_C$ and μ_D are not all equal, i.e. the mean yield is not the same for all four fertilizers

5% level of significance

$F = 19.4$; see Table 17.8

The critical value of F is 3.25, i.e. for $\alpha = 0.05$, $v_1 = 3$, $v_2 = 16$ (the d.f.s of the numerator and denominator of the F ratio), see Table C6(a)

Since $19.4 > 3.25$ we reject H_0, and conclude that there is a significant difference between the mean yields of the four fertilizers (5% level). □

17.5 FURTHER ANALYSIS FOLLOWING THE ANALYSIS OF VARIANCE FOR A COMPLETELY RANDOMIZED DESIGN

The conclusions of the example of the previous section may appear to be rather unsatisfactory since, having decided there are some significant differences, we may now wish to decide which fertilizer gives the best yield. We may be tempted to make lots of comparisons in pairs, but this is dangerous as we have already stated. Instead we will look at the mean yields of the fertilizers (treatments); see Table 17.9.

Table 17.9 The mean yields of four fertilizers

Fertilizer	*A*	B	C	D
Total yield	98	211	146	103
Number of plots	4	6	6	4
Mean yield	24.5	35.2	24.3	25.8

It is clear that fertilizer B gives the highest yield, and it seems likely that there is little to choose between A, C and D.

We can also obtain other useful estimates, for example:

(a) a 95% confidence interval for the mean effect of a treatment;
(b) a 95% confidence interval for the difference between the mean effects of two treatments;
(c) a 95% confidence interval for the within-treatment variance.

For (a) we use:
$$\bar{x}_i \pm \frac{t_{0.025,\nu}s}{\sqrt{n_i}}$$

For (b) we use:
$$\bar{x}_i - \bar{x}_j \pm t_{0.025,\nu}\, s\sqrt{\frac{1}{n_i}+\frac{1}{n_j}}$$

For (c) we use:
$$\frac{\nu s^2}{\chi^2_{0.025,\nu}} \quad \text{and} \quad \frac{\nu s^2}{\chi^2_{0.975,\nu}}$$

These are applications of results similar to those of Sections 10.3, 10.6 and 10.9 respectively. The various terms need explaining:

$\bar{x}_i$ is the mean response for treatment i
ν is the within-treatment d.f.
s^2 is the within-treatment mean square (remember mean square is really variance). In fact, s^2 is our 'pooled' estimate of the within-treatment variance, σ^2 say, which we have assumed is the same for all treatments.

n_i is the number of replications of treatment i

$\frac{s}{\sqrt{n_i}}$ is called the 'standard error' of the mean for treatment i.

Similarly, $s\sqrt{\frac{1}{n_i}+\frac{1}{n_j}}$ is called the 'standard error' of the difference between the means of treatments i and y.

Example 17.7

Following the ANOVA of Section 17.4:

(i) calculate a 95% confidence interval for the mean yield of fertilizer B;
(ii) calculate a 95% confidence interval for the difference between the mean yields of fertilizers B and D;
(iii) calculate a 95% confidence interval for σ^2, the within-treatment variance.

(i) The within-fertilizer mean square, $s^2 = 7.87$ (see Table 17.8) and is associated with $\nu = 16$ d.f.

For fertilizer B, the sample mean yield, $\bar{x}_B = 35.2$, and $n_B = 6$
Hence a 95% confidence interval for μ_B, the mean yield for fertilizer B, is given by:

$$35.2 \pm \frac{t_{0.025,16}\sqrt{7.87}}{\sqrt{6}}$$

$$35.2 \pm \frac{2.12\sqrt{7.87}}{\sqrt{6}}$$

$$32.8 \text{ to } 37.6 \qquad \text{(1 d.p.)}$$

(ii) For fertilizer D, $\bar{x}_D = 25.8$, $n_D = 4$
Hence a 95% confidence interval for $(\mu_B - \mu_D)$, the difference between the mean yields of fertilizers B and D, is given by:

$$(35.2 - 25.8) \pm 2.12\sqrt{7.87}\sqrt{\frac{1}{6} + \frac{1}{4}}$$

$$9.4 \pm 3.8$$

$$5.6 \text{ to } 13.2 \qquad \text{(1 d.p.)}$$

(iii) Our estimate of σ^2 is $s^2 = 7.87$ associated with $\nu = 16$ d.f. Hence a 95% confidence interval for σ^2 is given by:

$$\frac{16 \times 7.87}{\chi^2_{0.025,16}} \text{ to } \frac{16 \times 7.87}{\chi^2_{0.975,16}}$$

$$\frac{16 \times 7.87}{28.85} \text{ to } \frac{16 \times 7.87}{6.91}$$

$$4.4 \text{ to } 18.2 \qquad \text{(1 d.p.)}$$

□

17.6 TWO-WAY ANALYSIS OF VARIANCE FOR A RANDOMIZED BLOCK DESIGN

If, in the layout of 20 plots as shown in Fig. 17.1 we suspected that there were systematic factors other than fertilizer which might affect the yield of potatoes, we might be able to allow for this in the design. For example, if it was thought that drainage varied in the north–south direction, then it would be wise to divide the 20 plots into five *blocks*, each block consisting of a row of four plots running east–west, as in Fig. 17.2.

Then, instead of simply randomly allocating fertilizers to plots we allocate as follows:

> Allocate the fertilizers randomly to the plots in block 1 but with the restriction that each fertilizer must occur once and only once in the block. Repeat this procedure for the remaining four blocks.

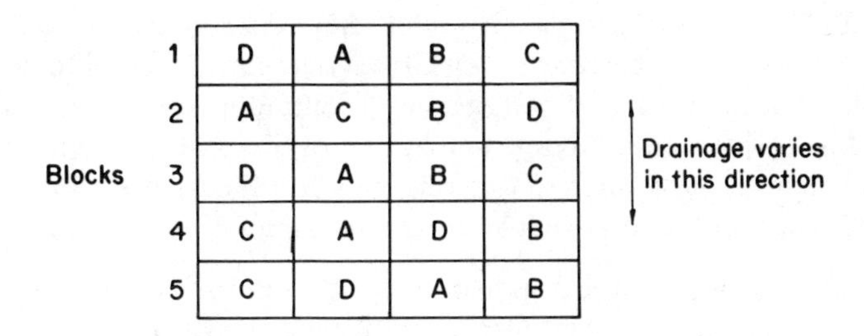

Fig. 17.2 20 plots divided in 5 blocks

The resulting allocation might be as in Fig. 17.2, which was obtained using random numbers (by letting 0 and 1 correspond to A, 2 and 3 to B, and so on. Numbers 8 and 9 are ignored, as are any numbers which correspond to a fertilizer which has already been allocated to a particular block).

This is an example of what is called a *randomized block design* (RBD). Because there are now two main sources of variation, fertilizers and blocks, the analysis which is performed is called *two-way* analysis of variance.

The general method of analysing the results of a randomized block experiment, and the concepts and assumptions of the analyses will now be described. Suppose we set out the results of a randomized block experiment consisting of bk experimental units (e.g. plots) on which k treatments are to be tested in b blocks as in Table 17.10, where x_{ij} is the response for the ith block and the jth treatment.

Table 17.10 The results of a randomized block design

	Treatment				
Block	*1*	*2*	...	*k*	*Total*
1	x_{11}	x_{12}	...	x_{1k}	B_1
2	x_{21}	x_{22}	...	x_{2k}	B_2
.	.			.	.
.	.			.	.
.	.			.	.
b	x_{b1}	x_{b2}	...	x_{bk}	B_b
Total	T_1	T_2	...	T_k	G, the grand total

We can consider that the total variation in the bk responses is attributable to three sources:

(i) The difference *between* treatments in their effect on the response.
(ii) The difference *between* blocks in their effect on the response.

(iii) The difference between plots even if they were in the same block and receiving the same treatment, which we refer to as the 'difference *within* blocks and treatments', although a more commonly used term is simply the 'residual'. Here 'residual' refers to that part of the total variation which is left over when the variation due to treatments and blocks has been subtracted. (Note a similar use of 'residual' in regression analysis; see Section 16.6.)

So we can consider that the response in each plot is made up of an element due to the treatment applied, an element due to the block in which the plot is situated, and a random element due to source (iii) above. The assumptions needed for the analysis of variance are:

1. The random element is normally distributed and has the same variance for all treatments and blocks (note a similar assumption in regression analysis; see Section 16.6).
2. The effect of treatments is the same for all blocks. Our assumption may be referred to as the assumption of 'no interaction between treatments and blocks'.

(If we cannot justify these assumptions a less powerful non-parametric test, Friedman's test, may be used instead.)

The calculations for the various entries of the ANOVA table for a RBD are as follows:

$$\text{Between-treatments SS} = \frac{T_1^2}{b} + \frac{T_2^2}{b} + \cdots + \frac{T_k^2}{b} - \frac{G^2}{bk}$$

$$\text{Between-blocks SS} = \frac{B_1^2}{k} + \frac{B_2^2}{k} + \cdots + \frac{B_b^2}{k} - \frac{G^2}{bk}$$

$$\text{Total SS} = \sum_{i=1}^{b} \sum_{j=1}^{k} x_{ij}^2 - \frac{G^2}{bk}$$

Residual SS = Total SS − Between-treatments SS − Between-blocks SS

Between-treatments d.f. = Number of treatments $-1 = k-1$

Total d.f. $= bk - 1$

Residual d.f. = Total d.f. − Between-treatments d.f. − Between-blocks d.f.

$$= (bk-1) - (k-1) - (b-1)$$

$$= (b-1)(k-1)$$

Notes

(i) notice that (as in Section 17.3) each time a total is squared it is divided by the number of observed values which contribute to that total.

(ii) notice that the usual '$n-1$' formula applies to the between-treatments d.f. and to the between-blocks d.f.
(iii) notice that the residual d.f. is the product of the d.f. for treatments and blocks.

These results are set out in an ANOVA table as in Table 17.11.

Table 17.11 ANOVA table for a RBD

Source of variation	*SS*	*d.f.*	*MS*	*F ratio*
Between treatments				
Between blocks				
Residual				
Total				

The mean square (MS) column is obtained by dividing the SS column by the d.f. column, and two F ratios are obtained:

$$\frac{\text{Between-treatments MS}}{\text{Residual MS}} \quad \text{and} \quad \frac{\text{Between-blocks MS}}{\text{Residual MS}}$$

These are the appropriate test statistics to test the null hypotheses that:

(i) the mean response is the same for all treatments;
(ii) the mean response is the same for all blocks.

The following numerical example illustrates the analysis of variance of a randomized block design.

Example 17.8

Suppose that the yields of potatoes for the design illustrated in Fig. 17.2 are as set out in Table 17.12.

Table 17.12 The responses for a RBD with four treatments and five blocks

	Fertilizer				
Block	*A*	*B*	*C*	*D*	*Total*
1	20	32	26	23	101
2	22	36	28	26	112
3	23	33	27	24	107
4	23	37	30	27	117
5	26	35	27	25	113
Total	114	173	138	125	550

Carry out an ANOVA for these data.

Between-fertilizers SS $= \dfrac{114^2}{5} + \dfrac{173^2}{5} + \dfrac{138^2}{5} + \dfrac{125^2}{5} - \dfrac{550^2}{5 \times 4}$

$= 15{,}518.8 - 15{,}125$

$= 393.8$

Between-blocks SS $= \dfrac{101^2}{4} + \dfrac{112^2}{4} + \dfrac{107^2}{4} + \dfrac{117^2}{4} + \dfrac{113^2}{4} - \dfrac{550^2}{5 \times 4}$

$= 15{,}163 - 15{,}125$

$= 38.0$

Total SS $= (20^2 + 32^2 + \cdots + 27^2 + 25^2) - \dfrac{550^2}{5 \times 4}$

$= 15{,}574 - 15{,}125$

$= 449.0$

Residual SS $= 449.0 - 393.8 - 38.0$

$= 17.2$

Between-treatment d.f. $= 4 - 1 = 3$

Between-blocks d.f. $= 5 - 1 = 4$

Total d.f. $= 5 \times 4 - 1 = 19$

Residual d.f. $= (5 - 1)(4 - 1) = 12$

Table 17.13 ANOVA table for Example 17.8

Source	*SS*	*d.f.*	*MS*	*F ratio*
Between fertilizers	393.8	3	131.27	91.80
Between blocks	38.0	4	9.50	6.64
Residual	17.2	12	1.43	
Total	449.0	19		

H_0: $\mu_A = \mu_B = \mu_C = \mu_D$, i.e. the mean yield is the same for all four fertilizers.

H_1: μ_A, μ_B, μ_C and μ_D are not all equal

5% level of significance

$F = 91.8$ (see Table 17.13)

The critical value of F is 3.49, i.e. for $\alpha = 0.05$, $\nu_1 = 3$, $\nu = 12$

Since $91.8 > 3.49$, we reject H_0, and conclude that there are significant differences between the mean yields of the four fertilizers (5% level.).

Of secondary importance, we can test another null hypothesis concerning blocks ($F = 6.64$ compared with a critical value of 3.26). The fact that the mean yields of the four blocks are significantly different confirms that we were right to include blocks in the design. By allowing for blocks in the ANOVA table, we have been able to get a more precise test for fertilizers, and a more precise estimate of the residual mean square (variance). □

17.7 FURTHER ANALYSIS FOLLOWING THE ANALYSIS OF VARIANCE FOR A RANDOMIZED BLOCK DESIGN

As we noted in Section 17.5, having rejected the null hypothesis that the mean yield is the same for all fertilizers, we may wish to refine this conclusion. We start by calculating the mean yield for each fertilizer using the data from Table 17.12, to obtain Table 17.14.

Table 17.14 The mean yields for four fertilizers

Fertilizer	*A*	*B*	*C*	*D*
Total yield	114	173	138	125
Number of plots	5	5	5	5
Mean yield	22.8	34.6	27.6	25.0

These means are spread out and it is not clear where the significant differences occur. One method which can be applied is to conclude that two treatment means are significantly different at the 5% level if they differ by more than

$$t_{0.025,\,\nu} s\sqrt{\frac{2}{b}},$$

where s^2 is the residual mean square, ν is the residual d.f.

For the data in Tables 17.13 and 17.14,

$$t_{0.025,\,\nu} s\sqrt{\frac{2}{b}} = 2.18 \times \sqrt{1.43} \times \sqrt{\frac{2}{5}} = 1.65.$$

Since all the means differ from each other by more than 1.65 we can conclude that there are significant differences between all pairs of fertilizer yields, with B being better than C, and so on.

17.8 SUMMARY

Two kinds of experimental design were considered:

(i) a completely randomized design (CRD) in which the allocation of treatments to individuals or experimental units is random.

(ii) a randomized block design (RBD) in which the allocation of treatments is random except for the constraint that each treatment must occur once and only once in each block. Blocks were introduced into the design to allow for suspected inherent natural variation.

Under certain assumptions the results of each type of experiment may be analysed initially using a method called analysis of variance. It was shown how the ANOVA of a CRD with two treatments was equivalent to an unpaired t test. Some further analysis following ANOVA was described.

APPENDIX TO CHAPTER 17

Formulae for sums of squares and degrees of freedom

The derivation of the formulae for the sums of squares and degrees of freedom for a completely randomized design (balanced case only) are set out below. If the number of replications per treatment is the same for all treatments, the design is said to be *balanced*. Let this number be n, so that the total number of observed responses is $N = nk$. Let x_{ij} be the ith response for the jth treatment, as in Table 17.15.

Table 17.15 The responses for a CRD

	Treatment			
	1	2	⋯	k
	x_{11}	x_{12}	⋯	x_{1k}
	x_{21}	x_{22}	⋯	x_{2k}
	⋮	⋮		⋮
	x_{n1}	x_{n2}	⋯	x_{nk}
Totals	T_1	T_2	⋯	T_k, $G = T_1 + T_2 + \cdots + T_k$
Mean	$\bar{x}_1$	$\bar{x}_2$	⋯	$\bar{x}_k$, $\bar{x} = G/N$

The difference between x_{ij} and $\bar{x}$, the overall mean, may be written as:

$$x_{ij} - \bar{x} = (x_{ij} - \bar{x}_j) + (\bar{x}_j - \bar{x})$$

Squaring and summing over all i and j gives:

$$\sum_{i=1}^{n} \sum_{j=1}^{k} (x_{ij} - \bar{x})^2 = \sum_{i=1}^{n} \sum_{j=1}^{k} (x_{ij} - \bar{x}_j)^2 + \sum_{i=1}^{n} \sum_{j=1}^{k} (\bar{x}_j - \bar{x})^2$$

The cross-product term disappears since

$$\sum_{i=1}^{n} \sum_{j=1}^{k} (x_{ij} - \bar{x}_j)(\bar{x}_j - \bar{x}) = \sum_{j=1}^{k} (\bar{x}_j - \bar{x}) \sum_{i=1}^{n} (x_{ij} - \bar{x}_j)$$

and
$$\sum_{i=1}^{n} (x_{ij} - \bar{x}_j) = 0,$$

for each j (the deviations of any group of observations from their mean is zero).

The above equation expresses the idea that:

$$\text{Total SS} = \text{Within-treatment SS} + \text{Between-treatment SS}$$

Now
$$\sum_{i=1}^{n} \sum_{j=1}^{k} (x_{ij} - \bar{x})^2 = \sum_{i=1}^{n} \sum_{j=1}^{k} x_{ij}^2 - N\bar{x}^2, \quad \text{using ideas of the Appendix to Chapter 15}$$
$$= \sum_{i=1}^{n} \sum_{j=1}^{k} x_{ij}^2 - \frac{G^2}{N}, \quad \text{since } \bar{x} = \frac{G}{N}$$

So, total sum of squares $= \text{(Sum of squares of all } N \text{ observations)} - \dfrac{G^2}{N}$

Also
$$\sum_{i=1}^{n} \sum_{j=1}^{k} (\bar{x}_j - \bar{x})^2 = \sum_{j=1}^{k} (\bar{x}_j - \bar{x})^2, \quad \text{summing over all } i$$
$$= n \sum_{j=1}^{k} (\bar{x}_j - \bar{x})^2$$
$$= n \sum_{j=1}^{k} x_j^2 - nk\bar{x}^2, \quad \text{using ideas of Appendix to Chapter 15}$$
$$= n \sum_{j=1}^{k} \left(\frac{T_j}{n}\right)^2 - nk\left(\frac{G}{N}\right)^2$$
$$= \frac{T_1^2}{n} + \frac{T_2^2}{n} + \cdots + \frac{T_k^2}{n} - \frac{G^2}{N}$$

The other term, the within-treatment sum of squares, may be found by difference, from the equation

$$\text{Total SS} = \text{Within-treatment SS} + \text{Between-treatment SS}$$

For the unbalanced case, i.e. when the number of replications per treatment is not constant but is, say, n_j for treatment j:

Total sum of squares $= \text{(Sum of squares of all } N \text{ observations)} - \dfrac{G^2}{N}$

Between-treatment sum of squares $= \left(\dfrac{T_1^2}{n_1} + \dfrac{T_2^2}{n_2} + \cdots + \dfrac{T_k^2}{n_k}\right) - \dfrac{G^2}{N}$

where
$$N = n_1 + n_2 + \cdots + n_k.$$

The total degrees of freedom may be allocated to 'between' and 'within' treatments. The rationale is as follows: when we calculate the between-treatments sum of squares we are essentially comparing the k treatment means, giving $(k-1)$ degrees of freedom. When we calculate the total sum of squares we are essentially comparing the N individual responses, giving $(N-1)$ degrees of freedom. The between-treatment degreess of freedom are then obtained using:

$$\text{Total d.f.} = \text{Within-treatment d.f.} + \text{Between-treatment d.f.},$$

i.e., Within-treatment d.f. $= (N-1)-(k-1) = N-k$

For the balanced case, when $N = nk$, the within-treatment d.f. becomes $k(n-1)$, and we can think of $(n-1)$ d.f. being contributed by each treatment.

For the unbalanced case, when $N = n_1 + n_2 + \cdots + n_k$, the within-treatment d.f. becomes $(n_1-1)+(n_2-1)+\cdots+(n_k-1)$, and we can think of (n_j-1) d.f. being contributed by the jth treatment, $j = 1$ to k.

WORKSHEET 17

Section 17.3

1. The following results are the weights of eggs (in grams) produced by eight hens fed on ordinary corn and eight fed on vitamin-enriched corn pellets. The allocation of type of corn to hen was done randomly.

Ordinary corn	31	32	34	33	32	32	31	33
Enriched corn	30	34	34	33	35	33	35	34

2. An experiment is to be made to compare the effects of two types of sleeping pill, A and B, on the length of sleep of patients. Compare the relative merits of using **either** a completely randomized design **or** a paired comparison design, assuming that 12 patients (= experimental units) are available for each type of design.

   Assume a completely randomized design is used and the following data, recording hours of sleep, are obtained:

   Perform an analysis of variance to decide whether there is a significant difference, at the 5% level, between the mean weight of eggs for the two types of corn. State any assumptions made.

   Now perform an unpaired t test on the above data, and compare your results with those of the analysis of variance.

A	9.4	8.3	8.6	7.6	7.4	8.0
B	10.9	8.3	9.5	8.2	7.3	9.1

   Test if there is a difference between the effects of the pills. State any assumptions that you make.

   (OLE)

Section 17.4

3. Below are the scores of 22 children taught by three different methods, A, B, C. The allocation of method to child was made randomly. Test, at the 5% level of significance, whether the mean scores for the three methods are equal. State any assumptions made.

A	116	117	138	100	125	130	134	124	114
B	132	137	131	108	111	130	140		
C	108	96	131	130	111	126			

►

4. Twenty samples were taken from a container of hydrogenated vegetable oil and five were given to each of the laboratory's four analysts to determine the melting point. Individual determinations are given below:

Analyst	*Melting point*				
A	93.60	94.64	96.30	93.62	93.51
B	96.44	96.53	98.38	97.00	97.63
C	92.57	94.01	94.49	93.29	95.87
D	95.55	95.90	94.25	95.80	96.21

Do you consider that there is any significant difference between the results of the analysts?

5. The percentage potassium in blood serum was determined by three different analysts using the same method. The following results were obtained:

Analyst	A	B	C
	0.015	0.012	0.019
	0.015	0.014	0.018
	0.017	0.016	0.020
	0.017	0.013	0.021
	0.018	0.010	0.019
	0.016	0.012	0.018

Use an analysis of variance technique to answer the following question:

Is there a significant variance among the average results obtained by each of the three analysts? Use a 1% level of significance.

Section 17.5

6. An experimenter was interested in comparing the effects of three kinds of reinforcement on the performance of 6-year-old girls in a visual discrimination task. The reinforcements, coded A, B and C, were:

A praise for correct responses;
B reproof for mistakes;
C silence.

Twelve girls were randomly chosen from a large number of 6-year-old girls in a certain town, and then randomly assigned to receive one of the three kinds of reinforcement. This procedure resulted in a balanced experiment with four girls per kind of reinforcement.

▶

The number of errors made by each girl in the visual discrimination task were as follows:

A	*B*	*C*
48	78	84
58	58	73
40	70	66
54	88	71

(a) Describe how you would have randomly assigned the kind of reinforcement for each girl.
(b) Use analysis of variance to test for significant differences between the mean number of errors for the three kinds of reinforcement.
(c) Is one kind of reinforcement significantly better or worse than the other two?

7. The systolic blood pressures (mm Hg) in the aorta for a number of patients each receiving one of four treatments were recorded as follows:

Treatment	*Measurements of pressure*				
A	120	127	115	113	121
B	125	130	119	126	
C	128	135	125	133	
D	116	122	115	118	121

(*Note*: $\sum x = 2209$, $\sum x^2 = 271{,}779$.)

(a) Obtain the analysis of variance table and use the table to draw conclusions about the treatments.
(b) Comment on the conditions which must prevail for the conclusions and the analysis to be appropriate.
(c) Obtain the standard errors for each treatment mean and the standard errors of the difference between any two treatment means. State how these values may be useful.

(IOS)

8. In an agricultural trial, four varieties of wheat were grown on 24 plots of land, 6 plots being used for each variety. The yields x are summarized in the table below.

Variety	*A*	*B*	*C*	*D*
$\sum x$	2.00	2.34	1.62	2.10
$\sum x^2$	0.71	0.94	0.48	0.75

Assume that these are independent random samples from four independent normally distributed random variables X, Y, Z, W which have the same variance σ^2 but whose means are unknown. ▶

(i) Find two-sided 98% confidence limits for σ.
(ii) Use the F distribution to test the hypothesis that the means of X, Y, Z, W are all equal.

(O&C)

9. In a horticultural experiment, three varieties of tomato plant are grown. The number n of plants of each variety and the yield x (in kg) of each plant are summarized in the table below.

Variety	n	$\sum x$	$\sum x^2$
Standard	8	95	1160
New A	6	92	1430
New B	6	76	1000

Assuming that the plant yields are normally distributed about means μ_S, μ_A, μ_B respectively and have the same variance σ^2 in all three varieties:

(a) estimate μ_S, μ_A, μ_B;
(b) estimate σ^2;
(c) use the F distribution to test the hypothesis $\mu_S = \mu_A = \mu_B$;
(d) find two-sided 95% confidence limits for μ_A.

(O&C)

Section 17.6

10. An experiment was conducted to decide whether the weal size from the intradermal injection of an antibody for guinea pigs depends on the method of preparation of the antibody. Four preparations were used, and each preparation was injected once into each of four guinea pigs. The preparation to be given to each of four sites on each animal was decided strictly at random.

The weal size data in coded units were as follows:

	Antibody preparation			
Guinea pig	*A*	*B*	*C*	*D*
1	46	56	62	48
2	53	63	57	58
3	58	65	61	53
4	61	62	65	57

Use analysis of variance to test the following hypotheses:

(i) the average weal size is the same for each antibody preparation;
(ii) the average weal size is the same for each guinea pig.

(*Hint*: Treat guinea pigs as 'blocks'.)

11. Concrete beams have to be made for the construction of a bridge. To make these beams, cement has to be mixed with gravel. There are five different types of cement and ▶

four batches of gravel. Each batch of gravel is sufficient for one mix with each type of cement. From each of the 20 combinations of cement and gravel, a test-beam is made and tested to destruction. The table below gives the breaking load in coded units for each test beam:

Gravel batch	*Cement type*				
	A	*B*	*C*	*D*	*E*
1	10	14	6	22	13
2	12	9	8	20	11
3	8	10	4	18	10
4	16	15	10	28	10

Use analysis of variance to decide whether there are any significant differences in mean breaking load for the five types of cement.

12. Information about the current state of a complex industrial process is displayed on a control panel which is monitored by a technician. In order to find the best display for the instruments on the control panel three different arrangements were tested by simulating an emergency and observing the reaction times of five different technicians. The results, in seconds, are given below:

Arrangement	*Technician*				
	P	*Q*	*R*	*S*	*T*
A	2.4	3.3	1.9	3.6	2.7
B	3.7	3.2	2.7	3.9	4.4
C	4.2	4.6	3.9	3.8	4.5

Carry out an analysis of variance and test for differences between technicians and between arrangements at the 5% significance level.

Currently arrangement C is used and it is suggested that this be replaced by arrangement A. Comment, briefly, on this suggestion and on what further information you would find useful before coming to a definite decision.

(AEB (1984))

Section 17.7

13. Carry out further analysis following the analysis of variance in Exercise 10 above, and state your conclusions.

14. Carry out further analysis following the analysis of variance in Exercise 11 above, and state your conclusions.

18

Quality control charts and acceptance sampling

▼

18.1 INTRODUCTION

Suppose we are manufacturers making items in a production process, and we may wish to control a particular variable. One way of doing this is to plot sample values of the variable on a *control chart*. In this chapter charts for both *continuous* and *discrete* variables will be discussed.

As an example of a continuous variable, suppose we make and pack flour in 3 kg bags. There is certain to be some variation in weight from bag to bag. We would wish to control the weight because if it is too low we may be breaking the law and if it is too high we may be losing profit. We can 'keep an eye' on the weight of flour by taking samples of bags of flour and plotting control charts both for the mean weight and also for the variation in weight.

As an example of a discrete variable, suppose we make light-bulbs which may be classified only as effective (when they light) or defective (when they fail to light). We would wish to control the fraction (proportion) of defective light-bulbs that we make. If this is too high, customers will complain about our bulbs. On the other hand, it may be expensive for us to maintain very low levels of defectives. We can 'keep an eye' on the fraction defective by taking samples of light-bulbs and plotting a control chart for the fraction of our sample which is defective.

By studying these control charts we can decide whether control may soon be lost because a plotted point lies outside *warning lines* or whether control has already been lost because a plotted point lies outside *action lines*.

Another type of quality control is called *acceptance sampling*. This concerns taking a sample of items from a large batch and counting the number of defective items in the sample. On the basis of the sample data, we decide either to accept the whole batch as being fit to leave the factory or to reject the whole batch as being unfit to leave the factory (in a borderline case we may decide to take a second sample before reaching a final decision). We may be said to be 'judging and sentencing' the batch.

As an example, suppose we make a large batch of 100 watt light-bulbs of the pearl mushroom type. We take a random sample of, say, 50 light-bulbs from the batch. We may then decide that we will accept the whole batch if there are fewer than 5 defectives, whereas if there are 5 or more defectives we will reject the batch.

18.2 CONTROL CHARTS FOR THE MEAN AND RANGE OF A CONTINUOUS VARIABLE

Recall from Section 9.5 that if a random variable, X, is normally distributed with mean μ and variance σ^2, then $\overline{X}$, the mean of samples of size n, has a normal distribution with mean μ and variance σ^2/n. It follows that:

(i) 95% of sample means will lie within the range $\mu \pm \dfrac{1.96\,\sigma}{\sqrt{n}}$.

(ii) 99.8% of sample means will lie within the range $\mu \pm \dfrac{3.09\,\sigma}{\sqrt{n}}$.

If we take samples of size n at intervals from our production process, we can, in principle, plot the sample means on a chart and note their position relative to four horizontal lines (see Fig. 18.1).

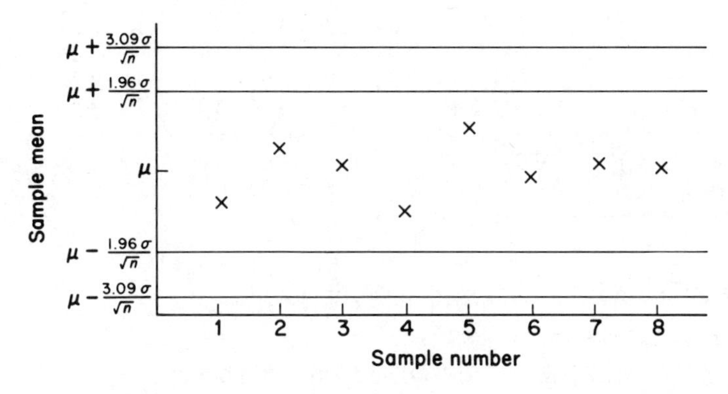

Fig. 18.1 Basis of a control chart for the sample mean

The problem is that we don't know μ and σ, the mean and variance of the variable we are trying to control. We will have to estimate μ and σ from sample data. If the means of the first few samples are $\bar{x}_1$, $\bar{x}_2$, and so on, then we can use $\bar{x}$, the mean of these sample means, as our estimate of μ. To estimate σ we use the fact (which we quote without proof) that the range of a sample size n taken from a normal distribution is proportional to σ, the constant of proportionality depending on n. If the ranges of the first few samples are w_1, w_2, and so on, then we use $k\bar{w}$ as our estimate of σ, where $\bar{w}$ is the mean range, and k depends on n (see Table 18.1).

Table 18.1 How k depends on n

Sample size, n	2	3	4	5	6	7	8	9	10
k	0.886	0.591	0.486	0.430	0.395	0.370	0.351	0.337	0.325

The reason for using this method of estimating σ, rather than the estimate described in Section 9.7, is that it is simpler and quicker to calculate the range of each sample.

Example 18.1

Ten samples of size four are taken from a process producing 3 kg bags of flour. The weights are as in Table 18.2. Estimate the mean and standard deviation of the weight of the bags. Plot a control chart for the sample mean, assuming that the weight of flour is normally distributed.

Table 18.2 The weights of 40 bags of flour

Sample number	*1*	*2*	*3*	*4*	*5*	*6*	*7*	*8*	*9*	*10*
	3.07	3.02	2.85	2.99	3.02	3.05	3.10	2.96	3.14	3.06
	3.02	2.98	3.06	3.14	3.07	2.95	3.00	3.02	3.10	2.98
	3.03	3.06	3.02	3.06	2.98	3.04	2.90	2.97	3.05	3.04
	2.99	2.96	2.97	3.02	3.06	3.07	2.95	3.10	3.08	3.10
Sample means ($\bar{x}_i$)	3.03	3.00	2.98	3.05	3.03	3.03	2.99	3.01	3.09	3.04
Sample range (w_i)	0.08	0.10	0.21	0.15	0.09	0.12	0.20	0.14	0.09	0.12

The overall mean is $\bar{x} = 3.025$ kg, our estimate of μ.

The mean range is $\bar{w} = 0.130$ kg.

For $n = 4$, $k = 0.486$, and $k\bar{w} = 0.063$ kg, our estimate of σ.

So our estimate of $\mu \pm \dfrac{1.96\,\sigma}{\sqrt{n}}$ is $3.025 \pm \dfrac{1.96 \times 0.063}{\sqrt{4}}$,

i.e. 2.96 and 3.09 (2 d.p.)

Horizontal 'warning' lines are drawn at values of 2.96 and 3.09.

Also our estimate of $\mu \pm \frac{3.09\,\sigma}{\sqrt{n}}$ is $3.025 \pm \frac{3.09 \times 0.063}{\sqrt{4}}$

i.e. 2.93 and 3.12 (2 d.p.)

Horizontal 'action' lines are drawn at values of 2.93 and 3.12. □

Figure 18.2 shows the warning and action lines for this example, and also the individual sample means (from Table 18.2).

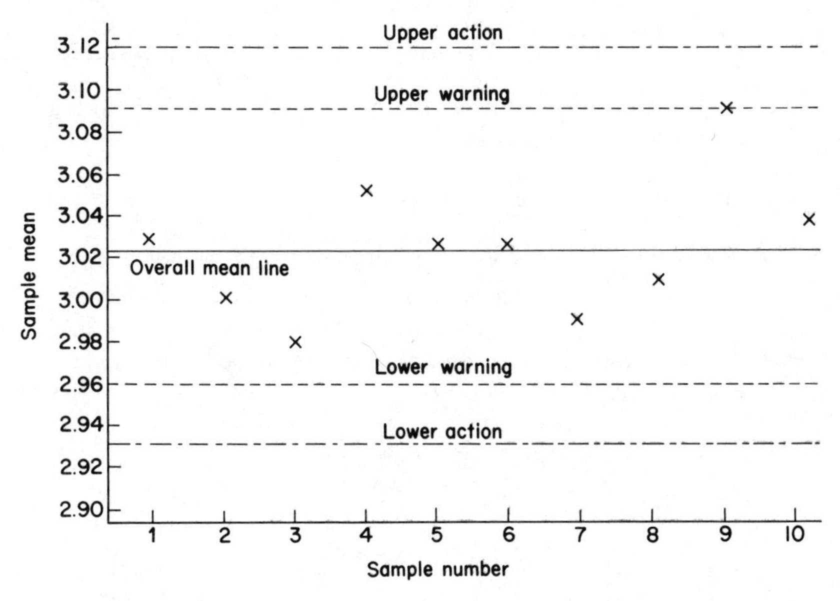

Fig. 18.2 A control chart for the sample mean weight

How should we interpret a control chart such as Fig. 18.2? Bearing in mind that, if μ and σ stay fixed at the values we estimate, only 1 in 20 (5%) of sample means should fall outside the warning lines, and only 1 in 500 (0.2%) of sample means should fall outside the action lines, we can use the following rules for interpreting the chart:

1. Control has been lost if any sample mean falls outside the action lines. For example, a point above the upper action line may indicate an increase in μ.
2. Control has been lost if two consecutive sample means fall outside the warning lines. For example, two consecutive sample means which fall below the lower warning line may indicate a decrease in μ.
3. Control has been lost if six consecutive sample means all fall above (or below) the overall mean line.

For Fig. 18.2, it appears that the process mean is under control, but it is advisable to plot a second chart, for the sample range, to see whether the process variability is also in control. For this chart, warning and action lines are drawn at values which are proportional to $\bar{w}$, the constants of proportionality being obtained from Table 18.3.

Table 18.3 Constants needed for plotting a control chart for the sample range

Sample size n	2	3	4	5	6	7	8	9	10
$D'_{0.975}$	2.809	2.176	1.935	1.804	1.721	1.662	1.617	1.583	1.555
$D'_{0.999}$	4.124	2.992	2.579	2.358	2.217	2.119	2.045	1.988	1.941

We multiply $\bar{w}$ by the two values corresponding to our sample size to obtain the warning and action lines respectively.

Example 18.2

Using the data in Table 18.1, plot a control chart for the sample range.

For these data, we already know that $\bar{w} = 0.130$. Selecting the constants from Table 18.3 for $n = 4$, we must draw lines at $1.935 \times 0.130 = 0.25$, and $2.579 \times 0.130 = 0.34$; see Fig. 18.3, on which the individual sample ranges (from Table 18.2) are also plotted.

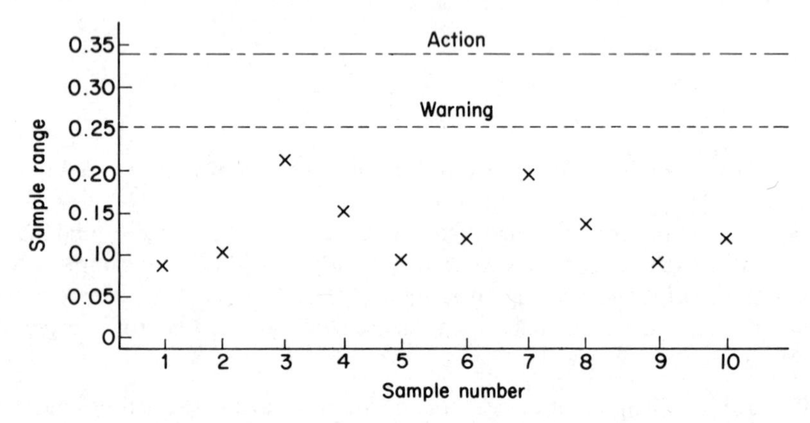

Fig. 18.3 A control chart for the sample range

How should we interpret a control chart such as Fig. 18.3? Bearing in mind that if σ stays fixed at the value we estimated, only one sample range in 40 should lie above the warning line, and that only one sample range in 1000 should lie above the action line, we can operate the following rules for interpreting the chart:

1. Control has been lost, in the sense that σ has increased, if any sample range falls above the action line.
2. Control has been lost, in the sense that σ has increased, if two consecutive sample ranges fall above the warning line.

Note that we do not have *lower* warning and action lines, since we are only interested in whether the variability has increased.

It is advisable to look at both control charts together. What can sometimes happen, for example, is that one 'maverick' value in a sample causes both an increase in variability (as indicated by the sample range chart) and a change in the mean (as indicated by the sample mean chart).

For Fig. 18.3, however, it appears that the variability in the weight of bags of flour is under control. □

18.3 CONTROL CHART FOR FRACTION DEFECTIVE

Let p denote the proportion or fraction of defective items in a production process for items which may be classified only as defective or effective. Using the normal approximation to the binomial (Section 7.3), we can state that, if $np > 5$ and $n(1-p) > 5$ (where n is the sample size):

(i) for 97.5% of samples the fraction defective will be less than $p + 1.96\sqrt{\dfrac{p(1-p)}{n}}$;

(ii) for 99.9% of samples the fraction defective will be less than $p + 3.09\sqrt{\dfrac{p(1-p)}{n}}$.

Since p is unknown, we estimate it from sample data. So if $\hat{p}$ is our estimate of p,

$$\hat{p} = \frac{\text{Number of defectives found}}{\text{Number of items inspected}}$$

A control chart for the fraction defective is a plot of fraction defective versus sample number with warning and action lines drawn at

$$\hat{p} + 1.96\sqrt{\frac{\hat{p}(1-\hat{p})}{n}} \quad \text{and} \quad \hat{p} + 3.09\sqrt{\frac{\hat{p}(1-\hat{p})}{n}}$$

Note that we do not have *lower* warning and action lines, since we are only interested in whether the fraction defective has increased.

Example 18.3

Light-bulbs are inspected at intervals. Each inspection consists of a sample of 100 bulbs which are tested and the number of defectives is recorded. Using the data ble 18.4, plot a control chart for the fraction defective, and comment on ree of control achieved.

Table 18.4 The number of defective light-bulbs in 10 samples of 100

Sample number	*1*	*2*	*3*	*4*	*5*	*6*	*7*	*8*	*9*	*10*
Number of defectives	12	20	14	6	13	7	20	28	16	14
Fraction defective	0.12	0.20	0.14	0.06	0.13	0.07	0.20	0.28	0.16	0.14

From this table, $\hat{p} = \dfrac{12+20+\ \ldots\ +16+14}{10 \times 100} = 0.150$

Since $n\hat{p}$ and $n(1-\hat{p})$ are both greater than 5, we can use the normal approximation to the binomial.

The warning line should be drawn at $0.15 + 1.96\sqrt{\dfrac{0.15 \times 0.85}{100}} = 0.22$

The action line should be drawn at $0.15 + 3.09\sqrt{\dfrac{0.15 \times 0.85}{100}} = 0.26$

Figure 18.4 shows the control chart based on these data.

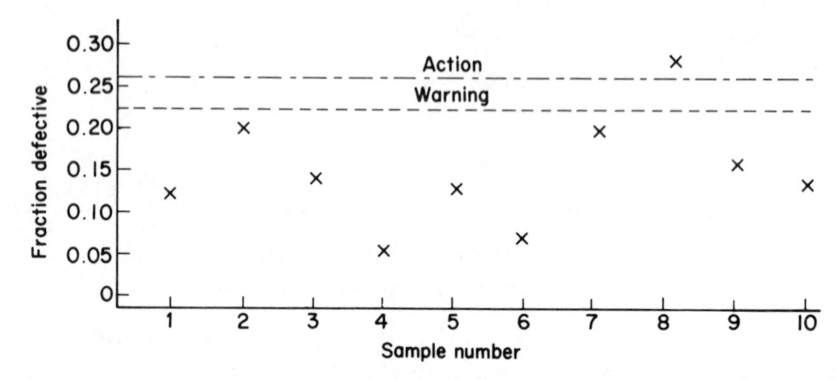

Fig. 18.4 A control chart for fraction defective

It appears that the fraction defective is not under control since one point is above the action line. □

18.4 ACCEPTANCE SAMPLING, A SINGLE SAMPLING PLAN

Suppose we manufacture items in large batches, and that these items may be classified only as defective or effective. From a sample of size n we will decide to accept the batch (as being fit to leave the factory), if there are only a small number of defectives in the sample, otherwise we will reject the batch. In other words, the basis of a *single sampling* plan is:

> Take a sample of n items and accept the batch if there are $\leqslant c$ defectives in the sample.

We can express the probability that a batch will be accepted, P_a, in terms of c and n, and the unknown fraction defective in the *batch*, p. So if X is the number of defectives, we know that:

$$X \sim B(n, p),$$

$$\therefore\ P_a = P(X \leqslant c),$$

i.e.

$$P_a = \sum_{x=0}^{c} \binom{n}{x} p^x (1-p)^{n-x}$$

If we plot P_a against p for a particular sampling plan, this plot is called the *operating characteristic* (OC) curve for the plan.

Example 18.4

Draw the OC curve for the following single sampling plan: take a sample of 50 items and accept the batch if there are $\leqslant 8$ defectives in the sample.

Here $$n = 50, \quad c = 8, \quad P_a = \sum_{x=0}^{8} \binom{50}{x} p^x (1-p)^{50-x}.$$

Since p can lie between 0 and 1, we calculate P_a for a number of values in this range using, in this case, tables of cumulative binomial probabilities (Table C.1). Before we do this, we know that if $p = 0$ there can be no defectives in the batch, and hence none in the sample, and so $P_a = 1$. Similarly, if $p = 1$ all items in the batch and sample are defective, and so $P_a = 0$. Other values are given in Table 18.5.

Table 18.5 P_a against p for a single sampling plan ($n = 50$, $c = 8$)

p	0	0.05	0.10	0.15	0.20	0.25	0.30
P_a	1	0.999	0.942	0.668	0.307	0.092	0.018

Note
P_a is very small for all p values greater than 0.3 for this plan.

Looking at Fig. 18.5 we can see that there is a high probability of accepting batches with $\leqslant 10\%$ defectives ($p \leqslant 0.1$) but this probability rapidly decreases as p increases.

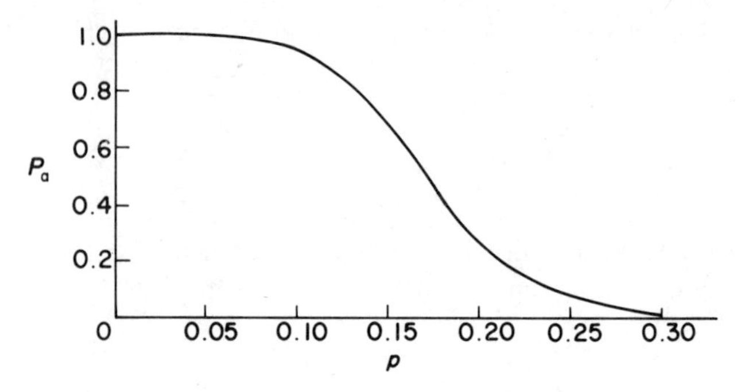

Fig. 18.5 The OC curve for a single sampling plan ($n = 50$, $c = 8$)

Ideally we can think of a sampling plan which will lead to certain acceptance of all batches having less than a certain critical fraction defective, p^* say, and certain rejection of all batches having more than p^* defective. The ideal OC curve would look like Fig. 18.6.

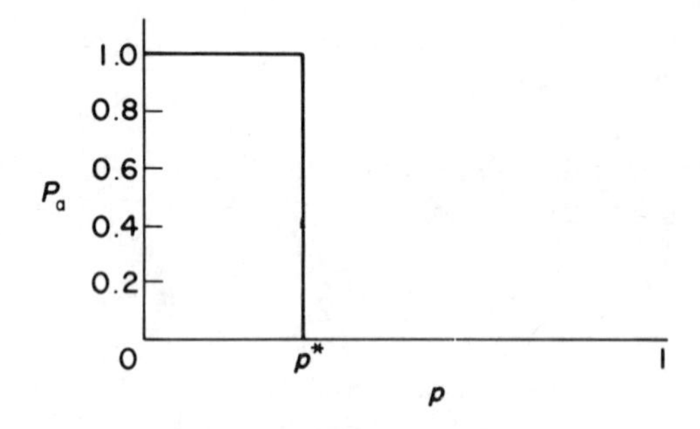

Fig. 18.6 The ideal OC curve □

18.5 ACCEPTANCE SAMPLING, A DOUBLE SAMPLING PLAN

It is clear that the OC curve, Fig. 18.5 (for the single sample plan of the previous section), does not have the ideal shape. In an attempt to get nearer to the ideal of Fig. 18.6, double sampling plans may be considered. The basis of a double sampling plan is as follows:

Take a first sample of n_1 items and:

(i) Accept the batch if there are $\leqslant c_1$ defectives in the sample.
(ii) Reject the batch if there are $\geqslant c_2$ defectives in the sample ($c_2 \geqslant c_1 + 2$).
(iii) Take a second sample of n_2 items if the number of defectives in the first samples lies between $(c_1 + 1)$ and $c_2 - 1)$ inclusive. Accept the batch if there are $\leqslant c_3$ defectives in total in the two samples, otherwise reject the batch.

Example 18.5

Draw the OC curve for the following double sampling plan:

Take a first sample of 20 items and:

(i) Accept the batch if there are $\leqslant 2$ defectives in the sample.
(ii) Reject the batch if there are $\geqslant 6$ defectives in the sample.
(iii) Take a second sample of 40 items if there are 3, 4 or 5 defectives in the first sample. Accept the batch if there are $\leqslant 9$ defectives in total, otherwise reject the batch.

Let X_1, X_2 denote the number of defectives in the first and second samples respectively. Then $X_1 \sim (20, p)$ and $X_2 \sim (40, p)$. The probability of accepting the batch is:

$$P_a = P(X_1 \leqslant 2) + P(X_1 = 3)\,P(X_2 \leqslant 6) + P(X_1 = 4)\,P(X_2 \leqslant 5) + P(X_1 = 5)\,P(X_2 \leqslant 4)$$

the last three terms expressing the idea that $X_1 + X_2 \leqslant 9$ for acceptance, if a second sample is needed.

We can calculate P_a for various values of p (see Table 18.6). When $p = 0$, $P_a = 1$, since there are no defectives in the batch or the sample. And, for example, when $p = 0.1$, using Table C.1,

$$\begin{aligned} P_a &= 0.677 + (0.867 - 0.677)0.901 + (0.957 - 0.867)0.794 \\ &\quad + (0.989 - 0.957)0.629 \\ &= 0.940 \end{aligned}$$

Comparing this table with Table 18.5 we see that the operating characteristic curves of the two plans would be almost identical.

Table 18.6 P_a against p for a double sampling plan

p	0	0.05	0.10	0.15	0.20	0.25	0.30
P_a	1	0.999	0.940	0.658	0.313	0.115	0.038

The question naturally arises as to which of the two plans is preferable. This question is answered in the following section. □

18.6 SINGLE VERSUS DOUBLE SAMPLING PLANS

How can we compare the relative merits of single and double sample plans? If the plans have the same OC curves, the only advantage of the double sampling plan may be that it requires fewer items to be sampled *on average*. In a single sampling plan the sample size, n, is always the same, but in a double sampling plan the sample size may be n_1 if only a first sample is needed or $n_1 + n_2$ if a second sample is needed. The number of items sampled is a random variable which we denote by N.

The average value of N in a double sampling plan is:

$$E(N) = n_1 + (\text{probability a second sample required})\, n_2$$

and so $E(N)$ depends on p, the fraction defective in the batch.

Example 18.6

Compare the average sample size for the single sampling scheme given in the example in Section 18.4 with the double sampling scheme given in the example in Section 18.5, noting that we have already shown that these plans have virtually the same OC curves.

For the single sample plan, $E(N) = 50$

For the double sampling plan,

$$E(N) = 20 + P(X_1 = 3, 4, 5) \times 40, \quad \text{where } X_1 \sim B(20, p)$$

For example, if $p = 0.1$,

$$E(N) = 20 + (0.989 - 0.677) \times 40 = 32.5;$$

see Table 18.7 for other values of $E(N)$.

The single and double sampling schemes may be compared graphically (see Fig. 18.7).

Table 18.7 The average number of items needed in a double sampling scheme

p	0	0.1	0.2	0.3	0.4	0.5
$E(N)$	20	32.5	43.9	35.2	24.9	20.8

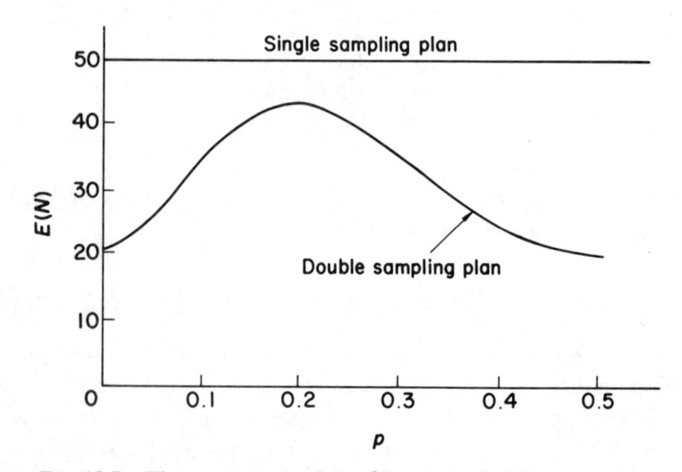

Fig. 18.7 The average number of items needed for two plans

We conclude that, although these plans have the same OC curves, the double sampling plan is, on average, more economical in terms of the average number of items which need to be sampled. □

18.7 SUMMARY

Control charts and acceptance sampling are two ways of controlling quality. For control charts, samples of product are taken at intervals. If the variable measuring quality is continuous, control charts for the sample mean and range are plotted. If the variable is discrete, a control chart for the fraction defective is plotted. Action and warning lines are drawn on the charts which enable the plotter to decide when control has been lost.

In acceptance sampling, a batch is accepted or rejected depending on the number of defectives found in items sampled from the batch. A single sampling scheme involves a fixed sample size, while a double sampling scheme may require a second sample to be taken. The latter scheme will require fewer items to be sampled on average, OC curves being 'equal'.

Section 18.2

1. The data below are the sample means and sample ranges for ten consecutive samples, each sample consisting of five measurements of a continuous random variable, X. Assuming X is normally distributed, plot control charts for the mean and range and comment on the degree of control.

Sample number	*1*	*2*	*3*	*4*	*5*	*6*	*7*	*8*	*9*	*10*
Sample mean	126.2	127.4	126.6	129.8	126.0	125.4	126.8	132.0	127.4	126.2
Sample range	8	6	7	6	8	7	6	19	6	7

2. The suspended solids content of a river downstream from the effluent outfall of a chemical plant was measured four times each month over a period of a year. Plot two control charts and comment on the state of control, given the following data, and assuming that the suspended solids content is normally distributed.

Month	*1*	*2*	*3*	*4*	*5*	*6*	*7*	*8*	*9*	*10*	*11*	*12*
Solids	60	70	80	60	60	80	30	60	50	100	110	50
content	60	20	50	30	20	60	40	10	50	90	80	110
(ppm)	20	40	30	50	80	30	70	60	40	80	100	120
	60	90	40	60	20	70	80	70	20	50	70	100

3. Quality control charts are used to aid control of the octane number of a gasoline product. Ten samples each of size three are taken as shown below. Assuming octane number is normally distributed, plot control charts for the sample mean and sample range. Discuss the state of control achieved.

Sample number	*1*	*2*	*3*	*4*	*5*	*6*	*7*	*8*	*9*	*10*
Octane number	93	88	89	86	91	92	88	86	84	82
	90	87	85	87	89	88	86	89	90	85
	87	87	84	85	92	89	90	89	87	84

4. Every half hour a bakery production inspector takes a random sample of five bread rolls from the production line and determines their masses in grams. The results for 10 consecutive samples are given below.

►

Sample number	*1*	*2*	*3*	*4*	*5*	*6*	*7*	*8*	*9*	*10*
Masses of the rolls (grams)	51.0	52.5	48.1	53.3	55.2	50.5	50.3	51.3	49.0	51.7
	52.6	49.4	52.9	46.6	52.4	48.2	52.3	52.7	49.9	49.1
	45.7	48.1	48.9	52.7	49.9	52.8	48.1	49.0	48.7	49.7
	52.3	50.1	47.4	47.3	53.0	49.5	49.2	52.8	49.0	49.0
	49.1	49.9	51.3	51.4	50.8	48.1	47.4	51.8	49.1	48.1

Calculate the mean and range for each of these 10 samples. Use these results to draw a control chart for the mean, showing 95% and 99.8% control lines and the control point corresponding to each sample.

At the Conference of Master Bakers, the manager of this bakery claims that, in the long run, less than 1 roll in 1000 leaves the above production line with a mass of 47 grams or less. In the light of the above results, is this statement correct? Justify your reply briefly.

(AEB (1983))

5. Reels of wire are wound automatically from a continuous source of wire. After each reel is wound the wire is cut and a new reel started. The aim is to wind 100 m onto each reel. The length of wire on 12 samples each of five reels was measured. The mean and range of each of the samples are given below:

Sample no.	*1*	*2*	*3*	*4*	*5*	*6*	*7*	*8*	*9*	*10*	*11*	*12*
Mean (m)	99.8	101.4	100.4	101.9	101.2	100.6	101.7	100.9	100.2	101.4	101.5	101.0
Range (m)	3.6	2.4	1.1	3.2	2.3	2.5	1.7	0.9	1.9	2.2	2.3	1.6

Assuming a normal distribution, use the information to estimate the standard deviation of the process and to calculate 95% and 99.8% control lines for the sample means. Draw a control chart for means and plot the 12 points.

Draw a control chart for ranges showing the upper 'action' and upper 'warning' limits given that these are respectively 2.34 and 1.81 times the mean sample range. Plot the 12 points and comment on the current state of the process. If the next sample measured 97.3, 100.6, 103.3, 100.1, 100.9 m, what action would you recommend and why?

Reels containing less than 95 m of wire are unacceptable to customers and reels containing more than 105 m make the process uneconomic. Comment on the ability of the process to produce reels which lie consistently within these limits.

(AEB (1984))

Section 18.3

6. The number of defectives found in 20 samples each of size 125 taken from a continuous process were as follows:

►

Sample number	1	2	3	4	5	6	7	8	9	10	11	12	13	14	15	16	17	18	19	20
Number of defectives	10	13	8	12	9	16	12	14	14	16	14	13	21	12	15	14	8	20	19	21

Estimate the fraction defective from these data, and plot a control chart for fraction defective showing the upper warning and action lines. Comment on the state of control.

7. Twelve samples each of approximately 1750 items were taken from a process. The actual numbers of items inspected in each sample and the number of defectives found were as follows:

Sample size	1780	1690	1610	1620	1690	1640	1650	1580	1610	1770	1660	1750
Number of defectives	180	150	130	110	250	120	170	120	180	150	150	90

Is the process in control? (*Hint*: in establishing control limits use the average sample size.)

Section 18.4

8. A random sample of 50 items is taken from a large batch of items and inspected. The batch is accepted if not more than two defectives are found. Draw the operating characteristic curve for this sampling plan.
   (a) What is the probability that a batch containing 12% defectives will be accepted?
   (b) What is the probability that a batch containing 5% defectives will be rejected?

9. A chemical company produces a large batch of components which is accepted or rejected according to the results of inspecting a sample of components. On inspection a component may be classified only as defective or effective. If the batch is accepted it will be sent out to one customer; otherwise it will be scrapped. It is suggested that a sample of 50 items should be taken and that the batch should be accepted if the sample contains at most 9 defectives.
   (a) What risk would the company take of rejecting a batch which contained only 10% defectives?
   (b) What risk would the customer take of being sent a batch which contained as many as 20% defectives?
   (c) Plot the OC curve for the sampling scheme, showing at least four plotted points.
   (d) What percentage of defectives must there be in a batch for it to have a 50% chance of being accepted?

10. Large batches of door hinges are produced at regular intervals by the Amalgamated Ironworks in Botherham. Let N be the number of hinges in each batch, and p be the ►

proportion of defectives. The company operates a single sample plan for its batch inspection by selecting 10 hinges at random and accepting the batch if there are two or less defectives, otherwise rejecting the batch.

(a) Show that the probability of accepting a batch is

$$(1-p)^8\,(1+8p+36p^2).$$

(b) Sketch the operating characteristic (OC) curve. (The values of this curve should be clearly indicated for two values of p other than 0 and 1.)

(c) When a batch is rejected it is not discarded, but instead subjected to 100% inspection and all defectives are replaced. Thus if a batch is rejected N hinges are inspected, whereas if the batch is accepted only 10 hinges are inspected. If $N = 200$, determine the expected amount of inspection in terms of p, the proportion of defectives. Evaluate your expression for $p = 0.1, 0.2, 0.3$, and comment upon the answers.

(AEB (1983))

Section 18.5

11. In a double sampling plan, a first sample of 25 items is taken and the batch is accepted if there are less than or equal to 5 defectives. The batch is rejected if there are 8 or more defectives. If there are 6 or 7 defectives a second sample of 40 items is taken and the batch is accepted if the total number of defectives found in the two samples is less than or equal to 8, otherwise the batch is rejected.

    Draw the OC curve. Calculate and plot the expected number of items sampled for $p = 0, 0.1, 0.2, 0.3, 0.4$ and 0.5.

12. When checking large batches of goods the following acceptance sampling plans have similar operating characteristics.

    Plan 1: Take a sample of size 50 and accept the batch if 3 or fewer defectives are found, otherwise reject it.

    Plan 2: Take a sample of size 30, accept the batch if zero or one defectives are found and reject the batch if 3 or more defectives are found. If exactly 2 defectives are found take a further sample of size 30. Accept the batch if a total of 4 or fewer defectives (out of 60) are found, otherwise reject it.

    Using the following table, or otherwise, verify that the two plans have similar probabilities of accepting a batch containing 5% defective.

    The table gives the probability of obtaining r or more defectives in n independent trials when the probability of a defective in a single trial is 0.05.

r	$n = 30$	$n = 50$
0	1.0000	1.0000
1	0.7854	0.9231
2	0.4465	0.7206
3	0.1878	0.4595
4	0.0608	0.2396

For the second plan, evaluate the expected number of items inspected each time the plan is used when the proportion defective in the batch is 0, 0.02, 0.05, 0.10 and 1.00. Sketch a graph of the expected number of items inspected against proportion defective in the batch.

What factors should be considered when deciding which of the two plans to use?

(AEB (1984))

Information on projects in statistics at A-level

▼

Only two A-level syllabuses in subjects containing statistics as a compulsory element require candidates to carry out projects. These are:

1. Pure Maths with Statistics 420 (London)
2. Statistics (Advanced) (JMB)

For their syllabus 420, the London Board require the candidate to carry out a total of five projects. These are not directly assessed but they must be brought to the examination room for reference in Paper 3. More information, including a list of possible project titles, may be obtained by writing to:

University of London Publications Office
52 Gordon Square
London WC1H 0PJ

For their Statistics (Advanced) syllabus, the JMB require the candidate to submit one project for assessment. This contributes a maximum of 15% to the overall assessment. More information, including a list of possible project titles, may be obtained from 'Notes of Guidance on A-level Statistics' by writing to:

Joint Matriculation Board
Manchester M15 6EU

Another publication, *Project Work in A-level Statistics*, which will be found useful mainly by candidates taking the JMB's Statistics (Advanced), is obtainable by writing to:

Centre for Statistical Education
25 Broomgrove Road
Sheffield S10 2BP

Appendix A
Answers to worksheets†

† In addition to the statement in the Acknowledgements that the accuracy of the answers given in this Appendix are solely the responsibility of the author, the following relates to the answers to those questions from past papers of the University of London School Examinations Board:

'The University of London School Examinations Board accepts no responsibility whatsoever for the accuracy or method of working in the answers given'.

WORKSHEET 1

1. (a) m.p.g., maximum m.p.h., number of cylinders, number of MOTs passed, make, colour.
   (b) weight, age, number of prizes won, number of tins of dog food eaten per week, type, hair colour.
   (c) income, average age, number in group, number of records in current charts, type of music, recording company.
   (d) annual profit, height of rig above sea-bed, number of men employed, number of barrels of oil filled per day, company owning rig, location (country).
   (e) balance-of-payments deficit, size (hectares), number of inhabitants, number of cathedrals, type of government, location (continent).
   (f) average charge for bed and breakfast, age, number of rooms, star-rating, type of location (city, small town, country), type of menu offered (e.g. *à la carte*).
   (g) pupil/teacher ratio, total floor area, number of pupils, number of teachers, type (state, independent), type (primary, secondary).
2. (a) 164.5–169.5; (b) guess 167 cm; (c) 167 cm, from cumulative frequency polygon; (d) 63 %; (e) 63 %.
3. (a) 1; (b) $1\frac{1}{2}$.
4. (a) BSc; (b) type of degree not quantifiable.
5. Shape identical to Fig. 1.6.
6. (a) (i) 11 %, (ii) 64 %, (iii) 25 %; (b) 11.8 minutes.
7.

Weight (g)	*Number of jars*
200 upto but not including 201	13
201 " " " " 202	27
202 " " " " 203	18
203 " " " " 204	10
204 " " " " 205	1
205 " " " " 206	1

8. (a) 62 %; (b) 52 %.
9. (b) too many different individual values.
   (c) (i) 21, (ii) 29. Average for highest scores in Division 2 is higher than for Division 1. (Averages to be discussed in Chapter 2.)

WORKSHEET 2

1. uncoded, $\bar{x} = \frac{590}{50} = 11.92$ minutes.
   coded with $c = 1$, $a = 11$; $\bar{x} = \frac{46}{50} + 11 = 11.92$ minutes.
   coded with $c = 1$, $a = 12$; $\bar{x} = -\frac{4}{50} + 12 = 11.92$ minutes.
   As expected, all three answers are the same.

2. (a) $\bar{x} = \frac{400}{200} = 2$; (b) expect 2 ($= \frac{1}{10} \times 20$) numbers out of 20 on average to be the digit 8. Hence, on this basis, can conclude numbers are random.

3. (a) $\bar{x} = \dfrac{n_1\bar{x}_1 + n_2\bar{x}_2}{n_1 + n_2}$, $\bar{x} = \dfrac{\sum n_i x_i}{\sum n_i}$, where the summations are from $i = 1$ to $i = k$.
   (b) $\bar{x}_1 = \frac{489}{22} = 22.33$, $\bar{x}_2 = \frac{615}{22} = 27.95$, $\bar{x} = 25.09$.

4. $\frac{206}{60} = 3.43$.

5. (a) $\Sigma(x_i - \bar{x}) = \Sigma x_i - \Sigma\bar{x} = \Sigma x_i - n\bar{x} = \Sigma x_i - \Sigma x_i = 0$.
   (b) $10 \times 4.3 - 62 = -19$.

6. ungrouped $\bar{x} = \dfrac{6730}{40} = 168.25$ cm; grouped $\bar{x} = \dfrac{1}{n}\sum f_i x_i = \dfrac{6725}{40} = 168.125$ cm.

   The difference between the two answers is small.

7.

Grade	F	E	D	C	B	A
Mark range	$\leqslant 48$	49–57	58–64	65–72	73–82	$\geqslant 83$

8. (a) mean = 150 km, median = 100 km, mode = 62.5 km; median preferred because of positive skewness (mean > median > mode).
   (b) mean = 1.95, median = 1, mode = 1; mean preferred because of approximate symmetry.

9. mean = £14,500, median = £13,310, mode = £12,710. Positive skewness since mean > median > mode.

10.

Number of members	*f/1000 group width*
less than 100	900
100–499	290
500–999	100
1,000–2,499	38.7
2,500–4,999	20.0
5,000–9,999	6.0
10,000–14,999	2.6
15,000–24,999	2.1
25,000–49,999	0.5
50,000–99,999	0.3
100,000–199,999	0.2

►

Median $= 500 + \frac{240 - 206}{256 - 206} \times (1000 - 500) = 840$ members; median preferred because of positive skewness which would be exhibited by histogram.

11. 5020 g, assuming growth at constant rate.
12. means are equal when $x_1 = x_2$.
13. (a) 95.1; (b) 95.1; (c) 95.1; (i) answers to (a) and (b) agree because the two methods are arithmetically identical; (ii) (b) and (c) agree coincidentally, usually these answers will not be the same.
14. 153.7.

WORKSHEET 3

1. (a) $\bar{x} = 1.04$, $s' = 0.0289$ kg; (b) $\bar{x} = 1.05$, $s' = 0.0289$ kg; (c) $\bar{x} = 1050$, $s' = 28.9$ g.

2. $s' = 1.534$ for uncoded and coded methods, whatever value of a is used.

3. $\bar{x} = \dfrac{25 \times 18.2 + 279}{40} = \dfrac{734}{40} = 18.35.$

   For first 25 observations, $\Sigma x_i^2 = 25(3.25^2 + 18.2^2) = 8545.$

   For all 40 observations, $\Sigma x_i^2 = 8545 + 5524 = 14{,}069$ and $s' = \sqrt{\dfrac{14{,}069}{40} - \left(\dfrac{734}{40}\right)^2}$

   $= 3.87.$

4. $\bar{x} = \dfrac{n_1 \bar{x}_1 + n_2 \bar{x}_2}{n_1 + n_2}$, $s' = \sqrt{\dfrac{n_1(s_1')^2 + n_2(s_2')^2}{n_1 + n_2} + \dfrac{n_1 n_2 (\bar{x}_1 - \bar{x}_2)^2}{(n_1 + n_2)^2}}.$

5. (a) $\bar{x} = \dfrac{66}{50} = 1.32$, $s' = \sqrt{\dfrac{168}{50} - \left(\dfrac{66}{50}\right)^2} = 1.27.$

   (b) 66.

   (c) $\dfrac{66 \times 100}{500} = 26.4\%$; expect $\dfrac{100}{4} = 25\%$.

6. $\bar{x} = 37.9$, $s' = 9.16$; (a) 78.3%; (b) 63.9%.

7. $\bar{x} = 6.49$, $s' = 1.714$, upper quartile $= 7$.

   Draw a line chart. Sample may not be representative because not all the population own a telephone. Also there could be variation in name length from one geographical area to another.

8. For histogram, use % frequency per 10 unit group width, for example. (i) median $= 7.64$ units; (ii) inter-quartile range $= 21.65 - 0.75 = 20.9$ units; (iii) mean $= 4.54$ units, assuming 51 or more means 51–65, for example; (iv) 0.085 or 8.5%.

   Interview period over preceding week because easier to remember over short period. For whole year, repeat the interview every month, say, using different random samples each time.

9. (i) mean $= 6.0$ (1 d.p.) miles, assuming that '30 miles and over' means '30 but under 45 miles'; (ii) median $= 3.54$ miles; (iii) inter-quartile range $= 6.45$ miles; (iv) 0.64 (64%), using mean $= 6.0$.

   Median is less than 1 mile for people travelling for education compared with 3.5 miles for people going to or from work. Explanation: schools serve local catchment areas, whereas those travelling to work may have better travel facilities such as cars and hence are more prepared to travel.

10. $1 - 0.827 = 0.173$; median $= £72.4$; semi-interquartile range $= \frac{1}{2}(90.9 - 58.9)$ $= £16.0$. Use % of men per £20 group width for histogram, for example.

11. $\bar{x} = 44.0$, $s' = 13.3$, median $= 44.0$, 3rd and 97th percentile.

12. (i) 32.06; (ii) 30.36; (iii) $33.93 - 26.83 = 7.1$; (iv) 0.135 (13.5%). ►

13. mean = 113.8, median = 115.1, standard deviation = 14.1; $S = -0.09$.
Other histogram would be slightly skewed to the right, instead of to the left. S is better than S' because it is independent of the units of the variable (i.e. dimensionless), whereas S' will have units (i.e. lb).

14. $s' = £6590$; inter-quartile range $= 16{,}895 - 10{,}733 = £6160$.

$$\text{Measure of skewness} = \frac{3(\text{mean} - \text{median})}{\text{standard deviation}} = \frac{3(14{,}500 - 13{,}310)}{6590} = 0.54$$

Although there is positive skewness, the measure of skewness is less than 1, and standard deviation could still be used as a measure of dispersion.

15. (a) $s' = 126.5$, inter-quartile range $= 225 - 50 = 175$;

$\dfrac{3(\text{mean} - \text{median})}{s'} = \dfrac{3(150 - 100)}{126.5} = 1.19$. Since this is > 1, there is marked positive skewness, so prefer the inter-quartile range.

(b) $s' = 1.52$, inter-quartile range $= 2 - 1 = 1$;

$\dfrac{3(\text{mean} - \text{median})}{s'} = \dfrac{3(1.95 - 2)}{1.52} = -0.1$. Since this is well inside the range -1 to $+1$, slight skewness only, prefer standard deviation.

16. (b) median = 53.67, inter-quartile range $= 63.3 - 42.4 = 20.9$, percentile range $= 72.0 - 29.5 = 42.5$. Median for sample 2 is larger.

(c) $\bar{x} = 52.25$, $s' = 15.57$, $\dfrac{3(\text{mean} - \text{median})}{s'} = -0.27$, small negative skewness.

17. (a) $\bar{x} = 36.7$, $s' = 11.2$ (using groups 10–19, 20–29, and so on). These are suitable measures of location and dispersion because histogram is almost perfectly symmetrical.
(b) 63.8%.

18. (a) If a number of observations are each multiplied by a constant, the mean and standard deviation should also be multiplied by the constant. If a number of observations are each multiplied by one constant and then a second constant is added to each, the new mean is obtained by multiplying the original mean by the first constant and adding the second constant, while the new standard deviation is obtained by multiplying the original standard deviation by the first constant.

19. (a) (i) $m + k$, s; (ii) km, ks; $\bar{x} = 11$, $s' = 4$, new marks are 35, 45, 55, 75, 90.
(b) For first set, sum of squares $= (5^2 + 17^2) \times 20 = 6280$
For second set, sum of squares $= (5^2 + 22^2) \times 20 = 10{,}180$

16,460

$$\text{Combined mean} = \frac{20 \times 17 + 20 \times 22}{40} = 19.5.$$

$$\text{Combined standard deviation} = \sqrt{\frac{16{,}460}{40} - 19.5^2} = 5.59.$$

►

20. $l = 1.25$, $k = -5$.

For w values; lower quartile $= 1.25 \times 40 - 5 = 45$, median $= 60$, upper quartile $= 70$. Inter-quartile range for x values is $60 - 40 = 20$, and for w values is $70 - 45 = 25$. Since $25 = 1.25 \times 20$, we have used only the value of l. Hence inter-quartile range for w is independent of k.

21. (a) $\bar{x} = \dfrac{n+1}{2}$, $s' = \sqrt{\dfrac{n^2-1}{12}}$.

(b) (i) $\bar{x} = 6.5$, $s' = 2.87$, $(s')^2 = 8.25$.
(ii) $\bar{x} = -5.5$, $s' = 2.87$, $(s')^2 = 8.25$.
(iii) $\bar{x} = -4.5$, $s' = 2.87$, $(s')^2 = 8.25$.

(c) (i) $\bar{x} = 51.5$, $s' = 28.87$, $(s')^2 = 833.25$.
(ii) $\bar{x} = -50.5$, $s' = 28.87$, $(s')^2 = 833.25$.
(iii) $\bar{x} = -49.5$, $s' = 28.87$, $(s')^2 = 833.25$.

WORKSHEET 4

1. No reason to suppose the two outcomes are equally likely.

2. The three outcomes are not equally likely. However, the four outcomes HH, HT, TH, TT, are equally likely. Hence $P(2 \text{ heads}) = \frac{1}{4}$, $P(1 \text{ head and } 1 \text{ tail}) = \frac{2}{4}$, $P(2 \text{ tails}) = \frac{1}{4}$.

3. Yes, but not a good estimate since 5 is not a large number.

4. (a) $\frac{3}{15}$; (b) $\frac{12}{15}$; (c) $\frac{8}{15}$; (d) $\frac{7}{15}$; (e) 1; (f) 0.

   In both cases the answers sum to a total of 1.

5. (a) $S = \{R1, R2, \ldots, R6, W1, \ldots, W6, B1, \ldots, B6\}$;
   (b) Yes, $\frac{1}{18}$;
   (c) (i) $\frac{6}{18}$, (ii) $\frac{3}{18}$, (iii) $\frac{9}{18}$, (iv) $\frac{6}{18}$, (v) $\frac{6}{18}$.

6. (a) $\frac{20}{40}$; (b) $\frac{16}{40}$; (c) $\frac{4}{40}$.

   $$\frac{20}{40} = \frac{\text{Area of rectangle on base 0–100}}{\text{Total area of histogram}}, \text{ and so on.}$$

   $$\text{Also } 1 = \frac{\text{Total area of histogram}}{\text{Total area of histogram}}$$

7. (a) 0.1, (b) $\frac{1}{3}$, (c) $\frac{1}{4}$; (i) No, since $P(B) \neq P(B|A)$; (ii) No, since $P(A \cap B) \neq 0$.

8. $P(A \cup B) = P(A) + P(B) - P(A)P(B)$, since A and B are independent.

   (a) $\frac{4}{7}$; (b) $\frac{3}{35}$; (c) $\frac{2}{7}$; (d) $\frac{3}{10}$.

9. $P(A \cup B) = P(A) + P(B)$, since A and B are mutually exclusive.

   (a) 0.2; (b) 0; (c) 0; (d) 0.

10. No. If A and B are mutually exclusive, $P(A \cap B) = 0$; whereas if A and B are statistically independent, $P(A \cap B) = P(A)P(B)$, which is > 0 if $P(A)$ and $P(B)$ are both > 0.

11. $P(A) = \frac{3}{36}$, $P(B) = \frac{10}{36}$: $P(C) = \frac{7}{36}$; (a) A and B are mutually exclusive, so are A and C; (b) no pairs are statistically independent.

12. (a) $P(A) = \frac{80}{100}$, $P(B) = \frac{85}{100}$, $P(A \cap B) = \frac{70}{100}$, $P(A \cup B) = \frac{95}{100}$, $P(B|A) = \frac{70}{80}$, $P(A|B) = \frac{70}{85}$;
    (b) (i) No, (ii) No.

13. (a) $A' \cup B'$ and $A \cap B$ are complementary events; $A' \cap B'$ and $A \cup B$ are complementary events.
    (b) (i) $P(A') = 0.6$, (ii) $P(B') = 0.7$, (iii) $P(A' \cup B') = 0.8$,
    (iv) $P(A' \cap B') = 0.5$, (v) $P(B'|A') = \frac{5}{6}$, (vi) $P(A'|B') = \frac{5}{7}$.

14. Yes, because $P(A \cap B) = P(A)P(B)$ leads to $P(A' \cap B') = P(A')P(B')$, using the results of Exercise 13, and $P(A') = 1 - P(A)$, $P(B') = 1 - P(B)$.

▶

15. No, unless A and B are complementary events. For A' and B' to be mutually exclusive we require $P(A' \cap B') = 0$. From Exercise 13, and using $P(A \cup B) = P(A) + P(B)$ for mutually exclusive events, we require $P(A) + P(B) = 1$.

16. (a) $\frac{34}{52}$; (b) $\frac{8}{52}$; (c) $\frac{16}{52}$; (d) 1; (i) '2' and '3'; (ii) 'numbered card' and 'picture card'.

17. (a) (i) $\frac{1}{28}$, (ii) $\frac{1}{28}$, (iii) $\frac{1}{28}$, (iv) $\frac{1}{28}$; (b) (i) $\frac{1}{4}$, (ii) $\frac{1}{2}$.

18. (i) $1 - p^2$; (ii) $1 - 2p + p^2$; machine A is more likely to fail when $0 < p < 1$; when $p = \frac{1}{2}$.

19. $p(1-p)^{r-1}$.

20. $P(A) = 0.2$, $P(B) = 0.1$, $P(A \cap B) = P(55) + P(66) = 0.02$. Since $P(A \cap B) = P(A)P(B)$, A and B are independent events.
For a fair die, $P(A) = \frac{1}{6}$, $P(B) = \frac{1}{6}$, $P(A \cap B) = \frac{2}{36}$. A and B are not independent.

21. (i) $\frac{1}{12}$; (ii) $\frac{3}{4}$; (iii) $\frac{1}{4}$.
Show $P(C) = \frac{1}{6}$ and hence $P(B \cap C) = P(B \cup C) - P(B) - P(C) = 0$.

22. (i) $\frac{1}{2}$; (ii) $\frac{1}{3}$; (iii) $\frac{2}{3}$; (iv) $\frac{4}{7}$.

23. (a) 0.336; (b) 0.452; (c) 0.188; (d) 0.024.
The total is 1 because the four events are mutually exclusive and exhaustive.

24. (a) $\frac{1}{8}$; (b) $\frac{3}{8}$; (c) $\frac{3}{8}$; (d) $\frac{1}{8}$.
P(eldest is a boy) $= \frac{2}{3}$.

25. (a) (i) 0.49, (ii) 0.42, (iii) 0.09; (b) (i) 0.21, (ii) 0.34.

26. (a) $\frac{76}{125}$; (b) 0.412.

27. (i) 0.095; (ii) 0.9033; (iii) Mr Smith. ($P(\geqslant 1$ defective) is 0.0975 for A and 0.0967 for B); (iv) 0.9029.

28. $P(R) = \frac{1}{2}$, $P(B \cap R) = \frac{11}{120}$, $P(B \cup R) = \frac{7}{12}$, $P(B|R) = \frac{11}{60}$.

29. (i) 0.12; (ii) 0.38; (iii) 0.3158; (iv) 0.24; (v) 0.7742.

30. (i) 0; (ii) $P(A \cup B) = P(A) + P(B)$. $P(A \cup B) = 1$ when A and B are also exhaustive. $A \cup B$ means 'either there is an orange in each of the n packets or there is a lemon in each of the n packets'. Probability of no refund $= (\frac{1}{2})^{n-1}$. Must buy at least 5 packets; 0.475; 0.7107.

31. (i) $\frac{2}{9}$; (iii) $\frac{2}{5}$; (iv) $\frac{5}{11}$; Similarly P (8 before 7) $= \frac{5}{11}$, P (9 before 7) $= \frac{2}{5}$, P (10 before 7) $= \frac{1}{3}$, P (win) $= \frac{2}{9} + 2(\frac{3}{36} \times \frac{1}{3} + \frac{4}{36} \times \frac{2}{5} + \frac{5}{36} \times \frac{5}{11}) = 0.4929$.

32. (a) (i) 0.0769, (ii) 0.0710, (iii) 0.0655, (iv) $(\frac{1}{13})(\frac{12}{13})^{n-1}$.
(b) (i) 0, (ii) 0.0059, (iii) 0.0109, (iv) $(n-1)\dfrac{12^{n-2}}{13^n}$.

33. $\dbinom{n}{r} p^r (1-p)^{n-r}$.

34. (a) $12! = 479{,}001{,}600$; (b) $2 \times \dfrac{11!}{12!} = \dfrac{1}{6}$; (c) $0.1364 = \frac{3}{22}$.

▶

With round table: (a) 11!, if specific places at the table do not matter (12! if they do); (b) 0.1818 = $\frac{2}{11}$; (c) 0.1455 = $\frac{8}{55}$.

35. $\binom{d}{x}$ is the number of ways of selecting x defectives from d defectives, and so on.

36. (i) 0.2545 = $\frac{14}{55}$; (ii) 0.1636 = $\frac{9}{55}$; (iii) 0.1091 = $\frac{6}{55}$, 0.3535 = $\frac{35}{99}$, 0.1636 = $\frac{9}{55}$.

37. (a) 224; (b) 425.

38. $P(1 \text{ boy}) = \dfrac{\binom{20}{4}\binom{15}{1}}{\binom{35}{5}} = 0.2239$; (i) 0.2665; (ii) 0.2092.

39. 0.2597.

40. $\frac{1}{35}$, $\frac{3}{7}$.

41. (i) 0.1493; (ii) $\frac{1}{17}$; P (exactly one woman deals a pair of queens) = 0.1812.

42. (a) 50,400; (b) (i) 180, (ii) 225.

43. (a) (i) $\frac{5}{18}$, (ii) $\frac{1}{6}$, (iii) 0.3571 = $\frac{5}{14}$, (iv) 0.5635 = $\frac{71}{126}$; (b) 18.

44. (a) $\frac{2}{7}$; (b) (i) $\frac{3}{4}$, (ii) $\frac{1}{4}$.

45. (a) 6.1%; (b) (i) 0.2623, (ii) 0.2459, (iii) 0.4918.

46. 0.1538, 0.9863. For the general population the diagnostic test improves the probability of detecting the disease from 0.01 to 0.1538. For the hospital population the improvement is from 0.8 to 0.9863.

47. (i) $\frac{1}{80}$; (ii) $\frac{3}{4}$; (iii) $\frac{9}{40}$; (iv) $\frac{1}{64}$; (v) $\frac{1}{6}$; (vi) $\frac{1}{14}$.

48. (i) $\frac{2}{5}$; (ii) $\frac{1}{2}$; (iii) $\frac{4}{5}$.

49. 0.400, 0.420, 0.400.

50. (a) Substituting the given result into Bayes' formula leads to $P(A|B) = P(A)$ (b) $P(C) = 0.310$. C and D are not independent, since $P(C|D) = 0.7$ so that $P(C) \neq P(C|D)$; $P(D|C') = 0.130$.

51. (a) $\frac{5}{12}$; (b) (i) 0.0965, (ii) $\frac{5}{11}$.

52. (i) $P(\text{white}) = \frac{1}{2}$, $P(\text{yellow}) = \frac{1}{8}$; (ii) $\frac{1}{3}$; (iii) $\frac{17}{64}$.

WORKSHEET 5

1. (a) discrete; (b) continuous; (c) discrete; (d) discrete; (e) continuous; (f) discrete; (g) discrete; (h) discrete; (i) continuous; (j) discrete.

2.

Number of married couples (x)	0	1	2
$p(x)$	0.6935	0.2972	0.0093

3.

Gambler's gain per bet (x)	3	-1
$p(x)$	$\frac{1}{6}$	$\frac{5}{6}$

$\mu = -£0.33$, $\sigma^2 = £^2 2.22$, $\sigma = £1.49$; £6

4.

Gain (x)	999	499	99	-1
$p(x)$	0.0001	0.0001	0.0010	0.9988

Expected gain $= -£0.75$

5. $E(x) = \frac{5}{3}$, $E(X^2) = \frac{10}{3}$, $\mu = \frac{5}{3}$, $\sigma^2 = \frac{5}{9}$.

6.

Number per car (x)	1	2	3	4	5
$p(x)$	0.550	0.200	0.150	0.075	0.025

$\mu = 1.825$, $\sigma^2 = 1.194$.

7.

x	0	1	2	3	4
$p(x)$	0.0625	0.2500	0.3750	0.2500	0.0625

Draw a graph similar to Fig. 5.1. $\mu = 2$, $\sigma^2 = 1$.

8. $k = \dfrac{1}{n}$, mean $= \dfrac{(n+1)}{2}$, variance $= \dfrac{(n^2-1)}{12}$

▶

9.

Number of red balls (x)	0	1	2
$p(x)$	$\left(\frac{n-r}{n}\right)^2$	$\frac{2r(n-r)}{n^2}$	$\left(\frac{r}{n}\right)^2$

$$\mu = \frac{2r}{n}, \sigma^2 = \frac{2r(n-r)}{n^2}.$$

10. p^3, $p^3(2-p^3)$, $p^3(2-p)^3$. Expected numbers are $2000p^3$, $1000p^3(2-p^3)$, $1000p^3(2-p)^3$.

p	0	0.5	1.0
$E(X_A)$	0	250	2000
$E(X_C)$	0	422	1000

Graph shows $E(X_C) > E(X_A)$, except when $p > 2 - 2^{1/3}$ $(= 0.74)$.

11. (i) (a) both E_1 and E_2 occur; (b) E_1 does not occur, but E_2 occurs;
(c) either E_1, or E_2, or both E_1 and E_2, occur;
(d) either E_1 or E_2 occurs, but not both.

(ii) 0.11;

x	0	1	2
$p(x)$	0.81	0.17	0.02

$\mu = 0.21$, $\sigma^2 = 0.21$.

12. $\frac{(n-4)(n-5)}{n(n-1)}$

13. $\mu = \frac{7}{6}$, $\sigma^2 = \frac{17}{36}$

Number of grandchildren (y)	0	1	2	3	4
$p(y)$	$\frac{28}{108}$	$\frac{33}{108}$	$\frac{31}{108}$	$\frac{12}{108}$	$\frac{4}{108}$

$E(Y) = \frac{49}{36}$. Assumption of independence may not be valid if: (i) there is inter-marrying in the village; (ii) if children from families with 1 (or 2) children are more likely to have families with 1 (or 2) children.

14. $\frac{6}{11}$; $\frac{11}{36}\left(\frac{5}{6}\right)^{2(n-1)}$, 5.54; 6 throws (assuming that the condition 'given that A wins eventually' does not apply to the last part of the question).

15. $\frac{ka}{(1-a)^2}$.

▶

16. $p(x) = p(1-p)^{x-1}$ for $x = 1, 2, \ldots$; $p(x) = 0$, otherwise, $\mu = \dfrac{1}{p}$, $\sigma^2 = \dfrac{(1-p)}{p^2}$.

17.

Number of fruits eaten (x)	1	2	3
$p(x)$	$\frac{35}{126}$	$\frac{55}{126}$	$\frac{36}{126}$

$\mu = 2.008$

18.

Number of correct words (x)	0	1	2	3
$p(x)$	$\frac{4}{35}$	$\frac{18}{35}$	$\frac{12}{35}$	$\frac{1}{35}$

$\dfrac{54+r}{35}$; 50

19. $G(t) = \dfrac{tp}{1-t(1-p)}$

20. $G(t) = \dfrac{t^3+2t^2+3t}{6}$

23. $k = e^{-2} = 0.1353$
 (i) $P(R=0) = 0.1353$, $P(R=1) = 0.1353$, $P(R=2) = 0.2030$, $P(R=3) = 0.1579$, $P(R=4) = 0.1410$;
 (ii) 3; (iii) 5; (iv) 3.

24. $k = 4$, $\mu = \frac{8}{15}$, $\sigma^2 = \frac{11}{225}$; $f(x)$ is a maximum when $x = 0.58$.

25. $\mu = 0$, $\sigma^2 = 2$, $P(|X| < \frac{1}{2}) = 0.3935$.

26. $\mu = 50$ minutes; (a) 0.6916; (b) 0.5617.

27. $k = \frac{3}{8}$, $E(X) = 0$, $\mathrm{Var}(X) = \frac{2}{5}$; (i) 0.1660; (ii) 0.5592.

28. (i) 0.85; (ii) 0.70; (iii) 0.55; 0.6, 0.9, 0.3, 100 hours; 0.648, 88 hours.

29. $\mu = 2.726$, $\sigma^2 = 0.1651$.

30. $k = 12$; $\mu = 0.6$, mode of $X = \frac{2}{3}$; $E(|X-\mu|) = 0.1659$.

31. (i) $A = 3$; (ii) $\mu = \frac{3}{2}$, $\sigma^2 = \frac{3}{4}$; (iii) 0.008.

32. (i) $F(x) = 2x^2 - x^4$, $0 \leqslant x \leqslant 1$; $F(x) = 0$, $x < 0$; $F(x) = 1$, $x > 1$.
 (ii) $F(x) = \dfrac{x}{27}(9+3x-x^2)$, $0 \leqslant x \leqslant 3$; $F(x) = 0$, $x < 0$; $F(x) = 1$, $x > 3$.

33. $k = \dfrac{3}{4}$; $f(x) = \dfrac{3x}{4}(2-x)$, $0 \leqslant x \leqslant 2$; $f(x) = 0$, elsewhere.

►

34. $1-e^{-kt}$, $t \geqslant 0$; 0, $t < 0$.
ke^{-kt}, $t \geqslant 0$; 0, $t < 0$.
$k = \frac{1}{2000}$, probability of a component lasting at least 4000 hours = 0.1353; 5000 hours.

35. $\mu = 1.25$, $\sigma^2 = 0.5375$; $P(X > 2) = 0.1852$.

36. $k = \frac{1}{2}$, median = 0.5858, inter-quartile range = $1 - 0.2679 = 0.7321$.

37. $k = \frac{2}{5}$.

$$F(z) = 0,\ z < 0;\ F(z) = 0.8(1 - \cos \tfrac{1}{2}z),\ 0 \leqslant z \leqslant \frac{2\pi}{3};$$

$$F(z) = 0.6 - 0.4\cos\left(z - \frac{\pi}{3}\right),\ \frac{2\pi}{3} \leqslant z \leqslant \frac{4\pi}{3};\ F(z) = 1,\ z > \frac{4\pi}{3}.$$

38. $\mu = \frac{2}{3}$, median $= 2 - \sqrt{2}$.

39. $k = \frac{1}{4}$; median $= \frac{5}{2}$, $\mu = \frac{31}{12}$.

40. $F(x)$, the distribution function is given. $F(10) = 1$ leads to $c = 0.659$; (a) 0.6887; (b) 0.5247.

41. (i) $A = \frac{1}{4}$; (ii) $\frac{9}{16}$; (iii) $m = 1.08$, $q^4 - 8q^2 + 4 = 0$; (iv) $\mu = \frac{16}{15}$; (v) $\sigma^2 = 0.1956$.

42. $\theta = 1.4427 = (\ln 2)^{-1}$, $\mu = 1.4427$, $\sigma^2 = 0.0827$; median $= \sqrt{2} = 1.414$, lower quartile, 1.189, upper quartile = 1.682.

43. $M(t) = \left(1 - \dfrac{t}{2}\right)^{-1}$, $\mu = \frac{1}{2}$, $\sigma^2 = \frac{1}{4}$

44. $M(t) = (1-t)^{-2}$, $\mu = 2$, $\sigma^2 = 2$.

45. $\mu = \dfrac{1}{\lambda}$, $\sigma^2 = \dfrac{1}{\lambda^2}$, $M_\mu(t) = e^{-t/\lambda}\left(1 - \dfrac{t}{\lambda}\right)^{-1}$, $\mu_1 = 0$, $\mu_2 = \dfrac{1}{\lambda^2}$.

Not surprised, since μ_1 is always zero. No coincidence, since $\sigma^2 = \mu_2$; see (5.18).

46. $\dfrac{1-\theta}{1-t} + \dfrac{\theta}{(1-t)^2}$, $\mu_3 = 2\theta^3 - 6\theta^2 + 6\theta + 2$.

47. (i) 110, 100; (ii) 0, 100; (iii) 1200, 10,000; (iv) 1080, 10,000; (v) 0, 1.

48. (b) (i) $\frac{1}{14}$; (ii) 480 p = £4.80.
(c) $n = 25$, maximum expected profit = 500 p = £5.00.

49. $P(x = r) = pq(p^r + q^r)$, $r = 0, 1, 2, \ldots$

$$G(t) = \frac{pq(2-t)}{(1-pt)(1-qt)};\ \text{variance of family size} = 2.$$

50. (i) 11.85; (ii) 0.3; (iii) 532 p = £5.32; (iv) 500 p = £5.00; (v) 12.
If he buys 10 magazines, mean monthly profit = $10x$;
If he buys 11 magazines, mean monthly profit = $10.85x - 10.5$;
$x \leqslant 12$ pence.

►

51. (i) $g(y) = \dfrac{by(y-1)}{5}$, $1 \leqslant y \leqslant 2$; $g(y) = 0$, otherwise.

(ii) $g(y) = \dfrac{6(y^2 - 3y + 2)}{5}$, $0 \leqslant y \leqslant 1$; $g(y) = 0$, otherwise.

(iii) $g(y) = \dfrac{3(\sqrt{y} + 1)}{5}$, $0 \leqslant y \leqslant 1$; $g(y) = 0$, otherwise.

(iv) $g(y) = \dfrac{3(\sqrt{1-y} + 1)}{5}$, $0 \leqslant y \leqslant 1$; $g(y) = 0$, otherwise.

(v) $g(y) = \dfrac{6 \log_e y (\log_e y + 1)}{5y}$, $1 \leqslant y \leqslant e$; $g(y) = 0$, otherwise.

(vi) $g(y) = \dfrac{6 \log_e y (\log_e y - 1)}{5y}$, $e^{-1} \leqslant y \leqslant 1$; $g(y) = 0$, otherwise.

52. $g(y) = \dfrac{5}{2\sqrt{y}}$, $98.01 \leqslant y \leqslant 102.01$; $g(y) = 0$, otherwise.

53. (i) $\frac{1}{8}$; (ii) $E(X^2) = 0.3125$, $\mu = 1.3125$; (iii) $g(y) = 2$, $1.0625 \leqslant y \leqslant 1.5625$; $g(y) = 0$, otherwise; $E(Y) = 1.3125$.

54. (i) $c = \frac{3}{4}$; (iii) $F(x) = \dfrac{3x^2}{4} - \dfrac{x^3}{4}$, $0 \leqslant x \leqslant 2$; $F(x) = 0, x < 0$; $F(x) = 1, x > 2$; (iv) R is 0 to 1; $k = 24$.

WORKSHEET 6

1. $p(x) = p$, $x = 1$; $p(x) = 1 - p$, $x = 0$; $p(x) = 0$, otherwise.
   $\mu = p$, $\sigma^2 = p(1-p)$, $G(t) = 1 - p + pt$.
   Note: The answer to this question may be obtained by treating the Bernoulli distribution as a binomial distribution with $n = 1$ trial.

2. (a) 0.0003; (b) 0.3277; (c) 0.9997.

3. (a) 0.0727; (b) 0.2692; (c) 0.3739; (d) 0.2308; (e) 0.0534; (i) 1.92, 0.9985; (ii) 2.08, 0.9985.

4. (a) 4; (b) 0; (c) (i) 0.032, (ii) 0.001.

5. 17.

6. (a) 0.2304; (b) 0.127; (c) 0.6582.

7. $\binom{n}{r} p^r (1-p)^{n-r}$; (a) 0.760; (b) 0.028; expected number = 0.27.

8. See Section 6.2; 0.086, 60, 90.

9. (i) 0.2503; (ii) 0.2061; (iii) 0.5.

10. (a) See Section 6.2; $n = 60$, $p = 0.4$; (b) 0.081, 0.009.

11. (a) (i) 0.368, (ii) 0.632; (b) (i) 0.018, (ii) 0.433 (using tables); (c) (i) 0.0183, (ii) 0.7476, (iii) 0.9817.

12. (i) 0.223; (ii) 0.191 (using tables); 2.34 days.

13. $\mu = \dfrac{a}{1 - e^{-a}}, \ \sigma^2 = \dfrac{a[1 - (a+1)e^{-a}]}{(1 - e^{-a})^2}.$

14. 0.0025; 78.3 days.

15. Mean = variance = μ; $\bar{x} = 5.35$, $s' = 4.56$. Since mean and variance are not approximately equal, not likely that all fish were hatched in the same season; 0.2193. Based on this probability one would expect $0.2193 \times 60 = 13$ fish to have less than 4 parasites. Since 29 fish were observed to have less than 4 parasites, the conclusion that fish were not hatched in the same season is supported.

16. mean = 7.6, variance = 4.7.

17. 0.0062; reject batch ($0.0011 > 0.001$); 20 samples may be insufficient to get a good estimate of the parameter of the Poisson distribution.

19. (a) $\mu = \dfrac{1-p}{p}, \ \sigma^2 = \dfrac{1-p}{p^2}.$

    (b) $G(t) = \dfrac{p}{1 - t(1-p)}$

    (c) (i) 0.0003; (ii) 0.0059; (iii) 0.9629.

20. (i) 5; (ii) 1.25.

21. $\mu = 6$, $\sigma^2 = 30$.

22. 0.6269.

23. rectangular with p.d.f. $f(x) = \frac{1}{12}$, $0 \leqslant x \leqslant 12$; $f(x) = 0$, otherwise.
mean = 6, variance = 12; $P(1 \leqslant x \leqslant 3) = \frac{1}{6}$.

24. (a) (i) $f(x) = \frac{1}{4}$, $2 \leqslant x \leqslant 6$; $f(x) = 0$, otherwise.
(ii) mean = 4, standard deviation = 1.155.
(iii) $F(x) = \dfrac{x-2}{4}$, $2 \leqslant x \leqslant 6$; $F(x) = 0$, $x < 2$; $F(x) = 1$, $x > 6$.
(b) (i) $k = \frac{3}{32}$; (ii) 1937.5 gallons; (iii) 0.1914.

25. (i) $\frac{2}{3}$; (ii) $\frac{2}{3}$; (iii) $\dfrac{x}{3}$; (iv) $\dfrac{x}{6} + \dfrac{1}{3}$; (v) 1.
$g(x) = \frac{1}{3}$, $0 \leqslant x \leqslant 2$; $g(x) = \frac{1}{6}$, $2 \leqslant x \leqslant 4$; $g(x) = 0$, elsewhere.
mean of $|X| = \frac{5}{3}$.

26. $M(t) = e^{\frac{1}{2}t^2}$.

27. (a) (i) 0.6826, (ii) 0.9544, (iii) 0.9974; (b) (i) 1.96, (ii) 1.64.

28. a, b^2; normal distribution.

29. (a) (i) 0.875, (ii) 0.375; (b) (i) 0.8849, (ii) 0.2119.

30. (a) (i) 47, (ii) 47, (iii) 906; expected cost = 8.54 p; (b) 0.049.

31. 21.645 kg, 24.735 kg.

32. (a) mean = 1000 hours, standard deviation = 120 hours; 0.0188.
(b) (i) 0.1056; (ii) 0.000124.

33. $M_\mu(t) = 1 + \sigma^2\left(\dfrac{t^2}{2!}\right) + 3\sigma^2\left(\dfrac{t^4}{4!}\right)$.

34. (i) 0.9409; (ii) 0.9991; (iii) 0.0583; 0.4814

35. (i) P, 0.0918; Q, 0.7495; R, 0.1587.
(ii) 42.3 cm; (iii) £188.52; (iv) £205.5.

36. (i) 0.8508; (ii) 0.7063; (iii) 0.3494; (iv) 0.0053; $z = -1.0477$.

37. (i) 0.8432; (ii) 0.0668; median = 102.42 mm.

38. (i) 15.87%; (ii) 84.13%; (iii) mean = 92.55, standard deviation = 13.29; (iv) mean = 93.25, standard deviation = 13.45. These agree quite well with the calculations in (iii), but agreement is not perfect because the distribution is not perfectly normal.

39. When $x = \mu$; mode = μ; median = μ.

40. (a) mean = 1000 hours; (b) 0.3012; (c) (i) 0.1666, (ii) 0.1647.

41. (a) $p(x) = \dfrac{e^{-0.02}0.2^x}{x!}$, $x = 0, 1, 2, \ldots$; $p(x) = 0$, otherwise; mean = variance = 0.2. ▶

(b) $f(t) = 0.25e^{-0.25t}$, $t \geqslant 0$; $f(t) = 0$, otherwise; mean = 4 min, variance = 16 min^2.

(c) (i) 0.1637, (ii) 0.3935

42. (i) 33.3 days; (ii) 23.1 days.

43. (a) 0.1859; (b) 39.35%.

44. (a) 0.2757; (b) 0.4313; (c) 0.25, 0.3012, 0.0753.

45. (i) 0.8318 min; (ii) 0.0821; (iii) 0.0341. Expected numbers are 34.08, 22.46, 14.81, 20.44. Poor model may be because mean time between successive arrivals changes through the working day. Suggest different model for each hour of the working day.

46. (a) 0.0821; (b) 0.4562; 6.16 seconds.

47. (a) $1/\lambda$; (b) $F(x) = 1 - e^{-\lambda x}$, $x \geqslant 0$; $F(x) = 0$, $x < 0$;
    0.301; median = 13.86 months; M = 4.46 months.

WORKSHEET 7

1. (a) (i) 0.2121, using $B(20, 0.08)$; (ii) 0.2166, using Poisson approximation. (b) (i) 0.1414, using $B(20, 0.08)$; (ii) 0.1378, using Poisson approximation. Good agreement (within 3%) in both cases.

2. 0.9596, using Poisson approximation with $a = 5$.

3. 0.6288, using Poisson approximation with $a = 4$ (assuming constant probability and independence).

4. (a) 0.0183; (b) 0.0003, using Poisson approximations with $a = 4$, $a = 8$ respectively.

5. 'n large, p small', although the one condition '$p < 0.1$' gives quite good approximate answers.

   (i) 0.2241; (ii) 0.577; P(both develop the chronic condition) $= 0.0049$; mean $= 0.14$.

6. (i) $1 - \binom{103}{101}(0.95)^{101}(0.05)^2 - \binom{103}{102}(0.95)^{102}(0.05)^1 - \binom{103}{103}(0.95)^{103}(0.05)^0$;
   (ii) 0.8874.

7. (a) 0.6778; (b) 0.0418; (c) 0.4325. (Normal approximation used in (b) and (c)).

8. 0.001.

9. 0.9738.

10. 0.9722.

11. 0.0853.

12. 0.0000 (4 d.p.)

13. (a) 0.1611; (b) 0.1586, using normal approximation; 0.1611 is correct.

14. (i) $\displaystyle\int_0^\infty 2xe^{-x^2}\mathrm{d}x = \left[-e^{-x^2}\right]_0^\infty = -0-(-1) = 1.$

    (ii) 0.5441.
    (iii) 0.7794.

15. (i) 0.194 (3 s.f.); (ii) 0.933 (3 s.f.); (iii) 0.986 (3 s.f.).

16. (a) (i) $p = \frac{1}{6}$, n large but unknown. Since np likely to be large, use normal approximation to Poisson;
    (ii) normal;
    (iii) Poisson, $a = 4.8$ per 6 seconds;
    (b) 0.7973, using normal approximation to binomial.
    (c) 0.2560, using Poisson approximation to binomial.

17. $P(R \leqslant 4) = \sum_{r=0}^{4} \binom{20}{r} p^r (1-p)^{20-r}$. When $p = \frac{1}{4}$, $P(R \leqslant 4) = 0.415$.

    0.0606, using normal approximation to binomial; smallest sample size is 408.
    $P(R = 4 | p = \frac{1}{4}) = 0.1896$, $P(R = 4 | p = \frac{1}{6}) = 0.2022$. Result supports '$p = \frac{1}{6}$' theory since this gives the larger probability.

▶

18. (i) $k = 2$; (ii) mean $= \frac{1}{3}$; (iii) $F(x) = 2x - x^2$, $0 \leqslant x \leqslant 1$; $F(x) = 0$, $x < 0$; $F(x) = 1$, $x > 1$. 0.795 (3 s.f.)

19. 'n large, p small', but '$p < 0.1$' alone gives quite good approximate answers. 0.00992; 0.323; 0.891 using normal approximation to Poisson or normal approximation to binomial, 0.888 using Poisson approximation to binomial.

20. $B(n, \frac{1}{6})$, normal approximation to binomial; $n = 402$.

21. (i) $\dfrac{1+4p}{5}$; (ii) $\dfrac{1-p}{1+4p}$ (iii) $8 + 32p$; (iv) 0.1292; (v) $n = 4$.

22. 0.6767; (a) 0.6406; (b) 0.6767. Poisson approximation is better because n large and p small. Normal approximation not so good because $np < 5$ ($n = 72$, $p = \frac{1}{36}$).

23. $n - 39 \geqslant 3.09n^{1/2}$ (not $n - 39 > 3.09n^{1/2}$ as stated in question).
Required value of n is 23, but this answer applies to P (less than 20 females), *not* to P (no more than 20 females).

24. $\frac{1}{3}$, mean $= \frac{20}{3}$, variance $= \frac{40}{9}$; pass mark $= 13$.

25. (i) 0.00992; (ii) smallest number $= 1151$; (iii) 0.594; (iv) 0.046.

26. 55.

27. (a) Poisson, with $a = 1.5$ per 30 seconds; 0.1912.
(b) (i) 0.159; (ii) $k = 112$.

28. (i) 0.05; (ii) 0.58; 0.07.

29. 0.5665; 0.067 using normal approximation to Poisson.

30. (a) (i) $P(X = 0) = 0.135$; (ii) $P(X \leqslant 4) = 0.947$.
(b) (i) 0.144; (ii) 0.111; (iii) 0.046 using normal approximation to Poisson.

WORKSHEET 8

1. $E(X) = 2.1$, $\text{Var}(X) = 0.63$, $E(Y) = 1.5$, $\text{Var}(Y) = 0.75$. Mean $= 3.6$, variance $= 1.38$; probability $= 0.5442$.

2. $E(U) = 2.8$, $\text{Var}(U) = 1.16$, $E(V) = -1.2$, $\text{Var}(V) = 1.16$; $E(UV) = -3.4$, $\text{Cov}(U, V) = -0.04$.

3. $k = \frac{1}{15}$, $E(X) = \frac{2}{3}$, $\text{Var}(X) = \frac{34}{45}$; $P(Y > 4) = \frac{1}{75}$, $R(Y) = \frac{4}{3}$, $\text{Var}(Y) = \frac{68}{45}$.

4. $E(X) = \frac{7}{2}$, $\text{Var}(X) = \frac{35}{12}$; $E(N_A + N_B - N_C) = \frac{7}{2}$, $\text{Var}(N_A + N_B - N_C) = \frac{35}{4}$, $E(|N_A - N_B|) = \frac{35}{18}$, $\text{Var}(|N_A - N_B|) = \frac{665}{324}$.

5. (i) $0.25e^{-1}$; (ii) $0.875e^{-1}$; (iii) $P(Z = 0) = \dfrac{3 + e}{4e}$; (iv) $E(Z) = 1$, $\text{Var}(Z) = 2$.

6. £24 per week, £^{2}204.4 per week.

7. (a) $\mu = 1000$, $\sigma = 119.9$ hours; (b) 0.0475; (c) 0.0017.

8. (a) 0.3015; (b) 0.1210.

9. (a) 0.5564; (b) 0.0000.

10. 0.8413, $X - Y \sim N(3, 2.5)$; 0.9713.

11. (a) (i) $N(\mu_x - \mu_y, \sigma_x^2 + \sigma_y^2)$; (ii) $N(3\mu_x, 9\sigma_x^2)$; (iii) $N(3\mu_x, 3\sigma_x^2)$.

    (b) 0.0668, 1553 units.

12. (i) $N(n\mu, n\sigma^2)$; (ii) $N(2\mu, 4\sigma^2)$; (iii) $N(0, 2\sigma^2)$; $N(8.5, 0.03^2)$, 0.7495, $N(0.5, 0.03^2)$.

13. 3, $\exp\left(\dfrac{\sigma^2 t^2}{2}\right)$, $k = \dfrac{1}{2\sigma}$.

14. (a) 0.1056; (b) $x = 2.6$; (c) 0.0124; (d) 7.

15. $\mu = 7.5478$, $\sigma = 0.2584$; 0.889; $\alpha = 0.0689$.

16. 0.0228, 4, 0.8510.

17. (a) (i) 0.4074, (ii) 0.3125, (iii) 0.3220; (b) 0.0062, 0.0062.

18. 0.423, 13.86 seconds.

19. (a) 0.125, (i) 0.713, (ii) 0.00055.

    (b) n large, p small (although as a reasonable working rule $p < 0.1$ may be used). 0.9496, assuming the year is not a leap year and that the probability of a birthday on New Year's Day is 1/365.

20. Poisson with mean $4n$. When n is large, use the normal approximation to the Poisson by putting $\mu = 4n$, $\sigma = \sqrt{4n}$, and using a continuity correction; 30 sessions.

21. mean $= \lambda_1 + \lambda_2$, variance $= \lambda_1 + \lambda_2$; (i) 0.014; (ii) 0.039, assuming that 'two men and two women' means 'exactly two men and exactly two women'; (iii) 0.030, assuming that 'four employees' means 'exactly four employees'; 0.019.

24. $\left(1 - \dfrac{t}{\lambda}\right)^{-n}$; mean $= \dfrac{n}{\lambda}$, variance $= \dfrac{n}{\lambda^2}$.

▶

25. p.g.f. of the distribution of $(X+Y)$ is $qt^2(1-pt)^{-2}$, a negative binomial distribution with parameters 2 and p. Probability function of $W = X+Y$ is $p(w) = \binom{w-1}{1}q^2p^{w-2} = (w-1)q^2p^{w-2}, w = 2, 3, \ldots$, where W is the number of tosses up to and including the second tail. Implication is that first tail is equally likely to occur on first, second, . . . , seventh toss (with probability $\frac{1}{7}$).

26. $g(t) = p_1t + p_2t^2 + \cdots + p_nt^n$. P.g.f. of $X-Y$ is $g(t)h\left(\frac{1}{t}\right) = g(t)\left[\frac{q_1}{t} + \cdots + \frac{q_n}{t^n}\right]$.

27. $\frac{2+t}{3}$; $\frac{1+t}{2}$; mean of $s = \frac{11}{6}$, variance of $s = \frac{17}{36}$.

28. (a)

Number of red (X)	Number of blue (Y) 0	1	2	Total
0	0	$\frac{2}{15}$	$\frac{1}{15}$	$\frac{3}{15}$
1	$\frac{3}{15}$	$\frac{6}{15}$	0	$\frac{9}{15}$
2	$\frac{3}{15}$	0	0	$\frac{3}{15}$
Total	$\frac{6}{15}$	$\frac{8}{15}$	$\frac{1}{15}$	1

(b) $E(X) = 1$, $\mathrm{Var}(X) = \frac{2}{5}$, $E(Y) = \frac{2}{3}$, $\mathrm{Var}(Y) = \frac{16}{45}$.

(c) $P(X=0|Y=1) = \frac{2}{8}$, $P(X=1|Y=1) = \frac{6}{8}$. mean $= \frac{3}{4}$, variance $= \frac{3}{16}$.

(d) $P(X=0$ and $Y=0) = 0$, $P(X=0) = \frac{3}{15}$, $P(Y=0) = \frac{6}{15}$. X and Y are not independent.

(e) $P(XY=0) = \frac{9}{15}$, $P(XY=1) = \frac{6}{15}$. $\mathrm{Cov}(X, Y) = -\frac{4}{15}$.

29. (a) $A = \frac{4}{24}$; (b) $\frac{9}{24}$; (c) $-\frac{1}{24}$, $\frac{61}{24}$; (d) $\frac{11}{24}$; (e) $\frac{5}{2}$; (f) $P(X=-1|Y=1) = \frac{4}{5}$, $P(X=0|Y=1) = \frac{1}{5}$.

30. (i) $X \sim B(3, 0.4)$, mean $= 1.2$, variance $= 0.72$; $Y \sim B(3, 0.2)$, mean $= 0.6$, variance $= 0.48$.

(ii) $P(X=2, Y=0) = 0.192$, $P(X=3, Y=0) = 0.064$, $P(X=1, Y=1) = 0.192$, $P(X=0, Y=2) = 0.048$, $P(X=0, Y=3) = 0.008$.

(iii) 0.352.

WORKSHEET 9

1. (a) advantages: (i) generally lead to unbiased estimates; (ii) generally lead to efficient estimates. Choose 25 3-digit numbers from the numbers 001 to 500, having given each pupil a different number in the range 001 to 500. Suppose first six numbers are 696, 848, 487, 419, 256, 743. Exclude 696, 848, 743 since they are outside range 1 to 500. Then 487, 419, 256 correspond to three pupils. Proceed until 25 pupils selected.
   (b) Observation point should be away from traffic-lights, cross-roads or other feature which might lead to dependent events. The mean number of vehicles per minute should be less than 30, otherwise normal distribution might be a better model.
   (c) (i) exponential; (ii) $B(5, \frac{1}{6})$; (iii) $B(n, \frac{1}{10})$, where n is the number of digits on the page. Could use Poisson approximation since n likely to be large and p is small, and better still a normal approximation since $np = n/10$ is likely to be greater than 30.

2. (i) 0.893; (ii) 5380 hours; (iii) 0.994.

3. $a_1 + a_2 + \cdots + a_n = 1$; $\frac{1}{n}$; $\text{Var}(\hat{\mu}) = \frac{\sigma^2}{n}$.

4. $\text{Var}(\overline{X}) = \frac{a}{n}$.

5. $\text{Var}(\overline{X}) = \frac{1-p}{np^2}$; Yes.

6. $\text{Var}(\overline{X}) = \frac{1}{n\lambda^2}$.

7. $k = 2$, $\frac{\theta^2}{3n}$.

8. $\text{Var}\left(\frac{3Y}{2}\right) = \frac{\theta^2}{8}$, $\text{Var}(2\overline{X}) = \frac{\theta^2}{6}$. Prefer $\frac{3Y}{2}$ because it has smaller variance, both estimators being unbiased.

9. Mean $= 3 - 5\theta$, variance $= 9\theta - 25\theta^2$; Estimate of θ is 0.27.

11. Mean $= \frac{\theta}{2}$, variance $= \frac{\theta^2}{12}$, $k = \frac{2}{n}$; estimate of θ is 2.7, which is unreasonable since one of the sample values is greater than 2.7 and $0 \leqslant x \leqslant \theta$.

12. $E(\overline{X}) = \theta$, $\text{Var}(\overline{X}) = \frac{\theta^2}{3n}$; $E(Y) = \frac{8\theta}{5}$; $E(T) = \theta$, so T is an unbiased estimator of θ; also $\text{Var}(T) = \frac{\theta^2}{24m}$. Since $n = 4m$, $\text{Var}(\overline{X}) > \text{Var}(T)$, so T preferred.

13. (a) 0.040; (b) 0.159; (c) 0.023.

14. $E(\hat{p}_2) = p^2\left[1 + \frac{4q}{3} + \frac{6q^2}{4} + \ldots\right] > p^2[1 + q + q^2 + \ldots]$

15. $f(x) = \frac{1}{a}$, $0 \leqslant x \leqslant a$; $f(x) = 0$, otherwise. $E(x) = \frac{a}{2}$, $V(x) = \text{Var}(x) = \frac{a^2}{12}$.

$E(\bar{x}) = \frac{a}{2}$, $V(\bar{x}) = \frac{a^2}{12n}$, $E(y) = \frac{na}{n+1}$, $V(y) = \text{Var}(y) = \frac{na^2}{(n+1)^2(n+2)}$.

Both $2\bar{x}$ and $\left(\frac{n+1}{n}\right)y$ are unbiased estimators of a. $\text{Var}(2\bar{x}) = \frac{a^2}{3n}$, $\text{Var}\left[\left(\frac{n+1}{n}\right)y\right]$ $= \frac{a^2}{n(n+2)}$. So $\left(\frac{n+1}{n}\right)y$ preferred since it has smaller variance for all $n > 1$.

16. (i) $E(X) = 2$, $\text{Var}(X) = 0.2$.
(ii) $P(M = 3) = P(M = 1) = 0.028$; $P(M = 2) = 1 - 2 \times 0.028 = 0.944$; $E(M) = 2$, $\text{Var}(M) = 0.056$.
(iii) $E(\bar{X}) = 2$, $\text{Var}(\bar{X}) = \frac{1}{15}$.

17. $E(X) = 3 - 3\theta$, $\text{Var}(X) = 5\theta - 9\theta^2$. $\text{Var}(p_1) = \frac{5\theta - 9\theta^2}{9n}$, $\text{Var}(p_2) = \frac{\theta(1-2\theta)}{2n}$

It follows that p_2 has the smaller variance, and hence smaller standard error (standard error is the square root of variance).

18. $N\left(\mu, \frac{\sigma^2}{10}\right)$, $N\left(\mu, \frac{\sigma^2}{3}\right)$; (i) m_2 is not an unbiased estimator, since $E(m_2) \neq \mu$. (ii) m_3 is an unbiased estimator, since $E(m_3) = \mu$; $\text{Var}(m_1) = \frac{\sigma^2}{10}$, $\text{Var}(m_3) = \frac{7\sigma^2}{18}$ hence m_1 is more efficient, since it has smaller variance.

20. (i) 8; (ii) 0.777, 0.013.

21. The first expression is the variance of the n observations, the second expression is an unbiased estimator of the population variance if we assume that the n observations were from a random sample.

33.09; 88,782.8; 15.1922.

22. (b) $\frac{n_1\sigma_2^2}{n_1\sigma_2^2 + n_2\sigma_1^2}$.

23. $w = \frac{\sigma_y^2}{\sigma_x^2 + \sigma_y^2}$, minimum variance $= \frac{\sigma_x^2\sigma_y^2}{\sigma_x^2 + \sigma_y^2}$, estimate $= 6.2$, standard error of estimate $= 0.1073$.

24. The distribution of the estimator of p in repeated sampling. Mean of this distribution is p.

$\frac{X}{n}$ is consistent because it is unbiased, and $\text{Var}\left(\frac{X}{n}\right) \to 0$ as $n \to \infty$.

$\text{Var}(\hat{p}_1) = \frac{3p(1-p)}{80}$, $\text{Var}(\hat{p}_2) = \frac{p(1-p)}{30}$. Since $\frac{3}{80} > \frac{1}{30}$, $\hat{p}_2$ is more efficient, relative efficiency $= \frac{8}{9}$.

WORKSHEET 10

1. (i) width increases; (ii) width increases; (iii) the effects are opposite, so we can't say without specific examples.

2. normal, μ, $\frac{\sigma^2}{n}$. A 98% confidence interval for the population mean, μ, is given by $\bar{x} \pm \frac{2.33\sigma}{\sqrt{n}}$ if σ is known, or $\bar{x} \pm \frac{ts}{\sqrt{n}}$ if σ is unknown; 98% of such intervals will actually contain the true value of μ;
   0.3108, 0.8664, 21.75 to 23.31.

3. 1.019 to 1.061.

4. (i) 110; (ii) 190.

5. 62, assuming weekly rent is normally distributed (although this assumption is not very important since $n(=62)$ is large).

6. (i) 32.02 to 35.60, assuming that the m.p.g. for the cars using Additive 1 is normally distributed;
   (ii) 34.03 to 37.63, assuming that the m.p.g. for the cars using Additive 2 is normally distributed;
   (iii) -4.35 to $+0.31$, assuming that the distributions of the m.p.g. for cars using Additives 1 and 2 are both normal and have the same variance.

7. -8.10 to 4.10.

9. (a) 0.26 to 1.44. Since this interval contains the value 1, the interval supports the claim. Assumptions: weight loss is approximately normally distributed, and the sample taken was randomly chosen from a population of women.
   (b) 0.29 to 1.03. Since this interval does not contain the value 0 and both limits are positive, the mean weight loss for women following diet A is significantly greater than the mean weight loss for women following diet B. Assumptions: the distributions of weight loss for women following diets A and B are both normal and have the same variance.

10. (a) 339.7 to 342.3. Since the interval contains the value 340, this result is consistent with the claim.
    (b) (i) 6.93 to 7.27, (ii) -1.46 to -1.14.

11. -0.11 to 1.53. Assumption: the difference between the hours of sleep with the drug and the inactive control is approximately normally distributed.

12. 1.58 to 9.84, using difference = (degree of corrosion for untreated half) − (degree of corrosion for treated half). Assumption: difference is normally distributed. Since the interval does not contain 0 and both limits are positive, the interval supports the claim.

13. (a) Histogram indicates approximate normality (fairly symmetrical with data bunched in the middle, unimodal);
    (b) (i) 4.42 to 4.68, (ii) 0.14 to 0.32;
    (c) 200.

▶

14. (a) 148, 1058;
    (b) 134.5 to 161.5, 645 to 2048;
    (c) 136.9 to 159.1.

15. 153.2 to 552.1. Since 250 lies inside the interval, but 100 does not, data support the claim that the variance is 250; 108.1 to 124.1.

16. (a) 0.8864; (b) (i) 2, (ii) -1.01 to 1.01.

17. Numbers taken from a set of numbers (population) so that each has the same chance of being selected in the sample. Number the lorries 01, 02, . . . , 92. Select twelve 2-digit random numbers from a table such as C.13, starting anywhere, reading in rows (for example), and ignoring 00 and any value above 92.

    (a) 0.72; (b) 0.01; (c) 0.27; 0.82 to 0.94. Since this interval does not contain the value 0.72 (the answer to (a)), it looks as though the vehicle tester was not choosing digits randomly.

18. When n large, distribution of $\hat{p}$ is approximately normal with mean p and variance $\dfrac{p(1-p)}{n}$; 0.08 to 0.13, which are approximate because we are using the normal approximation to the binomial and $\hat{p}$ is only an estimate of p; 769 to 1250.

19. (a) 0.106 to 0.144; (b) 0.079 to 0.121; (c) -0.003 to 0.053. Answer to (c) does not support the claim since the interval contains the value zero.

20. -0.12 to -0.06; Since both limits are negative, the claim that the newspaper is read by a higher proportion in the North-West is supported.

21. $B(n, p)$, $N\left[p, \dfrac{p(1-p)}{n}\right]$;

    (a) $B(100, p)$; estimate of p is 0.57;
    (b) $N(\mu, \sigma^2)$; estimate of μ is 0.57, estimate of σ^2 is 0.0025;
    (c) $N(\mu, \sigma^2)$; estimate of μ is 0.09, estimate of σ^2 is 0.0049.

    -0.048 to 0.228; since this interval contains zero, data support the claim that the probabilities are not significantly different.

22. (i) 385; (ii) $a = 0.804\lambda$, $b = 1.196\lambda$, 46 to 68.

23. $f(x) = \dfrac{1}{a-10}$, $10 \leqslant x \leqslant a$; $f(x) = 0$, otherwise; $E(X) = \dfrac{a+10}{2}$.

    Unbiased estimate of a is 32 (based on one observation). Unbiased estimate of a is 34 (based on 50 observations). 95% confidence interval for a is 25.77 to 42.23. All values lie between 0 and a, so $60 \leqslant a$. The contradiction is because the 95% confidence interval is based on Σx_i and Σx_i^2, and not on individual values.

24. (a) Criticism: method not random; it is possible that periodic problems may occur with period corresponding to every twentieth refrigerator. To obtain a random sample, give 100 refrigerators made in sequence the numbers 00, 01, . . . , 99. Then choose 5 of these numbers using 2-digit numbers from Table C.13.
    (b) 0.9452; £7790 to £8810.

WORKSHEET 11

1. (a) $H_0: \mu = 5$, $H_1: \mu > 5$, where μ is the population mean of the distribution of tar content per cigarette for the company's cigarettes.
   (b) $H_0: \mu = 5$, $H_1: \mu < 5$, where μ is the population mean tar content per cigarette for the new brand.
   (c) $H_0: \mu \neq 55$, $H_1: \mu \neq 55$, where μ is the population mean mark for all students who took the course over the past several years.

2. (a) $t = -2.35$, $t_{0.05,\,19} = 1.73$, Mean is not significantly higher than 5.
   (b) $t = -1.67$, $t_{0.01,\,99} = 2.37$. Mean is not significantly less than 5.
   (c) $t = 0.95$, $t_{0.025,\,19} = 2.09$, $t_{0.005,\,19} = 2.86$. Mean is not significantly different from 55 at either the 5% or the 1% levels of significance. With the 5% level, there is a 5% chance that we will wrongly reject H_0. With the 1% level, we have only a 1% chance of wrongly rejecting H_0, so we are less likely to reject H_0 the smaller the level of significance.

3. $H_0: \mu = 28$, $H_1: \mu > 28$, $t = 1.07$, $t_{0.05,\,9} = 1.83$. Mean mpg is not significantly greater than 28, assuming mpg for cars of this model is approximately normally distributed.

4. $H_0: \mu = 1$, $H_1: \mu \neq 1$, $t = 4.267$, $t_{0.05,\,10} = 1.81$. Mean weight of sugar is significantly different from 1.

5. (i) σ known; (ii) σ unknown and estimated from sample data, where σ is the population standard deviation.
   $t = -2.27$, $t_{0.05,\,8} = 1.86$. Mean is significantly lower than 100, assuming number of finished articles per hour is normally distributed.

6. $z = 2.236$, $z_{0.025} = 1.96$. Mean is significantly different from 2. 1.992 and 2.108; there is a 99% chance that the statement 'μ lies between 1.992 to 2.108' is correct. Alternative test would use $t = \dfrac{\bar{x} - \mu_0}{s/\sqrt{n}}$, $t_{0.025,\,4}$ and reject H_0 if $\dfrac{|\bar{x} - \mu_0|}{s/\sqrt{n}} > t_{0.025,\,4}$.

7. $z = 2.68$, $z_{0.05} = 1.64$. New lubricant has significantly increased spinning time. Assumptions are: (i) spinning time is normally distributed; (ii) the five spins can be thought of as a random sample of all spins using the new lubricant.

8. $z = -1.79$, $z_{0.025} = 1.96$. Mean is not significantly different from 1 lb. $z = -1.79$, $z_{0.05} = 1.64$. Mean is significantly less than 1 lb.

9. 0.015 (no interpolation). $z = -2.74$, $z_{0.025} = 1.96$. Mean weight is significantly different from 200.0 g.

10. $t = 2.84$, $t_{0.025,\,7} = 2.36$. There is a significant difference between the mean responses for the two treatments. Assumption is that the differences are approximately normally distributed.

11. $t = -1.85$, $t_{0.01,\,14} = 2.62$. Mean weight for hens receiving enriched corn is not significantly greater than for hens receiving ordinary corn. Assumption is that the variance in the weights is the same for both types of corn.

12. $t = -3.17$, $t_{0.025,\,398} = 1.96$. Mean incomes for the two regions are significantly different.

▶

13. $t = 2.10$, $t_{0.05,\,7} = 1.89$, $t_{0.01,\,7} = 3.00$. Mean score 'after' is significantly higher than mean score 'before' at the 5% level, but mean score 'after' is not significantly higher than mean score 'before' at the 1% level. In second test we run a smaller risk of wrongly rejecting H_0, so we are less likely to reject H_0. Assumption is that the differences are approximately normally distributed.

14. $t = 2.80$, $t_{0.05,\,9} = 1.83$. Mean hours of sleep are significantly greater 'with drug' than 'with control'.

15. $t = -1.81$, $t_{0.05,16} = 1.75$. Additive 2 gives significantly more mpg on average than Additive 1. Assumptions are: (i) mpg is normally distributed for cars using Additive 1, and also for cars using Additive 2; (ii) two distributions in (i) have the same variance.

16. $t = -0.79$, $t_{0.005,\,12} = 3.05$. Means are not significantly different.

17. $t = -2.07$, $t_{0.025,\,28} = 2.05$. Difference in mean height of men and women is significantly different from 12 cm. Assumptions are: (i) height of men in region is normally distributed, and height of women in region is normally distributed; (ii) the two distributions in (i) have the same variance.

18. (b) $t = 0.9956$, $t_{0.05,\,16} = 1.75$. Quenched steel is not significantly stronger than tempered steel, and vice versa. Conclusions depend on whether the distributions of strength for each type of steel are normal, and whether these distributions have the same variance. The allocation of the bars to either the quenching or the tempering treatment must be random for a properly designed experiment.

19. $t = -2.40$, $t_{0.05,\,8} = 1.86$. Mean for second sample is significantly greater, so it is reasonable to accept the statement.

20. (i) $t = 1.889$, $t_{0.025,\,11} = 2.20$. Means are not significantly different.
(ii) $t = 1.889$, $t_{0.05,\,11} = 1.80$. Method 1 gives significantly higher readings.

21. (i) $\bar{x} = 6.067$, $s^2 = 0.739$; (ii) $t = -2.66$, $t_{0.025,\,5} = 2.57$. Mean is significantly different from 7.
$t = -2.19$, $t_{0.025,\,7} = 2.36$. Means are not significantly different.

22. $t = 2.33$, $t_{0.025,\,9} = 2.26$. Means for the two methods of analysis are significantly different. Assumption is that the differences are normally distributed.

23. $t = 3.86$, $t_{0.05,\,7} = 1.89$. Mean number of eggs is significantly greater with the additive.

24. Assumptions are: (i) the distributions of mpg for cars tested with and without the additive are normal; (ii) the distributions in (i) have the same variance.

$t = 2.15$, $t_{0.05,\,18} = 1.73$. Mean mpg is significantly greater with the additive (5% level). Improved design would involve testing all 20 cars with the additive and also without the additive.

25. (a) $F = 1.78$, $F_{0.975,\,3,\,13} = 0.07$, $F_{0.025,\,3,\,13} = 4.36$. Variances are not significantly different.
(b) $t = 2.94$, $t_{0.005,\,16} = 2.92$. Mean survival times are significantly different.

26. $E(X_i^2) = \sigma^2 + \mu^2$, $E\left(\sum_{i=1}^{n} X_i\right) = n\mu$, $E\left[\left(\sum_{i=1}^{n} X_i\right)^2\right] = n\sigma^2 + n^2\sigma^2$.

90% confidence interval for standard deviation is 0.014 to 0.039; $F = 3.44$, $F_{0.05,\,6,\,3} = 8.94$.

27. (a) (i) 42.0 to 50.0, (ii) 35.2 to 50.8, (ii) 41.1 to 48.5.
    (b) $F = 1.88$, $F_{0.05, 51, 62} < 1.7$. Variance of the number of letters per day is significantly higher for the Principal's office and n_1, n_2 are large, so use $z = \dfrac{\bar{x}_1 - \bar{x}_2}{\sqrt{\dfrac{\sigma_1^2}{n_1} + \dfrac{\sigma_2^2}{n_2}}}$. $z = 2.46$, $z_{0.025} = 1.96$; there is a significant difference between the means.

28. (a) 226.8 to 232.4. Mean of counts per minute is significantly different from 200.
    (b) 155.0 to 271.2. Variance of counts per minute is not significantly different from 200.

29. 4.97 to 6.33, assuming temperature increase in approximately normally distributed. Mean temperature increase is not significantly different from 5°C. $t = 2.27$, $t_{0.05,7} = 1.98$; mean increase in temperature is significantly greater than 5°C.

30. (a) $\chi^2 = 50.6$, $\chi^2_{0.025, 49} = 70.2$, $\chi^2_{0.975, 49} = 31.6$. Variance in wing-length is not significantly different from 0.2.
    (b) $t = -7.00$, $t_{0.05, 49} = 1.68$. Mean wing-length is significantly less than 5 mm.

31. 0.9 mpg. $t = 2.45$, $t_{0.025, 268} = 1.96$. Increase is significantly different from zero; 4.7 p.

32. 52.5, 2.806; 52.02 to 52.98; (i) mean is significantly different from 52. (ii) $t = 2.11$, $t_{0.01,49} = 2.41$, mean is not significantly different from 52. $t = -0.996$, $t_{0.025,148} = 1.96$, means are not significantly different.

33. $s^2 = 41.5$; -3.61 to 15.61. Mean heights of boys and girls are not significantly different.

34. $\overline{X}_1 \pm 1.96\dfrac{\sigma}{\sqrt{n}}$, $\overline{X}_2 \pm 1.96\dfrac{\sigma}{\sqrt{n}}$, $\overline{X}_1 - \overline{X}_2 \sim N\left(\mu_1 - \mu_2, \dfrac{2\sigma^2}{n}\right)$;
    (i) 9.51 to 10.49, 10.51 to 11.49, reject H_0: $\mu_1 = \mu_2$; (ii) 0.0056.

35. (a) $\overline{X} > 1.10$; (b) (i) 0.09, (ii) 0.91.

36. (a) 20.49; (b) power values are 0.05, 0.17, 0.38, 0.64, 0.85, 0.96; (c) 20.35. Power at $\mu = 20.6$ is 0.88 (compared with 0.64); (d) 395.

37. (a) 8, 88.47; (b) mean of first eight values is 89.7. Since this is greater than 88.47, conclude that an increase in operating efficiency to 90 units has occurred.

38. 103.9, 0.03, 0.97.

40. $n = 19$, $h = 0.0249$; (a) 0.67; (b) $P(\text{stopping machine}|\mu = 2.55) = 0.99$, for example. $\sigma' = 0.085$. Control charts for sample mean and sample range.

41. $n = 5$, $m = 0.977$ kg. Reason for (A) is that it guarantees that bags with daily average mass as high as 1.02 kg will be rejected at most 5% of the time. Reason for (B) is that it guarantees that bags with daily mass as low as 0.94 kg will be accepted at most 5% of the time.

42. $t = -3.21$, $t_{0.025, 5} = 2.57$. Mean lifetime is significantly different from 12. Assumptions are: (i) distribution is normal; (ii) candles are a random sample from the manufacturer's output. 0.136.

WORKSHEET 12

1. (a) 0.172, 0.121; (b) 0.322.

2. (b) 0.042, 0.055.

3. 0.058, 0.608, $X \geqslant 16$; one-sided because the claim refers to a reduction in irritation, not simply a difference in irritation.

4. H_0: $p = 0.5$, H_1: $p > 0.05$; 0.021; 0.584.

5. H_0: $\mu = 55.26$, H_1: $\mu > 55.26$, 5% level, one-tail test, $z = 1.68$, $z_{0.05} = 1.64$, mean is significantly greater in 1981/82. Use middle result in Table 12.2, where x is the observed number of no-score draws in n games.

6. Decide new drug is significantly better if more than 8 of the 11 sleep for more than 3.2 hours; 0.1189.

7. 0.5; 0.110, 0.172, 0.205; 0.65, 0.267, 0.514; 0.237. Charles has highest type I error and lowest power, hence his is the least sensible test. Anne has lower type I error than Brian but also much lower power. Brian's test is most sensible, on balance.

8. $z = 2.0$, $z_{0.05} = 1.64$, new treatment is significantly better.

9. $z = 4.47$, $z_{0.005} = 2.58$, data do not support hypothesis.

10. $E(Y) = p$, $\text{Var}(Y) = \dfrac{p(1-p)}{n}$. $z = -5.96$, $z_{0.0005} = 3.29$.

    Reasons: (i) South-West not typical of England and Wales; (ii) letters may have all been posted on a particular day, not spread out through the week.

11. 0.821; $0.055 > 0.05$, so do not reject H_0; $z = 2.4$, $z_{0.05} = 1.64$, so this provides significant evidence.

12. (a) 0.077; (b) 0.911; $z = 2.126$, $z_{0.025} = 1.96$, there is a significant difference. Proportion of males in wild rabbits is greater than for rabbits bred in capacity.

13. 3600; $z = -1.706$, $z_{0.05} = 1.64$, data support researchers claim, assuming that the 800 interviewed are a random sample of the town's inhabitants.

14. $z = -1.633$, $z_{0.05} = 1.64$. Proportion is not significantly lower for those taking the drug.

15. $z = 1.11$, $z_{0.005} = 2.58$. Proportions of males and females are not significantly different.

16. $z = -1.136$, $z_{0.025} = 1.96$. Proportions of defectives for the two lines are not significantly different; 0.05 to 0.09 (2 d.p.).

17. $z = 1.92$, $z_{0.025} = 1.96$. Proportions are not significantly different.
    $z = -2.61$, $z_{0.025} = 1.96$. Proportion this year is significantly different from 0.1.

WORKSHEET 13

1. (a) $\chi^2 = 3.176$, $\chi^2_{0.05,\,2} = 5.99$, do not reject H_0 that examination results are independent of school; (b) $\chi^2 = 9.529$, $\chi^2_{0.05,\,2} = 5.99$, now reject H_0; (c) (b) is correct, χ^2 test is for frequencies, *not* percentages.

2. $\chi^2 = 5.556$, $\chi^2_{0.05,\,2} = 5.99$, do not reject H_0 that views of adults are independent of sex. $\chi^2 = 9.778$, $\chi^2_{0.01,\,2} = 9.21$, views of children and adults are significantly different.

3. $\chi^2 = 14.47$, $\chi^2_{0.05,\,4} = 9.49$, preferences are not independent of job type.

4. $\chi^2 = 72.94$, $\chi^2_{0.05,\,4} = 9.49$, place of marriage is not independent of length of courtship.

5. $\chi^2 = 18.31$, $\chi^2_{0.01,\,2} = 9.21$, observed frequencies are not consistent with the assumption of identical effects, a significantly higher proportion of patients do better on Drug A.

6. $a = 25$, $b = 5$, $c = -4$; $\chi^2 = 3.32$, $\chi^2_{0.05,\,2} = 5.99$. Proportions do not vary significantly.

7. $\chi^2 = 7.20$, $\chi^2_{0.05,\,2} = 5.99$. Queen's Cross has a significantly better record than Bakerloo.

8. $\chi^2 = 10.09$, $\chi^2_{0.05,\,1} = 3.84$. Proportion coming home is significantly greater in the North.

9. $\chi^2 = 2.91$, $\chi^2_{0.05,\,1} = 3.84$. Probability of burning out is not significantly diffferent for the two types of set.

10. Combine rows 1 and 2, and rows 3 and 4. $\chi^2 = 10.37$, $\chi^2_{0.05,\,1} = 3.84$. There is a significantly higher proportion of complications for mothers aged 30 or more.

11. $\chi^2 = 2.44$, $\chi^2_{0.05,\,1} = 3.84$. Data support the stated hypothesis.

13. (a) $\chi^2 = 4.32$, $\chi^2_{0.05,\,3} = 7.81$. Data support the stated hypothesis.
    (b) $\chi^2 = 13.6$, $\chi^2_{0.01,\,3} = 11.34$. Significantly more male births occur in spring.

14. (a) $\chi^2 = 17.9$, $\chi^2_{0.01,\,5} = 15.09$. (b) $\chi^2 = 30.56$, $\chi^2_{0.01,\,5} = 15.09$. Data do not support stated hypothesis.

15. $\chi^2 = 5.5$, $\chi^2_{0.05,\,2} = 5.99$. Data support hypothesis.

16. $\chi^2 = 7.36$, $\chi^2_{0.001,\,3} = 16.27$. Data support hypothesis.

17. $\chi^2 = 1.66$, $\chi^2_{0.05,\,5} = 11.07$. Data support hypothesis that the die is fair.

18. $\chi^2 = 0.49$, $\chi^2_{0.05,\,3} = 7.81$. Data support genetic theory.

19. $\chi^2 = 16.28$, $\chi^2_{0.05,\,7} = 14.07$. There is significant directional preference, birds preferring the three adjacent directions N, NE and E.

20. $\chi^2 = 0.22$, $\chi^2_{0.05,\,1} = 3.84$. Data support genetic theory.

21. (a) $\chi^2 = 8.76$, $\chi^2_{0.05,\,2} = 5.99$. Reject hypothesis of independence.
    (b) $\chi^2 = 0.97$, $\chi^2_{0.05,\,2} = 5.99$. Data support stated hypothesis.
    (c) Because of the difference in production times we cannot conclude that the number of defectives per production hour is the same for all three machines. It appears that, ▶

for the late shift, a defective is more likely to have come from A, but this would need to be confirmed by another test.

22. $\chi^2 = 7.30$, $\chi^2_{0.05,3} = 7.81$. Data support the stated hypothesis.

23. $\chi^2 = 4.43$, $\chi^2_{0.05,4} = 9.49$. Data for the first week support the hypothesis that the numbers borrowed are independent of the day of the week.

    $\chi^2 = 9.93$, $\chi^2_{0.05,4} = 9.49$. Data for the two weeks do not support the hypothesis that the numbers borrowed are independent of the day of the week.

24. (b) $\chi^2 = 60.49$, $\chi^2_{0.001,3} = 16.27$. Data do not support the stated hypothesis.

25. (a) $\chi^2 = 4.44$, $\chi^2_{0.05,4} = 9.49$. Data support the stated hypothesis.
    (b) $\chi^2 = 2.93$, $\chi^2_{0.05,3} = 7.81$. Data support the stated hypothesis.
    (c) Neither test *proves* that we have a binomial distribution; what they show is that the data are consistent with the two stated binomial distributions.

26. (a) $\chi^2 = 0.53$, $\chi^2_{0.01,2} = 9.21$. Data support the stated hypothesis.
    (b) $\chi^2 = 9.2$, $\chi^2_{0.05,1} = 3.84$. Proportions are significantly different, there are significantly more females.

27. $\chi^2 = 4.45$, $\chi^2_{0.05,5} = 11.07$. Data consistent with the stated distribution, i.e. that rates are equally likely to turn left or right.

28. (a) 0.25; (b) $\chi^2 = 5.86$, $\chi^2_{0.05,3} = 7.81$. Data consistent with $B(5, 0.25)$ distribution.

29. 0.4; $\chi^2 = 5.51$, $\chi^2_{0.05,3} = 7.81$, data consistent with $B(5, 0.4)$ distribution. 0.36 to 0.44.

30. binomial; p may not be (i) constant for all litters, (ii) 0.5.
    $\chi^2 = 12.9$, $\chi^2_{0.05,5} = 11.07$. Data do not support the $B(5, 0.5)$ hypothesis; p may not be constant for all litters.
    $\chi^2 = 1.51$, $\chi^2_{0.05,4} = 9.49$. Data support a $B(5, 0.46)$ hypothesis.

31. Expected frequencies are 7.6, 35.1, 64.8, 59.8, 27.6, 5.1.
    (a) $X_1^2 = 0.39$; (b) $X_2^2 = 1.51$.
    In the calculation of X_1^2, positive and negative differences tend to cancel out giving possibly small values of X_1^2 even when there are large differences between observed and expected frequencies. Small values of X_1^2 also occur when observed and expected frequencies are in good agreement. So X_1^2 is not discriminating. Also we have no way of testing either X_1^2 or X_2^2.

32. $\chi^2 = 5.83$, $\chi^2_{0.05,3} = 7.81$. Data support the hypothesis that the coin is fair.

33. $\chi^2 = 9.50$, $\chi^2_{0.05,4} = 9.49$. Data do not quite support the hypothesis of a Poisson distribution; more evidence needed.

34. (a) $\chi^2 = 1.03$, $\chi^2_{0.05,3} = 7.81$. Data support the hypothesis of a Poisson distribution.
    (b) mean = 1.5, variance = 1.67. It depends what 'approximately equal' means (which is why a χ^2 test is preferable!).

35. $\chi^2 = 7.16$, $\chi^2_{0.05,8} = 15.51$. Sample supports the stated belief.

36. Because of the lack of good guidance system, bombs falling randomly, i.e. number of hits per unit area is 'Poisson'.

mean = 0.9288, $\chi^2 = 1.27$, $\chi^2_{0.05,4} = 9.49$. Data consistent with Poisson distribution. Probability = 0.566.

37. Line chart preferred to bar chart for discrete data; mean = 1.2, variance = 1.2. Good agreement between observed and expected frequencies; 0.0907.

38. (i) 2.25; (ii) 10.5, 23.7, 26.7, 20.0, 11.3, 5.1, 2.7; (iii) $\chi^2 = 1.44$, $\chi^2_{0.05,4} = 9.49$, data support the hypothesis of a Poisson distribution.

39. (i) mean = 2.5, estimate of $p = 0.4$.
(ii) $\chi^2 = 0.35$, $\chi^2_{0.05,2} = 5.99$, data support stated theory.

40. The five points appear to lie on a vertical line corresponding to $a = 2$(?).

41. The five points appear to lie on a vertical line corresponding to $a = 1.4$(?).

42. (ii) $\chi^2 = 36.42$, $\chi^2_{0.05,3} = 7.81$. Data not consistent with a normal distribution.

43. (a) σ^2; (b) $\chi^2 = 9.42$, $\chi^2_{0.05,4} = 9.49$. Data support the hypothesis that the results are typical.

44. $\mu = 4.3$, $\sigma = 1.2$.

45. 0.8531; mean = 40.1, standard deviation = 0.7; 39.8 newtons.

46. There are six plotted points which appear to lie more or less on a straight line (or do they?). This exercise shows how subjective this graphical method is; the χ^2 test is preferable in this case.

WORKSHEET 14

1. $(\frac{1}{2})^5 > 0.025$, so H_0 never rejected. $n \leqslant 4$ for one-sided alternative.
2. (a) $0.0327 < 0.05$, significant increase in sales.
   (b) $t = 3.12$, $t_{0.05,10} = 1.81$, significant increase in sales.

   Normal probability plot looks non-linear, so the sign test is appropriate but the t test is not.
3. Sum of the first n positive integers.
4. (i) $0.0547 > 0.05$, coaching does not increase median mark significantly.
   (ii) $T = 6$, critical value of T is 10, coaching does increase median mark significantly.

   Prefer conclusion to (ii), since Wilcoxon is more powerful than the sign test. The alternative parametric test is the 'paired t test', which requires the assumption that the differences are approximately normally distributed. This assumption seems likely to be true from the results of a probability plot.
5. $T = 10.5$, critical value of T is 3. Median birthweights of first and second-born identical twins are not significantly different.
6. $T = 0$, critical value of T is 0. Median weights are significantly different.
7. A t test would be unsuitable because differences are negatively skew (rather than 'normal').

   (i) $0.172 > 0.025$, median weights of fish in the two populations are not significantly different; (ii) $T = 7.5$, critical value of T is 8, median weights are significantly different. Conclusion to (ii) preferred because Wilcoxon test is more powerful than the sign test.
8. Sign test simpler but Wilcoxon preferable here because it is more powerful. $T = 7.5$, critical value of T is 8. Data support the salesman's claim.
9. $0.006 < 0.025$, methods differ significantly in their effectiveness. Outline of the Wilcoxon test required, in which null hypothesis rejected if $T \leqslant 52$ (provided there are no zero differences). Wilcoxon preferred to sign test as in Exercises 7 and 8 above.
10. (i) because $R_1 + R_2$ is the sum of the first $(n_1 + n_2)$ positive integers.
    (ii) result follows after use of the hint and quoting answer to (i).
11. $U = 9$, critical value of U is 13. There is a significant difference in the median percentage mortality. An unpaired t test is an alternative test, but requires the assumptions that the distributions of percentage mortality for A and B are both normal and have equal variance.
12. $U = 25.5$, critical value of U is 7, median lengths of life are not significantly different.
13. $U = 7.5$, critical value of U is 6, median densities of liquid are not significantly different.
14. $U = 4.5$, critical value of U is 5 ($2\frac{1}{2}\%$ level, one-sided H_1), median number of errors is significantly less for Group 1. Group 2 might feel resentful and hence make more errors, so the experiment doesn't prove that tea-breaks reduce errors. Better to do an experiment in which all 12 clerks are measured for errors both before and after the ►

introduction of a tea-break. Also measure errors again after the novelty of the tea-break has worn off.

15. $U = 57.5$, critical value of U is 37, median weights of rats on diets A and B are not significantly different.

16. Scores for the 'Whisky group' look non-normal, so use Mann–Whitney U test. $U = 17$, critical value of U is 13. Alcohol did not significantly affect the average score.

17. $\chi^2 = 3.2$ if Yates's correction not used, $\chi^2 = 1.8$ with Yates's correction. Parametric test is unpaired samples t test, non-parametric test is Mann–Whitney U test. Relative merits: the χ^2 test is least preferable, since the magnitudes of the values are not used. The assumption of normality is probably not justified for the boys' pocket money (because of the 'outlier', 8.6). $U = 29.5$, critical value of U is 23, medians of pocket money for boys and girls are not significantly different.

WORKSHEET 15

1. The value of r is indeterminate since $\sum x_i y_i - \dfrac{\sum x_i \sum y_i}{n}$ and $\sum x_i^2 - \dfrac{\left(\sum x_i\right)^2}{n}$ are both zero.

2. $r = 0.694$, $t = 2.73$, $t_{0.05,8} = 1.86$, there is a significant positive correlation.

3. Choose 0.985, since scatter diagram indicates very high positive correlation. $t = 16.15$, $t_{0.05,8} = 1.86$, there is a significant positive correlation between the weights of first-born and second-born identical twins.

4. $r = -0.246$, $t = -1.56$, $t_{0.025,38} = 2.02$, not significant as expected from common sense.

5. $r = -0.246$, $z = -2.76$, $z_{0.025} = 1.96$, correlation coefficient is significantly different from 0.2.

6. -0.5178 to 0.0709, which does not contain 0.2 and so is consistent with previous answer.

7. (ii) $r = 0.7927$, $t = 4.11$, $t_{0.025,10} = 2.23$, correlation coefficient is significantly different from zero; (iii) 0.4015 to 0.9393; (iv) the observation $x = 43$, $y = 2.6$ is lower on the scatter diagram than expected from the general trend of the other observations.

8. $r_1 = -0.2673$, $r_2 = -0.1856$, $z = -0.229$, $z_{0.025} = 1.96$, the correlation coefficients are not significantly different.

9. $r_s = -0.9364$, critical value of r_s is 0.564, reject H_0 that the rankings are uncorrelated, and accept H_1 that high ranks for Judge 1 are associated with low ranks for Judge 2 (and vice versa).

10. $r_s = -0.7429$, critical value of r_s is 0.829, do not reject H_0 that the rankings of the two variables are uncorrelated. Disagree with researcher's conclusion; in any case we could not have concluded cause and effect even if we had rejected H_0.

11. $r_s = -0.5774$, critical value of r_s is 0.881, rankings of intelligence test score and lateral thinking score are uncorrelated.

12. Whisky group scores do not appear to be normally distributed; $r_s = 0.3155$, critical value of r_s is 0.738, sober and whisky ranks are uncorrelated.

13. If Pearson's r formula is applied to the ranks of two variables we get the formula for Spearman's r_s (also see Exercise 14). $r_s = 0.6$, which lies between the critical values of 0.648 and 0.564. Hence the result is significant at the 10 % level but not at the 5 % level.

14. Use $\sum_{i=1}^{n} x_i = \sum_{i=1}^{n} y_i = \dfrac{n(n+1)}{2}$ and $\sum_{i=1}^{n} x_i^2 = \sum_{i=1}^{n} y_i^2 = \dfrac{n(n+1)(2n+1)}{6}$.

    $r_s = -0.2167$, low negative correlation, critical value of r_s is 0.6, the ranks in event 1 and event 2 are not significantly positively correlated.

15. (i) points lie on a straight line with a positive gradient;
    (ii) ranks are in perfect agreement;

▶

For Overall and Mathematics, $r_s = 0.8374$, while for Overall and French, $r_s = 0.6643$. Hence Maths is the better predictor.

16. The advantage of rank correlation is that only ranks (not magnitudes) are required and we do not need to be able to assume the variables are normally distributed. Pearson's r is better (more powerful) if both variables are normally distributed.

    For head size and amount of hair, $r_s = -0.4303$, critical value = 0.564, not significant at the 10% level.
    For head size and amount of hair, $\tau = -0.2889$, critical value = 0.467, not significant at 10% level.

    For age and amount of hair, $r_s = -0.7091$, critical value = 0.564, significant at 10% level.
    For age and amount of hair, $\tau = -0.5111$, critical value = 0.467, significant at 10% level.

17. The value of the correlation coefficient is independent of the units in which X and Y are measured, provided b and d are positive.

18. (b) $r = -0.6304$, $t = -2.30$, $t_{0.025,8} = 2.31$, not quite significant at the 5% level;
    $r = 0.1471$, $t = 0.42$, $t_{0.025,8} = 2.31$, not significant at the 5% level.

19. (i) $E(W) = 0$, $\text{Var}(W) = 6\sigma^2$.

1. $y = 4.04 + 0.0357x$; when $x = 56$, $\hat{y} = 6.04$ g.

2. $y = 83.3 + 7.25x$; when $x = 10$, $\hat{y} = 155.8$; when $x = 0$, $\hat{y} = 83.3$, but this is extrapolation and so is an invalid prediction.

3. $y = 26.81 - 1.552x$; when $x = 3$, $\hat{y} = 22.15$; line should not be used for values of x much above 14 (maximum value in data).

4. $y = 2.98 + 4.8x$; $z = 1.039$, $z_{0.025} = 1.96$, slope is not significantly different from 3; 1.41 to 8.19.

5. $y = 1.1 + 1.09x$; 0.14 to 2.04; reject H_0: $\beta = 0$, since 99 % confidence interval does not contain zero.

6. (i) $w = 2.9 - 500x$; (ii) when $x = \frac{1}{250}$, $\hat{w} = 0.9$; (iii) -414 to -586.

7. $y = -1.998 + 0.9948x$; 0.92 to 1.07.

8. -2.34 to 4.54; do not reject H_0: $\alpha = 0$, since the confidence interval contains zero.

9. (i) $y = -71.2 + 3.5x$; (ii) when $x = 60$, $\hat{y} = 138.8$; (iii) 138.03 to 139.57.

10. (i) $\bar{y} = 10.4$; (ii) 13.33 to 16.67; (iii) 90 % confidence interval for $y_4 - y_1$ is 5.18 to 8.62.

11. $y = 51.2 + 0.032x$; (i) 50.92 to 51.48; (ii) 0.23 to 0.41.

12. $y = 57.2 - 0.48x$; (i) 47.15 to 48.05; (ii) 90 % confidence interval for $y_5 - y_{25}$ is 8.56 to 10.64.

13. $y = 7.059 + 1.309x$; 20.1, 33.2, 46.3; (a) estimate for $x = 20$ most reliable, being nearest centre of data; (b) estimate at $x = 30$ is least reliable, being outside the range of data.

14. If X denotes income, Y denotes savings, $y = -0.791 + 0.144x$, $x = 7.967 + 5.285y$; $r^2 = 0.8737^2 = 0.7633$; $bb' = 0.1444 \times 5.285 = 0.7632$.

15. $r = 0.9033$; $y = 87.4 + 4.46x$; when $x = 9$, $\hat{y} = 127.5$ cm; when $x = 30$, $\hat{y} = 221.2$ cm, but this is extrapolating far beyond the values of x in the data.

16. $x - \bar{x} = m'(y - \bar{y})$, where $m' = \dfrac{\sum(x_i - \bar{x})(y_i - \bar{y})}{\sum(y_i - \bar{y})^2}$; $r = 0.9977$; $y = -1.853 + 21.36x$; when $x = 2$, $\hat{y} = 40.87$; yes, since the two regression lines are virtually identical here.

17. (i) positive correlation; (ii) very little correlation; (iii) negative correlation; P is $(\bar{v}_1, \bar{v}_2)$; when $x = 6$, $\hat{y} = 6$ (nearest integer).

18. From scatter diagram, correlation coefficient will be positive and quite high (0.8 say). $x = -19.23 + 0.5794y$; when $y = 152.5$, $\hat{x} = 69.1$ inches; equation should be used to estimate values of x (height) only; $r = 0.78$.

19. (i) Since we want to predict x (marks in mathematics), use regression line of x on y; (ii) $x = 3.687 + 0.8222y$; when $y = 15$, $\hat{x} = 16$ (nearest integer); (iii) $r = 0.8001$; (iv) $r_s = 0.7455$.

20. $y = -12.9 + 8.3x$; $\log_{10} y = -0.436 + 0.4x$; when $x = 2.5$, $\hat{y} = 3.7$ cm.

►

21. (c) $\log_e y = 0.8321 + 1.6396 \log_e x$.

22. (ii) $y = 4.28 - \dfrac{2.48}{x}$; (iii) $x = 1$, $\hat{y} = 1.8$; $x = 4$, $\hat{y} = 3.66$.

23. $X = c + dx$, where $c = -17$, $d = 2$; $y = 128.5 + 4.7X$, $y = 9.4x + 48.6$.

24. (a) $M = 11.2 + 4.3\,T$; (b) $m = 2.15t + 89.7$; when $t = 16$, $\hat{m} = 124.1$, not reliable because of extrapolation outside range of t values in data.

25. (ii) scatter diagram shows non-linear trend; (iii) $y = 0.0124 + 1.438 \log_{10} x$; (iv) 3.23; equation should not be used for values of x much below 25 (minimum value in data).

26. Estimates of log α and log β are 2.104 and 1.144; estimates of α and β are 8.20 and 3.14; 1.110 to 1.178.

WORKSHEET 17

1. $F = 3.43$, $F_{0.05,1,14} = 4.61$, no significant difference between mean weights. $t = -1.85$, $t_{0.025,14} = 2.14$, same conclusion as with F test.

2. $F = 1.274$, $F_{0.05,1,10} = 4.96$, mean sleeping times for A and B are not significantly different.

3. $F = 1.02$, $F_{0.05,2,19} = 3.53$, mean scores for the three methods are not significantly different.

4. $F = 9.71$ $F_{0.05,3,16} = 3.25$, mean determinations of the four analysts are significantly different.

5. $F = 25.9$, $F_{0.01,2,15} = 6.36$, mean results of the three analysts are significantly different.

6. (a) Allot numbers 1, 2, . . . , 12, one to each girl. Choose four digits in the range 1 to 12 from random number tables, assign reinforcement A to corresponding girls. Choose four different digits and assign B, last four girls are assigned C.
   (b) $F = 7.89$, $F_{0.05,2,9} = 4.26$, mean number of errors for A, B, and C are significantly different.
   (c) A has significantly fewer errors than B or C (means are 50, 73.5, and 73.5 respectively).

7. (a) $F = 6.65$, $F_{0.05,3,14} = 3.36$, mean blood pressures for the four treatments are significantly different.
   (b) Variance within treatment is same for all treatments, blood pressure within each treatment is normally distributed.
   (c) For A and D, standard error of mean = 2.01, for B and C it is 2.25. Standard errors of differences between two means are 3.01 for A vs. B, A vs. C, B vs. D and C vs. D; 2.84 for A vs. D; 3.17 for B vs. C.

8. (i) 0.058 to 0.125; (ii) $F = 2.33$, $F_{0.05,3,20} = 3.10$, means of X, Y, Z, W are not significantly different.

9. (a) 11.9, 15.3, 12.7 kg (1 d.p.); (b) 5.2; (c) $F = 4.1$, $F_{0.05,2,17} = 3.60$, mean number of plants for the three varieties are significantly different. (i.e. reject H_0: $\mu_S = \mu_A = \mu_B$); (d) 13.4 to 17.3.

10. (i) $F = 5.1$, $F_{0.5,3,9} = 3.86$, average weal sizes for the four antibody preparations are significantly different;
    (ii) $F = 3.7$, $F_{0.05,3,9} = 3.86$, average weal sizes for the four guinea pigs are not significantly different.

11. $F = 26.9$, $F_{0.05,4,12} = 3.26$, mean breaking loads for the five types of cement are significantly different.

12. $F = 2.1$, $F_{0.05,4,8} = 3.84$, mean reaction times for the five technicians are not significantly different; $F = 10.1$, $F_{0.05,2,8} = 4.46$, mean reaction times for the three arrangements are significantly different.

13. Means for A, B, C and D are 54.5, 61.5, 61.3 and 54.0. For significance at the 5% level, ►

means must differ by more than $t_{0.025,\nu}\, s\sqrt{\frac{2}{b}} = 5.85$, so A and D differ significantly from B and C.

14. Means for A, B, C, D and E are 11.5, 12, 7, 22 and 11. Following method of Exercise 13, means must differ by more than 3.3. Hence, A, B, E have significantly greater breaking load than C and significantly smaller breaking load than D.

WORKSHEET 18

1. Sample mean chart; warning lines at 124.4 and 130.4, action lines at 122.6 and 132.1. Sample range chart; upper warning line at 14.4, upper action line at 18.9. Control lost at sample number 8, where mean and range are close to action line. A possible cause is one unusually high measurement.

2. Sample mean chart; warning lines at 36.2 and 83.8, action lines at 22.5 and 97.5. Sample range chart; upper warning line at 96.8, upper action line at 129.0. Control of range is good, but control of mean has been lost in months 11 and 12.

3. Sample mean chart; warning lines at 85.2 and 90.1, action lines at 83.7 and 91.5. Sample range chart; upper warning line at 8.1, upper action line at 11.1. Control of range is good, suspicion that mean has fallen by sample number 10.

4. Sample mean chart; warning lines at 48.4 and 52.0, action lines at 47.4 and 53.0. There are 2 values out of 50 below 47, and P ($\geqslant 2$ successes, for $n = 50, p = 0.001) = 0.0012$. Hence reject H_0: $p = 0.001$ so the statement is unlikely to be correct.

5. Sample mean chart; warning lines at 100.2 and 101.8, action lines at 99.7 and 102.3. Sample range chart; upper warning line at 3.9, upper action line at 5.0. Current state of process indicates reasonable control. Recommended action: tighter control on range. Since $\mu = 101$, $\sigma = 0.97$, virtually all reels contain between $(101 - 3 \times 0.97)$ and $(101 + 3 \times 0.97)$ m. Hence 95–105 m is well within the ability of the process.

6. Upper warning line at 0.168, upper action line at 0.200. Fraction defective appears to be increasing slowly.

7. Upper warning line at 0.1035, upper action line at 0.1114. Process is not in control, since two points lie above action lines.

8. (a) 0.0513; (b) 0.4595.

9. (a) 0.0245; (b) 0.4437; (c) when $p = 0.10$, $P_a = 0.976$, and so on; (d) 19.2%, using linear interpolation.

10. (b) When $p = 0.1$, $P_a = 0.9298$; when $p = 0.3$, $P_a = 0.3828$; (c) expected amount of inspection is 23.3, 71.2, 127.4 for $p = 0.1, 0.2, 0.3$. Amount of inspection climbs rapidly as p (proportion of defectives) increases.

11. 25, 26.2, 36.0, 37.8, 30.0, 25.8.

12. Probabilities of accepting a batch containing 5% defective is 0.7604 for Plan 1, 0.7636 for Plan 2. Expected numbers of items are 30, 33.0, 37.8, 36.8 and 30. Factors are: (i) how close OC curve is to ideal (see Fig. 18.6); (ii) which plan has lower average (expected) number of items inspected – clearly Plan 2 has in this case, since average for Plan 1 is 50.

Appendix B
Glossary of notation

Figures in brackets are chapter and section references.

ROMAN SYMBOLS

a	parameter, mean and variance of a Poisson distribution (6.3)
a, b	parameters of a rectangular (uniform) distribution (6.5)
a, b	intercept and slope (gradient) of sample regression line (16.1)
$A, B, \ldots$	events (4.2)
$B(n, p)$	binomial distribution with parameters n and p (6.2)
$\mathrm{Cov}(X, Y)$	covariance of random variables X and Y (8.2)
d	differences in paired samples data (10.7)
d	differences between paired ranks (15.8)
$D_1, D_2, \ldots, D_9$	first, second, . . . , ninth deciles (3.4)
E	expected frequency (13.1)
E	event (4.3)
E'	event complementary to event E (4.8)
$E(X)$	expectation of a random variable X (5.3)
$E[g(X)]$	expectation of $g(X)$, a function of X (5.3)
$f_1, f_2, \ldots$	frequencies with which the observations $x_1, x_2, \ldots$ occur (2.3)
$f(x)$	probability density function of a continuous random variable X (5.5)
$F(x)$	(cumulative) distribution function, $P(X \leqslant x)$, of a continuous random variable X (5.7).
F_{α, ν_1, ν_2}	α per cent point of the F distribution with ν_1, ν_2 degrees of freedom (Table C.6(a)–(d))
$G(t)$	probability generating function of a discrete random variable which can take values 0, 1, 2, . . . (5.4)
H_0	null hypothesis (11.2)
H_1	alternative hypothesis (11.2)
m	median of a continuous random variable (5.8)

$M(t)$	moment generating function about zero of a continuous random variable (5.9)
$M_\mu(t)$	moment generating function about the mean of a continuous random variable (5.9)
n	number of values in sample (sample size) (9.5)
n	number of observations (2.2)
n	a large number of trials (4.4)
n	number of trials in a binomial experiment (6.2)
$n!$	factorial n (or n factorial) (4.10)
${}^n\mathrm{P}_r$	the number of permutations of r objects taken from n objects (4.10)
$\binom{n}{r}$ or ${}^n\mathrm{C}_r$	the number of combinations of r objects taken from n objects (4.10)
$N(\mu, \sigma^2)$	normal distribution with mean, μ, and variance, σ^2 (6.6)
O	observed frequency (13.1)
p	the probability of success in each trial in a binomial experiment (6.2)
p	the probability of success in each trial in a geometric experiment (6.4)
$p(x)$	probability function, $P(X = x)$, of a discrete random variable X (5.2)
$P(A)$, $P(B)$, $P(E)$	probabilities of events A, B, E (4.3)
$P(E_1 \cap E_2)$	probability that both events E_1 and E_2 will occur (4.6)
$P(E_2 \mid E_1)$	probability that event E_2 will occur, given that E_1 has occurred (4.6)
$P(E_1 \cup E_2)$	probability that either or both of events E_1 and E_2 will occur (4.7)
$P_1, P_2, \ldots, P_{99}$	first, second, . . . , ninety-ninth percentiles (3.4)
$P(X = x)$	the probability that the random variable X takes the value x (5.2)
Poi (a)	Poisson distribution with parameter a (6.3)
P_{a}	probability that a batch will be accepted (18.4)
Q_1, Q_2, Q_3	lower quartile, median, upper quartile of a set of observations (3.4)
r	sample value of Pearson's product moment correlation coefficient (15.3)
r_{s}	Spearman's rank coefficient correlation (15.8)
s'	standard deviation of n observations (3.2)
s^2	unbiased estimate of σ^2, the variance of a normally distributed population (9.7)
S^2	(the name of) a random variable which is an unbiased estimator of σ^2, the variance of a normally distributed population (9.7)
S	outcome set (4.2)
$t_{\alpha, \nu}$	upper α percentage point for a t distribution with ν degrees of freedom (Table C.4 and 10.3)
T	Wilcoxon signed rank test statistic (14.3)
U	Mann–Whitney test statistic (14.4)
Var (X)	variance of the random variable X (5.3)
w	sample range (18.2)
$x_1, x_2, \ldots$	observations (2.1)
x	observed value of a random variable X (5.2)
$\bar{x}$	mean of a sample of observations (2.2)
$\bar{x}$	unbiased estimate of μ, the mean of a normally distributed population (9.6)
$\overline{X}$	(the name of) a random variable formed by taking the mean of $X_1, X_2, \ldots, X_n$ (9.5)

$X_1, X_2, \ldots, X_n$	(the name of) a sample of size n from the population of a random variable X (9.5)
X	(the name of) a random variable (5.2)
Y	(the name of) a random variable (5.10)
$\hat{y}$	predicted (mean) value of Y in regression analysis (16.3)
z	observed value of a random variable Z (6.6)
z_α	upper α percentage of point of standardized normal distribution (Table C.3(b) and 10.2)
Z	(the name of) a random variable having a standardized normal distribution (6.6)

GREEK SYMBOLS

α	significance level (11.3), Type I error (11.10)
α, β	intercept and slope (gradient) of population regression line (16.2)
β	Type II error (11.10)
θ	population parameter (9.4)
$\hat{\theta}$	point estimator of θ (9.4)
λ	parameter of an exponential distribution (6.7)
μ	mean value of a random variable (5.3)
μ	mean of a population (9.5)
μ	mean of a normal distribution (6.6)
ν	degrees of freedom (10.3)
ρ	Pearson product moment correlation coefficient for bivariate population (15.2)
σ, σ^2	standard deviation and variance of a random variable (5.3)
σ, σ^2	standard deviation and variance of a population (9.5)
σ, σ^2	standard deviation and variance of a normal distribution (6.6)
$\sum$	the operation of summing (2.2)
τ	Kendall's rank correlation coefficient (15.9)
$\phi(z)$	probability density function of a standardized normal variable, Z (6.6)
$\Phi(z)$	(cumulative) distribution function, $P(Z \leqslant z)$, of a standardized normal variable, Z (6.6)
χ^2	chi-squared distribution (10.8)
$\chi^2_{\alpha, \nu}$	α percentage point for a χ^2 distribution with ν degrees of freedom (Table C.5 and 10.9)

Appendix C
Statistical tables

CONTENTS

ACKNOWLEDGEMENTS

The author would like to thank the following authors and publishers for their kind permission to adapt from the following tables:

Tables C.1, 9, 10 and 13 from

Dunstan, F. D. J., Nix, A. B. J., Reynolds, J. F. and Rowlands, R. J. (1983) *Elementary Statistical Tables*, RND Publications, Cardiff.

Tables C.2, 3, 4, 5, and 6 from

Lindley, D. V. and Scott, W. F. (1984) *New Cambridge Elementary Statistical Tables*, Cambridge University Press, Cambridge.

Table C. 7 from

Runyon, R. P. and Haber, A. (1968) *Fundamentals of Behavioural Statistics*, Addison-Wesley, Reading, Mass.; based on values in Wilcoxon, F., Katti, S. and Wilcox, R. A. (1963) *Critical Values and Probability Levels for the Wilcoxon Rank Sum Test* and *the Wilcoxon Signed Rank Test*, American Cyanamid Co., New York; and Wilcoxon, F. and Wilcox, R. A. (1964) *Some Rapid Approximate Statistical Procedures*, American Cyanamid Co., New York.

Table C.8 from

Owen D. B. (1962) *Handbook of Statistical Tables*, Addison-Wesley, Reading, Mass.; based on values in Auble, D. (1953) Extended tables for the Mann–Whitney statistic. *Bulletin of the Institute of Educational Research at Indiana University*, 1:2.

Table C.11 from

Runyon, R. P. and Haber, A. (1968) *Fundamentals of Behavioural Statistics*, Addison-Wesley, Reading, Mass.; based on values in Olds, E. G. (1949) The 5 % significance levels of sums of squares of rank differences and a correction. *Annals of Mathematical Statistics*, **20**, 117–118 and Olds, E. G. (1938) Distribution of the sum of squares of rank differences for small numbers of individuals. *Annals of Mathematical Statistics*, **9**, 133–148.

Table C.12 from

Daniel, W. W. (1978) *Applied Nonparametric Statistics*, Houghton Mifflin Company, Boston, Mass.

TABLE C.1 CUMULATIVE BINOMIAL PROBABILITIES

The tabulated value is $P(X \leqslant r)$, where X has a binomial distribution with parameters n and p.

$p =$		0.05	0.10	0.15	0.20	0.25	0.30	0.35	0.40	0.45	0.50
$n = 5$ $r =$	0	0.774	0.590	0.444	0.328	0.237	0.168	0.116	0.078	0.050	0.031
	1	0.977	0.919	0.835	0.737	0.633	0.528	0.428	0.337	0.256	0.187
	2	0.999	0.991	0.973	0.942	0.896	0.837	0.765	0.683	0.593	0.500
	3	1.000	1.000	0.998	0.993	0.984	0.969	0.946	0.913	0.869	0.813
	4			1.000	1.000	0.999	0.998	0.995	0.990	0.982	0.969
$n = 10$ $r =$	0	0.599	0.349	0.197	0.107	0.056	0.028	0.013	0.006	0.003	0.001
	1	0.914	0.736	0.544	0.376	0.244	0.149	0.086	0.046	0.023	0.011
	2	0.988	0.930	0.820	0.678	0.526	0.383	0.262	0.167	0.100	0.055
	3	0.999	0.987	0.950	0.879	0.776	0.650	0.514	0.382	0.266	0.172
	4	1.000	0.998	0.990	0.967	0.922	0.850	0.751	0.633	0.504	0.377
	5		1.000	0.999	0.994	0.980	0.953	0.905	0.834	0.738	0.623
	6			1.000	0.999	0.996	0.989	0.974	0.945	0.898	0.828
	7				1.000	1.000	0.998	0.995	0.988	0.973	0.945
	8						1.000	0.999	0.998	0.995	0.989
	9							1.000	1.000	1.000	0.999
$n = 20$ $r =$	0	0.358	0.122	0.039	0.012	0.003	0.001	0.000	0.000		
	1	0.736	0.392	0.176	0.069	0.024	0.008	0.002	0.001	0.000	
	2	0.925	0.677	0.405	0.206	0.091	0.035	0.012	0.004	0.001	0.000
	3	0.984	0.867	0.648	0.411	0.225	0.107	0.044	0.016	0.005	0.001
	4	0.997	0.957	0.830	0.630	0.415	0.238	0.118	0.051	0.019	0.006
	5	1.000	0.989	0.933	0.804	0.617	0.416	0.245	0.126	0.055	0.021
	6		0.998	0.978	0.913	0.786	0.608	0.417	0.250	0.130	0.058
	7		1.000	0.994	0.968	0.898	0.772	0.601	0.416	0.252	0.132
	8			0.999	0.990	0.959	0.887	0.762	0.596	0.414	0.252
	9			1.000	0.997	0.986	0.952	0.878	0.755	0.591	0.412
	10				0.999	0.996	0.983	0.947	0.872	0.751	0.588
	11				1.000	0.999	0.995	0.980	0.943	0.869	0.748
	12					1.000	0.999	0.994	0.979	0.942	0.868
	13						1.000	0.998	0.994	0.979	0.942
	14							1.000	0.998	0.994	0.979
	15								1.000	0.998	0.994
	16									1.000	0.999
	17										1.000

Table C.1 (Contd.)

	$p =$	*0.05*	*0.10*	*0.15*	*0.20*	*0.25*	*0.30*	*0.35*	*0.40*	*0.45*	*0.50*
$n = 25$	$r =$ 0	0.277	0.072	0.017	0.004	0.001	0.000				
	1	0.642	0.271	0.093	0.027	0.007	0.002	0.000			
	2	0.873	0.537	0.254	0.098	0.032	0.009	0.002	0.000		
	3	0.966	0.764	0.471	0.234	0.096	0.033	0.010	0.002	0.001	0.000
	4	0.993	0.902	0.682	0.421	0.214	0.091	0.032	0.009	0.002	0.001
	5	0.999	0.967	0.839	0.617	0.378	0.193	0.083	0.029	0.009	0.002
	6	1.000	0.991	0.931	0.780	0.561	0.341	0.173	0.074	0.026	0.007
	7		0.998	0.974	0.891	0.726	0.512	0.306	0.154	0.064	0.022
	8		0.999	0.992	0.953	0.851	0.677	0.467	0.273	0.134	0.054
	9		1.000	0.998	0.983	0.929	0.811	0.630	0.425	0.242	0.115
	10			0.999	0.994	0.970	0.902	0.771	0.586	0.384	0.212
	11			1.000	0.999	0.989	0.956	0.875	0.732	0.543	0.345
	12				1.000	0.997	0.983	0.940	0.846	0.694	0.500
	13					0.999	0.994	0.975	0.922	0.817	0.655
	14					1.000	0.998	0.991	0.966	0.904	0.788
	15						0.999	0.997	0.987	0.956	0.885
	16						1.000	0.999	0.996	0.983	0.946
	17							1.000	0.999	0.994	0.978
	18								1.000	0.998	0.993
	19									1.000	0.998
	20										0.999
	21										1.000
$n = 30$	$r =$ 0	0.215	0.042	0.008	0.001	0.000					
	1	0.553	0.184	0.048	0.010	0.002	0.000				
	2	0.812	0.411	0.151	0.044	0.011	0.002	0.000			
	3	0.939	0.647	0.322	0.123	0.037	0.009	0.002	0.000		
	4	0.984	0.825	0.525	0.255	0.098	0.030	0.007	0.002	0.000	
	5	0.997	0.927	0.711	0.428	0.203	0.077	0.023	0.006	0.001	0.000
	6	0.999	0.974	0.847	0.607	0.348	0.160	0.059	0.017	0.004	0.001
	7	1.000	0.992	0.930	0.761	0.514	0.281	0.124	0.043	0.012	0.003
	8		0.998	0.972	0.871	0.674	0.431	0.225	0.094	0.031	0.008
	9		0.999	0.990	0.939	0.803	0.589	0.358	0.176	0.069	0.021
	10		1.000	0.997	0.974	0.894	0.730	0.508	0.292	0.135	0.049
	11			0.999	0.990	0.949	0.841	0.655	0.431	0.233	0.100
	12			1.000	0.997	0.978	0.916	0.780	0.578	0.359	0.181
	13				0.999	0.992	0.960	0.874	0.715	0.502	0.292
	14				1.000	0.997	0.983	0.935	0.825	0.645	0.428
	15					0.999	0.994	0.970	0.903	0.769	0.572
	16					1.000	0.998	0.988	0.952	0.864	0.708
	17						0.999	0.995	0.979	0.929	0.819
	18						1.000	0.999	0.992	0.967	0.900
	19							1.000	0.997	0.986	0.951
	20								0.999	0.995	0.979
	21								1.000	0.998	0.992
	22									1.000	0.997
	23										0.999
	24										1.000

Table C.1 (Contd.)

	$p=$	*0.05*	*0.10*	*0.15*	*0.20*	*0.25*	*0.30*	*0.35*	*0.40*	*0.45*	*0.50*
$n=40$	$r=$ 0	0.129	0.015	0.001	0.000						
	1	0.399	0.080	0.012	0.001	0.000					
	2	0.677	0.223	0.049	0.008	0.001	0.000				
	3	0.862	0.423	0.130	0.029	0.005	0.001	0.000			
	4	0.952	0.629	0.263	0.076	0.016	0.003	0.000			
	5	0.986	0.794	0.433	0.161	0.043	0.009	0.001	0.000		
	6	0.997	0.901	0.607	0.286	0.096	0.024	0.004	0.001	0.000	
	7	0.999	0.958	0.756	0.437	0.182	0.055	0.012	0.002	0.000	
	8	1.000	0.984	0.865	0.593	0.300	0.111	0.030	0.006	0.001	0.000
	9		0.995	0.933	0.732	0.440	0.196	0.064	0.016	0.003	0.000
	10		0.999	0.970	0.839	0.584	0.309	0.121	0.035	0.007	0.001
	11		1.000	0.988	0.912	0.715	0.441	0.205	0.071	0.018	0.003
	12			0.996	0.957	0.821	0.577	0.314	0.128	0.039	0.008
	13			0.999	0.981	0.897	0.703	0.441	0.211	0.075	0.019
	14			1.000	0.992	0.946	0.807	0.572	0.317	0.133	0.040
	15				0.997	0.974	0.885	0.695	0.440	0.214	0.077
	16				0.999	0.988	0.937	0.798	0.568	0.318	0.134
	17				1.000	0.995	0.968	0.876	0.689	0.439	0.215
	18					0.998	0.985	0.930	0.791	0.565	0.318
	19					0.999	0.994	0.964	0.870	0.684	0.437
	20					1.000	0.998	0.983	0.926	0.787	0.563
	21						0.999	0.993	0.961	0.867	0.682
	22						1.000	0.997	0.981	0.923	0.785
	23							0.999	0.992	0.960	0.866
	24							1.000	0.997	0.980	0.923
	25								0.999	0.991	0.960
	26								1.000	0.997	0.981
	27									0.999	0.992
	28									1.000	0.997
	29										0.999
	30										1.000

Table C.1 (Contd.)

	$p =$	*0.05*	*0.10*	*0.15*	*0.20*	*0.25*	*0.30*	*0.35*	*0.40*	*0.45*	*0.50*
$n = 50$ $r =$	0	0.077	0.005	0.000							
	1	0.279	0.034	0.003	0.000						
	2	0.540	0.112	0.014	0.001	0.000					
	3	0.760	0.250	0.046	0.006	0.000					
	4	0.896	0.431	0.112	0.018	0.002	0.000				
	5	0.962	0.616	0.219	0.048	0.007	0.001	0.000			
	6	0.988	0.770	0.361	0.103	0.019	0.002	0.000			
	7	0.997	0.878	0.519	0.190	0.045	0.007	0.001	0.000		
	8	0.999	0.942	0.668	0.307	0.092	0.018	0.002	0.000		
	9	1.000	0.976	0.791	0.444	0.164	0.040	0.007	0.001		
	10		0.991	0.880	0.584	0.262	0.079	0.016	0.002	0.000	
	11		0.997	0.937	0.711	0.382	0.139	0.034	0.006	0.001	0.000
	12		0.999	0.970	0.814	0.511	0.223	0.066	0.013	0.002	0.000
	13		1.000	0.987	0.889	0.637	0.328	0.116	0.028	0.004	0.000
	14			0.995	0.939	0.748	0.447	0.188	0.054	0.010	0.001
	15			0.998	0.969	0.837	0.569	0.280	0.095	0.022	0.003
	16			0.999	0.986	0.902	0.684	0.389	0.156	0.043	0.008
	17			1.000	0.994	0.945	0.782	0.506	0.237	0.077	0.016
	18				0.998	0.971	0.859	0.622	0.336	0.127	0.033
	19				0.999	0.986	0.915	0.726	0.447	0.197	0.059
	20				1.000	0.994	0.952	0.814	0.561	0.286	0.101
	21					0.997	0.975	0.881	0.670	0.390	0.161
	22					0.999	0.988	0.929	0.766	0.502	0.240
	23					1.000	0.994	0.960	0.844	0.613	0.336
	24						0.998	0.979	0.902	0.716	0.444
	25						0.999	0.990	0.943	0.803	0.556
	26						1.000	0.996	0.969	0.872	0.664
	27							0.998	0.984	0.922	0.760
	28							0.999	0.992	0.956	0.839
	29							1.000	0.997	0.976	0.899
	30								0.999	0.988	0.941
	31								1.000	0.995	0.967
	32									0.998	0.984
	33									0.999	0.992
	34									1.000	0.997
	35										0.999
	36										1.000

TABLE C.2 CUMULATIVE POISSON PROBABILITIES

The tabulated value is $P(X \leqslant r)$, where X has a Poisson distribution with mean a.

$a =$	0.5	1.0	1.5	2.0	2.5	3.0	3.5	4.0	4.5	5.0
$r =$ 0	0.607	0.368	0.223	0.135	0.082	0.050	0.030	0.018	0.011	0.007
1	0.910	0.736	0.558	0.406	0.287	0.199	0.136	0.092	0.061	0.040
2	0.986	0.920	0.809	0.677	0.544	0.423	0.321	0.238	0.174	0.125
3	0.998	0.981	0.934	0.857	0.758	0.647	0.537	0.433	0.342	0.265
4	1.000	0.996	0.981	0.947	0.891	0.815	0.725	0.629	0.532	0.440
5		0.999	0.996	0.983	0.958	0.916	0.858	0.785	0.703	0.616
6		1.000	0.999	0.995	0.986	0.966	0.935	0.889	0.831	0.762
7			1.000	0.999	0.996	0.988	0.973	0.949	0.913	0.867
8				1.000	0.999	0.996	0.990	0.979	0.960	0.932
9					1.000	0.999	0.997	0.992	0.983	0.968
10						1.000	0.999	0.997	0.993	0.986
11							1.000	0.999	0.998	0.995
12								1.000	0.999	0.998
13									1.000	0.999
14										1.000

$a =$	5.5	6.0	6.5	7.0	7.5	8.0	8.5	9.0	9.5	10.00
$r =$ 0	0.004	0.002	0.002	0.001	0.001	0.000	0.000	0.000	0.000	0.000
1	0.027	0.017	0.011	0.007	0.005	0.003	0.002	0.001	0.001	0.000
2	0.088	0.062	0.043	0.030	0.020	0.014	0.009	0.006	0.004	0.003
3	0.202	0.151	0.112	0.082	0.059	0.042	0.030	0.021	0.015	0.010
4	0.358	0.285	0.224	0.173	0.132	0.100	0.074	0.055	0.040	0.029
5	0.529	0.446	0.369	0.301	0.241	0.191	0.150	0.116	0.089	0.067
6	0.686	0.606	0.527	0.450	0.378	0.313	0.256	0.207	0.165	0.130
7	0.809	0.744	0.673	0.599	0.525	0.453	0.386	0.324	0.269	0.220
8	0.894	0.847	0.792	0.729	0.662	0.593	0.523	0.456	0.392	0.333
9	0.946	0.916	0.877	0.830	0.776	0.717	0.653	0.587	0.522	0.458
10	0.975	0.957	0.933	0.901	0.862	0.816	0.763	0.706	0.645	0.583
11	0.989	0.980	0.966	0.947	0.921	0.888	0.849	0.803	0.752	0.697
12	0.996	0.991	0.984	0.973	0.957	0.936	0.909	0.876	0.836	0.792
13	0.998	0.996	0.993	0.987	0.978	0.966	0.949	0.926	0.898	0.864
14	0.999	0.999	0.997	0.994	0.990	0.983	0.973	0.959	0.940	0.917
15	1.000	0.999	0.999	0.998	0.995	0.992	0.986	0.978	0.967	0.951
16		1.000	1.000	0.999	0.998	0.996	0.993	0.989	0.982	0.973
17				1.000	0.999	0.998	0.997	0.995	0.991	0.986
18					1.000	0.999	0.999	0.998	0.996	0.993
19						1.000	1.000	0.999	0.998	0.997
20								1.000	0.999	0.998
21									1.000	0.999
22										1.000

TABLE C.3 THE NORMAL DISTRIBUTION

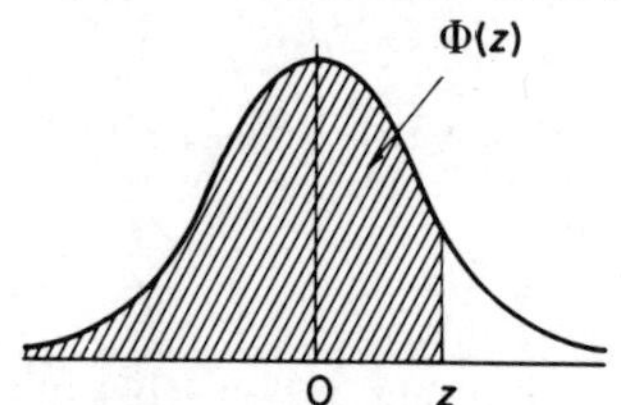

(a) *Distribution function.* The tabulated value is $\Phi(z) = P(Z \leqslant z)$, where Z is the standardized normal random variable, $N(0, 1)$.

z	*0.00*	*0.01*	*0.02*	*0.03*	*0.04*	*0.05*	*0.06*	*0.07*	*0.08*	*0.09*
0.0	0.5000	0.5040	0.5080	0.5120	0.5160	0.5199	0.5239	0.5279	0.5319	0.5359
0.1	0.5398	0.5438	0.5478	0.5517	0.5557	0.5596	0.5636	0.5675	0.5714	0.5753
0.2	0.5793	0.5832	0.5871	0.5910	0.5948	0.5987	0.6026	0.6064	0.6103	0.6141
0.3	0.6179	0.6217	0.6255	0.6293	0.6331	0.6368	0.6406	0.6443	0.6480	0.6517
0.4	0.6554	0.6591	0.6628	0.6664	0.6700	0.6736	0.6772	0.6808	0.6844	0.6879
0.5	0.6915	0.6950	0.6985	0.7019	0.7054	0.7088	0.7123	0.7157	0.7190	0.7224
0.6	0.7257	0.7291	0.7324	0.7357	0.7389	0.7422	0.7454	0.7486	0.7517	0.7549
0.7	0.7580	0.7611	0.7642	0.7673	0.7704	0.7734	0.7764	0.7794	0.7823	0.7852
0.8	0.7881	0.7910	0.7939	0.7967	0.7995	0.8023	0.8051	0.8078	0.8106	0.8133
0.9	0.8159	0.8186	0.8212	0.8238	0.8264	0.8289	0.8315	0.8340	0.8365	0.8389
1.0	0.8413	0.8438	0.8461	0.8485	0.8508	0.8531	0.8554	0.8577	0.8599	0.8621
1.1	0.8643	0.8665	0.8686	0.8708	0.8729	0.8749	0.8770	0.8790	0.8810	0.8830
1.2	0.8849	0.8869	0.8888	0.8907	0.8925	0.8944	0.8962	0.8980	0.8997	0.9015
1.3	0.9032	0.9049	0.9066	0.9082	0.9099	0.9115	0.9131	0.9147	0.9162	0.9177
1.4	0.9192	0.9207	0.9222	0.9236	0.9251	0.9265	0.9279	0.9292	0.9306	0.9319
1.5	0.9332	0.9345	0.9357	0.9370	0.9382	0.9394	0.9406	0.9418	0.9429	0.9441
1.6	0.9452	0.9463	0.9474	0.9484	0.9495	0.9505	0.9515	0.9525	0.9535	0.9545
1.7	0.9554	0.9564	0.9573	0.9582	0.9591	0.9599	0.9608	0.9616	0.9625	0.9633
1.8	0.9641	0.9649	0.9656	0.9664	0.9671	0.9678	0.9686	0.9693	0.9699	0.9706
1.9	0.9713	0.9719	0.9726	0.9732	0.9738	0.9744	0.9750	0.9756	0.9761	0.9767
2.0	0.9772	0.9778	0.9783	0.9788	0.9793	0.9798	0.9803	0.9808	0.9812	0.9817
2.1	0.9821	0.9826	0.9830	0.9834	0.9838	0.9842	0.9846	0.9850	0.9854	0.9857
2.2	0.9861	0.9864	0.9868	0.9871	0.9875	0.9878	0.9881	0.9884	0.9887	0.9890
2.3	0.9893	0.9896	0.9898	0.9901	0.9904	0.9906	0.9909	0.9911	0.9913	0.9916
2.4	0.9918	0.9920	0.9922	0.9925	0.9927	0.9929	0.9931	0.9932	0.9934	0.9936
2.5	0.9938	0.9940	0.9941	0.9943	0.9945	0.9946	0.9948	0.9949	0.9951	0.9952
2.6	0.9953	0.9955	0.9956	0.9957	0.9959	0.9960	0.9961	0.9962	0.9963	0.9964
2.7	0.9965	0.9966	0.9967	0.9968	0.9969	0.9970	0.9971	0.9972	0.9973	0.9974
2.8	0.9974	0.9975	0.9976	0.9977	0.9977	0.9978	0.9979	0.9979	0.9980	0.9981
2.9	0.9981	0.9982	0.9982	0.9983	0.9984	0.9984	0.9985	0.9985	0.9986	0.9986
3.0	0.9987	0.9987	0.9987	0.9988	0.9988	0.9989	0.9989	0.9989	0.9990	0.9990
3.1	0.9990	0.9991	0.9991	0.9991	0.9992	0.9992	0.9992	0.9992	0.9993	0.9993
3.2	0.9993	0.9993	0.9994	0.9994	0.9994	0.9994	0.9994	0.9995	0.9995	0.9995
3.3	0.9995	0.9995	0.9995	0.9996	0.9996	0.9996	0.9996	0.9996	0.9996	0.9997
3.4	0.9997	0.9997	0.9997	0.9997	0.9997	0.9997	0.9997	0.9997	0.9997	0.9998

Table C.3 (Contd.)

(b) *Upper percentage points.* The tabulated value is z_α, where $P(Z > z_\alpha) = \alpha$, so that $1 - \Phi(z) = \alpha$.

α	0.05	0.025	0.01	0.005	0.001	0.0005
z_α	1.64	1.96	2.33	2.58	3.09	3.29

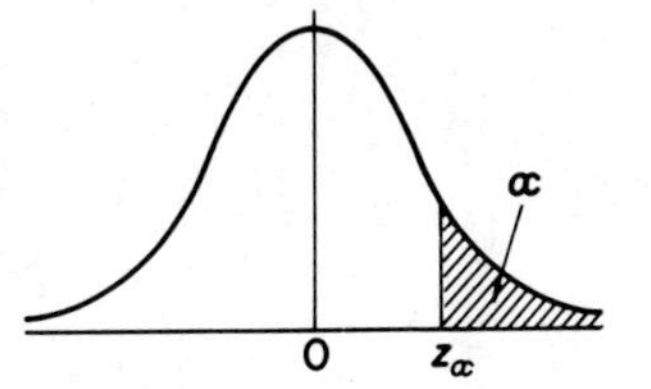

TABLE C.4 UPPER PERCENTAGE POINTS FOR THE t DISTRIBUTION

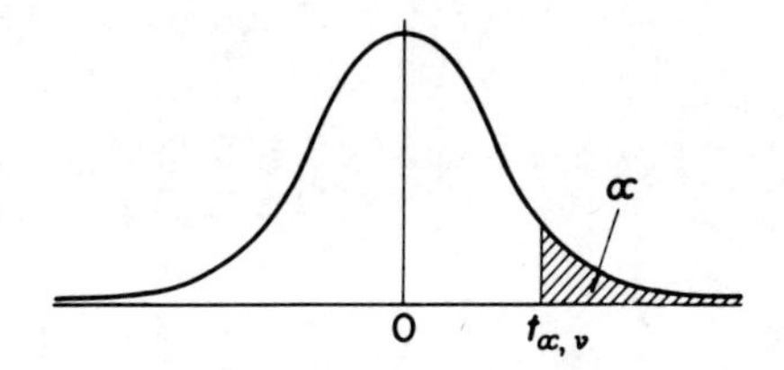

The tabulated value is $t_{\alpha,\nu}$, where $P(X > t_{\alpha,\nu}) = \alpha$, when X has the t distribution with ν degrees of freedom.

α		*0.05*	*0.025*	*0.01*	*0.005*	*0.001*	*0.0005*
$\nu =$	1	6.31	12.71	31.82	63.66	318.3	636.6
	2	2.92	4.30	6.96	9.92	22.33	31.60
	3	2.35	3.18	4.54	5.84	10.21	12.92
	4	2.13	2.78	3.75	4.60	7.17	8.61
	5	2.02	2.57	3.36	4.03	5.89	6.87
	6	1.94	2.45	3.14	3.71	5.21	5.96
	7	1.89	2.36	3.00	3.50	4.79	5.41
	8	1.86	2.31	2.90	3.36	4.50	5.04
	9	1.83	2.26	2.82	3.25	4.30	4.78
	10	1.81	2.23	2.76	3.17	4.14	4.59
	12	1.78	2.18	2.68	3.05	3.93	4.32
	14	1.76	2.14	2.62	2.98	3.79	4.14
	16	1.75	2.12	2.58	2.92	3.69	4.01
	18	1.73	2.10	2.55	2.88	3.61	3.92
	20	1.72	2.09	2.53	2.85	3.55	3.85
	25	1.71	2.06	2.48	2.79	3.45	3.72
	30	1.70	2.04	2.46	2.75	3.39	3.65
	40	1.68	2.02	2.42	2.70	3.31	3.55
	60	1.67	2.00	2.39	2.66	3.23	3.46
	120	1.66	1.98	2.36	2.62	3.16	3.37
	∞	1.64	1.96	2.33	2.58	3.09	3.29

TABLE C.5 PERCENTAGE POINTS FOR THE χ^2 DISTRIBUTION

The tabulated value is $\chi^2_{\alpha,\nu}$, where $P(X > \chi^2_{\alpha,\nu}) = \alpha$, when X has the χ^2 distribution with ν degrees of freedom.

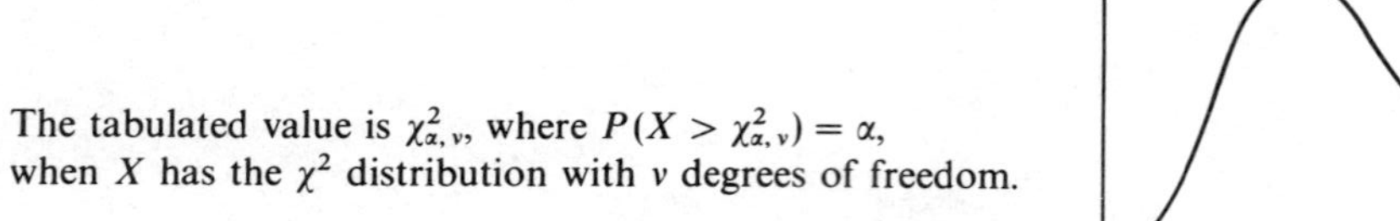

α	*0.999*	*0.995*	*0.99*	*0.975*	*0.95*	*0.9*	*0.1*	*0.05*	*0.025*	*0.01*	*0.005*	*0.001*
$\nu =$ 1	0.0^5157	0.0^4393	0.0^3157	0.0^3982	0.0^2393	0.0158	2.71	3.84	5.02	6.63	7.88	10.83
2	0.0^2200	0.0100	0.0201	0.0506	0.103	0.211	4.61	5.99	7.38	9.21	10.60	13.81
3	0.0243	0.0717	0.115	0.216	0.352	0.584	6.25	7.81	9.35	11.34	12.84	16.27
4	0.0908	0.207	0.297	0.484	0.711	1.06	7.78	9.49	11.14	13.28	14.86	18.47
5	0.210	0.412	0.554	0.831	1.15	1.61	9.24	11.07	12.83	15.09	16.75	20.52
6	0.381	0.676	0.872	1.24	1.64	2.20	10.64	12.59	14.45	16.81	18.55	22.46
7	0.599	0.989	1.24	1.69	2.17	2.83	12.02	14.07	16.01	18.48	20.28	24.32
8	0.857	1.34	1.65	2.18	2.73	3.49	13.36	15.51	17.53	20.09	21.95	26.12
9	1.15	1.73	2.09	2.70	3.32	4.17	14.68	16.92	19.02	21.67	23.59	27.88
10	1.48	2.16	2.56	3.25	3.94	4.86	15.99	18.31	20.48	23.21	25.19	29.59
12	2.21	3.07	3.57	4.40	5.23	6.30	18.55	21.03	23.34	26.22	28.30	32.91
14	3.04	4.08	4.66	5.63	6.57	7.79	21.06	23.68	26.12	29.14	31.32	36.12
16	3.94	5.14	5.81	6.91	7.96	9.31	23.54	26.30	28.85	32.00	34.27	39.25
18	4.90	6.26	7.02	8.23	9.39	10.86	25.99	28.87	31.53	34.81	37.16	42.31
20	5.92	7.43	8.26	9.59	10.85	12.44	28.41	31.41	34.17	37.57	40.00	45.31
25	8.65	10.52	11.52	13.12	14.61	16.47	34.38	37.65	40.65	44.31	46.93	52.62
30	11.59	13.79	14.95	16.79	18.49	20.60	40.26	43.77	46.98	50.89	53.67	59.70
40	17.92	20.71	22.16	24.43	26.51	29.05	51.81	55.76	59.34	63.69	66.77	73.40
50	24.67	27.99	29.71	32.36	34.76	37.69	63.17	67.50	71.42	76.15	79.49	86.66
60	31.74	35.53	37.48	40.48	43.19	46.46	74.40	79.08	83.30	88.38	91.95	99.61
100	61.92	67.33	70.06	74.22	77.93	82.36	118.5	124.3	129.6	135.8	140.2	149.4

TABLE C.6(a) 5 PER CENT POINTS OF THE F DISTRIBUTION

The tabulated value is $F_{0.05,\nu_1,\nu_2}$, where $P(X > F_{0.05,\nu_1,\nu_2}) = 0.05$ when X has the F distribution with ν_1, ν_2 degrees of freedom. The 95 per cent may be obtained using

$$F_{0.95,\nu_2,\nu_1} = \frac{1}{F_{0.05,\nu_1,\nu_2}}$$

$$\text{e.g. } F_{0.95,12,8} = \frac{1}{F_{0.05,8,12}} = \frac{1}{2.85} = 0.351$$

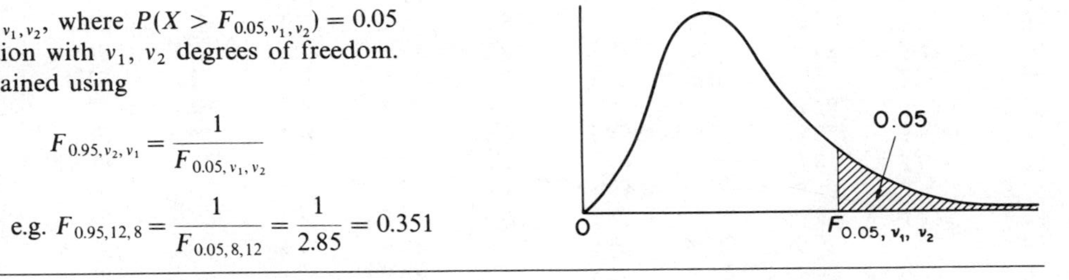

ν_2 \ ν_1 =	1	2	3	4	5	6	7	8	10	12	24	∞
1	161.4	199.5	215.7	224.6	230.2	234.0	236.8	238.9	241.9	243.9	249.1	254.3
2	18.5	19.0	19.2	19.2	19.3	19.3	19.4	19.4	19.4	19.4	19.5	19.5
3	10.1	9.55	9.28	9.12	9.01	8.94	8.89	8.85	8.79	8.74	8.64	8.53
4	7.71	6.94	6.59	6.39	6.26	6.16	6.09	6.04	5.96	5.91	5.77	5.63
5	6.61	5.79	5.41	5.19	5.05	4.95	4.88	4.82	4.74	4.68	4.53	4.36
6	5.99	5.14	4.76	4.53	4.39	4.28	4.21	4.15	4.06	4.00	3.84	3.67
7	5.59	4.74	4.35	4.12	3.97	3.87	3.79	3.73	3.64	3.57	3.41	3.23
8	5.32	4.46	4.07	3.84	3.69	3.58	3.50	3.44	3.35	3.28	3.12	2.93
9	5.12	4.26	3.86	3.63	3.48	3.37	3.29	3.23	3.14	3.07	2.90	2.71
10	4.96	4.10	3.71	3.48	3.33	3.22	3.14	3.07	2.98	2.91	2.74	2.54
12	4.75	3.89	3.49	3.26	3.11	3.00	2.91	2.85	2.75	2.69	2.51	2.30
15	4.54	3.68	3.29	3.06	2.90	2.79	2.71	2.64	2.54	2.48	2.29	2.07
20	4.35	3.49	3.10	2.87	2.71	2.60	2.51	2.45	2.35	2.28	2.08	1.84
24	4.26	3.40	3.01	2.78	2.62	2.51	2.42	2.36	2.25	2.18	1.98	1.73
30	4.17	3.32	2.92	2.69	2.53	2.42	2.33	2.27	2.16	2.09	1.89	1.62
40	4.08	3.23	2.84	2.61	2.45	2.34	2.25	2.18	2.08	2.00	1.79	1.51
60	4.00	3.15	2.76	2.53	2.37	2.25	2.17	2.10	1.99	1.92	1.70	1.39
∞	3.84	3.00	2.60	2.37	2.21	2.10	2.01	1.94	1.83	1.75	1.52	1.00

TABLE C.6(b) 2.5 PER CENT POINTS OF THE F-DISTRIBUTION

The tabulated value is $F_{0.025,\nu_1,\nu_2}$, where $P(X > F_{0.025,\nu_1,\nu_2}) = 0.025$, when X has the F distribution with ν_1, ν_2 degrees of freedom.

The 97.5 per cent point may be obtained using

$$F_{0.975,\nu_2,\nu_1} = \frac{1}{F_{0.025,\nu_1,\nu_2}}$$

$$\text{e.g.}\quad F_{0.975,12,8} = \frac{1}{F_{0.025,8,12}} = \frac{1}{3.51} = 0.285$$

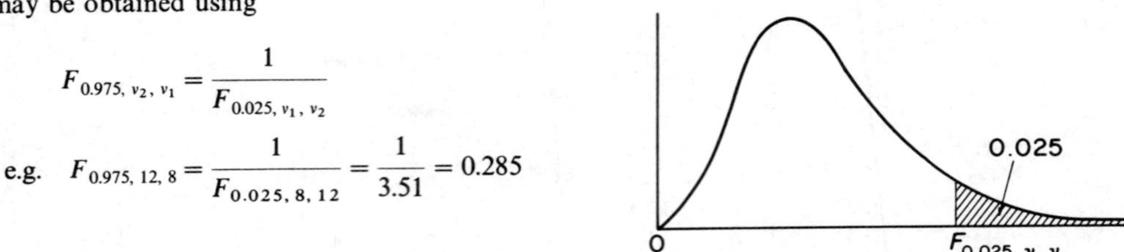

ν_1	1	2	3	4	5	6	7	8	10	12	24	∞
$\nu_2 =$ 1	648	800	864	900	922	937	948	957	969	977	997	1018
2	38.5	39.0	39.2	39.2	39.3	39.3	39.4	39.4	39.4	39.4	39.5	39.5
3	17.4	16.0	15.4	15.1	14.9	14.7	14.6	14.5	14.4	14.3	14.1	13.9
4	12.2	10.6	9.98	9.60	9.36	9.20	9.07	8.98	8.84	8.75	8.51	8.26
5	10.0	8.43	7.76	7.39	7.15	6.98	6.85	6.76	6.62	6.52	6.28	6.02
6	8.81	7.26	6.60	6.23	5.99	5.82	5.70	5.60	5.46	5.37	5.12	4.85
7	8.07	6.54	5.89	5.52	5.29	5.12	4.99	4.90	4.76	4.67	4.42	4.14
8	7.57	6.06	5.42	5.05	4.82	4.65	4.53	4.43	4.30	4.20	3.95	3.67
9	7.21	5.71	5.08	4.72	4.48	4.32	4.20	4.10	3.96	3.87	3.61	3.33
10	6.94	5.46	4.83	4.47	4.24	4.07	3.95	3.85	3.72	3.62	3.37	3.08
12	6.55	5.10	4.47	4.12	3.89	3.73	3.61	3.51	3.37	3.28	3.02	2.72
15	6.20	4.77	4.15	3.80	3.58	3.41	3.29	3.20	3.06	2.96	2.70	2.40
20	5.87	4.46	3.86	3.51	3.29	3.13	3.01	2.91	2.77	2.68	2.41	2.09
24	5.72	4.32	3.72	3.38	3.15	2.99	2.87	2.78	2.64	2.54	2.27	1.94
30	5.57	4.18	3.59	3.25	3.03	2.87	2.75	2.65	2.51	2.41	2.14	1.79
40	5.42	4.05	3.46	3.13	2.90	2.74	2.62	2.53	2.39	2.29	2.01	1.64
60	5.29	3.93	3.34	3.01	2.79	2.63	2.51	2.41	2.27	2.17	1.88	1.48
∞	5.02	3.69	3.12	2.79	2.57	2.41	2.29	2.19	2.05	1.94	1.64	1.00

TABLE C.6(c) 1 PER CENT POINTS OF THE F-DISTRIBUTION

The tabulated value is $F_{0.01, \nu_1, \nu_2}$, where $P(X > F_{0.01, \nu_1, \nu_2}) = 0.01$ when X has the F distribution with ν_1, ν_2 degrees of freedom.

The 99 per cent point may be obtain using

$$F_{0.99, \nu_2, \nu_1} = \frac{1}{F_{0.01, \nu_1, \nu_2}}$$

$$\text{e. g.} \quad F_{0.99, 12, 8} = \frac{1}{F_{0.01, 8, 12}} = \frac{1}{4.50} = 0.222.$$

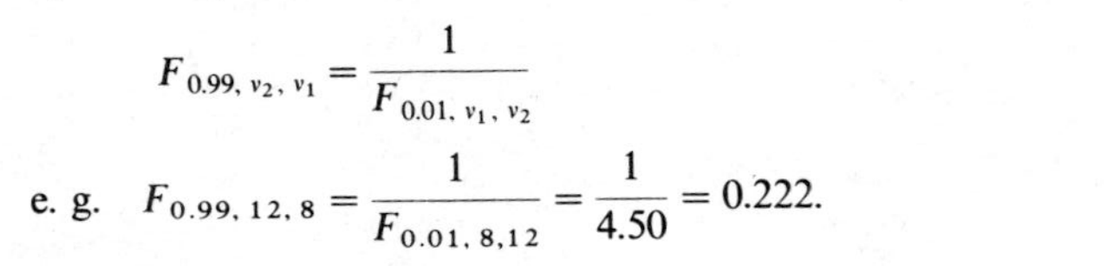

$\nu_1 =$		1	2	3	4	5	6	7	8	10	12	24	∞
$\nu_2 =$	1	4052	4999	5403	5625	5764	5859	5928	5981	6056	6106	6235	6366
	2	98.5	99.0	99.2	99.2	99.3	99.3	99.4	99.4	99.4	99.4	99.5	99.5
	3	34.1	30.8	29.5	28.7	28.2	27.9	27.7	27.5	27.2	27.1	26.6	26.1
	4	21.2	18.0	16.7	16.0	15.5	15.2	15.0	14.8	14.5	14.4	13.9	13.5
	5	16.3	13.3	12.1	11.4	11.0	10.7	10.5	10.3	10.1	9.89	9.47	9.02
	6	13.7	10.9	9.78	9.15	8.75	8.47	8.26	8.10	7.87	7.72	7.31	6.88
	7	12.3	9.55	8.45	7.85	7.46	7.19	6.99	6.84	6.62	6.47	6.07	5.65
	8	11.3	8.65	7.59	7.01	6.63	6.37	6.18	6.03	5.81	5.67	5.28	4.86
	9	10.6	8.02	6.99	6.42	6.06	5.80	5.61	5.47	5.26	5.11	4.73	4.31
	10	10.0	7.56	6.55	5.99	5.64	5.39	5.20	5.06	4.85	4.71	4.33	3.91
	12	9.33	6.93	5.95	5.41	5.06	4.82	4.64	4.50	4.30	4.16	3.78	3.36
	15	8.68	6.36	5.42	4.89	4.56	4.32	4.14	4.00	3.80	3.67	3.29	2.87
	20	8.10	5.85	4.94	4.43	4.10	3.87	3.70	3.56	3.37	3.23	2.86	2.42
	24	7.82	5.61	4.72	4.22	3.90	3.67	3.50	3.36	3.17	3.03	2.66	2.21
	30	7.56	5.39	4.51	4.02	3.70	3.47	3.30	3.17	2.98	2.84	2.47	2.01
	40	7.31	5.18	4.31	3.83	3.51	3.29	3.12	2.99	2.80	2.66	2.29	1.80
	60	7.08	4.98	4.13	3.65	3.34	3.12	2.95	2.82	2.63	2.50	2.12	1.60
	∞	6.63	4.61	3.78	3.32	3.02	2.80	2.64	2.51	2.32	2.18	1.79	1.00

TABLE C.6(d) 0.1 PER CENT POINTS OF THE F-DISTRIBUTION

The tabulated value is $F_{0.001,\nu_1,\nu_2}$, where $P(X > F_{0.001,\nu_1,\nu_2}) = 0.001$, when X has the F distribution with ν_1, ν_2 degrees of freedom.

The 99.9 per cent point may be obtained using

$$F_{0.999,\nu_2,\nu_1} = \frac{1}{F_{0.001,\nu_1,\nu_2}}$$

e.g. $F_{0.999,12,8} = \dfrac{1}{F_{0.001,8,12}} = \dfrac{1}{7.71} = 0.130.$

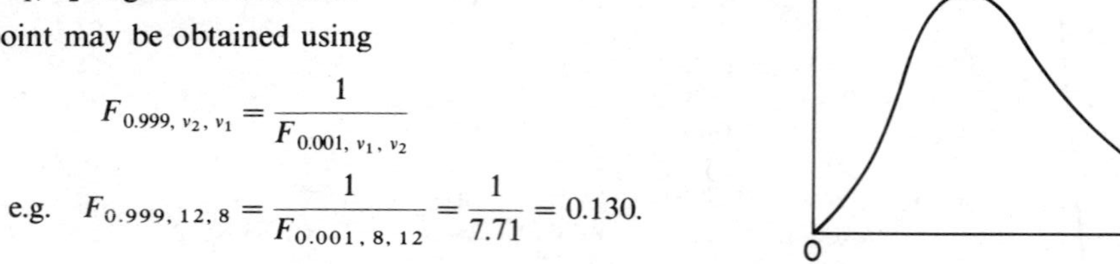

$\nu_1 =$	1	2	3	4	5	6	7	8	10	12	24	∞
$\nu_2 =$ 1*	4053	5000	5404	5625	5764	5859	5929	5981	6056	6107	6235	6366
2	999	999	999	999	999	999	999	999	999	999	1000	1000
3	167	149	141	137	135	133	132	131	129	128	126	124
4	74.1	61.3	56.2	53.4	51.7	50.5	49.7	49.0	48.1	47.4	45.8	44.1
5	47.2	37.1	33.2	31.1	29.8	28.8	28.2	27.7	26.9	26.4	25.1	23.8
6	35.5	27.0	23.7	21.9	20.8	20.0	19.5	19.0	18.4	18.0	16.9	15.8
7	29.3	21.7	18.8	17.2	16.2	15.5	15.0	14.6	14.1	13.7	12.7	11.7
8	25.4	18.5	15.8	14.4	13.5	12.9	12.4	12.1	11.5	11.2	10.3	9.33
9	22.9	16.4	13.9	12.6	11.7	11.1	10.7	10.4	9.89	9.57	8.72	7.81
10	21.0	14.9	12.6	11.3	10.5	9.93	9.52	9.20	8.75	8.44	7.64	6.76
12	18.6	13.0	10.8	9.63	8.89	8.38	8.00	7.71	7.29	7.00	6.25	5.42
15	16.6	11.3	9.34	8.25	7.57	7.09	6.74	6.47	6.08	5.81	5.10	4.31
20	14.8	9.95	8.10	7.10	6.46	6.02	5.69	5.44	5.08	4.82	4.15	3.38
24	14.0	9.34	7.55	6.59	5.98	5.55	5.23	4.99	4.64	4.39	3.74	2.97
30	13.3	8.77	7.05	6.12	5.53	5.12	4.82	4.58	4.24	4.00	3.36	2.59
40	12.6	8.25	6.59	5.70	5.13	4.73	4.44	4.21	3.87	3.64	3.01	2.23
60	12.0	7.77	6.17	5.31	4.76	4.37	4.09	3.86	3.54	3.32	2.69	1.89
∞	10.8	6.91	5.42	4.62	4.10	3.74	3.48	3.27	2.96	2.74	2.13	1.00

*Entries in the row $\nu_2 = 1$ must be multiplied by 100.

TABLE C.7 VALUES OF T FOR THE WILCOXON SIGNED RANK TEST

	Level of significance for one-sided H_1			
	0.05	0.025	0.01	0.005
	Level of significance for two-sided H_1			
n	*0.10*	*0.05*	*0.02*	*0.01*
5	0	–	–	–
6	2	0	–	–
7	3	2	0	–
8	5	3	1	0
9	8	5	3	1
10	10	8	5	3
11	13	10	7	5
12	17	13	9	7
13	21	17	12	9
14	25	21	15	12
15	30	25	19	15
16	35	29	23	19
17	41	34	27	23
18	47	40	32	27
19	53	46	37	32
20	60	52	43	37
21	67	58	49	42
22	75	65	55	48
23	83	73	62	54
24	91	81	69	61
25	100	89	76	68

TABLE C.8 VALUES OF U FOR THE MANN–WHITNEY U TEST

Critical values of U for the Mann–Whitney test for 0.05 (first value) and 0.01 (second value) significance levels for two-sided H_1, and for 0.025 and 0.005 levels for one-sided H_1.

n_2 \ n_1	1	2	3	4	5	6	7	8	9	10	11	12	13	14	15	16	17	18	19	20
1	–	–	–	–	–	–	–	–	–	–	–	–	–	–	–	–	–	–	–	–
	–	–	–	–	–	–	–	–	–	–	–	–	–	–	–	–	–	–	–	–
2	–	–	–	–	–	–	–	0	0	0	0	1	1	1	1	1	2	2	2	2
	–	–	–	–	–	–	–	–	–	–	–	–	–	–	–	–	–	–	0	0
3	–	–	–	–	0	1	1	2	2	3	3	4	4	5	5	6	6	7	7	8
	–	–	–	–	–	–	–	–	0	0	0	1	1	1	2	2	2	2	3	3
4	–	–	–	0	1	2	3	4	4	5	6	7	8	9	10	11	11	12	13	14
	–	–	–	–	–	0	0	1	1	2	2	3	3	4	5	5	6	6	7	8
5	–	–	0	1	2	3	5	6	7	8	9	11	12	13	14	15	17	18	19	20
	–	–	–	–	0	1	1	2	3	4	5	6	7	7	8	9	10	11	12	13
6	–	–	1	2	3	5	6	8	10	11	13	14	16	17	19	21	22	24	25	27
	–	–	–	0	1	2	3	4	5	6	7	9	10	11	12	13	15	16	17	18
7	–	–	1	3	5	6	8	10	12	14	16	18	20	22	24	26	28	30	32	34
	–	–	–	0	1	3	4	6	7	9	10	12	13	15	16	18	19	21	22	24
8	–	0	2	4	6	8	10	13	15	17	19	22	24	26	29	31	34	36	38	41
	–	–	–	1	2	4	6	7	9	11	13	15	17	18	20	22	24	26	28	30
9	–	0	2	4	7	10	12	15	17	20	23	26	28	31	34	37	39	42	45	48
	–	–	0	1	3	5	7	9	11	13	16	18	20	22	24	27	29	31	33	36
10	–	0	3	5	8	11	14	17	20	23	26	29	33	36	39	42	45	48	52	55
	–	–	0	2	4	6	9	11	13	16	18	21	24	26	29	31	34	37	39	42
11	–	0	3	6	9	13	16	19	23	26	30	33	37	40	44	47	51	55	58	62
	–	–	0	2	5	7	10	13	16	18	21	24	27	30	33	36	39	42	45	48

12	–	1	4	7	11	14	18	22	26	29	33	37	41	45	49	53	57	61	65	69
	–	–	1	3	6	9	12	15	18	21	24	27	31	34	37	41	44	47	51	54
13	–	1	4	8	12	16	20	24	28	33	37	41	45	50	54	59	63	67	72	76
	–	–	1	3	7	10	13	17	20	24	27	31	34	38	42	45	49	53	57	60
14	–	1	5	9	13	17	22	26	31	36	40	45	50	55	59	64	69	74	78	83
	–	–	1	4	7	11	15	18	22	26	30	34	38	42	46	50	54	58	63	67
15	–	1	5	10	14	19	24	29	34	39	44	49	54	59	64	70	75	80	85	90
	–	–	2	5	8	12	16	20	24	29	33	37	42	46	51	55	60	64	69	73
16	–	1	6	11	15	21	26	31	37	42	47	53	59	64	70	75	81	86	92	98
	–	–	2	5	9	13	18	22	27	31	36	41	45	50	55	60	65	70	74	79
17	–	2	6	11	17	22	28	34	39	45	51	57	63	69	75	81	87	93	99	105
	–	–	2	6	10	15	19	24	29	34	39	44	49	54	60	65	70	75	81	86
18	–	2	7	12	18	24	30	36	42	48	55	61	67	74	80	86	93	99	106	112
	–	–	2	6	11	16	21	26	31	37	42	47	53	58	64	70	75	81	87	92
19	–	2	7	13	19	25	32	38	45	52	58	65	72	78	85	92	99	106	113	119
	–	0	3	7	12	17	22	28	33	39	45	51	57	63	69	74	81	87	93	99
20	–	2	8	14	20	27	34	41	48	55	62	69	76	83	90	98	105	112	119	127
	–	0	3	8	13	18	24	30	36	42	48	54	60	67	73	79	86	92	99	105

TABLE C.9 THE FISHER z TRANSFORMATION

The table gives the values of the function $z(r) = \tanh^{-1} r$. For $r < 0$, the relationship $z(r) = -z(-r)$ may be used.

r	*0.00*	*0.01*	*0.02*	*0.03*	*0.04*	*0.05*	*0.06*	*0.07*	*0.08*	*0.09*
0.00	0.0000	0.0100	0.0200	0.0300	0.0400	0.0500	0.0601	0.0701	0.0802	0.0902
0.10	0.1003	0.1104	0.1206	0.1307	0.1409	0.1511	0.1614	0.1717	0.1820	0.1923
0.20	0.2027	0.2132	0.2237	0.2342	0.2448	0.2554	0.2661	0.2769	0.2877	0.2986
0.30	0.3095	0.3205	0.3316	0.3428	0.3541	0.3654	0.3769	0.3884	0.4001	0.4118
0.40	0.4236	0.4356	0.4477	0.4599	0.4722	0.4847	0.4973	0.5101	0.5230	0.5361
0.50	0.5493	0.5627	0.5763	0.5901	0.6042	0.6184	0.6328	0.6475	0.6625	0.6777
0.60	0.6931	0.7089	0.7250	0.7414	0.7582	0.7753	0.7928	0.8107	0.8291	0.8480
0.70	0.8673	0.8872	0.9076	0.9287	0.9505	0.9730	0.9962	1.0203	1.0454	1.0714
0.80	1.0986	1.1270	1.1568	1.1881	1.2212	1.2562	1.2933	1.3331	1.3758	1.4219

r	*0.000*	*0.001*	*0.002*	*0.003*	*0.004*	*0.005*	*0.006*	*0.007*	*0.008*	*0.009*
0.900	1.4722	1.4775	1.4828	1.4882	1.4937	1.4992	1.5047	1.5103	1.5160	1.5217
0.910	1.5275	1.5334	1.5393	1.5453	1.5513	1.5574	1.5636	1.5698	1.5762	1.5826
0.920	1.5890	1.5956	1.6022	1.6089	1.6157	1.6226	1.6296	1.6366	1.6438	1.6510
0.930	1.6584	1.6658	1.6734	1.6811	1.6888	1.6967	1.7047	1.7129	1.7211	1.7295
0.940	1.7380	1.7467	1.7555	1.7645	1.7736	1.7828	1.7923	1.8019	1.8117	1.8216
0.950	1.8318	1.8421	1.8527	1.8635	1.8745	1.8857	1.8972	1.9090	1.9210	1.9333
0.960	1.9459	1.9588	1.9721	1.9857	1.9996	2.0139	2.0287	2.0439	2.0595	2.0756
0.970	2.0923	2.1095	2.1273	2.1457	2.1649	2.1847	2.2054	2.2269	2.2494	2.2729
0.980	2.2976	2.3235	2.3507	2.3796	2.4101	2.4427	2.4774	2.5147	2.5550	2.5987
0.990	2.6467	2.6996	2.7587	2.8257	2.9031	2.9945	3.1063	3.2504	3.4534	3.8002

TABLE C.10 THE INVERSE FISHER z TRANSFORMATION

The table gives the values of the function $r(z) = \tanh z$. For $z < 0$, the relationship $r(z) = -r(-z)$ may be used.

z	0.00	0.01	0.02	0.03	0.04	0.05	0.06	0.07	0.08	0.09
0.00	0.0000	0.0100	0.0200	0.0300	0.0400	0.0500	0.0599	0.0699	0.0798	0.0898
0.10	0.0997	0.1096	0.1194	0.1293	0.1391	0.1489	0.1586	0.1684	0.1781	0.1877
0.20	0.1974	0.2070	0.2165	0.2260	0.2355	0.2449	0.2543	0.2636	0.2729	0.2821
0.30	0.2913	0.3004	0.3095	0.3185	0.3275	0.3364	0.3452	0.3540	0.3627	0.3714
0.40	0.3799	0.3885	0.3969	0.4053	0.4136	0.4219	0.4301	0.4382	0.4462	0.4542
0.50	0.4621	0.4699	0.4777	0.4854	0.4930	0.5005	0.5080	0.5154	0.5227	0.5299
0.60	0.5370	0.5441	0.5511	0.5581	0.5649	0.5717	0.5784	0.5850	0.5915	0.5980
0.70	0.6044	0.6107	0.6169	0.6231	0.6291	0.6351	0.6411	0.6469	0.6527	0.6584
0.80	0.6640	0.6696	0.6751	0.6805	0.6858	0.6911	0.6963	0.7014	0.7064	0.7114
0.90	0.7163	0.7211	0.7259	0.7306	0.7352	0.7398	0.7443	0.7487	0.7531	0.7574
1.00	0.7616	0.7658	0.7699	0.7739	0.7779	0.7818	0.7857	0.7895	0.7932	0.7969
1.10	0.8005	0.8041	0.8076	0.8110	0.8144	0.8178	0.8210	0.8243	0.8275	0.8306
1.20	0.8337	0.8367	0.8397	0.8426	0.8455	0.8483	0.8511	0.8538	0.8565	0.8591
1.30	0.8617	0.8643	0.8668	0.8692	0.8717	0.8741	0.8764	0.8787	0.8810	0.8832
1.40	0.8854	0.8875	0.8896	0.8917	0.8937	0.8957	0.8977	0.8996	0.9015	0.9033
1.50	0.9051	0.9069	0.9087	0.9104	0.9121	0.9138	0.9154	0.9170	0.9186	0.9201
1.60	0.9217	0.9232	0.9246	0.9261	0.9275	0.9289	0.9302	0.9316	0.9329	0.9341
1.70	0.9354	0.9366	0.9379	0.9391	0.9402	0.9414	0.9425	0.9436	0.9447	0.9458
1.80	0.9468	0.9478	0.9488	0.9498	0.9508	0.9517	0.9527	0.9536	0.9545	0.9554
1.90	0.9562	0.9571	0.9579	0.9587	0.9595	0.9603	0.9611	0.9618	0.9626	0.9633
2.00	0.9640	0.9647	0.9654	0.9661	0.9667	0.9674	0.9680	0.9687	0.9693	0.9699
2.10	0.9705	0.9710	0.9716	0.9721	0.9727	0.9732	0.9737	0.9743	0.9748	0.9753

Table C.10 (Contd.)

z	*0.00*	*0.01*	*0.02*	*0.03*	*0.04*	*0.05*	*0.06*	*0.07*	*0.08*	*0.09*
2.20	0.9757	0.9762	0.9767	0.9771	0.9776	0.9780	0.9785	0.9789	0.9793	0.9797
2.30	0.9801	0.9805	0.9809	0.9812	0.9816	0.9820	0.9823	0.9827	0.9830	0.9833
2.40	0.9837	0.9840	0.9843	0.9846	0.9849	0.9852	0.9855	0.9858	0.9861	0.9863
2.50	0.9866	0.9869	0.9871	0.9874	0.9876	0.9879	0.9881	0.9884	0.9886	0.9888
2.60	0.9890	0.9892	0.9895	0.9897	0.9899	0.9901	0.9903	0.9905	0.9906	0.9908
2.70	0.9910	0.9912	0.9914	0.9915	0.9917	0.9919	0.9920	0.9922	0.9923	0.9925
2.80	0.9926	0.9928	0.9929	0.9931	0.9932	0.9933	0.9935	0.9936	0.9937	0.9938
2.90	0.9940	0.9941	0.9942	0.9943	0.9944	0.9945	0.9946	0.9947	0.9949	0.9950
3.00	0.9951	0.9952	0.9952	0.9953	0.9954	0.9955	0.9956	0.9957	0.9958	0.9959
3.10	0.9959	0.9960	0.9961	0.9962	0.9963	0.9963	0.9964	0.9965	0.9965	0.9966
3.20	0.9967	0.9967	0.9968	0.9969	0.9969	0.9970	0.9971	0.9971	0.9972	0.9972
3.30	0.9973	0.9973	0.9974	0.9974	0.9975	0.9975	0.9976	0.9976	0.9977	0.9977
3.40	0.9978	0.9978	0.9979	0.9979	0.9979	0.9980	0.9980	0.9981	0.9981	0.9981
3.50	0.9982	0.9982	0.9982	0.9983	0.9983	0.9984	0.9984	0.9984	0.9984	0.9985
3.60	0.9985	0.9985	0.9986	0.9986	0.9986	0.9986	0.9987	0.9987	0.9987	0.9988
3.70	0.9988	0.9988	0.9988	0.9988	0.9989	0.9989	0.9989	0.9989	0.9990	0.9990
3.80	0.9990	0.9990	0.9990	0.9991	0.9991	0.9991	0.9991	0.9991	0.9991	0.9992
3.90	0.9992	0.9992	0.9992	0.9992	0.9992	0.9993	0.9993	0.9993	0.9993	0.9993

TABLE C.11 VALUES OF SPEARMAN'S r_s

	Level of significance for one-sided H_1			
	0.05	0.025	0.01	0.005
	Level of significance for two-sided H_1			
n	*0.1*	*0.05*	*0.02*	*0.01*
5	0.900	1.000	1.000	–
6	0.829	0.886	0.943	1.000
7	0.714	0.786	0.893	0.929
8	0.643	0.738	0.833	0.881
9	0.600	0.683	0.783	0.833
10	0.564	0.648	0.746	0.794
12	0.506	0.591	0.712	0.777
14	0.456	0.544	0.645	0.715
16	0.425	0.506	0.601	0.665
18	0.399	0.475	0.564	0.625
20	0.377	0.450	0.534	0.591
22	0.359	0.428	0.508	0.562
24	0.343	0.409	0.485	0.537
26	0.329	0.392	0.465	0.515
28	0.317	0.377	0.448	0.496
30	0.306	0.364	0.432	0.478

TABLE C.12 VALUES OF KENDALL's τ

	Level of significance for one-sided H_1			
	0.05	0.025	0.01	0.005
	Level of significance for two-sided H_1			
n	*0.1*	*0.05*	*0.02*	*0.01*
5	0.800	1.000	1.000	1.000
6	0.733	0.867	0.867	1.000
7	0.619	0.714	0.810	0.905
8	0.571	0.643	0.714	0.786
9	0.500	0.556	0.667	0.722
10	0.467	0.511	0.600	0.644
12	0.394	0.455	0.545	0.576
14	0.363	0.407	0.473	0.516
16	0.317	0.383	0.433	0.483
18	0.294	0.346	0.412	0.451
20	0.274	0.326	0.379	0.421
22	0.264	0.307	0.359	0.394
24	0.246	0.290	0.341	0.377
26	0.237	0.280	0.329	0.360
28	0.228	0.265	0.312	0.344
30	0.218	0.255	0.301	0.333

TABLE C.13 RANDOM DIGITS

The table gives 2500 random digits, from 0 to 9, arranged for convenience in blocks of 5.

87024	74221	69721	44518	58804	04860	18127	16855	61558	15430
04852	03436	72753	99836	37513	91341	53517	92094	54386	44563
33592	45845	52015	72030	23071	92933	84219	39455	57792	14216
68121	53688	56812	34869	28573	51079	94677	23993	88241	97735
25062	10428	43930	69033	73395	83469	25990	12971	73728	03856
78183	44396	11064	92153	96293	00825	21079	78337	19739	13684
70209	23316	32828	00927	61841	64754	91125	01206	06691	50868
94342	91040	94035	02650	36284	91162	07950	36178	42536	49869
92503	29854	24116	61149	49266	82303	54924	58251	23928	20703
71646	57503	82416	22657	72359	30085	13037	39608	77439	49318
51809	70780	41544	27828	84321	07714	25865	97896	01924	62028
88504	21620	07292	71021	80929	45042	08703	45894	24521	49942
33186	49273	87542	41086	29615	81101	43707	87031	36101	15137
40068	35043	05280	62921	30122	65119	40512	26855	40842	83244
76401	68461	20711	12007	19209	28259	49820	76415	51534	63574
47014	93729	74235	47808	52473	03145	92563	05837	70023	33169
67147	48017	90741	53647	55007	36607	29360	83163	79024	26155
86987	62924	93157	70947	07336	49541	81386	26968	38311	99885
58973	47026	78574	08804	22960	32850	67944	92303	61216	72948
71635	86749	40369	94639	40731	54012	03972	98581	45604	34885
60971	54212	32596	03052	84150	36798	62635	26210	95685	87089
06599	60910	66315	96690	19039	39878	44688	65146	02482	73130
89960	27162	66264	71024	18708	77974	40473	87155	35834	03114
03930	56898	61900	44036	90012	17673	54167	82396	39468	49566
31338	28729	02095	07429	35718	86882	37513	51560	08872	33717
29782	33287	27400	42915	49914	68221	56088	06112	95481	30094
68493	88796	94771	89418	62045	40681	15941	05962	44378	64349
42534	31925	94158	90197	62874	53659	33433	48610	14698	54761
76126	41049	43363	52461	00552	93352	58497	16347	87145	73668
80434	73037	69008	36801	25520	14161	32300	04187	80668	07499
81301	39731	53857	19690	39998	49829	12399	70867	44498	17385
54521	42350	82908	51212	70208	39891	64871	67448	42988	32600
82530	22869	87276	06678	36873	61198	87748	07531	29592	39612
81338	64309	45798	42954	95565	02789	83017	82936	67117	17709
58264	60374	32610	17879	96900	68029	06993	84288	35401	56317
77023	46829	21332	77383	15547	29332	77698	89878	20489	71800
29750	59902	78110	59018	87548	10225	15774	70778	56086	08117
08288	38411	69886	64918	29055	87607	37452	38174	31431	46173
93908	94810	22057	94240	89918	16561	92716	66461	22337	64718
06341	25883	42574	80202	57287	95120	69332	19036	43326	98697
23240	94741	55622	79479	34606	51079	09476	10695	49618	63037
96370	19171	40441	05002	33165	28693	45027	73791	23047	32976
97050	16194	61095	26533	81738	77032	60551	31605	95212	81078
40833	12169	10712	78345	48236	45086	61654	94929	69169	70561
95676	13582	25664	60838	88071	50052	63188	50346	65618	17517
28030	14185	13226	99566	45483	10079	22945	23903	11695	10694
60202	32586	87466	83357	95516	31258	66309	40615	30572	60842
46530	48755	02308	79508	53422	50805	08896	06963	93922	99423
53151	95839	01745	46462	81463	28669	60179	17880	75875	34562
80272	64398	88249	06792	98424	66842	49129	98939	34173	49883

Index